15th Conference of the Euro-American Consortium for Promoting the Application of Mathematics in Technical and Natural Sciences (AMiTaNS 2023)

Journal of Physics: Conference Series Volume 2675

Albena, Bulgaria
21 - 26 June 2023

Editor:

M.D. Todorov

ISBN: 978-1-7138-8771-3
ISSN: 1742-6588

Printed from e-media with permission by:

Curran Associates, Inc.
57 Morehouse Lane
Red Hook, NY 12571

Some format issues inherent in the e-media version may also appear in this print version.

Additional copies of this publication are available from:

Curran Associates, Inc.
57 Morehouse Lane
Red Hook, NY 12571 USA
Phone: 845-758-0400
Fax: 845-758-2633
Email: curran@proceedings.com
Web: www.proceedings.com

TABLE OF CONTENTS

Preface: Application of Mathematics in Technical and Natural Sciences 15th International Conference – AMiTaNS'23

The Euro-American Consortium for Promoting the Application of Mathematics in Technical and Natural Sciences organized its 15th annual conference of the series AMiTaNS (Application of Mathematics in Technical and Natural Sciences). The event took place during June 21-26, 2023 in the Congress Centre of Black-Sea resort of Albena, Bulgaria. Its aim was to foster existing and new personal contacts, scientific discussion as well as to provide a forum for the exchange of ideas and experience between specialists in the fields of Mathematics, Physics, Biology, Ecology, Economics, Computer Science and Technology. The event became a world conference covering 15 time zones between Chile and Nebraska through Texas, Indiana, Alabama, Florida, the UK, Europe, Cyprus, Russia, and China. The conference was managed from the Congress Centre in Albena where 40 participants and delegates were present in person and the other 40 attended the event via Zoom video platform.

Interest in this conference is growing year by year. This year we registered 70 talks and scientific reports, more than 80 participants – contributors and delegates from 16 countries around the world, representing the continents of Europe, North and South America, and Asia. About 18% of them attended the conference for the first time, with 50% of them being young scientists. The accepted talks were delivered and attended both on-location at the big hall and Corolla hall in Flamingo Grand hotel and on-line. Every day the morning sessions started by plenary lectures followed by contributed sessions on the following topics: Applied Analysis, Biomathematics, Continuum Mechanics, Mathematical Physics, Numerical Methods, Scientific Computing. As part of the program, leaders of their respective fields organized three special sessions: *Statistics & Modeling*, *Mathematical Methods in Reliability Theory*, *Algebraic and Geometric Aspects of Integrable Systems.* A number of very interesting presentations were delivered, which sparked stimulating scientific discussions both on- and off-line. The reader can find comprehensive and detailed information about the talks and reports including their records on the conference website at http://2023.eac4amitans.eu.

All contributed manuscripts were reviewed by anonymous independent referees via Morressier platform, and the approved revised manuscripts were accepted for publication in the present volume. The continued success of AMiTaNS in defiance of the extraordinary circumstances and challenges faced in the last few years undoubtedly enhances its reputation as an established annual meeting with an eminent and continuously growing core of contributors. Yet, the hybrid format turned out very flexible both for the organizers and contributors. It makes the conference accessible to all those who wish to attend either in person or remotely. We would like to thank the plenary speakers, who accepted our invitation and contributed to the conference through their elucidating lectures. We would also like to thank the special sessions organizers, hosts and co-hosts for the high scientific quality of the talks they selected, and for ensuring the smooth running of the conference in hybrid format. We are deeply indebted to the scores of anonymous referees all over the world, who approached their duties very professionally and thoroughly, and helped us to maintain high scientific standards. We highly appreciate the involvement of the members of the Program Committee and Steering Committee, who supported AMiTaNS from its inception, and worked tirelessly throughout the intensive labors of advertising, arranging, organizing, and reviewing. Special thanks go to our colleagues from the University of Essex at Colchester, UK, who worked with dedication so that parallel Zoom conference sessions could be held completely on-line.

We are very grateful to our publisher Institute of Physics, Bristol, UK for the helpful cooperation during the preparation of this volume.

Coordinator and *Zoom host* of AMiTaNS'23 and
Editor of the proceedings volume,
M.D. Todorov
Sofia, September 2023

Journal of Physics: Conference Series **2675** (2023) 011001 doi:10.1088/1742-6596/2675/1/011001

Program Committee

Organizer

Euro-American Consortium for Promoting the Application of Mathematics in the Technical and Natural Sciences Plc

Sponsor

– Ultra Quantum Inc., Huntsville, Alabama, USA

List of Participants

S. APOSTOLOV
St. Kliment Ohridski University of Sofia
BULGARIA

G. BOYADZHIEV
Institute of Mathematics and Informatics, BAS
Sofia
BULGARIA

Yu. BOZHKOV
Universidade Estadual de Campinas
Campinas - SP
BRAZIL

XUEDONG CHAI
China University of Mining and Technology
Xuzhou, Jiangsu
CHINA

YU-FEN CHANG
Western Norway University of Applied Sciences
Bergen
NORWAY

M. CHRISTOU
University of Nicosia
CYPRUS

I. CHRISTOV
Purdue University
West Lafayette, IN
USA

D. DANKOV
Institute of Mechanics, BAS
Sofia
BULGARIA

D. DANTCHEV
Institute of Mechanics, BAS
Sofia
BULGARIA
and Max-Planck Institute fuer Intelligente Systems
Stuttgart
GERMANY

A. DEMBINSKA
Warsaw University of Technology
POLAND

B. DIMITROV
Institute of Nuclear Research and Nuclear Energetics, BAS
& Institute for Advanced Physical Studies
Sofia
BULGARIA

Yu. DIMITROV
University of Forestry
and Institute of Mathematics and Informatics, BAS
Sofia
BULGARIA

V. DONEV
Technical University of Sofia , branch Sliven
BULGARIA

A. FEDOSEYEV
Ultra Quantum Inc.
Huntsville, AL
USA

I.R. GEORGIEV
Angel Kanchev University of Rousse
and Institute of Mathematics and Informatics, BAS, Sofia
BULGARIA

S. GEORGIEV
Angel Kanchev University of Rousse
and Institute of Mathematics and Informatics, BAS, Sofia
BULGARIA

V. GERDJIKOV
Institute for Nuclear Research and Nuclear Energy
and Institute of Mathematics and Informatics, BAS, Sofia
BULGARIA

A. GORONCY
Nicolaus Copernicus University
Torun
POLAND

G. GRAHOVSKI
University of Essex
Colchester
UK

Ya. GURIMSKAYA
Solestial Inc.
Tempe, AR
USA

T. GUROV
Institute for Information and Communication Technologies, BAS
Sofia
BULGARIA

S.-M. GUROVA
Institute for Information and Communication Technologies, BAS
Sofia
BULGARIA

V. GUSHCHIN
Institute for Computer Aided Design of RAS
Moscow
RUSSIA

S. HADJIIVANOV
Angel Kanchev University of Ruse
BULGARIA

R. HAYNATZKI
University of Nebraska Medical Center
Omaha, NE
USA

P. HOLDA
Warsaw University of Technology
POLAND

WEIPENG HU
Xi'an University of Technology
CHINA

B. IDIRIZOV
Angel Kanchev University of Ruse
BULGARIA

N. ILIEVA
Institute of Information and Communication Technologies
& Institute of Mathematics and Informatics, BAS
Sofia
BULGARIA

A. IVANOV
Paisii Hilendarski University of Plovdiv
BULGARIA

K. JASINSKI
Nicolaus Copernicus University
Torun
POLAND

P. JORDANOVA
Konstantin Preslavski University of Shumen
BULGARIA

A. KERZHAEV
Institute of Earthquake Prediction Theory and Mathematical Geophysics, RAS
Moscow
RUSSIA

A. KIRILOV
Institute of Information and Communication Technologies, BAS
Sofia
BULGARIA

M. KOLEVA
Angel Kanchev University of Ruse
BULGARIA

S. KOLEVA
Paisii Hilendarski University of Plovdiv
BULGARIA

O. KOROTUNOVA
Zaporizhia Polytechnic National University
Zaporizhia
UKRAINE

N. KUTEV
Institute of Mathematics and Informatics, BAS
Sofia
BULGARIA

M. LAZAROVA
Technical University of Sofia
Sofia
BULGARIA

ZHI QIANG LI
China University of Mining and Technology
Xuzhou
CHINA

E. LILKOVA
Institute of Information and Communication Technologies, BAS
1113 Sofia
BULGARIA

HONGYU LIU
City University of Hong Kong
Kowloon, Hong Kong SAR
CHINA

D. MENSEIDOV
Institute of Mechanics, BAS &
University of Telecommunications and Post
Sofia
BULGARIA

V. MIHOVA
Angel Kanchev University of Ruse
BULGARIA

I. MINKOVA
Angel Kanchev University of Ruse
BULGARIA
M. NIKOLO
Saint Louis University, MO
USA

N. NIKOLOV
RWTH Aachen University
GERMANY

G. NIKOLOVA
Institute of Mechanics, BAS
Sofia
BULGARIA

M. NIKOLOVA
Angel Kanchev University of Ruse
BULGARIA

O. OLEVSKYI
Oles Honchar National University
Dnipro
UKRAINE

V. OLEVSKYI
Ukrainian State University of Chemical Technology
Dnipro
UKRAINE

Tz. OSTROMSKY
Institute of Information and Communication Technologies, BAS
Sofia
BULGARIA

J. OUAZZANI
ArcoFluid Consulting LLC
Orlando, FL
USA

N. PAPANICOLAOU
University of Nicosia
CYPRUS

XUBIAO PENG
Beijing Institute of Technology
CHINA

A. POHJONEN
University of Oulu
FINLAND

ZHIJUN (GEORGE) QIAO
University of Texas Rio Grande Valley
Edinburg, TX
USA

K. RACHWAL
Warsaw University of Technology
POLAND

E. REYES
Universidad de Santiago de Chile
Santiago
CHILE

V. RUKAVISHNIKOV
Computing Center of FEB of the RAS
Khabarovsk
RUSSIA

G. SHYSHKANOVA
Zaporizhzhia Polytechnic National University
Zaporizhzhia
UKRAINE
and Humboldt-Universitaet zu Berlin
GERMANY

M. SIDDIQUY
Institute of Business Administration
Karachi
PAKISTAN

D. SLAVCHEV
Institute of Information and Communication Technologies, BAS
Sofia
BULGARIA

M. STEHLIK
Johannes Kepler University in Linz
AUSTRIA,
& Universidad de Valparaíso
CHILE

M. STOENCHEV
Technical University of Sofia
BULGARIA

M. TODOROV
Technical University of Sofia
BULGARIA

V. TODOROV
Institute of Information and Communication Technologies, BAS&
Institute of Mathematics and Informatics, BAS
Sofia
BULGARIA

V. TRANEVA
Prof. Asen Zlatarov University
Bourgas
BULGARIA

A. VASSILEV
St. Kliment Ohridski University of Sofia
BULGARIA

V. VATCHEV
University of Texas, Rio Grande Valley
Rancho Viejo, TX
USA

R. VEJAR
Universidad de La Serena
CHILE

L. VULKOV
Angel Kanchev University of Ruse
BULGARIA

P. YANEVA
Paisii Hilendarski University of Plovdiv
BULGARIA

JIN-JIE YANG
China University of Mining and Technology
Xuzhou, Jiangsu
CHINA

A.YORDANOVA
Trakia University
Stara Zagora
BULGARIA

HONGYU ZHANG
China University of Mining and Technology
Xuzhou, Jiangsu
CHINA

T. ZHELYAZOV
Technical University of Sofia
BULGARIA

Peer Review Statement

All papers published in this volume have been reviewed through processes administered by the Editors. Reviews were conducted by expert referees to the professional and scientific standards expected of a proceedings journal published by IOP Publishing.

- **Type of peer review:** Single Anonymous
- **Conference submission management system:** Morressier
- **Number of submissions received:** 86
- **Number of submissions sent for review:** 53
- **Number of submissions accepted:** 48
- **Acceptance Rate (Submissions Accepted / Submissions Received × 100):** 55.8
- **Average number of reviews per paper:** 1.0416666666666667
- **Total number of reviewers involved:** 32
- **Contact person for queries:**
  Name: Michail Todorov
  Email: mditod@gmail.com
  Affiliation: EAC4AMiTaNS - Mathematical Modelling and Numerical Methods

Building Machine Learning Models to Improve Hypertension Diagnosis

Roman Haynatzki[1, a)], **Thomas Windle**[1], **John Windle, MD**[1]

[1]*Biomedical Informatics Program, University of Nebraska Medical Center,* 42nd and Emile St, Omaha, NE 68198, USA

[a)] Corresponding author: rhaynatzki@unmc.edu

Abstract. Hypertension (HT) is a major risk factor for heart disease, stroke, and kidney disease. A diagnosis of HT is based on two or more blood pressure (BP) readings taken on two or more visits, and such diagnosis may vary depending on the threshold BP values utilized by specific guidelines. Our hypertension identification approach has been recently proposed and validated for HT diagnosis and could be used as ground truth in modeling of latent HT diagnosis. Our approach to modeling of latent HT diagnosis with high precision will leverage analytics to "Big Data", such as electronic health records (EHRs). In this work, we will review the time complexity underlying the classical ML methods XGBoost (XGB) and Artificial Neural Networks (ANN). In particular, we compare the XGB and ANN to leverage their strengths. The performance of all algorithms for diagnosing HT will be characterized using the area under the curve (AUC) approach on a big EHR longitudinal dataset. Predictor variable importance and model interpretability will be assessed using the Shapley values approach.

1. INTRODUCTION AND BACKGROUND

Hypertension (HT), or high blood pressure, is a serious issue in the US for several reasons. HT is highly prevalent, affecting nearly half of adults [14]. This means that millions of Americans are at risk of developing serious health problems related to HT. There are many effective treatments for HT, but low diagnosis rates (below 66% [15]) lead to poor management, excess cardiovascular morbidity, and death. HT is a major risk factor for heart disease, stroke, and kidney disease. Heart disease is the leading cause of death in the US and was responsible for approximately 670,000 deaths, which is about 1 in every 5 deaths in 2017. HT can damage organs such as the eyes and the brain and can contribute to the development of dementia. The costs associated with HT are also very high. According to the American Heart Association (AHA), the annual cost of treating HT in the US is estimated to be between $131 billion and $198 billion. HT disproportionately affects certain groups of people, including African Americans, Hispanic Americans, and older adults. As a result, these groups are also more likely to experience complications related to HT. While HT is a serious issue, it is also largely preventable through lifestyle changes such as a healthy diet, regular exercise, avoiding tobacco products, weight control, and taking HT medications regularly. Unfortunately, up to 75% of hypertensive patients are not treated to target, often because they are unaware of their condition or do not take it seriously. Approximately one in three hypertensive patients is undiagnosed. HT is sometimes referred to as the "silent killer" because it often has no symptoms or warning signs. Thus, many people may not be aware that they have high blood pressure because they do not see a healthcare provider regularly and their blood pressure is not checked on a regular basis. Others may not be aware of the risks associated with HT or may be hesitant to seek medical care. This project addresses the important clinical problem of accurate HT diagnosis, with the potential of improving the timing of preventative interventions and treatment of HT.

The overarching objective of this paper is to model HT diagnosis with high precision that will leverage Artificial Intelligence and Machine Learning (AI/ML) advanced analytics to "Big Data", such as electronic health records (EHRs). We have utilized the UNMC Research Instance of CRANE database that includes demographics, clinical diagnosis, medications, patient reported measures and outcomes, and over 40 other variables per patient. Additionally, the models (XGBoost, or XGB, and Partial Logistic Artificial Neural Network, or PLANN) that we are using are on a new dataset that has not been previously analyzed by these methods.

1.1 Hypertension Identification Approach

Given that antihypertensive medications can mask the presence of hypertension, relying solely on high blood pressure measurements might not accurately identify truly hypertensive patients. To address this challenge and establish a definitive ground truth, we consulted with five medical professionals. They advised that while individuals with a systolic blood pressure of 130 or above, or a diastolic blood pressure of 80 or above, can be considered hypertensive, a comprehensive definition should also include past medical history. Specifically, a patient (i) taking antihypertensive medications who also (ii) has a previous ICD diagnosis of hypertension can be confidently classified as hypertensive. However, if only one of (i) or (ii) is met, it's possible the patient either successfully managed their hypertension without medication or is using antihypertensive drugs for a different medical condition. In building our models, we aimed to ensure robustness and sought to determine which variables most significantly predict the presence of HT.

For ideal patient evaluations, emphasis is placed on the reliability and accuracy of all measurements. It's assumed that the data gathered, including readings of systolic and diastolic blood pressure, previous diagnoses of HT, and medication history, are precise and reliable. Accurate measurements ensure healthcare professionals have a comprehensive and exact view of the patient's health status, which aids in making informed decisions about their treatment plans.

However, discrepancies can arise in patient data, which can present challenges. A patient might have been misdiagnosed with HT previously, or their blood pressure might have been recorded inaccurately. Such inconsistencies, whether due to human error, equipment malfunction, or other reasons, can distort our understanding of the patient's condition and might lead to unsuitable or even harmful treatment approaches.

In summary, accuracy in all facets of patient evaluation, diagnosis, and HT management is of paramount importance. Healthcare professionals should be vigilant about potential data inconsistencies and contemplate their implications when formulating treatment plans. By recognizing and addressing these potential discrepancies, we can aim to provide optimal and effective care for each patient, thereby enhancing overall health outcomes.

2. METHODS

Machine learning (ML) and *artificial intelligence* (AI) are algorithmic analyses of *large* and *complex* data. ML is a subset of AI used to model data, namely, *prediction/estimation* of HT probabilities. AI, in the form of mathematical or statistical algorithms for prediction, refers to the application of statistical techniques to analyzing data and making predictions or inferences. It involves using mathematical models and statistical methods to identify patterns, correlations, and trends in large datasets, and then using those patterns to make predictions about future or unseen data.

Thus, the focus is the correct implementation of the ML algorithm:

- tuning/finding suitable hyperparameters;
- selecting the best set of hyperparameters;
- running the algorithm in real time;
- creating an ensemble model by combining multiple weak learners;
- improving overall prediction accuracy by sequentially addressing the weaknesses of previous models;
- learning complex patterns and relationships in data;
- solving classification, regression, or other complex tasks;
- adapting to new inputs through training and adjusting weights;
- reducing overfitting; and
- (in case of categorical response variable) classifying instances based on the likelihood of belonging to a particular class.

2.1 SHAP (SHapley Additive exPlanations) Importance Values

SHAP values quantify the contribution of each feature (aka, variable) to the prediction of a machine learning model. SHAP values provide local (at patient level) explanations by estimating the contribution of each feature for a particular patient and considering all possible feature combinations and averaging their effects, allocating the total effect of a feature among all possible feature subsets. However, it's important to note that SHAP values are suited for explaining the contributions of features after models have been finalized, and not for model selection.

The Shapley values were originally proposed as a method for fairly distributing the "value" of a coalition (a group of players working together) among its members [4]. In our case, these SHAP values are being looked at for different features, or variables, instead of players. The larger the SHAP importance value for a given feature (e.g., age), the more this feature contributes to the presence of HT. Given that the value function

$$val_x(S) = \int \hat{f}(x_1, \dots, x_p) d\mathbb{P}_{x \notin S} - E_X\left(\hat{f}(X)\right)$$

and linear model

$$\hat{f}(X) = \beta_0 + \beta_1 x_1 + \dots + \beta_p x_p$$

for HT risk where $x_1, x_2, \dots, x_p$ are the p features' values, β_j refers to the weight of feature j and S refers to a coalition, or subset of features, drawn from $\{x_1, x_2, \dots, x_p\}$. Additionally, $|S|$ denotes the number of features in subset S. The formula used to calculate the Shapley value for each feature j is as follows [11], [13]:

$$\varphi_j(val) = \sum_{S \subseteq \{1,\ldots,p\} \setminus \{j\}} \frac{|S|!\,(p - |S| - 1)!}{p!} \left(val(S \cup \{j\}) - val(S)\right)$$

$$= \frac{1}{\text{number of features}} \sum_{\text{coalitions including feature } j} \frac{\text{marginal contribution of } j \text{ to coalition}}{\text{number of coalitions excluding } j \text{ of this size}}.$$

The SHAP value implements the Shapley value formulas. SHAP incorporates computational optimizations and approximations to efficiently compute feature attributions for high-dimensional datasets and complex models. This scalability makes SHAP applicable to real-world machine learning tasks, where the number of features and instances can be significant. The number of features comes in play when presenting the number of ways to choose a subset of features from a set of features, and the resulting binomial coefficients contribute to the computational complexity of calculating the Shapley values using the original formula. The SHAP framework addresses this issue by introducing several techniques to reduce the computational burden while calculating the Shapley values efficiently. They include coalition reduction where SHAP employs sampling or other approximation methods to estimate the Shapley values without exhaustively evaluating all possible coalitions. This reduces the computational complexity significantly. Also, SHAP employs consistency and linearity properties, thus ensuring that the feature attribution methods have desirable characteristics like local accuracy, consistency, and missingness, making SHAP importance values more interpretable and efficient.

2.2 AUC Value

The area under the ROC curve (AUC) shown in Figure 1 below is a metric used to evaluate the performance of a binary classification model, and in our case, we are classifying the patient being truly hypertensive vs not truly hypertensive. AUC represents the overall quality of the model's predictions by measuring the trade-off between its true positive rate (sensitivity) and false positive rate (1 - specificity) across different classification thresholds.

The ROC curve as shown below is created by plotting the true positive rate (TPR) against the false positive rate (FPR) at various threshold settings. The AUC is then computed as the area under this curve, ranging from 0 to 1. An AUC value closer to 1 indicates a better classifier with higher discriminatory power, while a value of 0.5 represents a random or non-informative classifier. The image below shows a graphical representation of the ROC curve with values as was obtained from our models [19]. The AUC is a valuable metric for assessing a model's performance, capturing both its sensitivity and specificity, especially when the dataset size is sufficiently large. However, a cautionary note is warranted for imbalanced datasets. For instance, in a scenario where 95% of patients are labeled as hypertensive (the majority class), a model making nearly constant predictions can misleadingly achieve a high AUC value. In such contexts, an AUC value exceeding 0.95 should be viewed with suspicion, as it may indicate the model isn't genuinely discriminating between classes as effectively as the score suggests.

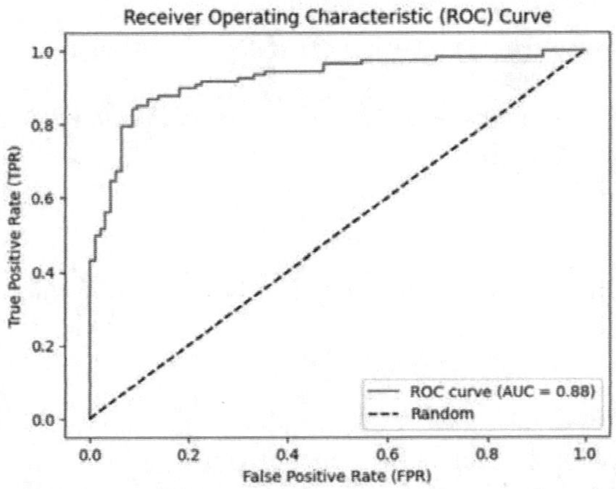

FIGURE 1. Receiver Operating Characteristic (ROC) Curve with an AUC (Area Under the Curve) value of 0.88, indicating strong predictive accuracy.

2.3 XGBoost Method

The Extreme Gradient Boosted trees (XGBoost) is a relatively recent ensemble method for constructing trees [10]. Trees in other ensemble algorithms are created in the conventional manner, i.e. either using Gini Impurity or Entropy. But XGBoost introduces a new metric, called *similarity score*, for node selection and splitting.

A very general (nonmathematical) description of the steps involved in creating a Decision Tree using the similarity score is shown below:

1) Create a single-leaf tree.
2) For the first tree, compute the average of the target variable as prediction and calculate the residuals using the desired loss function. For subsequent trees, the residuals come from prediction made by the previous tree.
3) Calculate the similarity score using the following formula:

$$Similarity\ Score = \frac{Gradient^2}{Hessian + \lambda}$$

   where the Hessian is equal to the number of residuals; Gradient^2 = squared sum of residuals; λ is a regularization hyperparameter.
4) Using the similarity score, we select the appropriate node. The higher the similarity score, the more homogeneity.
5) Using the similarity score, we calculate the *Information gain*. Information gain gives the difference between old similarity and new similarity and thus tells how much homogeneity is achieved by splitting the node at a given point. It is calculated using the following formula:

$$Information\ Gain = Left\ Similarity + Right\ Similarity - Similarity\ for\ Root$$

6) Create the tree of desired length using the above method. Pruning and regularization would be done by varying the regularization hyperparameter.

7) Predict the residual values using the Decision Tree you constructed.
8) The new set of residuals is calculated using the following formula:

$$New\ Residuals = Old\ Residuals + \rho\ \Sigma\ Predicted\_Residuals$$

where ρ is the learning rate.
9) Go back to step 1 and repeat the process for all the trees.

The actual XGBoost algorithm is mathematically involved [9] and the details are omitted here.

2.4 Neural Networks Method

Partial Logistic Artificial Neural Network (*PLANN*) with a single output node uses, as inputs, the prognostic factors, to increase the predictive ability of the model [1-2]. An artificial neural network refers to the implementation of neural networks in the context of artificial intelligence and machine learning. PLANN is a subset of neural networks that focuses on learning from data to perform specific tasks. The PLANN consists of three main types of layers: (i) input, (ii) hidden, (iii) output. The input layer(s) is where the data is entered into the network. Each node in the input layer represents a feature/attribute of the input data. The hidden layer(s) is sandwiched between the input and output layers. The hidden layer(s) performs computations on the input data. Each node in the hidden layer takes the weighted sum of the inputs from the previous layer, adds a bias term, and then applies an activation function to produce an output. The activation function is usually a non-linear function, such as a sigmoid function. The output layer is the final layer of the neural network and produces the network's prediction or output.

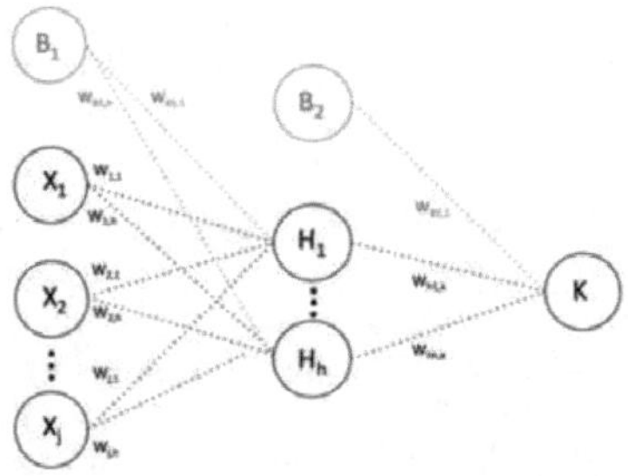

FIGURE 2. PLANN: Single hidden layer feed forward NN (nodes are uni-directionally connected), where X = inputs, H = hidden nodes, K = output node, B = bias nodes.

In our case, PLANN provides the continuous probability, $P(HT{=}1)$, of being truly hypertensive using as transformation function, for both input and output layers, the logistic (sigmoid) function:

$$f(\eta) = \frac{1}{1 + e^{-\eta}}$$

where η is the linear combination of the input covariates X_i (i = 1, 2, ..., p) with weights w_i of the input-hidden layer. PLANN is estimated using the maximum likelihood approach, and the contribution to the log-likelihood for each patient is calculated over all intervals that that patient is at risk for HT. The *output node* is a large target vector with kth component 0 if the patient is not truly hypertensive, and 1 if the patient is truly hypertensive. Therefore, for each patient, PLANN estimates the probability that a patient is truly hypertensive, and this can be converted into a yes/no categorical variable. Converting a continuous variable ranging from 0 to 1 into a binary type of variable is commonly done using a thresholding approach. The threshold chosen will determine the cutoff point for the binary classification. For example, if we want to classify values greater than or equal to 0.5 as "1" and values less than 0.5 as "0", we would set the threshold at 0.5. We would then apply thresholding, where for each value of the continuous variable, we compare it to the chosen threshold. If the value is greater than or equal to the threshold, we assign a value of "1" (or "yes"). If the value is less than the threshold, assign a value of "0" (or "no").

2.5 Comparing Efficiency XGB vs PLANN: Time Complexity

Despite utilizing a supercomputer, the computational time required for AI/ML models can extend multiple days when dealing with significantly large datasets. Consequently, it becomes imperative to assess the time complexity of the two specific models to understand the reasons for their disparate performance. Upon initial observation, the XGBoost model exhibits a runtime of approximately ten minutes when executed on our dataset with the aid of a supercomputer. On the other hand, the PLANN model's execution on the same dataset takes several hours, even with access to a supercomputer. Understanding the underlying factors contributing to this discrepancy is crucial, and thus we have undertaken an analysis of the time complexity for both models. It is important to acknowledge that comparing these two models poses challenges due to their fundamentally distinct algorithms. Nevertheless, our goal is to provide an explanation for the observed differences in their respective running times.

When assessing algorithms' time complexities, the constant multiplier preceding the 'n' term is often less significant in comparison, where 'n' is the size of the problem. Common time complexities include constant (O(1)), logarithmic (O(log n)), linear (O(n)), quasilinear (O(n log n)), polynomial (O(n^a)), exponential (O(a^n)), and factorial (O(n!)). Typically, algorithms with lower complexities are expected to run faster, while those with higher complexities will run slower. In the context of our comparison between XGBoost and PLANN, the explanation from the original paper by Tianqi Chen et al [9] reveals that XGBoost's time complexity is essentially O(n log n) due to its greedy nature. A greedy algorithm efficiently performs operations at each step without revisiting and re-optimizing previous decisions. Regarding PLANN, the analysis indicates that its forward and backpropagation processes contribute to a polynomial time complexity of O(n^a). This suggests a relationship between the dataset size (n) and the degree of the polynomial (a) in determining the algorithm's computational efficiency. Notably, there are numerous resources available for analyzing the time complexity of neural network algorithms. However, even a conservative estimate of O(n^3) for neural networks would significantly increase their runtime compared to XGBoost's O(n log n) complexity. In conclusion, understanding the time complexities of both XGBoost and PLANN aids in explaining the differences in their respective running times. XGBoost's greedy nature leads to a faster execution, while PLANN's polynomial complexity accounts for its comparatively longer runtime [3]. It is also important to note that Convolutional Neural Networks (CNN) are more complex neural networks than PLANN, designed to process grid-structured data, such as images. The complexity of these networks arises from the use of convolutional layers, which apply a convolution operation to the input, passing the result to the next layer[18].

2.6 Interpreting Model Prediction

Both *accuracy* and *interpretability* are important properties of models. The ability to correctly *interpret* a model:

- increases user’s trust;
- provides insight into how a model may be improved; and

- supports understanding of the process being modeled.

Complex (nonlinear) models are, on average, more precise but less interpretable than simple models. Even for large data, despite having lower accuracy, simple models (e.g., *linear* or *logistic regression*) are sometimes preferred for their ease of interpretation. The area under the ROC curve (*AUC*) is often used to assess the predictive performance of the models and for model selection purposes.

The *SHAP* (SHapley Additive exPlanations) values are used as importance values for predictor variables (aka features) for a particular prediction resulting from a given ML algorithm. [4-6]

2.7 Dataset and Software

We used the language and environment for statistical computing Python in all calculations. We used a new dataset to determine the effects leading to HT in cardiac patients. This dataset was extracted from the CRANE EHR Database, on cardiac patients that were treated by faculty members in the Department of Internal Medicine, from January 2017 until December 2021. To the best of our knowledge, this dataset has not previously been analyzed using the methods in this paper. These data were generated by an *observational study* with the main outcome variable being presence of HT (or lack thereof) and what factors influence the presence of HT the most. The dataset originally consisted of more than 500,000 patients. It was also of interest to note that of the truly hypertensive, approximately 55% are obese and of those that are not truly hypertensive, only 42% are obese.

The variables recorded at the time of entry include demographic characteristics as well as presence of diabetes, HT medications, and prior HT diagnosis. Also, many clinical and biochemical variables that had been recorded both at the time of entry and at the follow-up visits, were not included in the current dataset. Some of the variables included in the analyzed dataset are as follows:

ID = Patient id (integer)
HTMEDICATION = 1, if patient uses HT medication for any reason; 0, otherwise
ICD10CODE = 1, if patient has been previously ICD-10 diagnosed with HT; 0, otherwise
AGE = Patient's age in years at the last valid measurement
GENDER = 0, if female; 1, if male
SYSTOLIC = Integer representing the patient's systolic blood pressure
DIASTOLIC = Integer representing the patient's diastolic blood pressure
WT = Patient's weight in pounds
HT = Patient's height in inches
BMI = Patient's BMI
RACE = Patient's race (1, if American Indian; 2, if Asian; 3, if African American; 4, if Pacific Islander; 5, if White; 6, if Multiple; 7, otherwise)
ETHNICITY = Patient's ethnicity (0, if patient identifies as Hispanic; 1, otherwise).

3. RESULTS

Multiple datasets were merged before performing the actual data analysis. There were datasets for the patient vitals (e.g., systolic blood pressure and diastolic blood pressure; SBP, DBP), for the demographics, for all of the HT medications used by patients, and for the HT diagnosis per patient. Some observations were removed, including those that were outside of the 2017-2021 timeframe, those with missing blood pressure (SBP or DBP), those with obviously spurious blood pressure measurements of either greater than 300 SBP or greater than 150 DBP, as well as those aged 17 and younger or those aged 86 and older. Additionally, those that have kidney disease or a kidney transplant were deleted. The cleaned dataset comprised a significant subset of the initial observations, focusing on around 85,000 patients from the original 500,000. For the machine learning algorithms, the data is randomly split into two: 80% for training and 20% for validation, as illustrated in Figure 3.

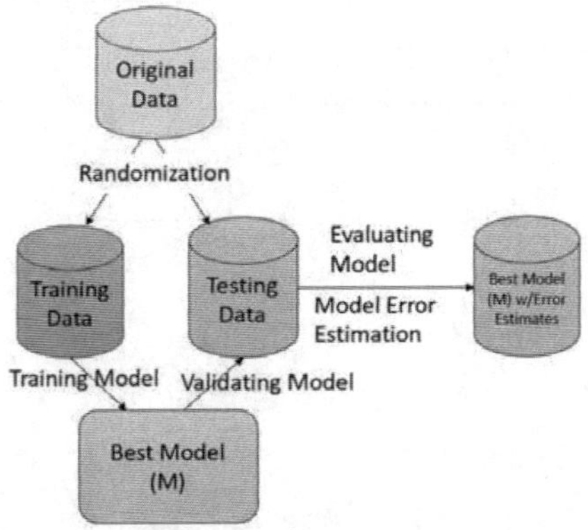

FIGURE 3. A diagram of tuning and testing of a machine learning algorithm in the general case.

For each of the two models (XGBoost and PLANN) used, the initial model was tuned based on the training data. This training dataset is used to create the best model, and the resulting model/algorithm was then validated on the testing data. Based on the predictor values in the testing data, the algorithm calculated the response variable (i.e., HT presence) for every observation within the testing dataset. The true values of HT presence in the testing dataset are known, so the predicted responses are compared to the actual responses and the AUC was calculated. The real data results from the comparison of the two different approaches using the AUC and SHAP Values are shown in Tables 1 and 2 below. According to those tables, it is conclusive that both the PLANN and XGBoost models had AUC values of almost 0.9; hence, performed very well. From both the XGBoost and PLANN model, the Shapley importance values of the features "Weight", "BMI", and "Age" have a relatively high impact on the model's predictions of true HT compared to other features. This implies that changes in weight, BMI, and age have a strong influence on the predicted outcome. Additionally, from Figure 5, we can see that the risk for HT increases with weight. This is a result that we have expected to obtain according to literature searches. For example, in literature, we can find that a reduction of 20 pounds can reduce the systolic blood pressure by 5-20 units (mmHg) [8]. What needs to be discussed with medical professionals is the finding in Figure 4 in the clinical sense. The data shows that the risk for HT is higher for heavier patients between the ages of 18 and 60, but these results flip after the age of 60. In other words, after the age of 60, it seems that heavier patients have a lower risk of HT. This phenomenon is discussed in the mathematical/statistical sense in the discussion section.

Method	AUC
XGBoost	0.888
Neural Nets (PLANN)	0.885

TABLE 1. AUC values for both models on the real dataset: XGBoost and PLANN.

Importance Values	PLANN	XGB
Weight	0.125	0.071
BMI	0.103	0.032
Age	0.085	0.030
Height	0.054	0.005
Sex	0.048	<0.001
Hispanic	0.009	<0.001

TABLE 2. SHAP values for both models on the real dataset: XGBoost and PLANN.

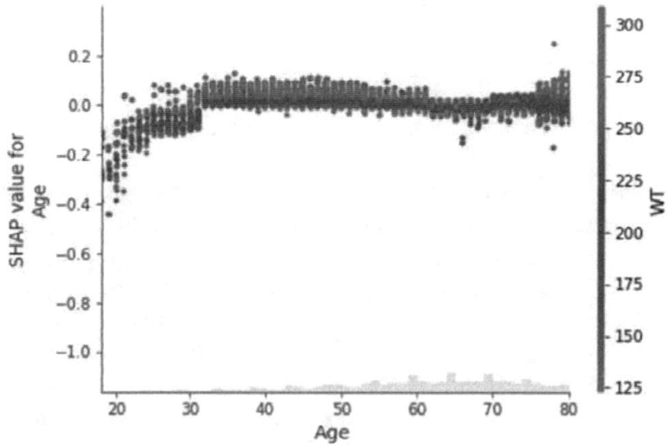

FIGURE 4. Relationship between the age, weight, and HT risk of a patient.

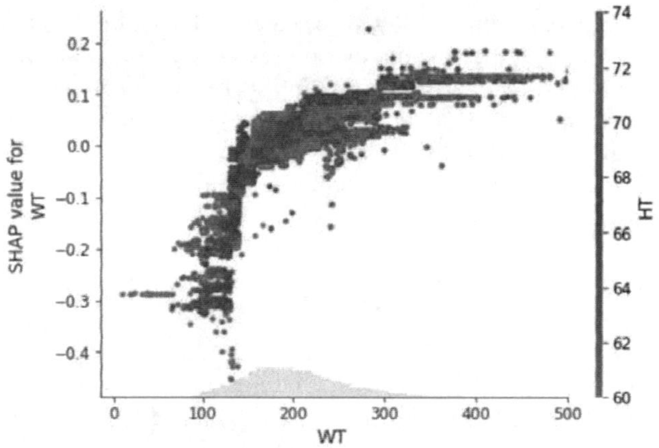

FIGURE 5. Relationship between the height, weight, and HT risk of a patient.

4. DISCUSSION AND CONCLUSIONS

We applied and compared the performance of two *Artificial Intelligence/Machine Learning* (AI/ML) approaches (XGBoost and PLANN) to a large-sized cardiac dataset. AI/ML approaches are popular for modeling large, complex data. For this dataset, the PLANN and XGBoost performed similarly well, with AUC value approximately 0.9. The dataset we are working with may not have complexity of patterns or structure that require the neural network model. In such cases, models for more structured data like XGBoost, can perform well. For example, an X-Ray or MRI scan with data in the form of pixels is unstructured. Both models, XGBoost and PLANN capture well the complexity of the data patterns.

Even though BMI (Body Mass Index) is a derived metric that depends on weight and height, it can still be useful to include both weight and BMI as features in AI/ML models such as XGBoost or PLANN. The reason behind this lies in the way these algorithms handle input features. Both XGBoost and PLANN are capable of learning complex, non-linear relationships between the input features and the target variable. They don't strictly require features to be uncorrelated, which is a key assumption in some other models, such as linear regression. While weight and BMI are indeed correlated (since BMI is calculated from weight and height), they provide different perspectives on an individual's physical condition. Including both of these variables allows the model to potentially capture more intricate patterns in the data.

An example of this is that atrial fibrillation (AF) and HT share a complex interplay, with substantial epidemiological evidence suggesting a strong correlation between the two conditions. HT is a well-established risk factor for the development of AF [16]. It contributes to structural and functional changes in the heart, such as left ventricular hypertrophy, stiffness of the heart chambers, and alterations in the heart's electrical system, which in turn can lead to the development of AF. Furthermore, the presence of AF can exacerbate HT by disrupting the normal rhythm and flow of blood in the heart, which can lead to a cycle of worsening of both conditions. Interestingly, studies have shown that taller individuals have a higher risk of AF [12]. A possible explanation lies in the fact that taller individuals have larger atria (the upper chambers of the heart), which could make them more susceptible to developing AF. Larger atria mean more cells and a greater chance for abnormal electric pathways to develop, which can disrupt the heart's rhythm. Furthermore, the larger blood volume that comes with greater height could also place a larger workload on the heart, potentially contributing to the development of HT. Despite greater height leading to lower BMI at a constant height, these factors can lead to an increased risk of AF and HT. This highlights that it is even beneficial to include both height and BMI in models that can handle such correlated variables.

It is important to also discuss the fact that the less heavy patients over the age of 60 have a higher risk of HT, which seems counterintuitive. A possible explanation is as follows. It is plausible that patients above the age of 60 have a lower risk for HT since this dataset is based on the data from a medical center, and this sample may not fully represent the broader population. The observed lower risk in older, heavier patients could be attributed to several factors. Firstly, the medical center's older patient population may consist of individuals who are more health-conscious, have better access to healthcare, and adhere to regular medical check-ups. This demographic bias could result in an overrepresentation of individuals who are actively managing their health and thus exhibit lower rates of HT. Secondly, it is also essential to consider the potential for survivor bias in the dataset. If individuals who were overweight or obese and had high blood pressure earlier in life had higher mortality rates, the surviving older population may be skewed toward individuals who had a more favorable response to weight and blood pressure. This could lead to a misleading observation that heavier individuals have lower blood pressure after the age of 60. However, it is vital to recognize that this observation is based on a limited and possibly biased sample from a medical center, and the actual prevalence of HT in the general population may differ significantly. To draw more conclusive findings, broader population-based studies would be necessary, considering a more diverse range of individuals from various demographics and regions.

There have been other such studies with similarly intriguing results potentially due to this form of sampling bias. A UCLA Survival Analysis study [7], conducted using data from the real Worcester Heart Attack Study dataset, has yielded intriguing and somewhat unexpected findings. As a reputable source, UCLA analyzed various features (eg, age, BMI, gender) to determine their impact on the risk of death in cardiac patients. Surprisingly, the study revealed that individuals with a BMI of 25 who experienced a 5-unit increase showed a statistically significant reduction in the death risk. Also, patients with a BMI of 30 and a 5-unit increase in BMI did not exhibit a statistically significant change in the risk of death. These results are thought-provoking and warrant discussion with medical professionals to explore potential medical reasons behind this phenomenon. However, from a statistical perspective, it's important to acknowledge that this study focused on cardiac patients rather than the general population, which could account for some of the unexpected outcomes observed. Further examination and consultation with medical experts are necessary to gain deeper insights into these findings.

5. REFERENCES

[1] Biganzoli E, Boracchi P, Mariani L, Marubini E. Feed forward neural networks for the analysis of censored survival data: a partial logistic regression approach. *Stat Med*, 17(10):1169-1186. (1998)
[2] Biganzoli E, Boracchi P, Marubini E. A general framework for neural network models on censored survival data. *Neural Netw*. 15(2):209–18. (2002)
[3] Fredenslund, L. *Computational Complexity Of Neural Networks*. Lunalux.io. https://lunalux.io/computational-complexity-of-neural-networks/. (2023)
[4] Shapley LS. A value for *n*-person games. In: Kuhn HW, Tucker AW (eds.) *Contribution to the Theory of Games* (AM-28) 2:307-318. (1953)
[5] Lipovetsky S, Conklin M. Analysis of regression in game theory approach. *Applied Stochastic Models in Business and Industry* 17(4):319-330. (2001)
[6] Hausken K, Mohr M. (2001) The value of a player in *n*-person games. *Soc Choice Welfare* 18: 465-483.
[7] *Introduction to Survival Analysis in SAS*. Stats.oarc.ucla.edu. https://stats.oarc.ucla.edu/sas/seminars/sas-survival/. (2021)
[8] Zelman, K. M., MPH, RD, & LD. *Take Charge of Your Blood Pressure*. WebMD. https://www.webmd.com/hypertension-high-blood-pressure/features/take-charge-of-your-blood-pressure. (2004)
[9] Chen T, Guestrin C. XGBoost: A Scalable Tree Boosting System. KDD '16: Proceedings of the 22nd ACM SIGKDD International Conference on Knowledge Discovery and Data Mining, August 2016. pp 785-794.
[10] Haynatzki, R. Prediction of Survival Models: A Comparison Between Machine Learning and Cox Regression. (2021)
[11] Ichiishi, Tatsuro. *Game Theory for Economic Analysis*. New York: Academic Press. pp. 118–120. ISBN 0-12-370180-5. (1983)
[12] Rosenberg, M. A., Patton, K. K., Sotoodehnia, N., Karas, M. G., Kizer, J. R., Zimetbaum, P. J., Chang, J. D., Siscovick, D., Gottdiener, J. S., Kronmal, R. A., Heckbert, S. R., & Mukamal, K. J. (2012). The impact of height on the risk of atrial fibrillation: the Cardiovascular Health Study. *European Heart Journal*, *33*(21), 2709–2717. https://doi.org/10.1093/eurheartj/ehs301
[13] Molnar, C. 5.9 Shapley Values | Interpretable Machine Learning. In *christophm.github.io*. Lulu.com. https://christophm.github.io/interpretable-ml-book/shapley.html. (2019)
[14] Centers for Disease Control and Prevention (CDC). "High Blood Pressure Facts." Centers for Disease Control and Prevention, 23 Jan. 2023, www.cdc.gov/bloodpressure/facts.htm.

[15] Mills, Katherine T., et al. "Global Disparities of Hypertension Prevalence and Control." Circulation, vol. 134, no. 6, 9 Aug. 2016, pp. 441–450, www.ncbi.nlm.nih.gov/pmc/articles/PMC4979614/, https://doi.org/10.1161/circulationaha.115.018912.
[16] Dzeshka, M. S., Shantsila, A., Shantsila, E., & Lip, G. Y. H. (2017). Atrial Fibrillation and Hypertension. *Hypertension*, *70*(5), 854–861. https://doi.org/10.1161/hypertensionaha.117.08934
[17] Google Developers. (2019). Classification: ROC Curve and AUC | Machine Learning Crash Course. Google Developers. https://developers.google.com/machine-learning/crash-course/classification/roc-and-auc.
[18] Smeda, Kousai. *Understand the architecture of CNN*. Medium; Towards Data Science. https://towardsdatascience.com/understand-the-architecture-of-cnn-90a25e244c7. (2019)
[19] Shabani, F., Kumar, L., & Ahmadi, M. Assessing Accuracy Methods of Species Distribution Models: AUC, Specificity, Sensitivity and the True Skill Statistic. *Global Journal of Human-Social Science: B Geography, Geo-Sciences, Environmental Science & Disaster Management*, *18*(1). https://researchnow.flinders.edu.au/en/publications/assessing-accuracy-methods-of-species-distribution-models-auc-spe. (2018)

Temporal-causal modeling of air pollution in the city of Plovdiv, Bulgaria: a case study

A V Ivanov, S G Gocheva-Ilieva and M P Stoimenova-Minova

Department of Mathematical Analysis, Faculty of Mathematics and Informatics, University of Plovdiv Paisii Hilendarski, 24 Tzar Asen Street, 4000 Plovdiv, Bulgaria
E-mails: aivanov@uni-plovdiv.bg, snow@uni-plovdiv.bg, mstoimenova@uni-plovdiv.bg

Corresponding author's e-mail: aivanov@uni-plovdiv.bg

Abstract. High levels of air pollution are dangerous to human health, which is a current problem for densely populated cities worldwide. Studying this problem can help detect pollutants' time dependencies on basic meteorological measurements and other factors for future prediction and elaborate corresponding alarms when official upper pollution limits are exceeded. In this work, time-causal models based on previous daily time observations and meteorological measurements in the city of Plovdiv, Bulgaria, are applied. Vector-type temporal-causal models are constructed and analyzed for carbon dioxide (CO2), nitrogen dioxide (NO2), sulfur dioxide (SO2), and fine dust particles below size 10, 2.5, and 1 micron (PM10, PM2.5, and PM1), respectively. Pollution levels are predicted seven days ahead.

1. Introduction

Among the most important problems in the modern world is environmental pollution. A complex mix of natural and anthropogenic factors, related to each other, determines the degree of pollution of air, water, soil and all ecological systems. Ambient air pollution in urban areas is one of the biggest environmental health risks. High, as well as lower, but constant levels of the concentration of harmful emissions in the air, lead to the complication of diseases in cases of stroke, heart disease, lung cancer, as well as in chronic and acute respiratory diseases, Alzheimer's disease and others [1-3]. According to the World Health Organization (WHO), the combined effects of ambient air pollution and household air pollution are associated with 6.7 million premature deaths annually [4].

In recent years, many researchers have focused their research in the field of statistical processing of empirical data to analyze, model and predict concentrations of harmful outdoor air pollutants. Research mainly concerns individual pollutants such as PM_{10} – fine particulate matter with diameter below 10 microns, with diameter below 2.5 microns ($PM_{2.5}$), NO – nitrogen monoxide, NO_2 – nitrogen dioxide, SO_2 – sulfur dioxide, ground level ozone (O_3), CO – carbon monoxide and more. For this purpose, both classic and the most avant-garde computer-based methods and algorithms with machine learning (ML) are developed and applied. Classical methods such as multiple linear regression (MLR), Principal component analysis (PCA), stochastic autoregressive integrated moving average (ARIMA), seasonal ARIMA, numerical simulation modeling, gaseous air contamination modeling and others are used, for example, in [5-10]. The authors of [5] analyze and model the tropospheric ozone in Warsaw depending on its precursors for selected heat waves using PCA and MLR. Stochastic Box-Jenkins ARIMA and MLR models of PM10 are developed in [6,7], gaseous air pollution is studied in [8] for some Bulgarian cities. Computational fluid dynamics model capable of simulating the dispersion of pollutants in a large urban environment is reported in [9]. In [10], supercomputer-based numerical simulations using the finite volume method and semi-implicit difference schemes are employed for short-term prediction of extreme meteorological conditions and atmospheric air quality.

A strong alternative to standard statistical and numerical methods for air pollution research are statistical models based on ML techniques. This field covers numerous variants of Neural network (NN), Support Vector Machine (SVM), decision trees, Random forest and many others. Results in this field are published in [11-15].

Paper [11] applies a wavelet model to decompose meteorological data to improve feature selection and designs a radial basis function long short-term memory (LSTM) model for modeling $PM_{2.5}$ pollution data. Authors of [12] develop a new hybrid model for air quality prediction based on a reed deer metaheuristic optimized LSTM Deep Learning network coupled with ARIMA models. Sánchez et al. [13] combine SVM regression and multilayer perceptron (MLP) to model nitrogen oxides (NOx), CO, SO_2, O_3 and PM_{10}. Models of each individual pollutant with predictors being the remaining pollutants were constructed and the relative influence of the predictors in the models was assessed. In [14], ensemble decision tree models using random forest and CART Ensemble and Bagging are built, then the stacking paradigm is applied through the MARS method to improve the PM_{10} forecasts.

In addition to the study and modeling of different pollutants, modeling the relationships between multiple pollutants is also an important task. In practice, there is a mutual influence between the components of the air, determined by chemical reactions, weather conditions and the complex dynamics of emissions. There are several different types of methods in this field. A vector type of multivariate ARIMA is implemented in [15], where 3- and 4 dimensional ARIMA models are built.

A more preferred method for studying mutual relationships between different factors is time-series Granger causality analysis [16]. It is applied in various fields. Among them is the study of the links between air pollution and economic indicators [17], causal dynamics between CO_2 emissions, health expenditures and economic growth [18], causal effects of air pollution on health [19,20], and other. In particular, although a small number of publications, have used Granger causality to study interdependencies between air pollutants' time series. In [21] the authors develop a methodology for discovering causal links between the ozone variations, nitrogen oxides and local meteorological conditions, using Positive Matrix Factorization (PMF), k–means cluster analysis and the Granger causality test. In [22] a quantitative evidence is presented to verify the relative impacts from stationary emission sources to a number of air quality stations applying time-series Granger causality. Classification models based on causality tests and different machine learning (ML) algorithms are reported in [23]. The author considered the causality influence of the industrial pollutants NO_2, SO_2, CO and O_3 and meteorological variables on $PM_{2.5}$. Zhu et al. [24] proposed a spatio-temporal Granger causality approach which integrates different urban dynamic big data to select the most influential dataset for air quality estimation. An analysis of the main methods and results in time series modeling related to air pollution can be found in recent review papers [25,26].

The purpose of this study is to investigate the mutual relationships between the main air pollutants using vector-type temporal-causal analysis. Modeling is conducted with empirical data for nitrogen dioxide (NO_2), sulfur dioxide (SO_2), carbon dioxide (CO_2), and particular matter (PM) with size under 10, 2.5, and 1 micron (PM_{10}, $PM_{2.5}$, and PM_1), and meteorological factors. The goal is to identify the causal relationships between these pollutants, to identify the problematic pollutants, as well as to demonstrate the possibilities for evaluating short-term forecasts for future pollution. The approach is used for local data for the North-western part of the city of Plovdiv, Bulgaria. But in the same way, it can easily be applied to any urban area, both locally and on a much more general scale, depending on the available data.

The study was carried out using IBM SPSS Statistics [27,28].

2. Air quality standards, data and methods

In this section, we briefly present the data used, the national and European standards for air quality, as well as the methods used in the present study.

2.1. Air quality polices and standards

As a member of the European Union, legal regulations have been adopted and are in force in Bulgaria, fully in line with European environmental policies, norms and standards [29]. Ambient air quality is systematically monitored and controlled by the national Executive Environment Agency (ExEA) [30] and European Environment Agency [31]. Especially, the European Directive 2008 [32] sets the following daily mean limit value for NO_2 in outdoor air: 120 μg/m³ (micrograms per cubic meter) equivalent to 0.062 ppm (parts per million); for SO_2 the maximum allowable 24-hour mean concentration is 125 μg/m³ (0.047 ppm), which should not be exceeded more than three times per calendar year.

The levels of the main air pollutants in Bulgaria in the larger cities are measured by 36 automated measuring stations. The data is collected and processed online at short intervals and reported in different aggregated form to ExEA and the European Agency. Two state automated and certified measuring stations have been installed in Plovdiv. Plovdiv is characterized by relatively high values of particulate matter PM_{10}, while the other type of particulate matter - PM_1 and $PM_{2.5}$ are not measured in the mentioned stations.

2.2. Study area and data

The data used for modeling are measured in the city of Plovdiv, Bulgaria. Plovdiv is the second-largest city in Bulgaria. It is located in the Upper Thracian valley, on both banks of the Maritsa River, on seven syenite hills. It has a population of over 360 000 as of 2023 in the greater metropolitan area. The climate is transitional-continental with an average annual temperature of 12.3°C. During the summer months, the average maximum temperatures are within 30-31°C, with maximum daily temperatures over 40 and 42°C also being reached. The average annual minimum temperature is 6.5°C. In Plovdiv, weak winds prevail (0 - 5 m/s), with winds up to 1 m/s being up to 95% during the year, which at low temperatures helps to retain harmful emissions in the air [33]. Other conditions for increased levels of air pollutants are intensive car traffic, various combustion processes from household and industry, road maintenance and others in the Plovdiv District. In particular, NO_2 and SO_2 mainly are results of combustion of fossil fuels, coil at power plants and the automobile traffic.

In this study we examine average daily data for six air pollutants, namely NO_2, SO_2, CO_2, PM_{10}, $PM_{2.5}$, PM_1, collected by Alphasense automatic monitoring station with optical sensors. The station is located on the roof of Plovdiv University, New building. The period of collected data is from 11 April 2021 to 24 March 2023. The main descriptive statistics of the pollutants' time series are given in Table 1.

Table 1. Descriptive statistics of the air pollutants' variables.

Statistic	NO2 (ppm)	SO2 (ppm)	CO2 (ppm)	PM10 (ppm)	PM2.5 (ppm)	PM1 (ppm)
N	713	713	713	713	713	713
Mean	0.383	0.420	196.35	28.09	9.39	5.87
Std. Dev.	0.102	0.018	93.61	109.33	15.34	7.50
Variance	0.010	0.0003	8763.42	11953.8	235.60	56.26
Minimum	0.191	0.386	2.89	0.68	0.68	0.51
Maximum	0.565	0.476	366.30	1434.86	174.97	62.15

From Table 1 it can be seen that for the studied period the average 24-hour concentration of NO_2 particles is 0.383(ppm), equal to 741 $(\mu g / m^3)$. This value exceeds the permissible limit for the European Union, as well as of the WHO [4,31,32] of 120 $(\mu g / m^3)$ =0.062(ppm) about 6 times. Also, Table 1 shows that the average daily SO2 limit of 125 $(\mu g / m^3)$=0.047(ppm) is actually 1117.3 $(\mu g / m^3)$ thus exceeding the limit by more than 8.9 times. We will consider these two pollutants problematic for the given area of Plovdiv. In addition to pollutants, the monitoring station also measures the following meteorological time series: Temperature (°C), Humidity (%), Pressure (Pa) and Luminosity (Luxe). They are also included in the constructed models.

2.3. Brief description of temporal-causal modeling

Temporal causal modelling (TCM) is a statistical approach based on Granger causality test [16]. It attempts to discover key causal relationships in vector time series data. This approach differs from traditional time series modeling where you must explicitly specify the predictors for a target series. Since temporal causal modeling typically involves building models for multiple related time series, the result is referred to as a model system.

In the simplest case, let us assume as given two stationary time series, $X=\{X_1, X_2, ..., X_t, ..., X_N\}$ and $Y=\{Y_1, Y_2, ..., Y_t, ..., Y_N\}$. with values at the same time points $t, t=1,2,...,N$. We consider the symmetric autoregressive processes of order q

$$\begin{aligned} Y_t &= a_0 + \sum_{k=1}^{q} a_k Y_{t-k} + \sum_{k=1}^{q} b_k X_{t-k} + \varepsilon_t, \quad \varepsilon_t \in N(0,\sigma_1^2) \\ X_t &= c_0 + \sum_{k=1}^{q} c_k X_{t-k} + \sum_{k=1}^{q} d_k X_{t-k} + \delta_t \quad \delta_t \in N(0,\sigma_2^2), \end{aligned} \tag{1}$$

where $t=q+1,...,N$, a_0, a_k, b_k, c_0, c_k, d_k, $k=0,1,...,q$ are constant coefficients, q is the number of lags.

The Granger test examines the following two null hypotheses sequentially:

$$\begin{aligned} H_0^{(Y)} &: b_1 = b_2 = ... = b_q = 0, \\ H_0^{(X)} &: d_1 = d_2 = ... = d_q = 0. \end{aligned} \tag{2}$$

In essence, the hypothesis $H_0^{(Y)}$ tests whether the time series *X* Granger cause *Y*, and reciprocally $H_0^{(X)}$ checks if *Y* Granger causes *X*. A statistically significant test result with small *p*-value (for example $p \le 0.05$) means that the null hypothesis is false and should be rejected. A *p*-value greater than 0.05 has come to refer to the "null" causal effect. Hypotheses (2) can be tested with F-test, LM or t-test on lagged values of X and *Y*. It should be noted that the result of the tests depends on the number of lags q.

In the multivariate case, let us have an m-dimensional time series matrix $\mathbf{X} = \left(X^{(1)}, X^{(2)}, ..., X^{(m)}\right)$ with $X^{(j)} = \left\{X_1^{(j)}, X_2^{(j)}, ..., X_N^{(j)}\right\}$, $j=1,2,...,m$. The TCM implies that all these time series are stationary and standardized them before processing them. The TCM explores the autoregressive process of the type

$$X_t^{(j)} = \mathbf{A}_0 + \mathbf{A}_1 X_{t-1}^{(j)} + ..., \mathbf{A}_q X_{t-q}^{(j)} + \varepsilon_{X,t}, \quad t=q+1,...,N, \quad ,j=1,2,...,m \tag{3}$$

where $\mathbf{A}_j$ are matrices of coefficients, $\varepsilon_{X,t}$ is a random error with standard normal distribution. The null hypothesis for the Granger causality test implies that a variable $X^{(j)}$ does not cause $X^{(i)}$, $(i \ne j)$

through the check $\mathbf{A}_i = \mathbf{0}$. The TCM algorithm then builds the autoregressive time series model for each target $X^{(i)}$ and includes only those inputs that have a causal relationship with the target $X^{(j)}$. This is done for selected or all vectors from the matrix **X** at a given number of lags q. A valid TCM model can be further used to obtain predictive models that allow adequate prediction of future values not used in building the models. This approach differs from traditional time series modeling where you must explicitly specify the predictors for a target series. Since temporal causal modeling typically involves building models for multiple related time series, the result is referred to as a model system. In this study the TCM engine of IBM SPSS Statistics was implemented [27].

2.4. Statistical measures to evaluate model performance

The quality of constructed TCM models was evaluated by the following standard statistical measures: root mean square error (RMSE), root mean squared percent error (RMSPE), mean absolute percentage error (MAPE), and coefficient of determination (R^2), given by the expressions:

$$\mathrm{RMSE} = \sqrt{\frac{1}{N}\sum_{t=1}^{N}\left(Y_t - \hat{Y}_t\right)^2}, \quad \mathrm{RMSPE} = 100\sqrt{\frac{1}{N}\sum_{t=1}^{N}\left(\frac{Y_t - \hat{Y}_t}{Y_t}\right)^2},$$

$$\mathrm{MAPE} = \frac{1}{N}\sum_{t=1}^{N}\left|\frac{Y_t - \hat{Y}_t}{Y_t}\right|, \quad R^2 = \frac{\sum_{t=1}^{N}\left(\hat{Y}_t - \bar{Y}\right)^2}{\sum_{t=1}^{N}\left(Y_t - \bar{Y}\right)^2}, \tag{4}$$

where $\hat{Y}_t$ is the value predicted by the model at time t, $\bar{Y}$ is the mean value of Y, N is the sample size. We consider that models are better when the coefficient of determination is close to 1 and the error measures are small.

Model selection with multiple nested valid models was performed by using the Akaike information criterion (AIC) and Schwarz's Bayesian information criterion (BIC), where the minimum value of the criterion is selected [34].

Along with these measures, for the reliability of the time series models, their residuals were examined, which should also be small.

3. Results

3.1. Initial data processing and verification of TCM application conditions

To conduct a TCM analysis, we have six time series of pollutants from Table 1 and four meteorological variables. Before moving on to building the models, we conducted their initial processing.

The studied data has the size of data samples N=713. Building autoregressive models does not allow for missing data. In our case, the original raw data had less than 15% missing data, which was previously replaced by the linear interpolation method.

Then the sequence plots of the variables are represented. In this study, the main results for variables NO2 and SO2, corresponding to NO_2 and SO_2 pollutants were obtained. Their line plots are shown in Figure 1. From the behavior of the graphs for both variables, no trend is observed. Besides the graphic study, their ACFs and partial ACF (partial autocorrelation functions) were also plotted. In the same way, the remaining variables were examined and no trends were established. We can conclude that all variables are stationary. In the general case, if some of the variables has a trend, then the trend should be removed, for example, by differentiating (once or repeatedly) to obtain a stationary time series. The stationary is a condition for applying TCM.

Figure 1 shows a clear annual cyclic recurrence. During the winter periods, the values of NO2 and SO2 are the greatest, while during the summer months, low values are observed. From here it can be concluded that the influence of weather conditions is significant. In the studied time period, the amplitudes for both variables have similar behavior.

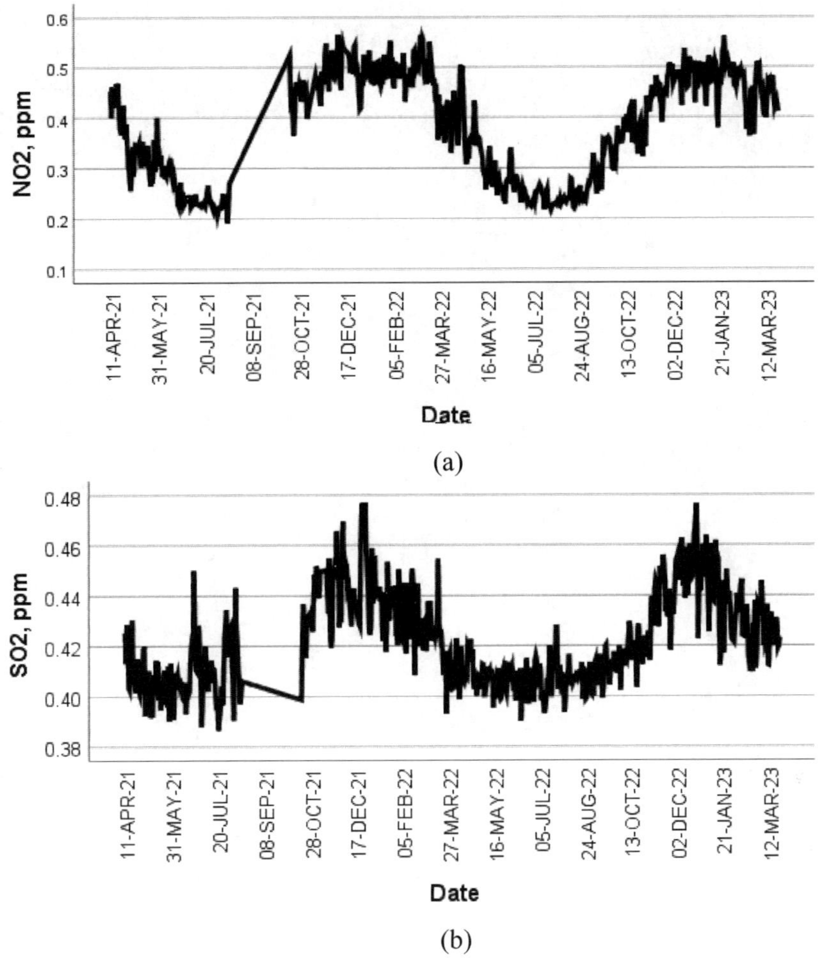

Figure 1. Line plots of the investigated problematic variables: (a) NO2, (b) SO2.

3.2. Results from the TCM

In this section, we present the causal relations obtained by applying the TCM method.

3.2.1. Building the overall TCM system model

To build TCM models we use data for N=713 days described in section 2.2. For model validation and forecasting, the data for the next 7 days, chronologically after these 713, were used as out of data sample to N1=720.

Since TCM typically involves building models for multiple related time series, the result is referred to as a model system. Our model system includes all ten variables – for the time series of the six pollutants and the four meteorological variables. Each of these variables can be a target and all the others are in the role of predictors. Since the system is of autoregressive type, an important parameter is the maximum number of lags that consider the influence of the time series back from the current moment t. In the built models, we have chosen a maximum number of lags equal to 7 days. In formula (1), this corresponds to $q = 7$. The causal relations between variables are calculated on the basis of correlations and their statistical significance is reported. In the resulting TCM system, only those variables remain that have statistically significant coefficients at a given level α. We apply the case $\alpha = 0.05$ as an upper significance limit.

Figure 2 shows the resulting TCM system of all causal relations between all ten variables. The arrows in the circle show the direction and strength of influence of the predictor on the target. A thicker arrow indicates a stronger effect. Figure 2 shows that there are only eight variables left in the model. For example, for NO2 the strongest causal links stem from SO2 and PM2.5. Causal links have been established for SO2 with Pressure and PM2.5.

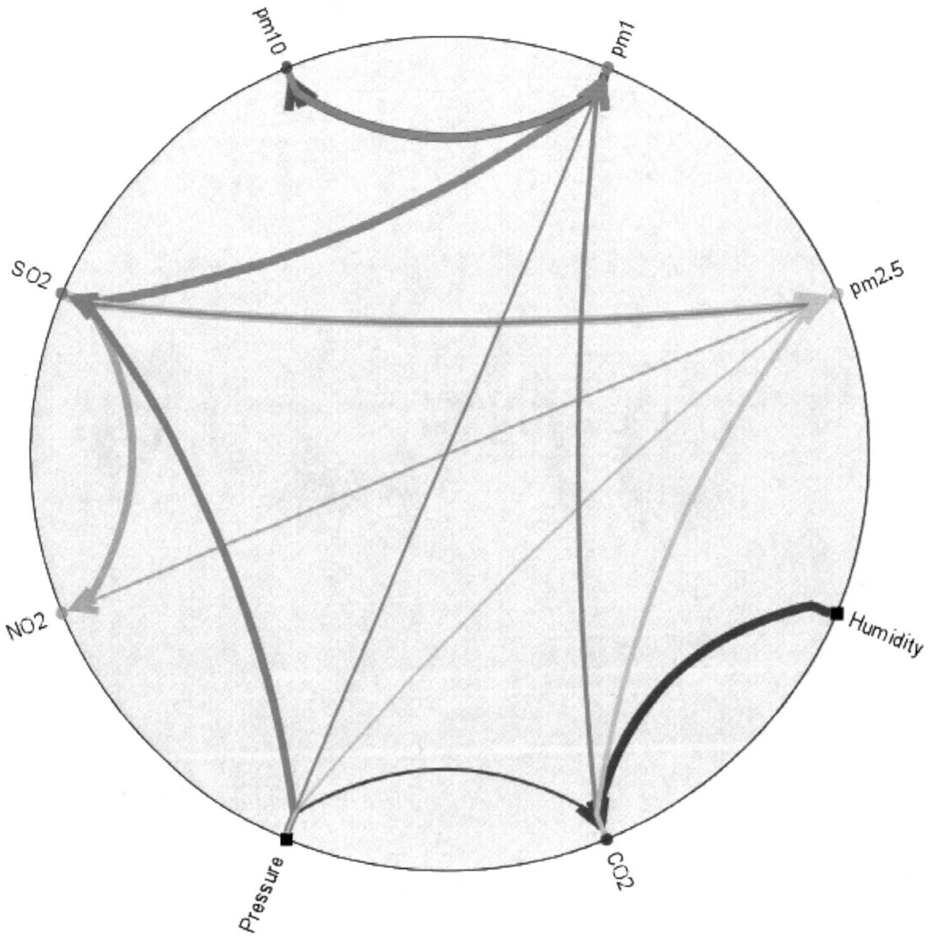

Figure 2. Graphical representation of the causal relations between the time series in the model system.

In more detail, the causal impact of the predictors on NO2 and SO2 is displayed in Figure 3 and Figure 4, respectively. Variables that affect the targets of interest are referred to as causes. For example, Figure 3 shows that we have the following causal links:

- the strongest: Pressure $\rightarrow$ SO2 $\rightarrow$ NO2;
- weaker: Pressure $\rightarrow$ SO2 $\rightarrow$ PM2.5 $\rightarrow$ NO2, Humidity $\rightarrow$ CO2 $\rightarrow$ PM2.5 $\rightarrow$ NO2;
- the weakest: Luminosity $\rightarrow$ NO2.

From the impact diagram on Figure 4, we similarly have the causal relations

- the strongest: Pressure → SO2;
- weaker: Humidity → CO2 → PM2.5 → SO2;
- the weakest: Luminosity → SO2.

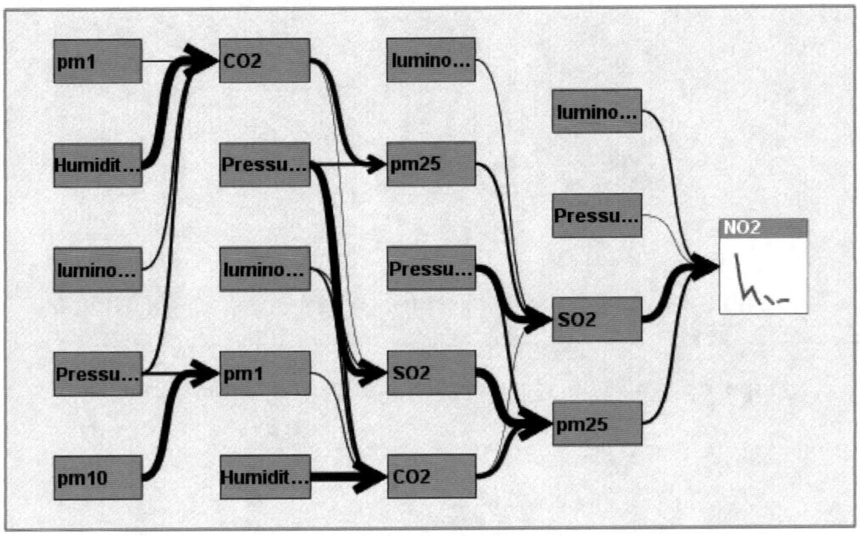

Figure 3. Impact diagram of causes of time series on NO2.

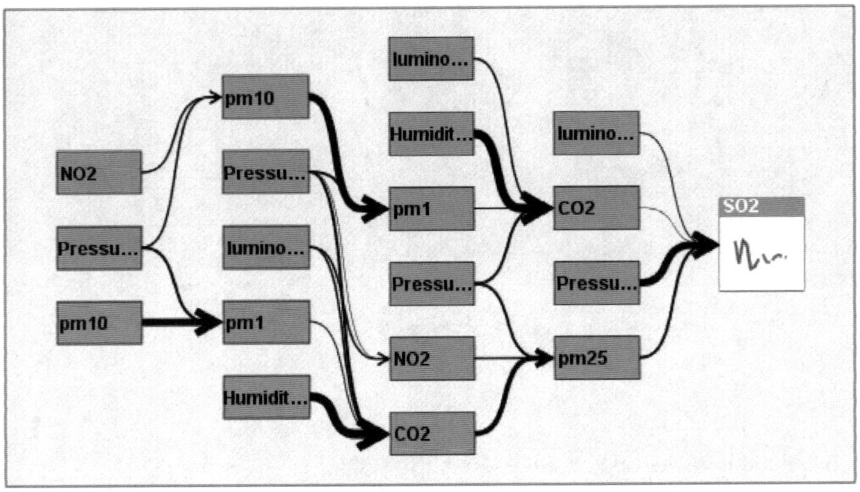

Figure 4. Impact diagram of causes of time series on SO2.

Conversely, the effects of NO2 and SO2 as predictors on the other variables are illustrated in the diagrams of Figures 5 and 6 respectively. Figure 5 shows the effects of NO2 - strongest on PM10 and weaker on PM2.5. Figure 6 displays the impact diagram of the effects of SO2 – weaker on NO2 and weakest on CO2. In the impact diagrams of causes or effects we need to toggle in the output item that displays the impact diagram.

Impact diagrams for causes and effects were also built for all pollutant variables.

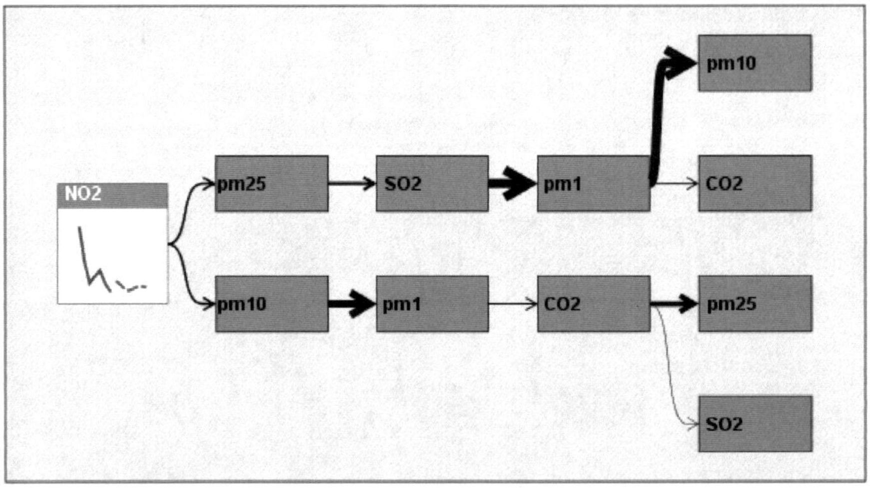

Figure 5. Impact diagram of effects of NO2 on other variables.

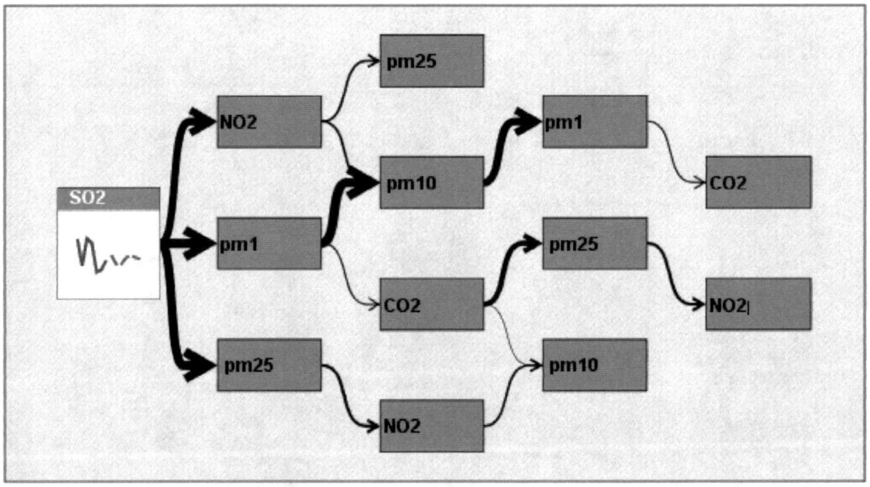

Figure 6. Impact diagram of effects of SO2 on other variables.

3.2.2. Model evaluation

The obtained statistical indicators calculated from formulas (4) for the top causal models generated for each of the pollutants are presented in Table 2. The main criterion for choosing the best model is the coefficient of determination R^2, which is the default criterion for quality evaluation of the causal models. A larger value of R^2 is assumed to correspond to a better model. Also in Table 2, the values of the AIC and BIC are presented, which are used for choosing the best model. It is necessary to recall that these criteria are applied to nested models and the model with their smallest possible values is selected.

In Table 2, the models are arranged in descending order relative to R^2. A best model with R^2=0.98 was obtained for CO2, i.e. it explains about 98% of the data. The next model is for NO2, which fits the measured data with 94%, followed by the model for SO2 - with a fit of 79%. The lowest and unsatisfactory result of R^2=0.39 was obtained for PM10.

Table 2. Fit statistics for the top causal models of the air pollutants' variables.

Model target	RMSE	RMSPE	MAPE	AIC	BIC	R^2
CO2 (ppm)	12.14	0.40	0.11	3559.74	3.723.89	0.98
NO2 (ppm)	0.02	0.06	0.05	-5177.66	-5013.51	0.94
SO2 (ppm)	0.01	0.02	0.01	-6677.03	-6512.89	0.79
PM1 (ppm)	4.72	1.45	0.65	2226.96	2391.10	0.62
PM2.5 (ppm)	10.80	1.84	0.88	3395.48	3559.63	0.53
PM10 (ppm)	86.72	7.00	2.87	6336.44	6500.59	0.39

3.2.3. Predictive performance of the models

The quality of fit of the NO2 and SO2 models to the measured data is depicted in Figures 7 and 8, respectively. At the far right are the predicted values for the 7 days retained for model verification. The graphs show a very good fit between the model predicted values and the measured ones

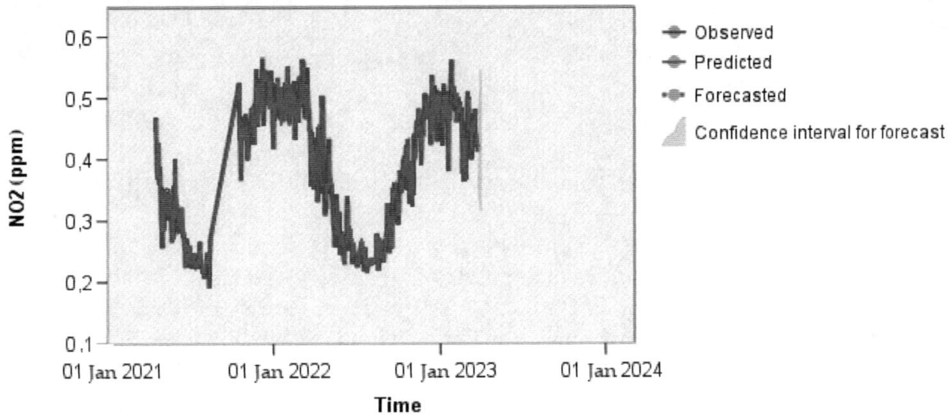

Figure 7. Observed and predicted values for NO2.

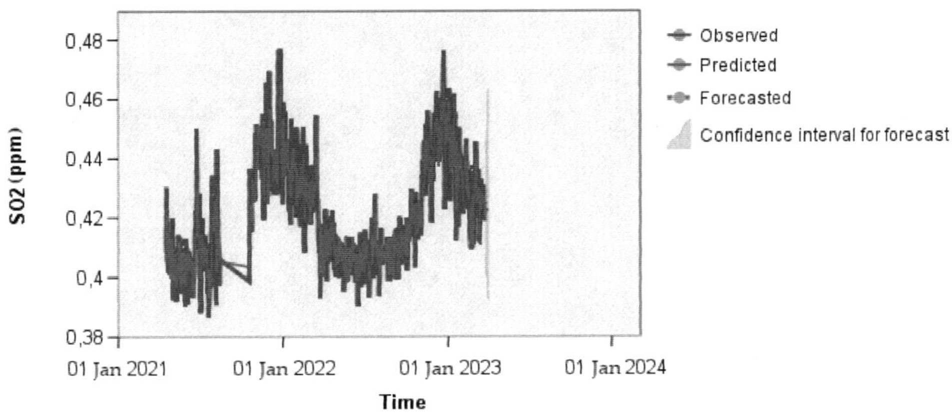

Figure 8. Observed and predicted values for SO2.

3.2.4. *Diagnosis of residuals*

The quality of the constructed models and their adequacy can be verified by the behavior of their residuals $\varepsilon_t = Y_t - \hat{Y}_t$ (see also (4)). Figure 9 shows the residuals of the built TCM models of NO2 and SO2 for the entire investigated time period. Deviations from zero are small, within 5% of the variables' mean (see Table 1). It can be concluded that the models built are adequate and can be used for predictions.

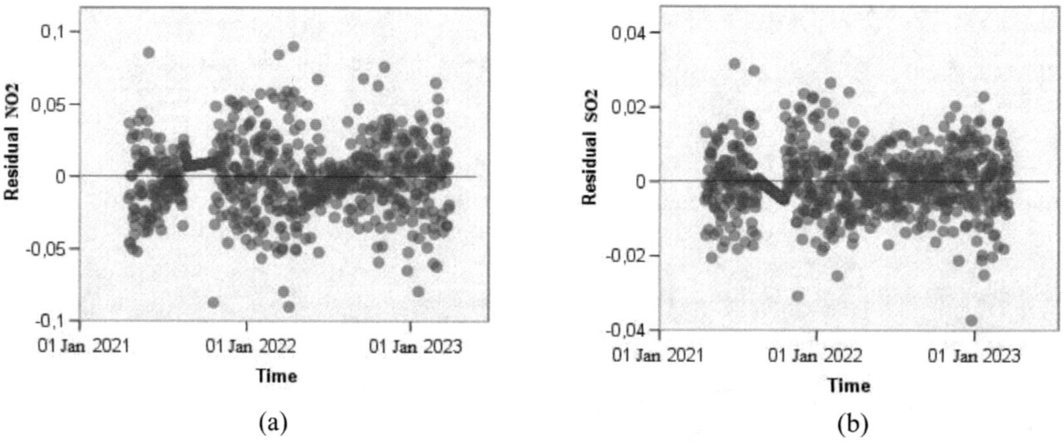

Figure 9. Residual plots by time for problematic variables: (a) NO2, (b) SO2.

In addition, Kolmogorov-Smirnov (K-S) tests were performed to check the normality of residuals. The corresponding values of the K-S statistics are: for CO2: 0.086, for NO2: 0.074, for SO2: 0.050; for PM1: 0.184; for PM2.5: 0.212, and for PM10: 0.268. All K-S tests are significant. This suggests that formal normality of the residuals was not achieved. This is explained by the presence of a large number of outliers. Instead, the behavior of the ACF of the model residuals is a more important property [35]. The ACFs of residuals for the six investigated variables are shown in Figure 10. In general, the residuals are within the confidence intervals, from which it can be concluded that the models are adequate.

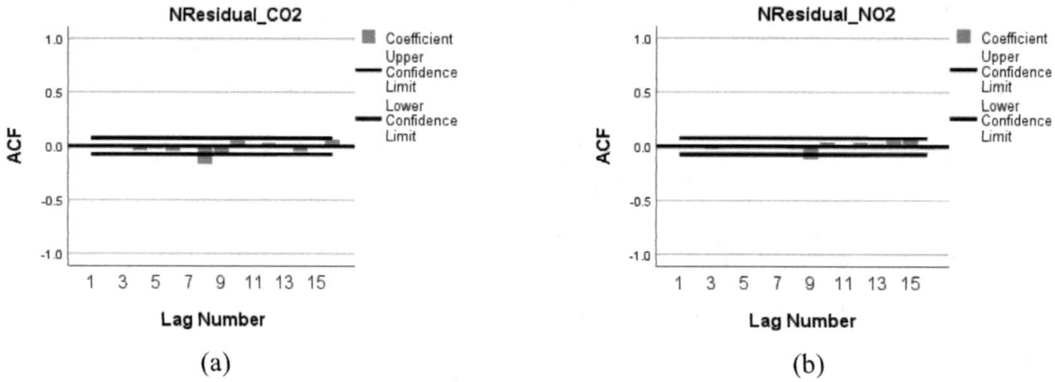

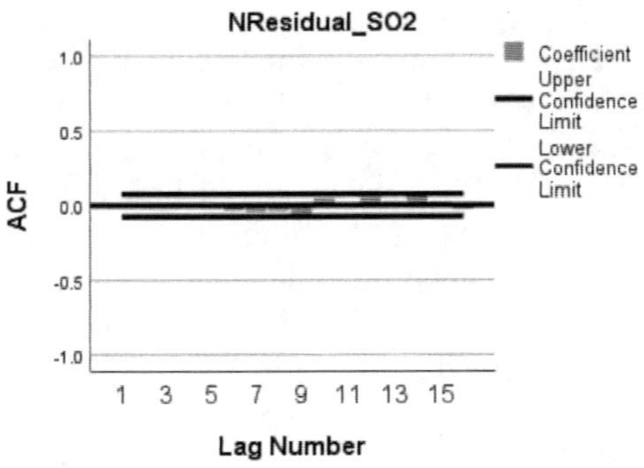

(c)

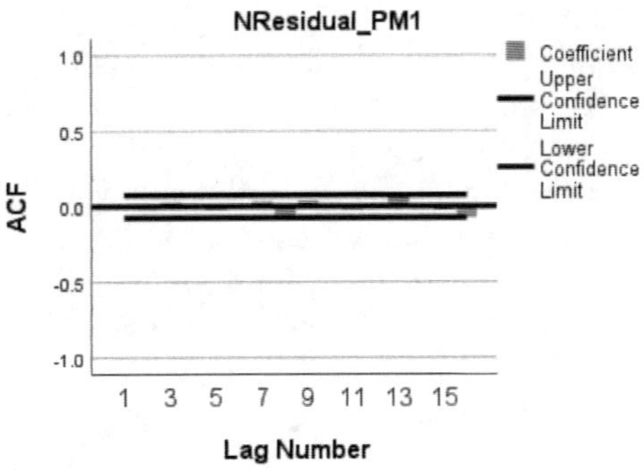

(d)

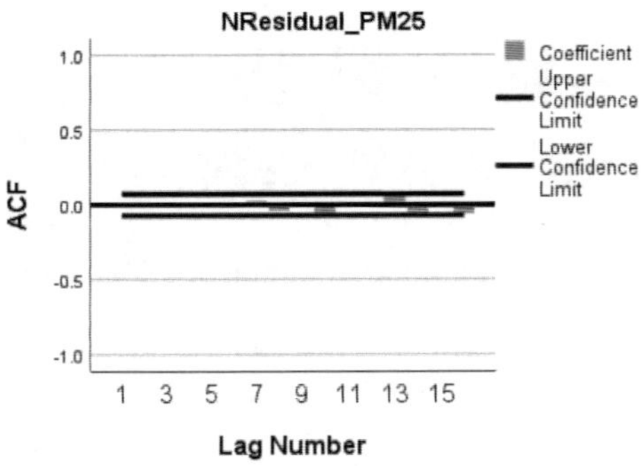

(e)

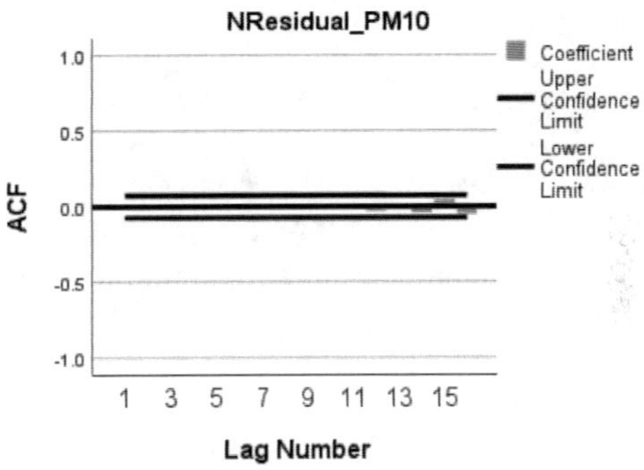

(f)

Figure 10. ACF of the residuals for all variables.

3.3. Application of TCM models to forecast out of data values
Verification of the models was performed to predict pollution 7 days after the data used in the modeling process. Figure 11 presents the forecasted values of NO2 and SO2 against those measured with lower and upper limits of confidence levels. To obtain these values, all of the predictors are forecasted in the built TCM model from the previous subsection 3.2. Good forecasts are observed for the two pollutants.

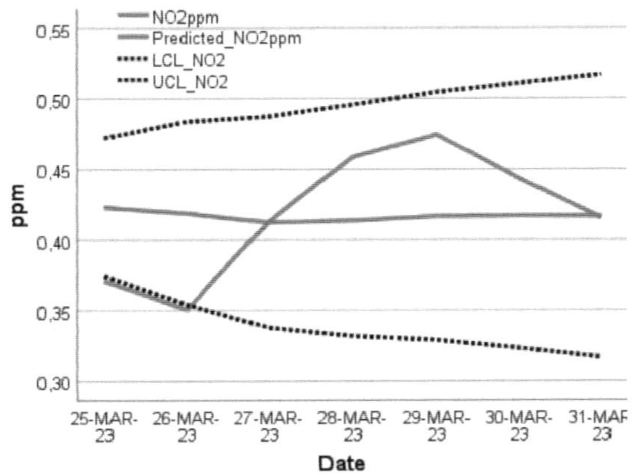

(a)

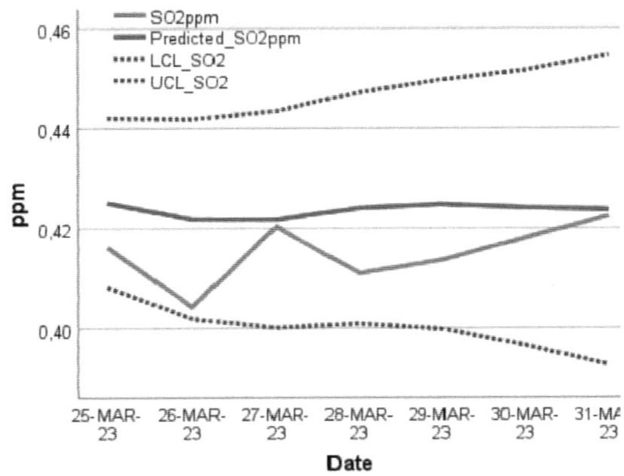

(b)

Figure 11. Forecasting for 7 days ahead with lower and upper confidence limits for: (a) NO2, (b) SO2.

4. Conclusion

The main purpose of the study in this paper was to reveal the Granger causal relationships between all measured time series, to identify the problematic pollutants with exceeded levels, as well as to demonstrate the possibilities of the TCM method for short-term forecasts. For this purpose, average daily measurements of six pollutants were used (NO2, SO2, CO2, PM10, PM2.5 and PM1) and four meteorological variables were used (Temperature, Pressure, Humidity, and Luminosity) for the period from 11th April 2021 to 24th March 2023 or 713 days. It was found that the pollutants NO2 and SO2 exceeded average daily values, respectively 6 and 8.9 times the national and European permissible limits. The overall TCM system vector type model was built. The impact diagrams were obtained. The latter visualize mainly the results for NO2 and SO2. The highest results achieved according to the coefficient of determination are: model fit of 98% for CO2, 94% for NO2, and 79% for SO2. The TCM system model was applied to predict all the variables included in it, then 7 days ahead forecasts were generated.

We can conclude that TCM provides a convenient and effective vector method for studying and modeling causal relationships, as well as for forecasting time series in the field of air pollution.

Acknowledgement

This study has emanated from a research conducted with the financial support of the Bulgarian National Science Fund (BNSF), Grant number KP-06-IP-CHINA/1.

References

[1] Shahrbaf M A, Akbarzadeh M A, Tabary M and Khaheshi I 2021. Air pollution and cardiac arrhythmias: A comprehensive review *Current Problems in Cardiology* **46**(3) 100649. https://doi.org/10.1016/j.cpcardiol.2020.100649

[2] Papadogeorgou G, Kioumourtzoglou M A, Braun D and Zanobetti A 2019 Low levels of air pollution and health: Effect estimates, methodological challenges, and future directions *Current Environmental Health Reports* **6** 105–115 https://doi.org/10.1007/s40572-019-00235-7

[3] Schwartz J, Bind M A and Koutrakis P 2017 Estimating causal effects of local air pollution on daily deaths: effect of low levels *Environmental Health Perspectives* **125** 23–29 http://dx.doi.org/10.1289/EHP232

[4] World Health Organization 2021 WHO global air quality guidelines: particulate matter (PM2.5 and PM10), ozone, nitrogen dioxide, sulfur dioxide and carbon monoxide. Available online at: https://apps.who.int/iris/handle/10665/345329

[5] Rozbicka K, Majewski G, Rogula-Kozłowska W and Rozbicki T 2020 Tropospheric ozone assessment in urban environment–Warsaw case study of selected heat waves *Journal of Atmospheric and Solar-Terrestrial Physics* **209** 105418. https://doi.org/10.1016/j.jastp.2020.105418

[6] Veleva E, Filipova M and Zheleva I 2022 Statistical study of particulate matter (PM10) air contamination in the city of Vidin, Bulgaria *AIP Conference Proceedings* **2522**(1) 050014-–050014-11. https://doi.org/10.1063/5.0101245

[7] Minkova I, Filipova M and Zheleva I 2023 Statistic study of particulate matter (PM10) air contamination in the city of Svishtov, Bulgaria *AIP Conference Proceedings*, AMITANS 2022 Conference Proceedings (to be published)

[8] Nikolova M, Filipova M and Zheleva I 2023 Statistic study of gaseous air contamination in the city of Ruse, Bulgaria *AIP Conference Proceedings* AMITANS 2022 Conference Proceedings (to be published)

[9] Fernández-Pacheco V M, Álvarez-Álvarez E, Blanco-Marigorta E and Ackermann T 2023 CFD model to study PM10 dispersion in large-scale open spaces *Scientific Reports* **13**(1) 5966 https://doi.org/10.1038/s41598-023-33144-9

[10] Starchenko A V, Danilkin E A, Prokhanov S, Kizhner L and Shelmina E 2022. A supercomputer-based modeling system for short-term prediction of urban surface air quality *Supercomputing Frontiers and Innovations* **9**(1) 17–31. https://doi.org/10.14529/js220102
[11] Chen Y C and Li D C 2021 Selection of key features for PM2.5 prediction using a wavelet model and RBF-LSTM *Applied Intelligence* **51** 2534–2555. https://doi.org/10.1007/s10489-020-02031-5
[12] Gunasekar S, Kumar G and Agbulu G P 2022 Air quality predictions in urban areas using hybrid ARIMA and metaheuristic LSTM *Computer Systems Science & Engineering* **43**(3) 1271–1284. https://doi.org/10.32604/csse.2022.024303
[13] Sánchez A S, Nieto P G, Fernández P R, del Coz Díaz J J and Iglesias-Rodríguez F J 2011 Application of an SVM-based regression model to the air quality study at local scale in the Avilés urban area (Spain) *Mathematical and Computer Modelling* **54**(5-6) 453–1466. https://doi.org/10.1016/j.mcm.2011.04.017
[14] Gocheva-Ilieva S, Ivanov A and Stoimenova-Minova M 2022. Prediction of daily mean PM10 concentrations using random forest, CART Ensemble and Bagging Stacked by MARS *Sustainability* **14**(2) 798. https://doi.org/10.3390/su14020798
[15] Gocheva-Ilieva S, Stoimenova M, Ivanov A, Voynikova D and Iliev I 2016 Stochastic univariate and multivariate time series analysis of PM2.5 and PM10 air pollution: a comparative case study for Plovdiv and Asenovgrad, Bulgaria In: Proceedings of 8th Conference of the Euro-American Consortium for Promoting the Application of Mathematics in Technical and Natural Sciences (AMiTaNS'2016), Albena, Bulgaria, 22–27 June 2016, Todorov M D (ed) *AIP Conference Proceedings* Melville, NY **1773**(1), 110004. http://dx.doi.org/10.1063/1.4965008
[16] Granger C W 1969 Investigating causal relations by econometric models and cross-spectral methods *Econometica: Journal of the Econometric Society* **37**(3) 424–438. https://doi.org/10.2307/1912791
[17] Kumar V and Chander R 2016 Foreign direct investment and air pollution: Granger causality analysis *IOSR Journal of Business and Management* **2**(20) 12–17.
[18] Chaabouni S, Zghidi N, Ben Mbarek M 2016 On the causal dynamics between CO2 emissions, health expenditures and economic growth *Sustainable cities and society* **22** 184–191. https://doi.org/10.1016/j.scs.2016.02.001
[19] Bind M-A 2019 Causal modeling in environmental health Annu Rev Public Health 40(1):23–43. https://doi.org/10.1146/annurev-publhealth-040218-044048
[20] Schwartz J, Bind M A and Koutrakis P 2017 Estimating causal effects of local air pollution on daily deaths: effect of low levels *Environmental health perspectives* **125**(1) 23–29. https://doi.org/10.1289/EHP232
[21] Sfetsos A and Vlachogiannis D 2013 An analysis of ozone variation in the Greater Athens Area using Granger Causality *Atmospheric Pollution Research* **4**(3) 290297. https://doi.org/10.5094/APR.2013.032
[22] Chan C H, Juang J Y, Chu T H, Mao C H and Huang S Y 2023 A novel evaluation of air pollution impact from stationary emission sources to ambient air quality via time-series Granger Causality In: Wen T H, Chuang T W, Tipayamongkholgul M (eds) *Earth Data Analytics for Planetary Health. Atmosphere, Earth, Ocean & Space* 33–53 (Singapore: Springer). https://doi.org/10.1007/978-981-19-8765-6_2
[23] Wang B 2019 Applying machine-learning methods based on causality analysis to determine air quality in China *Pol J Environ Stud* **28**(5) 3877–3885. https://doi.org/10.15244/pjoes/99639
[24] Zhu J Y, Sun C. and Li V O K 2017 An extended spatio-temporal granger causality model for air quality estimation with heterogeneous urban big data *IEEE Transactions on Big Data* **3**(3) 307–319. https://doi.org/10.1109/TBDATA.2017.2651898
[25] Bai L, Wang J, Ma X and Lu H 2018 Air pollution forecasts: an overview *Internacional Journal of Environmental Research and Public Health* **15**(4) 780. https://doi.org/10.3390/ijerph15040780

[26] Rybarczyk Y and Zalakeviciute R 2018 Machine learning approaches for outdoor air quality modelling: A systematic review *Applied Sciences* **8**(12) 2570. https://doi.org/10.3390/app8122570
[27] IBM SPSS Statistics. Available online at: https://www.ibm.com/products/spss-statistics
[28] Yordanova L, Kiryakova G, Veleva P, Angelova N and Yordanova A 2021 Criteria for selection of statistical data processing software *Proceedings of the IOP Conference Series: Materials Science and Engineering* **1031**(1) 012067. https://doi.org/10.1088/1757-899X/1031/1/012067
[29] Executive Environment Agency, Bulgaria. Available online at: http://eea.government.bg/bg/soer/2014/air/kachestvo-na-atmosferniya-vazduh
[30] European Environment Agency. Available online at: https://www.eea.europa.eu//publications/status-of-air-quality-in-Europe-2022
[31] Air Quality Standards. European Commission. Environment. Available online at: https://environment.ec.europa.eu/topics/air/air-quality/eu-air-quality-standards_en
[32] Directive 2008/50/EC of the European Parliament and of the Council of 21 May 2008 on Ambient Air Quality and Cleaner Air for Europe 2008 *Official Journal of the European Union* L 152/1
[33] Plovdiv. Available online at: https://en.wikipedia.org/wiki/Plovdiv
[34] Konishi S and Kitagawa G 2008 *Information Criteria and Statistical Modeling* (New York: Springer)
[35] Livieris I, Stavroyiannis S, Pintelas E and Pintelas P 2020 A novel validation framework to enhance deep learning models in time-series forecasting *Neural Computing and Applications* **32**(23) 17149–17167. https://doi.org/10.1007/s00521-020-05169-y

Stochastic modelling of daily air pollution in Burgas, Bulgaria

S K Koleva, S G Gocheva-Ilieva and H N Kulina

Department of Mathematical Analysis, Faculty of Mathematics and Informatics, University of Plovdiv Paisii Hilendarski 24 Tzar Asen Street, 4000 Plovdiv, Bulgaria
E-mails: stkoleva@uni-plovdiv.bg, snow@uni-plovdiv.bg, kulina@uni-plovdiv.bg

Corresponding author's e-mail address: stkoleva@uni-plovdiv.bg

Abstract. Exceeding the norms and limits of atmospheric air pollution causes enormous damage to the population's health and the environment. Determining the factors affecting air quality is a current task in a local, regional, and global scale. In this study, we use daily time series data for the main air pollutants in Burgas, Bulgaria – O_3, NO, NO_2, CO, SO_2, and PM_{10}, to analyze, model, and forecast these levels depending on meteorological factors. For this purpose, the stochastic ARIMA method and ARIMA with transfer functions are applied. Results are obtained for univariate and multivariate time series. Particular attention is paid to the concentrations of the secondary pollutant ground-level ozone (O_3), which are modelled as a function of all variables considered. Results were evaluated using root mean square error, mean absolute percentage errors, and the coefficient of determination. Short-term forecasts have been obtained for seven days ahead. Model accuracy up to 84% has been established.

1. Introduction

The quality of the air we breathe and its pollution are among the most pressing topics to date. There are 36 automated and certified air control stations in Bulgaria and the main air pollutants in the larger cities of the country are monitored systematically. For the whole country, monitoring is carried out by the Executive Environment Agency, European and national norms and standards, and World Health Organization documents [1-3]. Although in recent years, harmful emissions into the ambient air have decreased, some systematic exceedances of air pollution for several cities in Bulgaria are still observed. It should be clarified that air pollution for each region depends to a significant extent not only on its weathering and geographical and climatic characteristics but also on a large number of anthropogenic factors. Such factors are the results of human activity, including various production processes, road traffic, emissions of harmful emissions from households, and others. The characteristics and harmful effects of air pollutants on human health are described, for example, in [1-3].

The availability of a huge volume of empirical measurement data in the field of air pollution enables their analysis and statistical modelling to derive trends and dependences, as well as to predict the future state of ambient air quality. A large number of studies on data processing on atmospheric pollutants have been published in the scientific literature. At the same time, in Bulgaria, such studies are relatively few. To analyse and model concentrations of the main air pollutants, such as O_3 – ground-level ozone, CO – carbon monoxide, NO – nitrogen monoxide, SO_2 – sulfur dioxide, NO_2 – nitrogen dioxide, PM_{10} – fine particulate matter below 10 microns and others different statistical and numerical methods were applied. The classical linear stochastic ARIMA (Auto Regressive Integrated Moving Average) approach and its variants [4] have been used in [5-9]. In [5], a stationary stochastic ARMA/ARIMA modelling approach has been considered to forecast the daily mean air pollutants O_3, CO, NO, and NO_2

concentration. The models were calibrated by means of different information criteria, such as the Akaike Information Criterion (AIC), Hannon–Quinn Information Criterion (HIC), Bayesian Information criterion (BIC), etc. and the plots of the Autocorrelation function (ACF), and Partial autocorrelation function (PACF). Guarnaccia et al. [6] built univariate seasonal ARIMA (SARIMA) models for predicting CO concentration levels in a district of Monterrey, Mexico and concluded that the use of hourly data allowed the achievement of reliable prediction especially on a short time of 24-hour period. The authors of [7] apply ARIMA to model, analyse and forecast SO_2 and NO_2 time series values, using average daily measurements. Statistical studies on PM_{10} pollution in two Bulgarian cities - Pernik and Ruse, are presented in [8,9]. A two-step SARIMA approach was developed in [10] for PM_{10} concentration prediction. To achieve a real-world effect, meteorological variables are first predicted with one-dimensional SARIMA models, then the pollutant data is modelled depending on the historical and newly calculated predictive values of the predictors.

In recent years, cutting-edge approaches based on machine learning (ML) algorithms have been actively used to model and predict empirical ambient air data. Many authors combine or compare their results with ARIMA and other classical statistical approaches, hybrid models are also developed. Wavelet analysis with energy spectrograms, combined with the multilayer feed-forward back propagation NN algorithm and compared with ARIMA are reported in [11] to forecast PM_{10} monthly levels in selected regions of Chennai. Discrete Haar wavelet transform and ARIMA models were coupled to improve ozone prediction in [12]. Principal Component Analysis (PCA) and path seeker technique are implemented in [13] to model the CO concentration. Paper [14] deals with the Random Forest algorithm for analysis and forecasting of PM_{10} pollution. The authors of [15] construct hybrid Elman ANN and ARIMA models using empirical data of SO_2. Boosted trees method with regularized regression from ensemble learning group are applied for studying ground-level ozone and PM_{10} in [16]. In [17] the authors develop several ML models of O_3 and PM_{10}, based on hourly data. The best performance is obtained using the stochastic Gradient Boosting method, representative of ensemble tree-based methods, demonstrating computational efficiency and robustness to overfitting. More results on ML methods and case studies in the field of air pollutants could be found in the review article [18].

The aim of this study, based on the collected empirical data for the city of Burgas are as follows: 1) To apply univariate ARIMA methods for stochastic regression modelling and forecasting separately the concentrations of the six air pollutants O_3, SO_2, NO, NO_2, CO and PM_{10}; 2) To create multivariate ARIMA models of each of these time series, depending on meteorological time series; 3) To construct the multivariate ARIMA model of tropospheric ozone; 4) To evaluate the models' performance and their residuals; 5) To validate the models by short-term predictions of pollution.

The study was carried out using IBM SPSS Statistics [19,20].

2. Data and methods

2.1. Description of data and initial data processing

This section presents the data and the results from the initial analysis performed on them.

2.1.1. Study area

Burgas is the second largest city on the Bulgarian Black Sea coast. Here is the largest chemical and oil refinery in Southeastern Europe, which is also the largest employer in Bulgaria, the international and second busiest Bulgarian airport, as well as the most significant Bulgarian port and the only oil port in the country. The city is located on the westernmost point of the Black Sea. The climate is humid subtropical with wet and continental influence. The environment area is urbanized, with high building density, intensive car traffic and industrial activity in the municipality of Burgas. Three estuarine lakes are located on the territory of the city. A specific influence is also exerted by the sea breeze, whose circulation has a direct impact on the climate and the dispersion of atmospheric pollutants.

2.1.2. Data
In this article, we explore data on six major air pollutants for the city of Burgas for the period from the 1 January 2019 to 31 March 2023 with average daily data or for N=1551 days. In particular, we are looking for mathematical models describing the behaviour of O_3, μg/m^3 (ground level ozone concentrations), CO, mg/m^3 (carbon monoxide), NO, μg/m^3 (nitrogen monoxide), SO_2, μg/m^3 (sulfur dioxide), NO_2, μg/m^3 (nitrogen dioxide), PM_{10}, μg/m^3 (fine particulate matter with diameter below 10 microns) in relation to meteorological data.

The following seven meteorological variables are used to build the multivariate models: MaxT, °C (maximum daily mean air temperature), MinT, °C (minimum daily average air temperature), Cloud, % (cloudiness), Humidity, % (relative air humidity), Precipi, mm (precipitation), Speed, m/s (wind speed), Pressure, mbar (ambient air pressure). The data were recorded from the automated measured station situated in Dolno Ezerovo, a corner of the city of Burgas. The GPS coordinates of the station Dolno Ezerovo are 42°31′05.99" N 27°22′22" E. Data were retrieved from the official sites [21,22].

The maximum number of missing values for each initial time series is less than 2%. In the analyses the missing values were replaced using linear interpolation. Several extreme values of pollutants were found and replaced by the next largest. The basic descriptive statistics of time series of pollutants and meteorological variables are given in Table 1.

Table 1. Descriptive statistics of the examined data of air pollutants and meteorological data for the city of Burgas.

Variable	Mean	Median	Skewness	Kurtosis
O3, μg/m^3	44.52	45.88	-0.14	-0.59
NO, μg/m^3	4.90	4.94	1.38	4.28
NO2, μg/m^3	14.10	13.17	0.72	0.45
CO, mg/m^3	0.31	0.25	1.48	2.89
SO2, μg/m^3	10.85	10.63	0.66	1.51
PM10, μg/m^3	33.79	30.21	1.47	3.12
MaxT, °C	17.59	17.00	0.01	-1.05
MinT, °C	10.49	10.00	-0.08	-0.98
Cloud, %	0.38	0.32	0.57	-0.86
Humidity, %	0.71	0.71	-0.57	1.77
Precipi, mm	1.70	0	5.01	31.48
Speed, m/s	1.47	1.31	1.56	3.80
Pressure, mbar	1010.3	1010.6	-33.62	1256.29

The next Figure 1 shows the sequences plots of the initial time series of the examined air pollutants. There are pronounced seasonal changes. Higher concentrations of pollution for SO_2, NO, NO_2, CO and PM_{10}, are observed in winter period while in summer the levels drop. This behaviour is opposite for ozone. This indicates that weather conditions have a strong influence on pollutants' emissions. The remaining causes of increased concentrations of harmful emissions, including car traffic, seasonal combustion processes from households and manufacturing enterprises, and others, are relatively constant and do not change sharply over time, compared to meteorological changes. Their influence is stochastically comprised in the PM_{10} values. This will make it easier for us to create models to predict pollutant concentrations for a short period of time ahead based on the meteorological conditions.

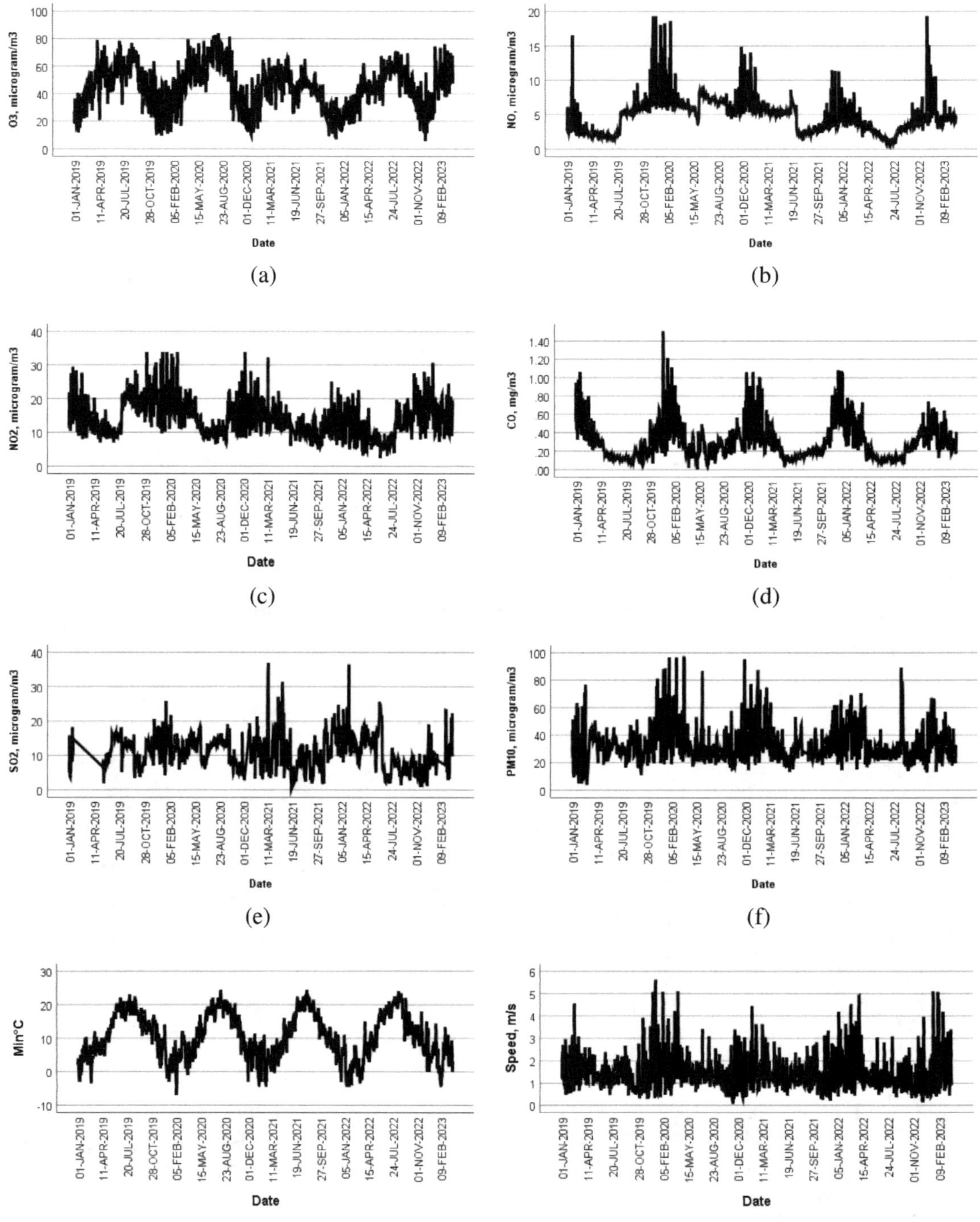

Figure 1. Sequence plots of the investigated variables: (a) O3, (b) NO, (c) NO2, (d) CO, (e) SO2, (f) PM10, (g) Tmin, (h) Speed.

2.1.3. Methods used

We will apply the Box-Jenkins ARIMA methodology [4]. The general form of the ARIMA model of a given time series variable Y is represented as ARIMA (p, d, q) where p is the number of autoregressive terms, d is the number of differences, q is the number of moving average terms. The main requirements for building an adequate ARIMA model include normal or close to normal distribution, linear dependence, absence of missing values and stationarity [4,23].

- Univariate ARIMA

When the time series is stationary, i.e. the probability distribution does not depend on time, so its mean and variance are constant, the univariate AR process of order p is described by a difference equation of order p of the type:

$$Y_t = \phi_1 Y_{t-1} + \phi_2 Y_{t-2} + \ldots + \phi_p Y_{t-p} + a_t = \left(\sum_{j=1}^{p} \phi_j B^j \right) Y_t + a_t, \quad t = p+1, \ldots, N \tag{1}$$

where $(\phi_1, \phi_2, \ldots, \phi_p)$ are constant coefficients (model parameters), a_t. is a stochastic term (random error, white noise) for each time t, under the assumption $a_t \sim WN\left(0, \sigma^2\right)$, B is the backward (lag) operator $BY_t = Y_{t-1}$.

If the process contains mostly systematic random fluctuations a_t. around some fixed level, then it is defined as a stochastic MA process. The MA model with moving averages is of order q if, for each t, it is represented by a difference equation of order q of the type

$$Y_t = a_t - \theta_1 a_{t-1} - \theta_2 a_{t-2} - \ldots - \theta_q a_{t-q} = \left(1 - \sum_{j=1}^{q} \theta_j B^j \right) a_t, \quad t = q+1, \ldots, N \tag{2}$$

where $\theta_1, \theta_2, \ldots, \theta_q$ are constant parameters.

In the general case, ARIMA (p,d,q) is written as

$$\left(1 - \sum_{j=1}^{p} \phi_j B^j \right) (1-B)^d Y_t = \left(1 - \sum_{j=1}^{q} \theta_j B^j \right) a_t + c \,. \tag{3}$$

In (3) c is a constant,

- Multivariate ARIMA

When the dependent time series Y is considered according to other time series $X_{1t}, X_{2t}, \ldots, X_{kt}$ which values are known in the same time period, a multidimensional ARIMA could be used. It is called ARIMA with transfer functions (TF) [4]. The ARIMA/TF model is written as:

$$(1-B)^d Y_t = \frac{MA}{AR} a_t + \sum_{i=1}^{k} \left(\frac{Num_i}{Den_i} (1-B)_i^{d_i} B^{b_i} X_{it} \right) + \mu \tag{4}$$

where μ is constant, B^{b_i} is a delay member with a positive integer lag b_i, MA and AR, Num_i and Den_i have the form of the difference polynomials with constant coefficients:

$$AR = \left(1 - \sum_{j=1}^{p} \phi_j B^j \right), \; MA = \left(1 - \sum_{j=1}^{q} \theta_j B^j \right),$$

$$Num_i = \left(\sum_{j=0}^{u} \omega_{ij} B^j \right), \; Den_i = \left(1 - \sum_{j=1}^{v} \gamma_{ij} B^j \right) \tag{5}$$

To obtain statistically correct ARIMA or ARIMA/TF models, the significance of its coefficients must be achieved at a set level α (usually 0.05).

2.1.4. Data transformation
The basic assumptions for ARIMA analysis include the normal or close to normal distribution and stationarity of the time series involved [23]. As it is seen from Table 1, the coefficients of skewness and kurtosis of the considered time series are different from zero, therefore the normality condition is violated. To stabilize the variance and improve the distribution to normality the Yeo-Johnson power transformation was used by the expressions [24]:

$$trY = YJ(\lambda, Y) = \begin{cases} \left\{(Y+1)^{\lambda} - 1\right\}/\lambda & Y \geq 0,\ \lambda \neq 0 \\ \log(Y+1) & Y \geq 0,\ \lambda = 0 \\ -\left\{(-Y+1)^{2-\lambda} - 1\right\}/(2-\lambda) & Y < 0,\ \lambda \neq 2 \\ -\log(-Y+1) & Y < 0,\ \lambda = 2 \end{cases}, \quad \lambda \in [-2, 2] \tag{6}$$

where Y is the original variable, trY is its transformed variable, and λ is a real parameter. The parameter λ is tuned by a stepwise procedure using some statistical tools or normality test.

Table 2 shows the obtained values for λ, and Skewness and Kurtosis of the transformed pollutant's variables. With the small values of Skewness and Kurtosis, it can be concluded that the distributions of the transformed variables are close to a normal distribution. The box plots of the standardized transformed variables are presented in Figure 2. It is observed the improvement of the distribution, however, the outliers exist, and could not be neglected.

Table 2. Statistics of the transformed variables of air pollutants.

Transformed pollutant variable	Parameter λ	Skewness	Kurtosis
trO3	1.2	0.018	-0.600
trNO	0.6	0.665	1.248
trNO2	0.2	-0.014	-0.201
trCO	-1.8	0.437	-0.564
trSO2	0.8	0.352	0.708
trPM10	-0.2	-0.429	2.724

2.1.5. Statistical measures to assess the quality of models
The quality of models is evaluated by standard statistical indicators: coefficient of determination, root mean square error, mean absolute percentage error (R^2, RMSE, MAPE), calculated by the expressions:

$$R^2 = \frac{\sum_{t=1}^{N}\left(\hat{Y}_t - \bar{Y}\right)^2}{\sum_{t=1}^{N}\left(Y_t - \bar{Y}\right)^2}, \quad RMSE = \sqrt{\frac{1}{N}\sum_{t=1}^{N}\varepsilon_t^2}, \quad MAPE = \frac{100}{N}\sum_{t=1}^{N}\left|\frac{\varepsilon_t}{Y_t}\right|, \quad \varepsilon_t = Y_t - \hat{Y}_t . \tag{7}$$

where $\hat{Y}_t$ is the value predicted by the model at time t, $\bar{Y}$ is the mean value of Y, N is the sample size. We will look for models for which the coefficient of determination is close to 1 and the errors are small.

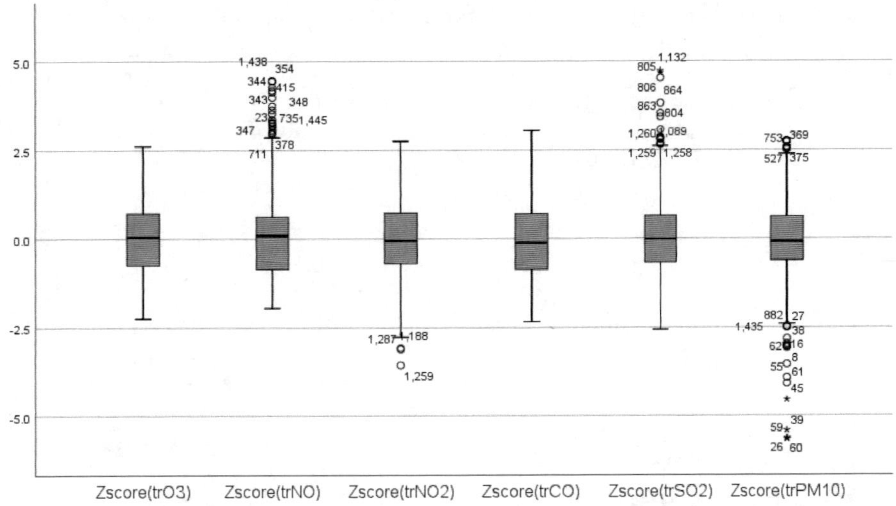

Figure 2. Box plots of the standardized transformed variables of air pollutants.

3. Results and discussion

Using the developed methodology, univariate and multivariate models of transformed air pollutant variables have been constructed. Univariate models are a special case when there are no predictors.

All models were built using data for N=1544 days. The data of the last 7 days will be used for model validation and forecasting the pollution level.

3.1. Building and evaluating the univariate ARIMA models

For each variable of the six pollutants, separate univariate ARIMA models were built using their corresponding transformed variables. The process of constructing an ARIMA model requires five basic steps to correctly determine the parameters p, d, q.

1) The corresponding sequence plots of the ACF and PACF are constructed and examined to reveal the presence of trends (*d*). The lags outside the permissible values are inspected to find initial values of *p* and *q*.

2) In the presence of a PACF coefficient with a value close to 1 or -1, an augmented Dickey-Fully (ADF) test is conducted to specify whether it is a trend or not and to determine the exact value of the parameter *d*.

3) An ARIMA (p,d,q) is constructed with selected parameters.

4) The statistical significance of the parameters is checked. In our case, the significance level should be Sig.<0.05. Insignificant parameters are removed and the values of *p*, *q* are adjusted. The parameter selection procedure can be repeated many times.

5) A detailed analysis of the residuals of the model is carried out to establish its statistical adequacy. The residual values of the model are investigated to be within the appropriate confidence interval using their autocorrelation functions (ACF). It is also recommended to check the residuals with the Ljung-Box portmanteau test or other appropriate statistical tests. For this purpose, such a test must be insignificant, i.e., Ljung-Box Sig. >0.05. This means that the residuals of the ARIMA model do not contain autocorrelation.

When obtaining several different adequate ARIMA models, the simpler one is chosen or the one that has the best statistical indicators, in this case the attention is on the indicators of (7).

The next Figure 3 shows the univariate ARIMA predictions for transformed variables.

Table 3. Statistics of univariate models of transformed variables of air pollutants.

Transformed variable	Univariate ARIMA model	R^2	RMSE	MAPE	Ljung-Box Sig.
trO3	ARIMA(1,0,6)	0.756	16.442	20.610	0.058
trNO	ARIMA(1,0,5)	0.775	0.577	10.842	0.051
trNO2	ARIMA(1,0,10)	0.611	0.376	8.370	0.314
trCO	ARIMA(1,0,21)	0.812	0.036	16.602	0.210
trSO2	ARIMA(1,0,13)	0.715	1.582	20.163	0.246
trPM10	ARIMA(1,0,3)	0.477	0.131	3.819	0.533

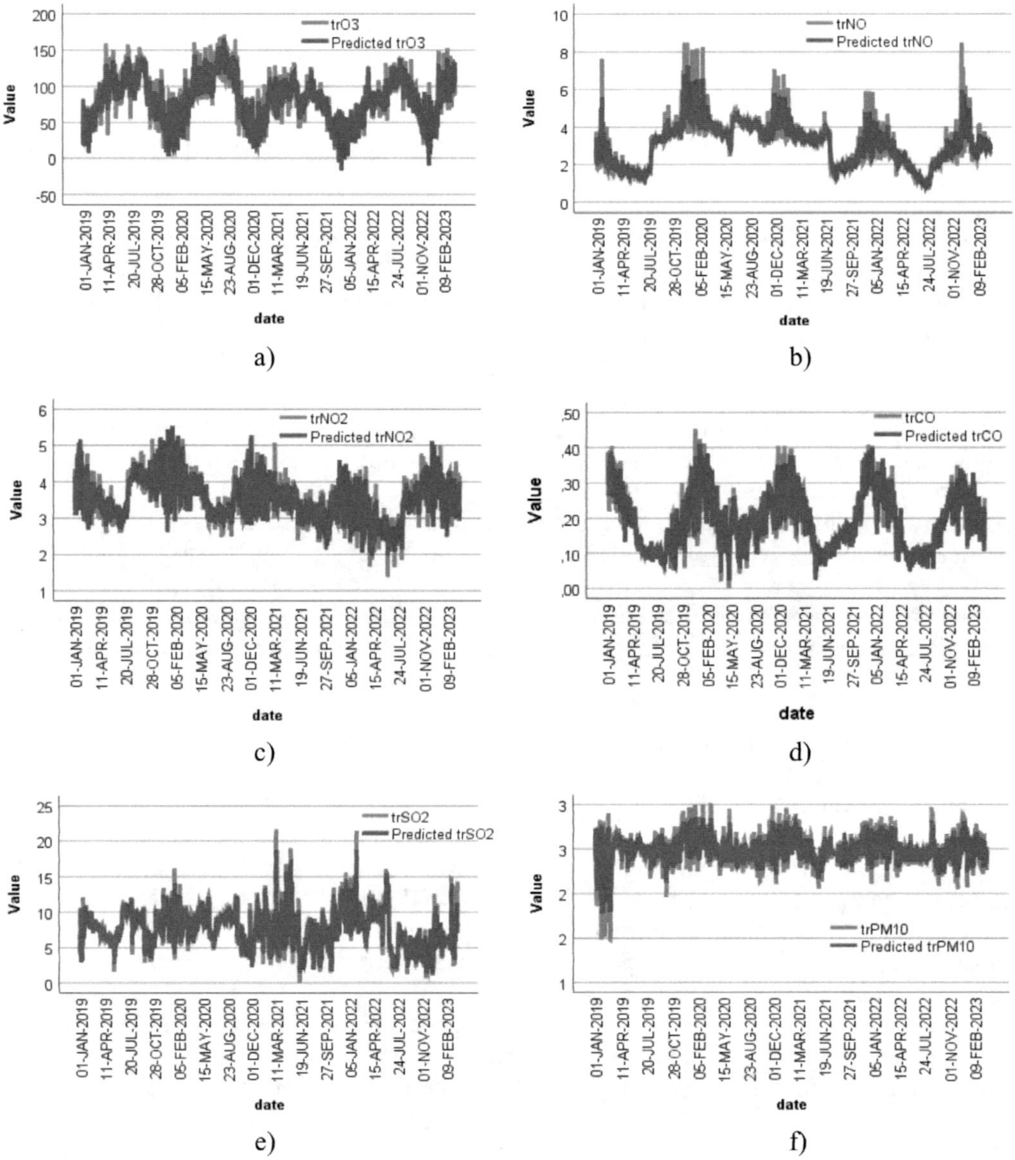

Figure 3. Line plots of the transformed versus modelled values of the six studied air pollutants.

3.2. Building the multivariate ARIMA models

ARIMA/TF models are built for each pollutant separately. The seven meteorological time series from Table 1 were used as predictors. The basic statistics of the selected models are given in Table 4. The comparison with the statistics of the univariate models shows that all models with transfer functions have improved values over the univariate models - both with a larger data matching rate (R^2) and lower RMSE values, as well as the MAPE drops.

Table 4. Summary statistics of the built air pollutant's ARIMA/TF models (for transformed variables).

Transformed dependent variable	ARIMA model	R^2	RMSE	MAPE	Ljung-Box Sig.	Predictors
trO3	ARIMA/TF (1,0,4)	0.840	13.356	16.004	0.368	MinT, Speed, Humidity, Precipi, trNO, trNO2, trCO, trSO2, trPM10
trCO	ARIMA/TF(1,0,3)	0.854	0.032	15.853	0.178	MinT, Speed, Cloud, Pressure
trNO	ARIMA/TF(1,0,5)	0.808	0.535	10.766	0.096	MaxT, Cloud, Humidity, Pressure
trSO2	ARIMA/TF(1,0,13)	0.726	1.552	19.774	0.359	MaxT, Pressure, Humidity
trNO2	ARIMA/TF(1,0,10)	0.749	0.303	6.693	0.097	MinT, Speed, Cloud, Pressure
trPM10	ARIMA/TF(1,0,3)	0.531	0.124	3.606	0.522	MaxT, MinT, Speed, Cloud

3.3. Study on the residuals of the multivariate ARIMA model of ozone

We have paid particular attention to the secondary air pollutant ground-level ozone which was modelled as a function of all the variables considered. 84% coefficient of determination was achieved. In this subsection, we present the results of the residuals diagnostics of the model ARIMA/TF (1,0,4) of trO3. Figure 4 illustrates the ACF and PACF of the residuals of the ARIMA/TF ozone model for 24 lags. It snows a lack of serial correlation. Figure 5 shows the box plot of the residuals and Figure 6 shows the histogram of the residuals of this model. We will add that the formal Kolmogorov-Smirnov test to check the normality of the residuals is small with a statistic equal to 0.035, df=1544, and is significant due to the presence of outliers. In general, we could conclude that the distribution of the residuals is close to normal.

The model performance in terms of the coefficient of determination R^2 is illustrated in Figure 7.

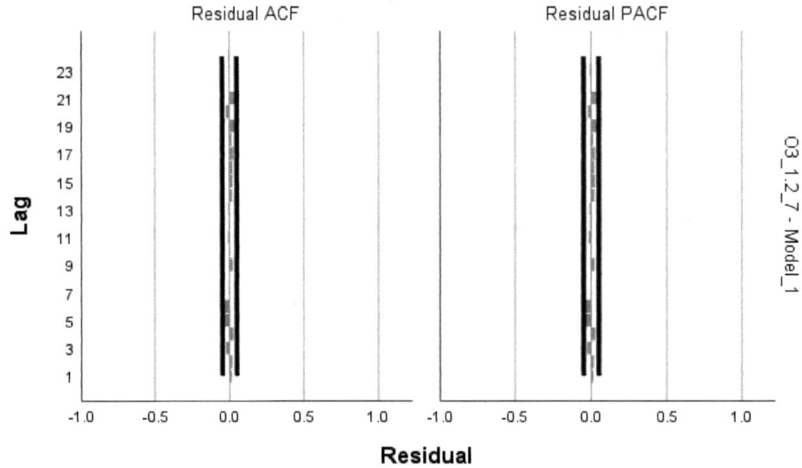

Figure 4. ACF and PACF figures of the residuals of the ARIMA/TF model of trO_3_7.

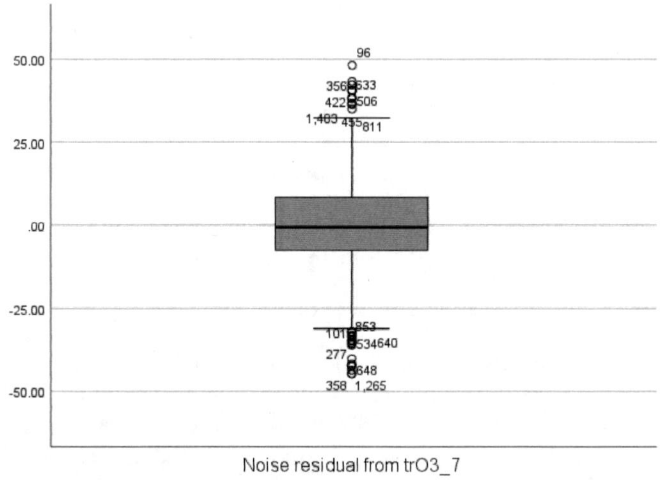

Figure 5. Box plot of model residuals for ozone.

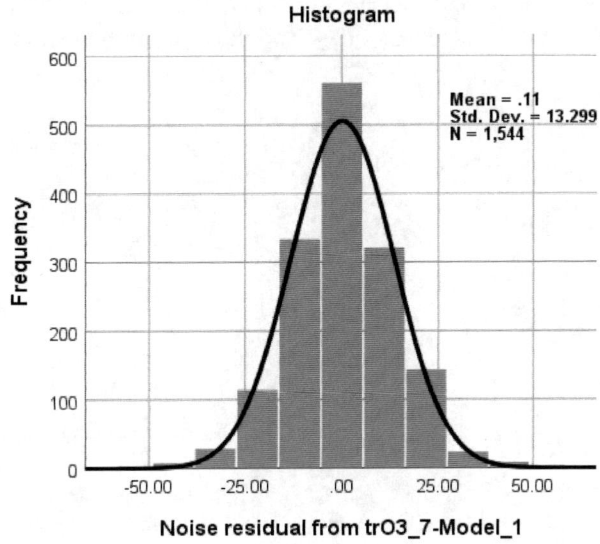

Figure 6. Histogram of ARIMA/TF ozone model residuals

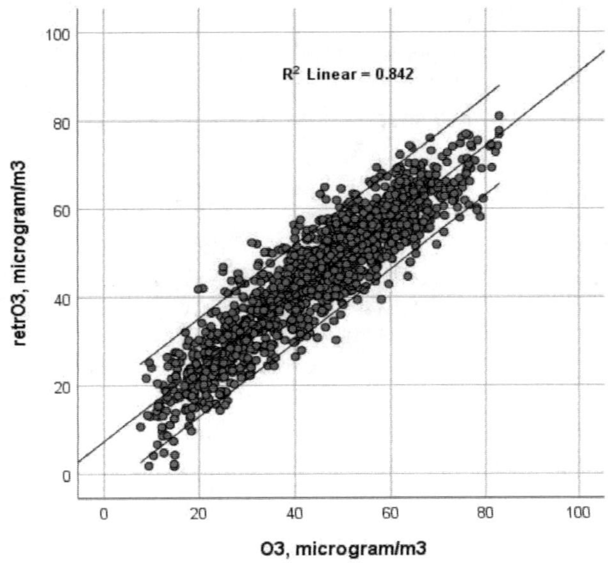

Figure 7. Scatterplot for achieved R^2=84% model fit to data obtained by the retransformed predicted values from the ozone model with a 95% confidence interval.

3.4. Application of ARIMA/TF models to predict future concentrations

To validate the model, from all available data of the dependent variable (trO3) the last few values could be deleted. In our case, these are the data for the last 7 days. To obtain the forecasts, it is necessary to know the values of the predictors (i.e. meteorological variables and the other pollutants). In the real case, the synoptic weather forecasts could be used and the required values of the pollutant variables must be predicted. The O_3 forecasts were obtained without the participation of the last 7 ozone values in the modeling procedure. In Figure 8. a comparison is shown of the observations and model ARIMA/TF

(1,0,4) predictions for the previous 7 days (on the left side of the vertical line) and the observations and forecasted pollutant data for the validated last 7 days (on the right of the vertical line).

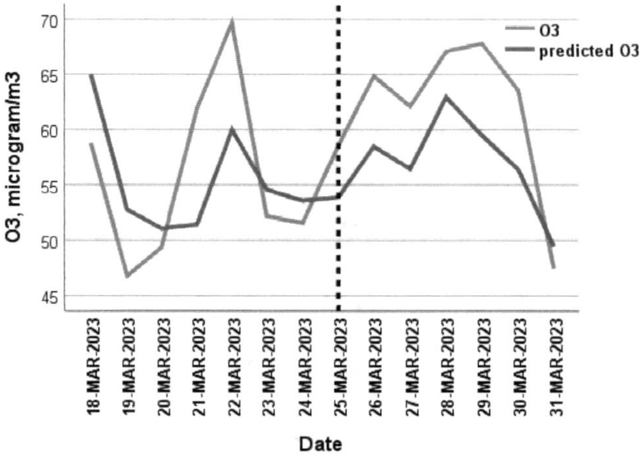

Figure 8. Observed versus predicted ozone values using the ARIMA/TF model for the last 7 days (March 18 to 24, 2023, to the left of the vertical line) and observed versus forecasts using the ARIMA/TF model for the last 7 days (March 25 to 31, 2023, to the right of the vertical line).

The conducted analysis and diagnostics of the model residuals in this subsection give us reason to conclude that the constructed ARIMA/TF(1,0,4) ozone model (see Table 4) is statistically valid and capable of predicting concentrations in a short period ahead.

4. Conclusion

In this study, we investigated average daily data for six air pollutants and seven meteorological variables for the city of Burgas, Bulgaria over a period of 4 years and 3 months. The Box-Jenkins stochastic method for time series analysis was applied for modelling. Univariate ARIMA and multivariate ARIMA/transfer function models were built and statistically investigated. In the preprocessing stage, we used the Yeo-Johnson transformation to stabilize variance and improve the distribution of data. Applying the ARIMA/TF allowed to build models reviling the impact of the given 7 meteorological variables on the considered 6 air pollutants separately. Statistical analyses were carry out to evaluate the model performance and conduct the analyses of the model's residuals, thus validated the model adequacy and usefulness. Especially, the ozone model was built and 84% of data fit was achieved. The model was applied for 7-days ahead forecasting.

More sophisticated modeling approaches are planned as future work, such as ML, discrete wavelet transform, singular spectrum analysis, or others to improve the performance of models and forecasts.

Acknowledgement

This study was supported by the Bulgarian National Science Fund, Grant KP-06-N52/9.

References

[1] Executive Environment Agency, Bulgaria, http://eea.government.bg/bg/soer/2014/air/kachestvo-na-atmosferniya-vazduh
[2] European Environment Agency, https://www.eea.europa.eu//publications/status-of-air-quality-in-Europe-2022
[3] Air Quality Standards. European Commission. Environment, http://ec.europa.eu/environment/air/quality/standards.htm

[4] Box G E P, Jenkins G M and Reinsel G S 1994 *Time Series Analysis, Forecasting and Control*, 3rd edn (New Jersey: Prentice-Hall, Inc)
[5] Kumar U and Jain V K 2010 ARIMA forecasting of ambient air pollutants (O3, NO, NO2 and CO) *Stochastic Environmental Research and Risk Assessment* **24**(5) 751–760
[6] Guarnaccia C, Breton J G C, Breton R M C, Tepedino C, Quartieri J and Mastorakis N E 2018 ARIMA models application to air pollution data in Monterrey, Mexico *AIP Conference* Proceedings **1982**(1) 020041
[7] Gourav, Rekhi J K, Nagrath P and Jain R 2020 Forecasting air quality of delhi using arima model *Lecture Notes in Electrical Engineering* **612** 315–325 (Singapore: Springer)
[8] Stoimenova M P 2016 Stochastic modeling of problematic air pollution with particulate matter in the city of Pernik, Bulgaria *Ecologia Balkanica* **8**(2) 33–41
[9] Veleva E, Filipova M and Zheleva I 2022 Statistical study of particulate matter (PM10) air contamination in the city of Vidin, Bulgaria *AIP Conference Proceedings* **2522**(1) 050014-1–050014-11
[10] Gocheva-Ilieva S, Ivanov A and Iliev I 2019 Exploring key air pollutants and forecasting particulate matter PM10 by a two-step SARIMA approach *AIP Conference Proceedings* **2106** 020004
[11] Angelena J P, Raj A S, Viswanath J and Muthuraj D 2021 Evaluation and forecasting of PM10 air pollution in Chennai district using Wavelets, ARIMA, and Neural Networks algorithms *Pollution* **7**(1) 55–72
[12] Salazar L, Nicolis O, Ruggeri F, Kisel'ák J and Stehlík M 2019 Predicting hourly ozone concentrations using wavelets and ARIMA models *Neural Computing and Applications* **31** 4331–4340
[13] Ivanov A, Voynikova D, Gocheva-Ilieva S, Kulina H and Iliev I 2015 Using principal component analysis and general path seeker regression for investigation of air pollution and CO modeling *AIP Conference Proceedings* **1684** 1–11
[14] Ivanov A, Gocheva-Ilieva S and Stoimenova-Minova M 2022 Random forest regression for statistical modeling and forecasting of PM10 *AIP Conference Proceedings* **2522**(1) 100005
[15] Sánchez A B, Ordóñez C, Lasheras F S, de Cos Juez F J and Roca-Pardiñas J 2013 Forecasting SO2 pollution incidents by means of Elman artificial neural networks and ARIMA models *Abstract and Applied Analysis* **2013** Art ID 238259
[16] Ivanov A, Gocheva-Ilieva S and Stoimenova M 2020 Hybrid boosted trees and regularized regression for studying ground ozone and PM10 concentrations *AIP Conference Proceedings* **2302** 060005
[17] Krylova M and Okhrin Y 2022 Managing air quality: Predicting exceedances of legal limits for PM10 and O3 concentration using machine learning methods *Environmetrics* **33**(2) e2707
[18] Bai L, Wang J, Ma X and Lu H 2018 Air pollution forecasts: an overview *Internacional Journal of Environmental Research and Public Health* **15**(4) 780
[19] IBM SPSS Statistics, https://www.ibm.com/products/spss-statistics
[20] Yordanova L, Kiryakova G, Veleva P, Angelova N and Yordanova A 2021 Criteria for selection of statistical data processing software *Proceedings of the IOP Conference Series: Materials Science and Engineering* **1031**(1) 012067
[21] Regional Inspectorate of Environment and Water – Burgas, https://www.eea.government.bg/kav/reports/air/qReport/92/01#
[22] World Weather API and Weather Forecast, https://www.worldweatheronline.com/
[23] Wilks D S 2011 *Statistical Methods in the Atmospheric Sciences* 3rd edn (Amsterdam: Elsevier)
[24] Yeo I K and Johnson R A 2000 A new family of power transformations to improve normality or symmetry *Biometrika* **87**(4) 954–959

Mathematical Modeling of Carbon Nanoparticle Distribution in Polymer Composites for High Electrical Conductivity: Implications for Green Technologies

Dzhihan Menseidov[1,2]

[1] Bulgarian Academy of Sciences, Institute of Mechanics
Acad. G. Bontchev St., bl. 4, 1113 Sofia, Bulgaria
[2.] University of telecommunications and post-Sofia, Acad. Stefan Mladenov 1,

email: menseidov@imbm.bas.bg

Abstract: A mathematical framework has been formulated to elucidate the conductivity behavior of polymer composites infused with carbon nanoparticles. This linear model, while simplistic, provides a foundational understanding of how factors like nanoparticle morphology, concentration, and surface properties, in conjunction with polymer matrix attributes such as viscosity and surface tension, influence the spatial distribution of these nanoparticles. The overarching aim is to harness the potential of electrically conductive composites in bolstering green technological advancements in sectors like road transport, solar energy, batteries, and environmental conservation. The model underscores the pivotal role of high electrical conductivity in ensuring the efficient, sustainable, and environmentally benign operation of these technologies. This mathematical construct emerges as an instrumental tool, paving the way for the strategic design and optimization of electrically conductive composites tailored for diverse green technology applications.

Keywords: Mathematical framework, Conductivity behavior, Nanoparticle morphology, Green technological advancements, Strategic design.

1. Introduction

As the technological landscape undergoes a transformative shift, the imperative for materials that are both sustainable and efficient becomes paramount. Electrically conductive composites, especially those enriched with carbon nanoparticles embedded within a polymer matrix, stand out as potential game-changers in the realm of green technologies. These composites are characterized by their elevated electrical conductivity, a trait indispensable for the seamless functioning of green technologies. Yet, the path to fully unlocking their capabilities is riddled with intricacies.

The intricate interplay between the carbon nanoparticles' attributes, such as their size, form, concentration, and surface characteristics, and the inherent properties of the polymer matrix, like viscosity and surface tension, governs their distribution within the matrix. Attaining an ideal nanoparticle dispersion, which is the linchpin for amplifying electrical conductivity, is a nuanced endeavor.

This is where our mathematical framework comes to the fore. By encapsulating all these multifaceted variables, our model offers predictive insights into the spatial arrangement of nanoparticles within the

polymer matrix. Beyond just theoretical understanding, this model serves as a strategic compass, guiding the optimization of the dispersion process to maximize electrical conductivity.
In this discourse, we embark on a deep dive into the genesis and ramifications of our mathematical framework. We elucidate how this model, despite its linear nature, can be a potent asset in the design and fine-tuning of electrically conductive composites, thereby propelling the progress of green technologies. We beckon you to accompany us on this intellectual odyssey, where the confluence of mathematical rigor and materials science crafts the blueprint for a sustainable technological future.

2. The state of research and the properties of carbon nanoparticles and polymer composites

The field of carbon nanoparticle and polymer composites has seen significant research interest due to their potential applications in creating electrically conductive materials. In particular, polymers such as Poly(lactic acid) (PLA) and Polyvinylidene fluoride (PVDF) have been a focus due to their unique properties when combined with carbon nanoparticles.

2.1. Properties of Carbon Nanoparticles

Carbon nanoparticles, such as carbon nanotubes and graphene, offer a variety of attributes that make them ideal for enhancing the electrical conductivity of polymer composites. Their high aspect ratio, unique electronic properties, and high thermal conductivity contribute to their effectiveness. Carbon nanoparticles can create conductive networks within the polymer matrix, enabling the transfer of charges and thus increasing the electrical conductivity of the composite[1].
Moreover, their size and shape can be controlled to a large extent, allowing them to be custom-tailored to the needs of specific applications. For instance, the nanoparticle's shape can influence its dispersion within the polymer matrix, which in turn impacts the overall electrical conductivity of the composite[2].

2.2 PLA and PVDF Polymer Matrix

Poly(lactic acid) (PLA) and Polyvinylidene fluoride (PVDF) are two commonly used polymers in the formation of composites with carbon nanoparticles. PLA is a bioactive and biodegradable thermoplastic made from renewable materials like corn starch or sugar cane. It possesses good mechanical and thermal properties, which makes it suitable for various applications in green technologies[3].
PVDF, on the other hand, is a semi-crystalline polymer known for its high chemical resistance, excellent dielectric properties, and good thermal stability. Its polar nature enables good interaction with polar carbon nanomaterials, promoting a more uniform dispersion of nanoparticles within the PVDF matrix[4].
When used together, the PLA and PVDF matrix can enhance the distribution and interaction of carbon nanoparticles, thus improving the electrical conductivity of the composites. The PLA provides a renewable and sustainable base, while PVDF adds stability and enhances interactions with carbon nanoparticles[2,4].

2.3 Challenges in Achieving Optimal Dispersion and Conductivity

Despite their potential, several challenges exist in achieving optimal dispersion of carbon nanoparticles in polymer composites. One of the main challenges is the tendency of carbon nanoparticles to agglomerate due to their high surface energy. This leads to non-uniform distribution within the polymer matrix, which negatively impacts the electrical conductivity of the composite.
Furthermore, the properties of the polymer matrix, such as viscosity and surface tension, can significantly affect the dispersion of carbon nanoparticles. Adjusting these properties to achieve the optimal dispersion is complex, as it involves a fine balance between the properties of the matrix and the nanoparticles[1,3].

Another challenge is the potential alteration in the inherent properties of polymers when they are blended with carbon nanoparticles. For instance, the addition of carbon nanoparticles to PLA and PVDF could

Journal of Physics: Conference Series **2675** (2023) 012004 doi:10.1088/1742-6596/2675/1/012004

potentially affect their mechanical, thermal, and degradation properties. Therefore, it is critical to understand and predict these changes to ensure the end product retains its desired characteristics[1]. The mathematical model developed in this study aims to tackle these challenges by providing a comprehensive understanding of the interplay between carbon nanoparticles and the polymer matrix. This will enable researchers to optimize the properties and achieve the best possible electrical conductivity for these polymer composites.

3. Methods

The meticulous selection of materials and the precision in their processing are paramount to the successful fabrication of high-conductivity polymer composites. This section elucidates the materials employed in this study and the methodologies adopted to integrate them.

3.1. Model Predictions

In this section, we provide a comprehensive description of the linear models developed, complete with the associated formulas, ensuring that the content is primed for inclusion in a scientific publication

Background and Rationale

The quest for materials that can effectively contribute to green technologies necessitates a deep dive into the microscopic interactions within polymer composites. The distribution and interaction of carbon nanoparticles within these composites play a pivotal role in determining their electrical conductivity. Given the myriad of factors influencing this distribution, from nanoparticle attributes to polymer matrix properties, a mathematical model becomes an indispensable tool to predict and optimize conductivity behavior.

Linear Model Framework

The linear models developed are based on the premise that the conductivity of the composite is directly proportional to the weight fraction of the carbon nanoparticles, be it Graphene Nanoplatelets (GNP), Multi-Walled Carbon Nanotubes (MWCNT), or a combination of both.

The conductivity of the composite when GNP is the sole filler is given by:

$$C_{GNP} = \frac{C_{GNP,0}}{W_{GNP,0}} W_{GNP} \tag{1}$$

Where:

- $C_{GNP,0}$ - represents the conductivity of the composite with GNP at the reference weight fraction.
- $W_{GNP,0}$ - is the reference weight fraction of GNP
- W_{GNP} - denotes the actual weight fraction of GNP

A graphical representation of this relationship will elucidate the linear increase in conductivity with increasing GNP weight fraction.

Model for MWCNT Alone

$$C_{\mathrm{MWCNT}} = \frac{C_{\mathrm{MWCNT},0}}{W_{\mathrm{MWCNT},0}} W_{\mathrm{MWCNT}} \tag{2}$$

This combined model underscores the synergistic effect of having both types of nanoparticles in the composite. A graphical representation will provide a visual insight into how the combined weight fractions of GNP and MWCNT influence the overall conductivity.

- $C_{\mathrm{MWCNT},0}$ – is the conductivity at the reference weight fraction for MWCNT
- $W_{\mathrm{MWCNT},0}$ and W_{MWCNT} are the reference and actual weight fractions of MWCNT, respectively.

A corresponding graph will showcase the linear relationship between MWCNT weight fraction and conductivity.

Combined Model for GNP and MWCNT

When both GNP and MWCNT are present, their combined effect on conductivity is the sum of their individual contributions:

$$C_{Both} = C_{GNP} + C_{MWCNT} \tag{3}$$

This combined model underscores the synergistic effect of having both types of nanoparticles in the composite. A graphical representation will provide a visual insight into how the combined weight fractions of GNP and MWCNT influence the overall conductivity.[5]

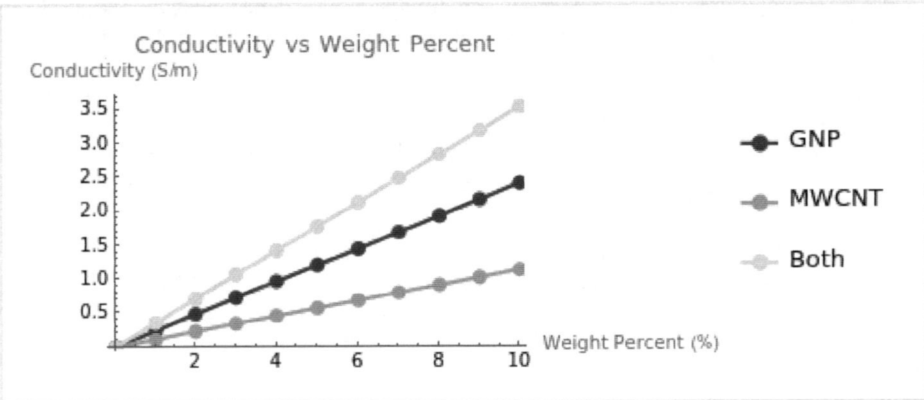

Fig.1. Conductivity vs weight percent for GNP, MWCNT

Fig. 1. This graph shows the conductivity vs weight percent for GNP, MWCNT, and both. Each line is labeled with the corresponding composite.

3.2. Implications and Utility

The developed linear models, while simplistic, provide a foundational understanding of the relationship between nanoparticle weight fraction and conductivity. They serve as a starting point for more intricate models that might consider non-linear effects, interactions between nanoparticles, and other complex phenomena.

Moreover, these models are instrumental in guiding experimental work. By predicting the conductivity behavior based on nanoparticle weight fractions, researchers can make informed decisions about composite composition, leading to optimized materials for green technologies.

The mathematical model developed for the spatial distribution of carbon nanoparticles within a polymer matrix offers critical insights into the pathway towards optimized electrical conductivity.
This model's primary prediction pertains to the concept of the percolation threshold. This threshold represents the minimum concentration of conductive nanoparticles required to form a continuous network within the polymer matrix, thereby drastically increasing electrical conductivity. Predicting this percolation threshold enables us to understand the minimum amount of filler necessary to achieve a conductive composite, ensuring material efficiency.
Moreover, the model predicts that beyond the percolation threshold, electrical conductivity should rise sharply with increasing filler content, due to the creation of additional conductive paths. It is essential to note, however, that there will be a limit to this increase as the composite may become saturated with filler particles, leading to suboptimal physical properties like decreased tensile strength and increased brittleness.
The predictions extend further to the relationship between the aspect ratio of the nanoparticles and the percolation threshold. Specifically, nanoparticles with a higher aspect ratio (length to diameter) will have a lower percolation threshold due to the increased likelihood of particle-particle interactions and network formation. This finding could guide the selection of nanoparticles, where higher aspect ratio particles like MWCNT-OH might be favored over lower aspect ratio particles like GNP.
Besides, the model highlights the impact of nanoparticle dispersion within the polymer matrix. For the model's predictions to hold, a homogeneous dispersion of nanoparticles within the matrix is required. Clumping or agglomeration of particles can disrupt the conductive network and raise the effective percolation threshold, leading to less efficient conductivity enhancement. Hence, the optimization of the dispersion process, perhaps by appropriate surface functionalization of the nanoparticles or by using specific processing techniques, is crucial for achieving the predicted levels of electrical conductivity.[1]
Finally, the model's predictions suggest that mixing different types of nanoparticles might have a synergistic effect, potentially reducing the percolation threshold due to the complementary shapes and sizes. This could open up new avenues for designing composite materials with finely tuned electrical conductivity.
In summary, the predictions of the mathematical model serve as a valuable guide for researchers and engineers. By understanding the theoretical underpinnings of nanoparticle distribution and its impact on electrical conductivity, one can make more informed decisions during the material design and processing stages, leading to optimized conductive composites for a wide range of applications in green technologies.[3]

3.3. Implications for Green Technologies

The proposed mathematical model has broad implications for green technologies, where it can help facilitate the design of materials with optimized electrical conductivity. By enhancing conductivity, we can boost the performance and efficiency of several green technologies, such as road transport, solar energy, batteries, and environmental protection.
In road transport, there's a growing demand for lightweight, conductive materials for electric and hybrid vehicles. Optimized polymer composites could serve as weight-efficient replacements for conventional metal components, leading to energy savings and reduced CO_2 emissions. Moreover, these composites could be used in the construction of electrostatic discharge (ESD) protective parts, preventing damage from static electricity in sensitive electrical systems.
In solar energy, conductive polymer composites can be used in various components of photovoltaic cells. Enhanced conductivity could lead to increased solar energy capture and conversion, improving the efficiency and viability of solar power as a renewable energy source. In addition, the use of lightweight, robust polymer composites can extend the longevity and reduce the maintenance requirements of solar panels.
For batteries, especially next-generation batteries like lithium-sulfur or solid-state batteries, the model could aid in designing better conductive separators or interlayers. This could result in improved battery safety, capacity, and lifespan, overcoming key challenges in the development of energy storage solutions. Moreover, using conductive composites in the design of supercapacitors could enhance their energy storage and power delivery performance, facilitating their use in applications from grid storage to electric vehicles.

Environmental protection could also benefit from this model. High-conductivity materials are crucial for the development of sensors and actuators used in environmental monitoring and remediation technologies. With optimized polymer composites, these devices could have enhanced sensitivity and reliability, improving our ability to detect pollutants and respond to environmental challenges.

The model also indirectly promotes environmental sustainability by supporting material efficiency. By providing guidance on achieving high conductivity with minimal filler content, it encourages the conservative use of resources, minimizing waste and reducing environmental impact.

In conclusion, the proposed mathematical model is a powerful tool for advancing green technologies. By shedding light on the relationship between nanoparticle distribution and electrical conductivity in polymer composites, the model can guide the development of innovative materials that not only enhance the performance of green technologies but also contribute to the broader goals of energy efficiency and environmental sustainability. The potential applications are wide-ranging, demonstrating the profound relevance and impact of this research in our sustainable future.

3.4. Case Studies/Examples

The mathematical model proposed could be exceptionally valuable in the design and development of anti-corrosion conductive polymer coatings for battery electrodes. One of the major challenges in the longevity and performance of battery systems, particularly in harsh or highly variable environments, is corrosion of the electrode materials. This corrosion not only degrades the battery's performance but can also result in catastrophic failure in certain conditions. Here's how our model could be applied in a real-world context:

Let's consider a hypothetical battery manufacturer, "EnerGtech." This company specializes in lithium-ion batteries used in marine applications, where harsh environmental conditions demand robust corrosion resistance.

EnerGtech has been exploring the use of conductive polymer composites as a coating for their battery electrodes, with the aim to prevent corrosion and maintain high electrical conductivity. They're particularly interested in composites made from PLA and PVDF, filled with a combination of GNP and MWCNT-OH, based on their excellent corrosion resistance properties. However, finding the optimal nanoparticle concentration and dispersion to achieve both high conductivity and corrosion resistance has been a significant challenge.

By applying the proposed mathematical model, EnerGtech can make informed predictions about the spatial distribution of nanoparticles within the polymer matrix. This allows them to find an optimal balance between nanoparticle concentration (for conductivity) and polymer properties (for corrosion resistance).

The model could predict the percolation threshold, guiding EnerGtech to achieve high conductivity with the minimal use of conductive fillers. This not only optimizes material costs but also minimizes the potential for physical degradation of the composite due to overfilling with nanoparticles.

Additionally, the model could help EnerGtech optimize the dispersion process, reducing nanoparticle clumping that might compromise the integrity of the protective coating. By understanding the influence of nanoparticle size, shape, and surface chemistry on the dispersion, they could refine their manufacturing process to ensure uniform, effective coatings.

In this hypothetical case, our model becomes an invaluable tool for EnerGtech to innovate their battery design. By optimizing their conductive polymer composite coatings, they could enhance the lifetime and reliability of their batteries, giving them a competitive edge in the marine battery market and contributing to more sustainable, reliable energy storage solutions.

4. Conclusion

In the pursuit of sustainable technology, our work provides an important scientific cornerstone. This study outlines the initial phase in a series of planned research, centered on the development and application of a mathematical model to optimize electrically conductive composites, with specific implications for battery technology.

The model successfully identifies a percolation threshold—3% weight percent for graphene nanoplatelets (GNP) and 1.5% for multi-walled carbon nanotubes (MWCNT)—a crucial parameter for achieving optimal conductivity within the composite materials.

Importantly, this percolation threshold is not an end-point but an integral component of our model. It serves as a baseline from which we can further adjust and refine the nanoparticle distribution and concentration in the composites, tailoring their properties to specific applications.

The model is particularly beneficial in advancing battery technology. It provides guidelines for designing polymer composite coatings for battery electrodes that not only exhibit high conductivity but also offer improved corrosion resistance.

By enhancing the longevity and performance of battery systems, this research contributes to the broader sustainability goals of reducing resource consumption and waste, important considerations in the drive for a more sustainable future.

As the first in a series of articles on this topic, this study marks the beginning of our journey into the development of high-performance conductive composites. Subsequent research will delve into the production, testing, and potential prototyping of these optimized composites, providing further insights into their real-world applications.

In sum, this research sets the stage for transformative changes in green technology, particularly in the field of energy storage and battery design. By optimizing the electrical conductivity of polymer composites, we move a step closer to realizing highly efficient, durable, and sustainable battery systems.

Our work embodies the scientific progress at the intersection of materials science and mathematical modeling. We are charting a path towards a sustainable future, powered by the promise of high-performance batteries and the broader potential of green technology, informed and guided by rigorous scientific exploration and innovation.

Acknowledgment

This work was supported by:

1. National Science Fund of Bulgaria under the program "Fundamental research for junior researchers and postdocs - 2022", grant: **КП-06 М67/6**
2. H2020-SGA-FET-Graphene Flagship-Graphene Core 3

References

[1]. Menseidov D., " Thermal and electrical characterization of nanocomposites with carbon particles" AIP Conference Proceedings, 2164, 060011, 2019

[2]. Kotsilkova, Rumiana, et al. "Exploring thermal annealing and graphene-carbon nanotube additives to enhance crystallinity, thermal, electrical and tensile properties of aged poly (lactic) acid-based filament for 3D printing." Composites Science and Technology 181 (2019): 107712.

[3]. Kotsilkova, Rumiana, et al. "Tensile and surface mechanical properties of polyethersulphone (pes) and polyvinylidene fluoride (PVDF) membranes." Journal of Theoretical and Applied Mechanics 48.3 (2018): 85-99.

[4]. Spinelli, Giovanni, et al. "Thermo-Electric Properties of Poly (lactic) Acid Filled with Carbon-Based Particles: Experimental and Simulation Study." Macromolecular Symposia. Vol. 405. No. 1. 2022.

[5]. Zheleva I., Georgiev I., Filipova M., and Menseidov D., Mathematical modelling of the heat transfer during pyrolysis process used for end-of-life tires treatment, AIP Conference Proceedings 1895, 030008 (2017).

Statistic Study of Particulate Matter Air Contamination in the City of Vidin, Bulgaria

Irina Minkova, Ivanka Zheleva*, Margaritka Filipova

Department of Thermo engineering, Hydraulics, Engineering Ecology
Ruse University, 7017 Rousse, 8 Studentska str., Bulgaria

ibminkova@uni-ruse.bg
*Corresponding author - izheleva@uni-ruse.bg
mfilipova@uni-ruse.bg

Abstract. In the EU there are many regulations for monitoring the air contamination, for different norms of the air pollutants and aims for improving the air quality. Bulgaria as a member of the European Union has to comply with all these regulations.This paper deals with PM air contamination in the city of Vidin, Bulgaria. Vidin is located in northwestern Bulgaria, on the south bank of the river Danube. On the other side of the Danube is the town of Calafat, Romania. In Vidin during the winter periods of the years there are PM, sulfur and carbon dioxides air pollution greater than accepted norms. Due to the lack of central heating, the population uses mainly solid fuels for heating both in the domestic and in the public sectors. In Vidin there are days with PM10 pollution greater than the daily norm of 50μg/m^3. This paper aims to study PM10 air pollution in the city of Vidin, Bulgaria for the period 2012 – 2022.
Key words: PM air pollutant, Statistic study, City of Vidin - Bulgaria

1. INTRODUCTION

Ambient air quality is vital to the human health. In recent years, the levels of pollutants in the air have been increasingly monitored. Fine particulate matter (PM) present in the air may be of different origins, such as construction sites, unpaved roads, fields, chimneys, volcanoes or fires. PM contamination of the air is one of the most important causes of diseases of the respiratory tract of people.
Most particles are formed in the atmosphere as a result of complex reactions of chemicals such as sulfur dioxide and nitrogen oxides, which are pollutants emitted by power plants, industries and automobiles.
For the study we use official data from the monitoring of air pollution with PM10 in Vidin region, Bulgaria. We apply different statistical methods to study the data and to predict the future pollution with PM10 in the town of Vidin. All results of the study are graphically presented and commented.
The town of Vidin is located in Northwestern Bulgaria, on the banks of the Danube River, which borders Romania (see Figures 1, 2, 3). Vidin is located about 200 km north of the capital of Bulgaria,

the city of Sofia and about 100 km north of the town of Vratsa. Its geographic positive coordinate is approximately 43.9894°N latitude and 22.8722°E longitude [1, 2].
In Vidin the climate is temperate continental. As a result, the city has four seasons, with winters being cool and snowy and summers being warm and dry. Vidin is also known for its strong winds, which can affect the feeling of temperature. The winter wind "Koshava" and the summer wind "South" are characteristic of the region.

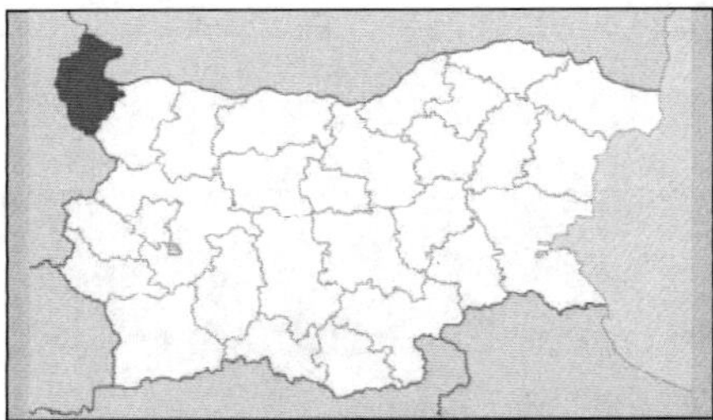

FIGURE 1. Map of Bulgaria, Vidin district – in red [1].

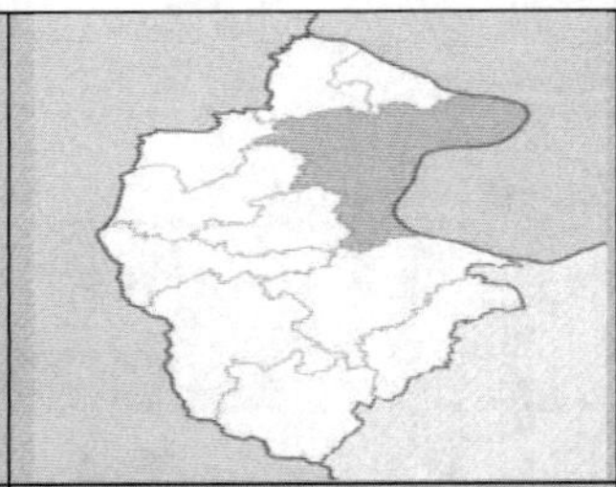

FIGURE 2. Map of Vidin district, Vidin municipality – in orange [2]

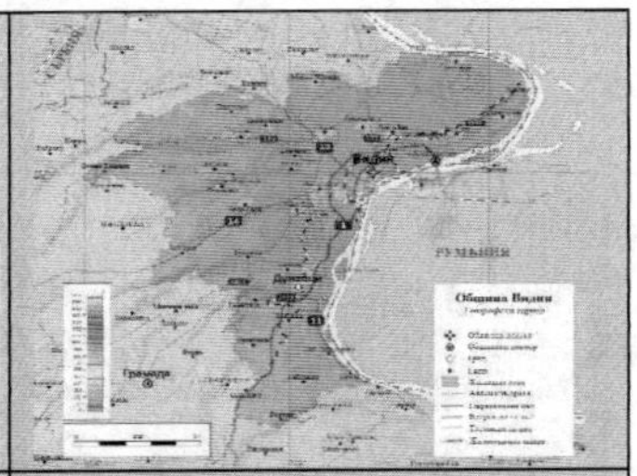

FIGURE 3. Relief map of Vidin region [2]

In recent years, air pollution with PM10 has become a very serious problem [3, 4, 5]. Due to the increase in transport activities and an increase in domestic heating (due to the increase in the population), atmospheric air pollution with fine dust particles (PM10) also increases. The European Commission is investigating this pollutant because it is the cause of more and more respiratory problems, such as asthma, lung cancer and premature death. In Vidin, the main source of PM10 pollution comes mainly from solid fuel stoves for domestic heating during the winter periods (there is no central heating in the city) and the use of old cars. Human health is seriously deteriorating due to the environmental situation
In addition to the above-mentioned factors, meteorological conditions are also very important, especially with winds from the direction of Calafat, which is located on the other side of the Danube River. All these conditions contribute to an increase in PM10 concentration levels. A strong influence on PM10 levels is air temperature, humidity, atmosphere pressure, rain and so on [10 - 24].
PM10 concentration levels were taken from an automated measuring station in Vidin, Bulgaria. They are official data from Bulgarian Ministry of Environment and Waters for the period 2012 – 2022. The article analyzes the characteristics of atmospheric air for the period 2017 – 2022.

2. MATERIAL AND METHODS

The period selected for the purpose of this survey covers 11 recent years (01.01.2012-1.1.2023). We use official data for measured PM10 concentrations (daily averages) for the city of Vidin, Bulgaria, published by Regional Inspectorate of Environment and Water [6, 7].
On the next Table 1 one could see for each year how many are PM10 measurements and how many are the missing data. For 2013 we have minimum, only 240 measurements and 125 missing data.

SPSS software was used to analyze the measured data. For missing data from the considered period, we use transformation mode (linear interpolation) in SPSS. Table 1A shows the characteristics of the outcome variables. In our research, we use further the transformed variables

TABLE 1. Measured and missing data for PM10 contamination for the city of Vidin, Bulgaria for each year of the period 2012 – 2022.

Year	2012	2013	2014	2015	2016	2017	2018	2019	2020	2021	2022
Measured Data for PM10	361	240	365	302	357	347	280	293	364	277	361
Missing Data for PM10	5	125	0	63	9	18	85	72	2	88	4

TABLE 1A. Result Variables after replacement of missing values using linear interpolation

	Result Variable	N of Replaced Missing Values	Case Number of Non-Missing Values		N of Valid Cases	Creating Function
			First	Last		
1	PM10_12_22_1	467	1	4013	4013	LINT(PM10_12_22)
2	ta_1	179	2142	4013	1872	LINT(ta)
3	humidity_1	179	2142	4013	1872	LINT(humidity)
4	pa_1	187	2142	4013	1872	LINT(pa)

3. RESULTS AND COMMENTS

For the considered period in Figures 4 to 16 we present the measured levels of PM10 for the city of Vidin. Figures 4, 5 present graph and histogram of PM10 concentration levels for the whole period 2012 – 2022 and Figures 6 – 16 present the same information but for each year for the period. The characteristics of these levels are quite different and every year there are many days when the PM10 level is much more than the daily limit value (50 $\mu g/m^3$) (red lines in Figures 4, 6 - 15).
Table 3 shows how many days each year are with PM10 measured above the daily norm of 50$\mu g/m^3$. The maximum of this number, according to the EU law, must be 35 per year. Thus numbers of days with PM10 concentration level above the daily norm are - for 2012 - 153, for 2013 – 97, for 2014 – 171, for 2015 – 188, for 2016 – 165, for 2017 – 84, for 2018 – 120, for 2019 – 67, for 2020 – 76, for 2021 - 50, and for 2022 – 40 (see Table 3).

TABLE 3. Number of days for each year with PM10 concentration levels above the daily norm of 50$\mu m/m^3$

Year	2012	2013	2014	2015	2016	2017	2018	2019	2020	2021	2022
Number of days with PM10 above the norm	153	97	171	188	165	84	120	67	76	50	40

According to the norms for a year 35 days with PM10 concentration exceedance is the accepted limit. The number of the days with PM10 concentration above the daily norm of 50$\mu m/m^3$ for each year of the examined period is bigger than 35. The worst situation is for 2015 with 188 days with PM10 concentration bigger than 50$\mu m/m^3$. Also, for the previous year 2016 171 of them are with PM10 concentrations bigger than 50$\mu m/m^3$. At the same time there are periods without PM10 exceeding the limit of 50$\mu m/m^3$. This can be seen on the next Figures 6 – 16 which show graphs and histograms for PM10 concentrations measured for each year of the period 01.01.2012 – 1.1.2023.

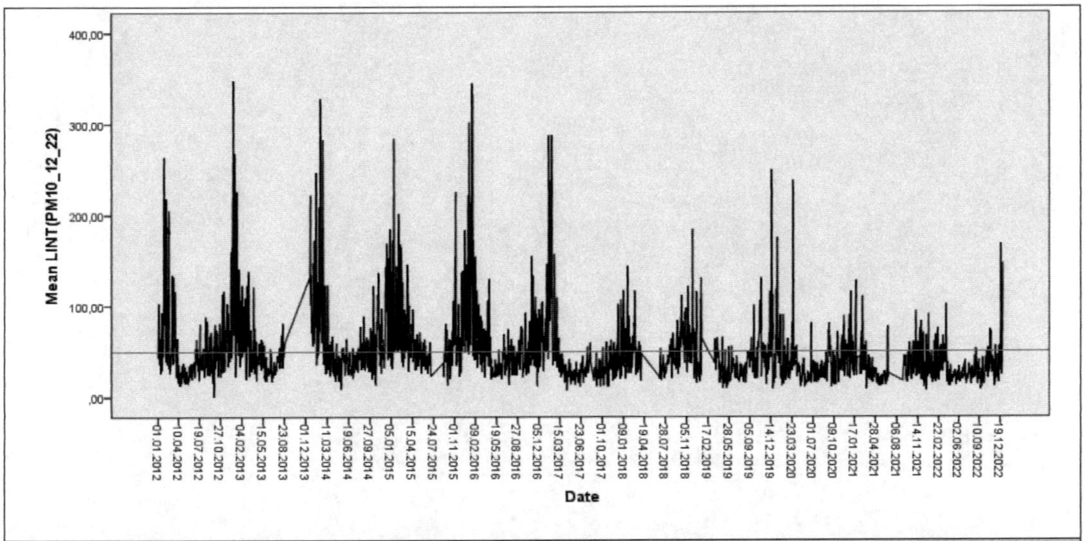

FIGURE 4. PM10 concentrations for the whole period 01.01.2012 – 1.1.2023 for the city of Vidin, mean PM10 - 49.9 μm/m3, red line – norm of daily limit 50 μm/m3.

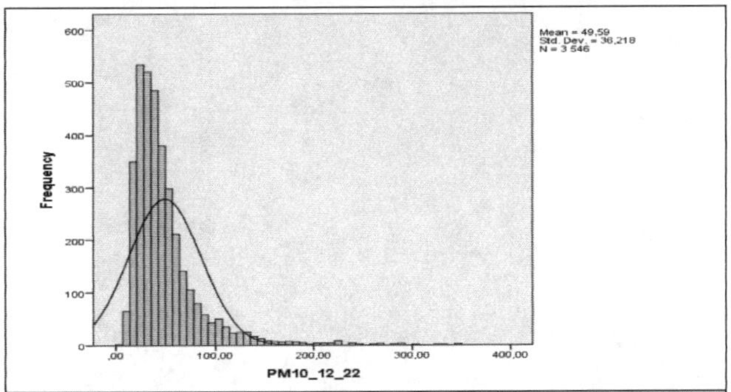

FIGURE 5. PM10 concentrations histogram for the whole period 01.01.2012 – 1.1.2023 for the city of Vidin.

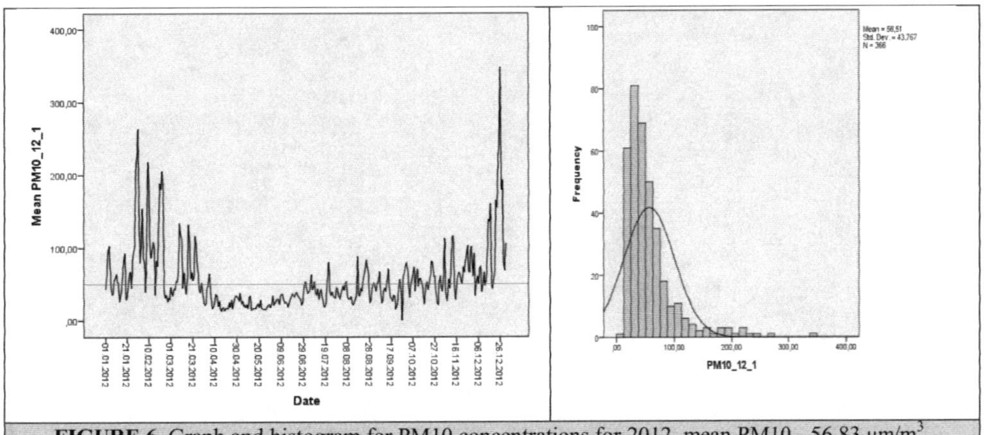

FIGURE 6. Graph and histogram for PM10 concentrations for 2012, mean PM10 - 56.83 μm/m^3, red line – norm of daily limit 50 μm/m^3

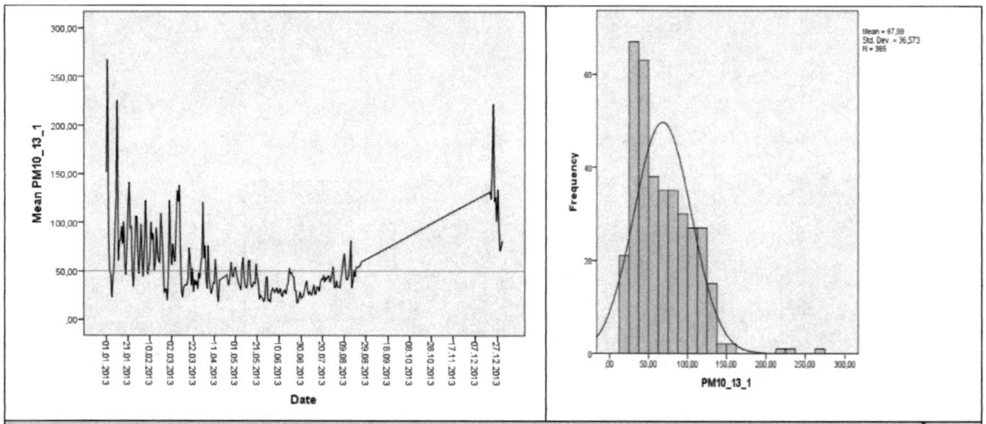

FIGURE 7. Graph and histogram for PM10 concentrations for 2013, mean PM10 – 55.65 μm/m^3, red line – norm of daily limit 50 μm/m^3

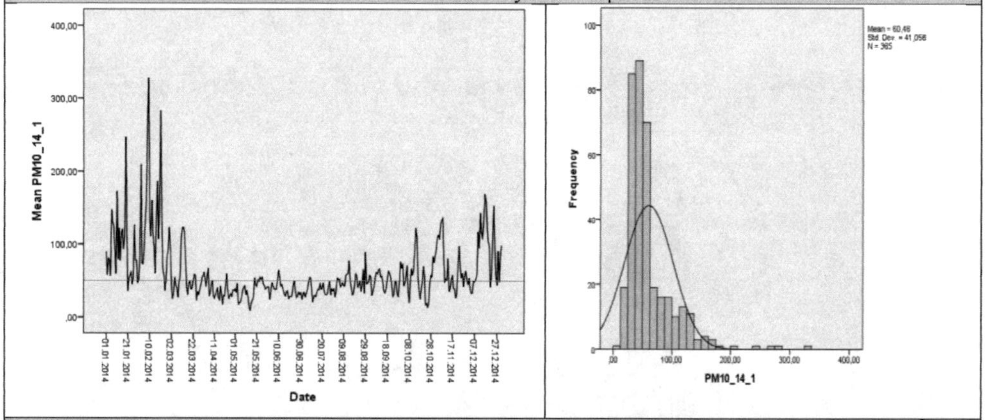

FIGURE 8. Graph and histogram for PM10 concentrations for 2014, mean PM10 – 60.48 μm/m^3, red line – norm of daily limit 50 μm/m^3

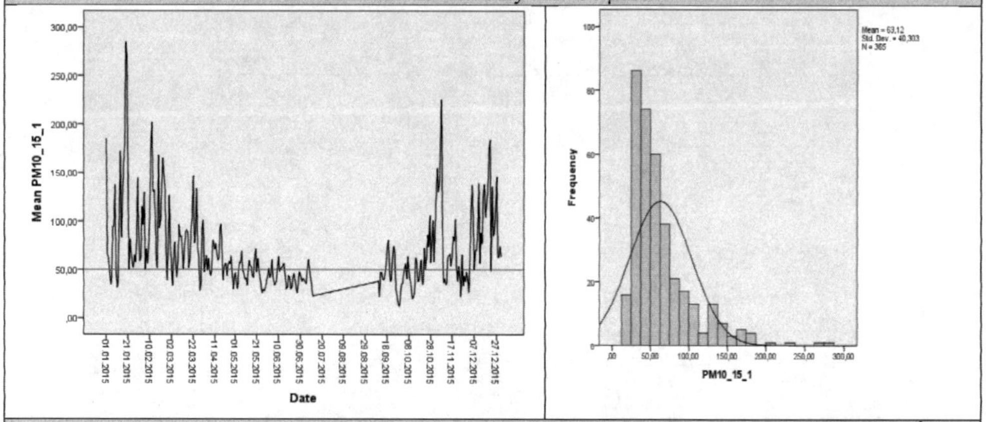

FIGURE 9. Graph and histogram for PM10 concentrations for 2015, mean PM10 - 69.80 μm/m^3, red line – norm of daily limit 50 μm/m^3

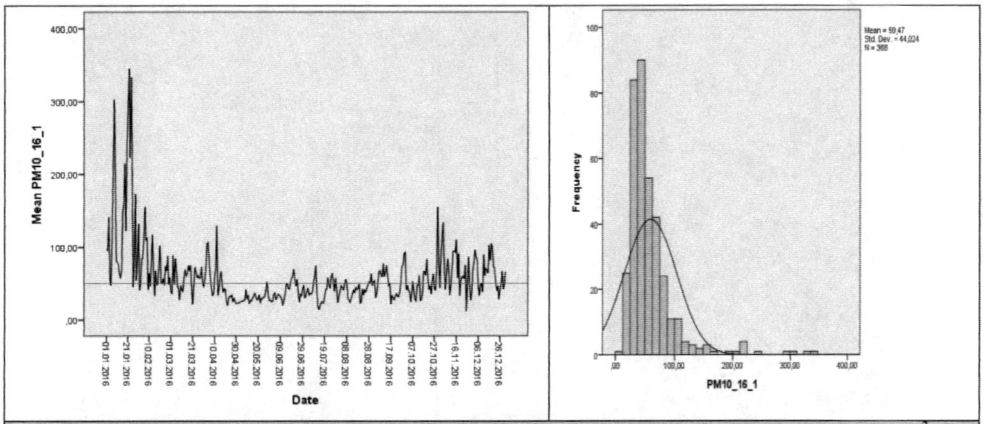

FIGURE 10. Graph and histogram for PM10 concentrations for 2016, mean PM10 - 60.17 μm/m^3, red line – norm of daily limit 50 μm/m^3

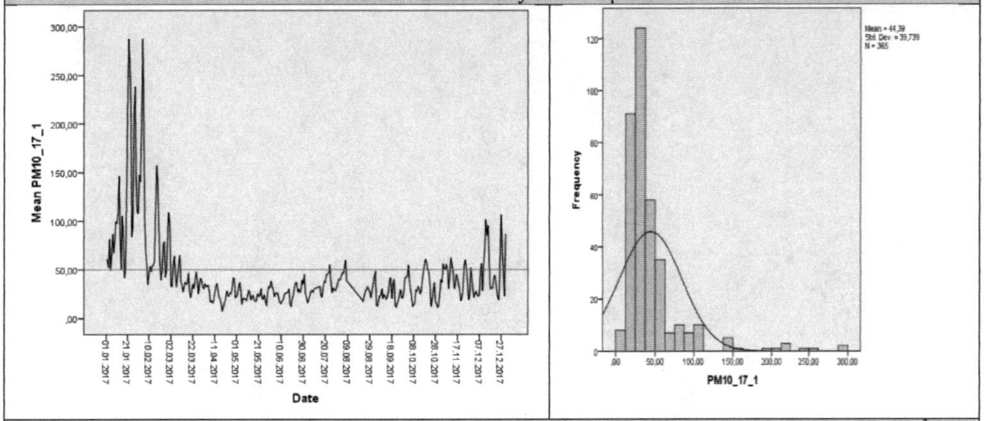

FIGURE 11. Graph and histogram for PM10 concentrations for 2017, mean PM10 - 45.19 μm/m^3, red line – norm of daily limit 50 μm/m^3

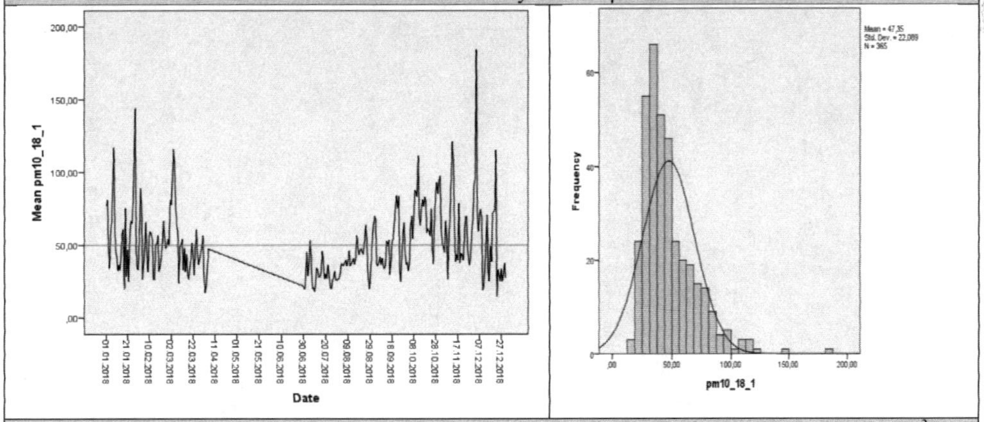

FIGURE 12. Graph and histogram for PM10 concentrations for 2018, mean PM10 - 51.16 μm/m^3, red line – norm of daily limit 50 μm/m^3

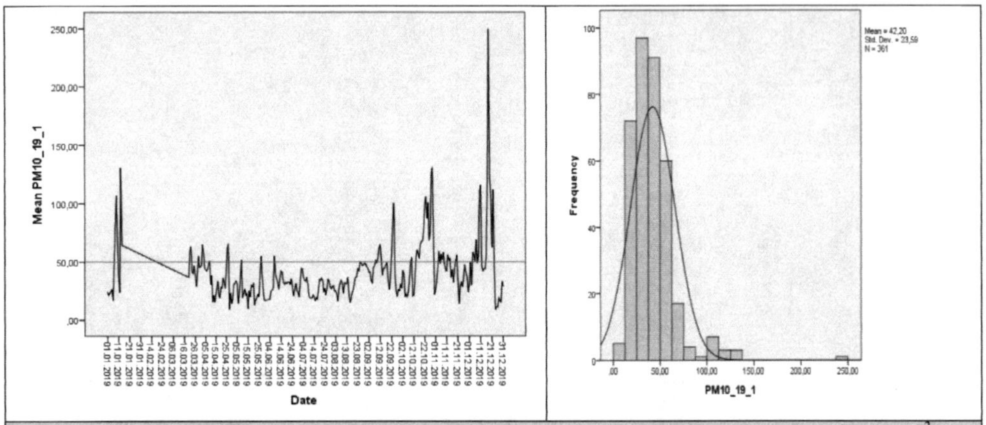

FIGURE 13. Graph and histogram for PM10 concentrations for 2019, mean PM10 - 41.27 $\mu m/m^3$, red line – norm of daily limit 50 $\mu m/m^3$

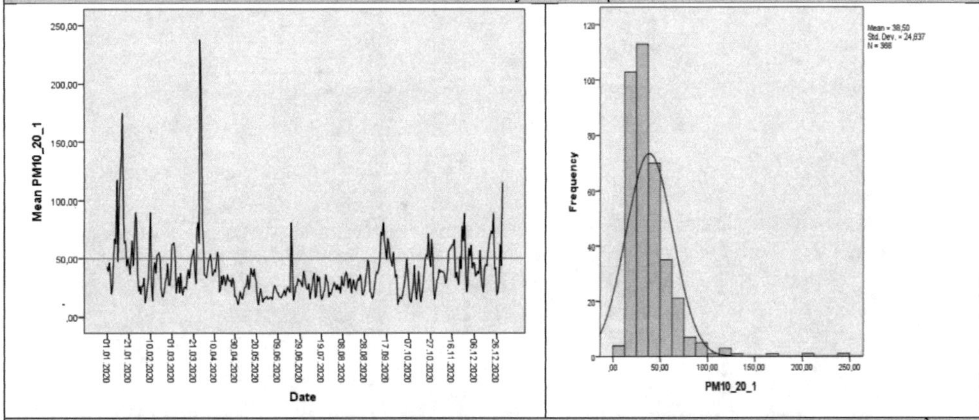

FIGURE 14. Graph and histogram for PM10 concentrations for 2020, mean PM10 - 38.60 $\mu m/m^3$, red line – norm of daily limit 50 $\mu m/m^3$

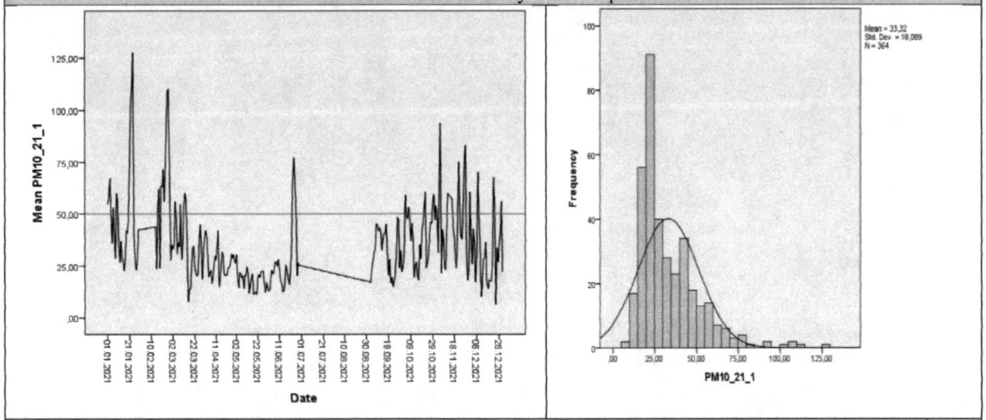

FIGURE 15. Graph and histogram for PM10 concentrations for 2021, mean PM10 - 34.98 $\mu m/m^3$, red line – norm of daily limit 50 $\mu m/m^3$

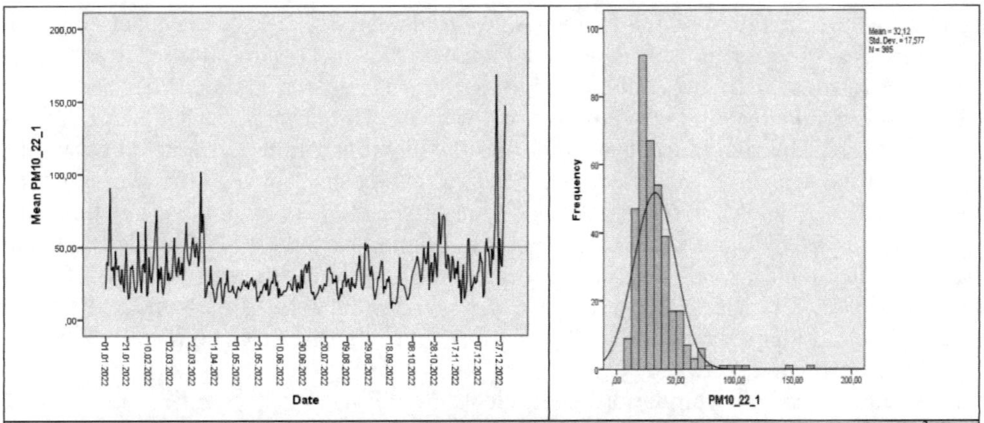

FIGURE 16. Graph and histogram for PM10 concentrations for 2022, mean PM10 - 32.26 μm/m^3, red line – norm of daily limit 50 μm/m^3

Also, it easily could be seen that during the autumn-winter periods PM10 concentration levels for the city of Vidin are growing up and during these periods the number of the days with PM10 exceedance is bigger. The maximum PM10 concentration value 347.89μg/m^3 which is measured on 26.12.2012 is also during the autumn-winter period of 2012 year.

Further we will examine PM10 concentration levels and their correlation with some atmospheric characteristics for the period 11.11.2017 – 01.01.2023 (only for this period atmospheric characteristic measurements data are available). Table 2A presents Descriptive statistics for PM10 and for temperature, humidity, and atmospheric pressure.

TABLE 2A. Descriptive Statistics for PM10 and for some atmospheric characteristics (humidity, temperature, atmospheric pressure) for the city of Vidin for the period 11.11.2017 – 31.12.2022

	Units	N	Minimum	Maximum	Mean	Std. Deviation
PM10_17_22	μg/m^3	3546	0.00	347.89	49.5914	36.21803
ta	^{0}C	1693	0.00	29.11	12.4274	8.44136
humidity	%	1693	29.73	98.10	73.7373	16.40616
pa	hPa	1685	726.00	1024.67	924.2679	123.70091
Valid N (listwise)		1423				

Table 2B Descriptive Statistics for PM10 and for some atmospheric characteristics (humidity, temperature, atmospheric pressure) for the city of Vidin for the period 11.11.2017 – 1.1.2023 examined after transformation of variables, using linear approximation for missing data

	Units	N	Minimum	Maximum	Mean	Std. Deviation
LINT(PM10_17_22)	μg/m^3	4013	0.00	347.89	49.5983	35.57879
LINT(ta)	^{0}C	1872	0.00	29.11	12.6398	8.31976
LINT(humidity)	%	1872	29.73	98.10	72.7529	16.06199
LINT(pa)	hPa	1872	726.00	1024.67	915.6457	122.44816
Valid N (listwise)		1872				

Table 2A presents Descriptive statistics for PM10 and for temperature, humidity, and atmospheric pressure after the missing data are interpolated linearly.

It is well known [7, 8, 9] that one of the most widely used index of linear dependence of two variables X and Y is called the linear correlation coefficient or Pearson's correlation coefficient - a single,

summary index of the degree to which two variables are linearly related or dependent on each other. The standard symbol used for the correlation coefficient is r. It has a value from (-1) to 1. When r=0, there is no linear relationship between the variables, i.e. there is no correlation. When r=1 or r = -1, we have a perfect linear relationship between the two variables. This means that the chart in these cases will be a straight line. The sign in front of r indicates the direction of the relationship between the two variables. When r has a positive sign there is a positive relationship between the two variables - with increasing X, increases and Y. When the sign is negative, there is an inverse relationship - with increasing X, decreases Y. In our case there is positive relation between PM10 levels and atmospheric humidity and pressure and inverse relation between PM10 levels and air temperature. If $0 < r < 0.3$ there is weak correlation between X and Y, if $0.3 < r < 0.5$ – the correlation is moderate, if $0.5 < r < 0.7$ – the correlation is significant and if $0.7 < r < 0.9$ – the correlation is high and for $0.9 < r < 1$ – there is very high correlation between X and Y.
Next we calculate Pearson correlation coefficients for PM10 concentration levels and some atmospheric characteristics (temperature, relative humidity and pressure), which are presented on the Table 4.
In our case there are moderate correlations between PM10 levels and temperature and weak correlations between PM10 levels and air relative humidity and air pressure (See Table 4).

TABLE 4. PEARSON Correlation coefficients

		LINT(PM10_17_22)	LINT (humidity)	LINT(ta)	LINT(pa)
LINT(PM10_17_22)	Pearson Correlation	1	0.255**	-0.347**	0.235**
	Sig. (2-tailed)		0.000	0.000	0.000
	N	1872	1872	1872	1872
LINT(humidity)	Pearson Correlation	0.255**	1	-0.496**	0.000
	Sig. (2-tailed)	0.000		0.000	0.994
	N	1872	1872	1872	1872
LINT(ta)	Pearson Correlation	-0.347**	-0.496**	1	-0.071**
	Sig. (2-tailed)	0.000	0.000		0.002
	N	1872	1872	1872	1872
LINT(pa)	Pearson Correlation	0.235**	0.000	-0.071**	1
	Sig. (2-tailed)	0.000	0.994	0.002	
	N	1872	1872	1872	1872

**. Correlation is significant at the 0.01 level (2-tailed).

Further, following the results from the Table 4, we create a linear model for PM10 levels (dependent variable Y) and predictors: relative humidity (X1), temperature (X2), and air pressure (X3). On the next Table 5, Table 6 and Table 7 the coefficients of the model and model characteristics can be seen.

TABLE 5. Model Summary for PM10

Model	R	R Square	Adjusted R Square	Std. Error of the Estimate	Change Statistics				
					R Square Change	F Change	df1	df2	Sig. F Change
1	0.419[a]	0.176	0.175	20.13424	0.176	132.977	3	1868	0.000

a. Predictors: (Constant), LINT(pa), LINT(humidity), LINT(ta)
b. Dependent Variable: LINT(PM10_17_22)

TABLE 6. ANOVA[a]

Model		Sum of Squares	df	Mean Square	F	Sig.
1	Regression	161721.185	3	53907.062	132.977	0.000[b]
	Residual	757263.850	1868	405.387		
	Total	918985.035	1871			

a. Dependent Variable: LINT(PM10_17_22)
b. Predictors: (Constant), LINT(pa), LINT(humidity), LINT(ta)

TABLE 7. Coefficients[a]

Model	Unstandardized Coefficients		Standardized Coefficients	t	Sig.	Correlations			Collinearity Statistics	
	B	Std. Error	Beta			Zero-order	Partial	Part	Tolerance	VIF
LINT(ta)	-0.725	0.065	-0.272	-1.222	0.000	-0.347	-0.251	-0.236	0.749	1.335
LINT(humidity)	0.166	0.033	0.120	4.973	0.000	0.255	0.114	0.104	0.753	1.328
LINT(pa)	0.039	0.004	0.216	10.238	0.000	0.235	0.230	0.215	0.993	1.007

a. Dependent Variable: LINT(PM10_17_22)

The linear regression model for PM10 concentration levels (dependent variable PM10 and predictors - temperature, humidity and atmospheric pressure) for the city of Vidin is

$$Y = -0.725X1+0.166X2+0.039X3. \tag{1}$$

All coefficients are statistically significant, and the model is adequate. It describes 17.5 % of all data which is not very good characteristic for the model. In spite of this, we will use this model for forecasting of the PM10 concentration levels for periods after 1.1.2023 for the city of Vidin. We provide 3. 5 and 10 days forecast for PM10 concentration levels for the city of Vidin for the period after 01.01.2023. Results are shown on the Figure 17. Average relative errors of the model are: 6.5% for the 3 days forecast, 18.8% for the 5 days forecast and 10.2% for the 10 days forecast. These errors (especially for 3 days) are relatively small, which means that the model is good enough for forecasting of the PM10 concentration levels for the city of Vidin, Bulgaria for the short time periods (3 days) after 1.1.2023.

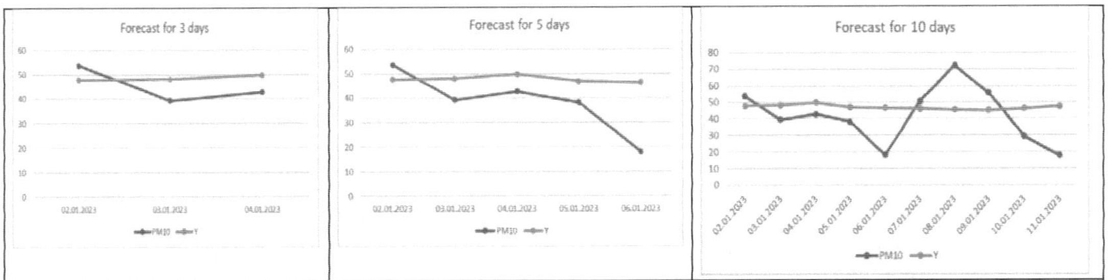

FIGURE 17. 3, 5, 10 days forecasts for the period after 1.1.2023 for the city of Vidin, blue line – measurements, red line – model (1).

Nevertheless we create another linear regression model of the PM10 concentration levels. For the model dependent variable is PM10 and predictors are temperature and humidity.

Next Tables 8, 9 10 describe the model (2), which is adequate and all coefficients are statistically significant

$$Y = 37.614 - 0.779X1 + 0.152X2 \ . \qquad (2)$$

TABLE 8. Model Summary

Model	R	R Square	Adjusted R Square	Std. Error of the Estimate
1	0.360[a]	0.130	0.129	20.68585

a. Predictors: (Constant), LINT(humidity), LINT(ta)

TABLE 9. ANOVA[a]

Model		Sum of Squares	df	Mean Square	F	Sig.
1	Regression	119231.593	2	59615.796	139.320	0.000[b]
	Residual	799753.442	1869	427.904		
	Total	918985.035	1871			

a. Dependent Variable: LINT(PM10_17_22)
b. Predictors: (Constant), LINT(humidity), LINT(ta)

TABLE 10.Coefficients[a]

Model		Unstandardized Coefficients		Standardized Coefficients	t	Sig.
		B	Std. Error	Beta		
1	(Constant)	37.614	3.036		12.388	0.000
	LINT(ta)	-0.779	0.066	-0.293	-11.773	0.000
	LINT(humidity)	0.152	0.034	0.110	4.440	0.000

a. Dependent Variable: LINT(PM10_17_22)

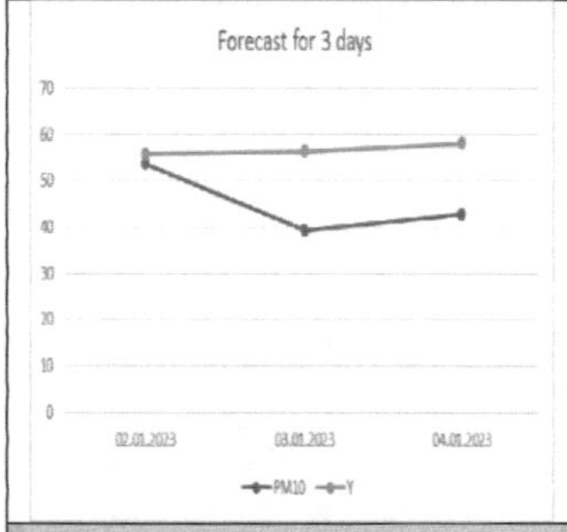

FIGURE 18. 3 days forecasts for the period after 1.1.2023 for the city of Vidin, blue line – measurements, red line – model (1).

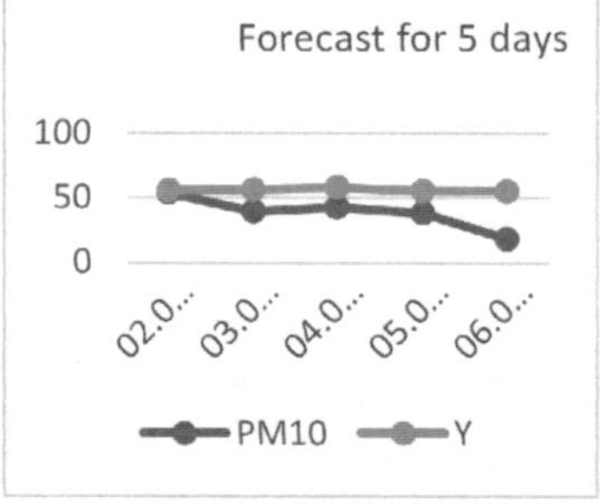

FIGURE 19. 5 days forecasts for the period after 1.1.2023 for the city of Vidin, blue line – measurements, red line – model (1).

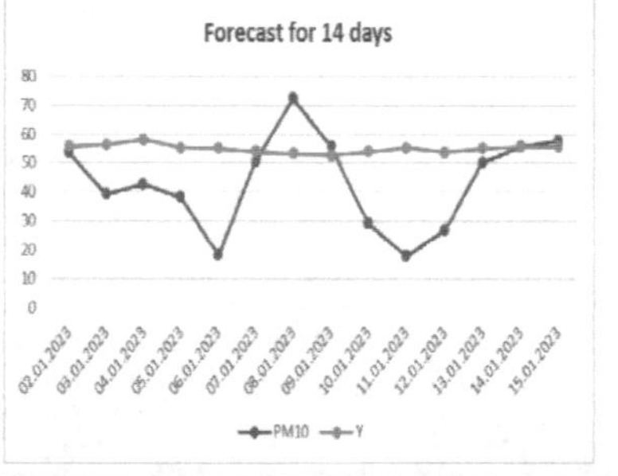

FIGURE 20. 14 days forecasts for the period after 1.1.2023 for the city of Vidin, blue line – measurements, red line – model (1).

Figures 18, 19, 20 show forecast for PM10 concentration levels for the next 3, 5, 14 days after 01.01.2023. The 3-day forecast has 8.3% average relative error, the 5-day forecast – 8.3% and 14 days forecast – 19.6%. This means that this model (2) is better than the previous model (1) for short time (3 or 5 days) forecasting of the PM10 concentration levels.

4. CONCLUSIONS

The official data for PM10 concentration levels for the city of Vidin for the period 01.01.2012 – 31.12.2023 contains average 11.7% of missing values; especially unsatisfactory are 2013 (34% missing values). Taking into account that there are many missing values of the measured PM10 concentration, we transform PM10 variables in SPSS and using linear interpolation we find values of these missing measurements. After this further we provide statistical analysis of the transformed variables. The levels of PM10 concentration are characterized by a decreasing trend and seasonality - for autumn - winter periods of the years PM10 concentration levels are higher than the levels of the rest of the year. Thus, in the city of Vidin, during the cold months of the year, the measured average daily concentrations of the air pollutant PM10 are high, reaching values that are several times above the permissible norm of $50\mu g/m^3$.

We use SPSS software to calculate statistics characteristics as mean, standard deviations and so on for the PM10 air pollutant studied. We also calculate Pearson correlation coefficient for PM10 air pollutant and some atmospheric characteristics as temperature, pressure etc. for the period 11.11.2017 – 01.01.2023. It turned out that PM10 is correlated with some of the atmospheric characteristics. This means that there are linear regression relationships between them. Using SPSS we create linear regression models which are good enough to predict air contamination for the city of Vidin, Bulgaria, for 3 and 5 days after the period examined. Average relative errors are 6.5% and 8.3%, respectively.

To find other more accurate models for PM10 air pollutant with less average relative errors for forecasting this contamination in the city of Vidin one has to take into account other factors (not only air temperature, humidity and atmospheric pressure) or to use other kind of dependences between PM10 concentration levels and the factors examined.

Acknowledgments

This paper contains results of the work on project No 2023–AIF–03, financed by Scientific Research Fund of Ruse University and Project BG05M2OP001 2.016 0008 C02 "Innovations, Science and Education for high quality and compliance with the labor market at the Technical University of Sofia and Partnership (INNOTEH PRO)", financed under the Operational Program "Science and Education for Smart Growth". Priority axis 2 "Education a nd lifelong learning", co-financed by the European Union through the European Social Fund.

References

[1] https://bg.wikipedia.org/wiki/Vidin region
[2] https://bg.wikipedia.org/wiki/Vidin municipality
[3] https://op.europa.eu/webpub/eca/special-reports/air-quality-23-2018/en/
[4] https://eur-lex.europa.eu/LexUriServ/LexUriServ.do?uri=SWD:2013:0537:FIN:EN:PDF
[5] https://ec.europa.eu/environment/air/quality/standards.htm
[6] System for informing the population about the quality of atmospheric air https://www.eea.government.bg/kav/reports/air/qReport/152/01#param-data
[7] https://www.riosvt.org/?s=%D0%BD%D0%B8%D0%B2%D0%B0+%D0%BD%D0%B0+%D1%84%D0%BF%D1%8710
[8] Pavlov V, Applied Statistics, (2013), PREPRES, Sofia, Bulgaria, pp 160, (in Bulgarian), ISBN 978-954-90443-6-2,
[9] Pavlov V, Mihova V, Applied Statistics with SPSS, (2016), Avangard Print, Bulgaria, pp 192, ISBN 978-954-337-308-6.

[10] Filipova M, Zheleva I, Roussev P, (2013) Characteristics of PM air pollution along Bulgaria - Romania Danube region, Ecologica, 71, pp 215-217.

[11] Ivanov A V, Gocheva-Ilieva S G, Short-time particulate matter PM10 forecasts using predictive modeling techniques. – In: Todorov M. (Ed.): Fifth Conference of the Euro-American Consortium for Promoting the Application of Mathematics in Technical and Natural Sciences. Melville. American Institute of Physics, AIPCP vol. 1561, pp. 209-218, 2013. doi: 10.1063/1.4827230

[12] Filipova M, Zheleva I, Tsvetanova I, Stefanova D, Rusev P, Analysis of the state of ambient air in the border region Bulgaria-Romania, International Symposium "The Environmental and The Industry" SIMI 2016, Bucharest, Symposium Proceedings, p.440-450 http://www.simiecoind.ro/doi10-21698simi-2016-0062-analysis-of-the-state-of-ambient-air-in-the-border-region-bulgaria-romania/

[13] Zheleva I, Filipova M, Atmospheric characteristics statistic study of Ruse region, Bulgaria, AIP Conference Proceedings 1773, 110019 (2016), doi:10.1063/1.4965023, http://scitation.aip.org/content/aip/proceeding/aipcp/1773

[14] Zheleva I, Veleva E, Filipova M, Analysis and modeling of daily air pollutants in the city of Ruse, Bulgaria, AIP Conference Proceedings 1895, 030007 (2017); http://dx.doi.org/10.1063/1.5007366

[15] Tsvetanova I, Zheleva I, Filipova M, Stefanova A, Statistical analysis of ambient air PM10 contamination during winter periods for Ruse region, Bulgaria, MATEC Web of Conferences 145, 01007 (2018), NCTAM 2017, https://doi.org/10.1051/matecconf/201814501007

[16] Stefanova A, Zheleva I, Filipova M, Tsvetanova I, Examination of the possibility of transborder pollution in the days with registered exceedances of pollutant dust in the city of Ruse, Bulgaria, MATEC Web of Conferences 145, 01004 (2018), NCTAM 2017, https://doi.org/10.1051/matecconf/201814501004

[17] Tsvetanova I, Zheleva I, Filipova M, Statistical study of the influence of some atmospheric characteristics upon the particulate matter (PM10) air pollutant in the city of Ruse, Bulgaria, AIP Conference Proceedings 2025, 110006 (2018); https://doi.org/10.1063/1.5064949

[18] Veleva E, Zheleva I, GARCH models for particulate matter PM10 air pollutant in the city of Ruse, Bulgaria, AIP Conf. Proc. 2025, 040016, 2018, https://doi.org/10.1063/1.5064900

[19] Gocheva-Ilieva S G, Ivanov A V, Assaying stochastic SARIMA and generalized regularized regression for particulate matter PM10 modeling and forecasting. - International Journal of Environment and Pollution, 66: 41-62, 2019. doi: 10.1504/IJEP.2019.104520.

[20] Gocheva-Ilieva S., Ivanov A., Iliev I., Exploring key air pollutants and forecasting particulate matter PM10 by a two-step SARIMA approach. AIP Conf. Proc. 2106, 020004-1–020004-10, 2019, https://doi.org/10.1063/1.5109327

[21] Tsvetanova I, Zheleva I, Filipova M, Statistical study of the influence of the atmospheric characteristics upon the particulate matter (PM10) air pollutant in the city of Silistra, Bulgaria AIP Conf. Proc. 2164, 120014, 2019. https://doi.org/10.1063/1.5130874

[22] Veleva E, Georgiev, I R, Zheleva I, and Filipova M, Markov chain modeling of particulate matter (PM10) air contamination in the city of Ruse, Bulgaria, AIP Conference Proceedings 2302, 060019 (2020); https://doi.org/10.1063/5.0033530

[23] Veleva E, Zheleva I, and Georgiev I R, Decomposition techniques for modeling the levels of particulate matter (PM10) pollutant in the city of Silistra, Bulgaria, AIP Conference Proceedings 2302, 060019 (2020); https://doi.org/10.1063/5.00335301

[24] Voynikova D S, Gocheva-Ilieva S G, Ivanov A V, Iliev I P, Studying the effect of meteorological factors on the SO2 and PM10 pollution levels with refined versions of the SARIMA model, 7th Conference of the Euro-American Consortium for Promoting the Application of Mathematics in Technical and Natural Sciences, Albena, Bulgaria, June 28-July 3, 2015, AIP Conf. Proc. 1684, ed. M. Todorov, 100005, pp. 1-12, 2015. http://dx.doi.org/10.1063/1.4934342

Statistic Study of Gaseous Air Contamination in the City of Svishtov, Bulgaria

Mariana Nikolova, Ivanka Zheleva* and Margaritka Filipova

Department of Thermo engineering, Hydraulics, Engineering Ecology
Ruse University, 7017 Rousse, 8 Studentska str., Bulgaria

mnikolova@uni-ruse.bg,
*Corresponding author - izheleva@uni-ruse.bg
mfilipova@uni-ruse.bg

Abstract. Polluted air has a negative impact on human health and plant and animal life. There are two types of sources of air pollution: human activity (transport, heating, industry and others) and natural phenomena such as natural fires, volcanic activity, deforestation and others. Substances that are the most common air pollutants are dust particles, carbon and nitrogen dioxides, carbon and nitrogen oxides, hydrocarbons; aldehydes; radioactive substances, heavy metals, sulfur dioxide, ozone, etc. Gaseous air pollutants of primary concern in urban environments include sulfur dioxide, nitrogen dioxide, and carbon monoxide. Sulfur dioxide is a colorless gas that is a product of volcanic eruptions and various industrial processes. It's usually released when coal and oil are burned. Nitrogen oxides, especially nitrogen dioxide, are poisonous gases with a pungent, suffocating odor. There are many regulations aimed at air pollution and pollution monitoring for all European countries. Bulgaria, as part of the EU, must comply with all these regulations. There are still many places and periods of time in Bulgaria over the years with air pollution greater than daily norms. The present study is dedicated to the study of air pollution with gaseous substances in the area of Svishtov, Bulgaria. The town of Svishtov is located in Northern Bulgaria, on the banks of the Danube River, where the southernmost point of the river along its entire length is located. A statistical analysis of the level of air pollution in Svishtov based on official data from the monitoring station in the city is presented. The measurements cover the period 2021 - 2023. All research results are presented graphically and commented.

Introduction

Svishtov is situated in northern central Bulgaria on the right bank of the Danube river in its southernmost point, across from Zimnicea, Romania. The town lies 237 kilometres north-east of the capital Sofia, and 251 kilometres west of port Varna - the major of Black Sea. Other important cities in the region are Pleven 80 kilometres away, and Ruse 90 kilometres away (see Figures 1, 2, 3).
The area of Svishtov has a moderate continental climate. According to the climatic zoning of the Republic of Bulgaria, the Svishtov's area falls into the Northern climatic region of the Danube plain. Due to the great openness of the Danube plain, during the cold part of the year, continental air masses from the north and northeast freely invade here, as a result of which the winter is relatively cold. In normally cold winters average monthly temperatures drop to –14.1°C. During the cold half-year, the lowest measured temperature is –28.0°C. The first negative temperatures appear in November and are maintained until March. The average daily air temperature passes 0°C during the first days of December and the last days of February. In the middle of March, it rises above 5°C and follows a

smooth sinusoidal course that has a maximum in the summer months. During the month of July - August, the average night-time temperature is above 23°C and can reach 35°C. The average maximum temperature in summer reaches 29.1°C. The total annual amount of precipitation reaches 543 mm. Maximum rainfall occurs in the months of May - June and minimum in September. Snow cover in the region is observed for about 40 days. The relative humidity of the air is approximately 73%. The proximity of the Danube river favors the retention of moisture in the air, which, combined with the course of minimal temperature inversions, favors the formation of fogs. The average annual number of foggy days is 39. West-southwest and east-northeast winds dominate in Svishtov. The cloud cover in the area has a maximum during the winter months. The region is characterized by a good duration of sunshine (greater than 2,000 hours/year).

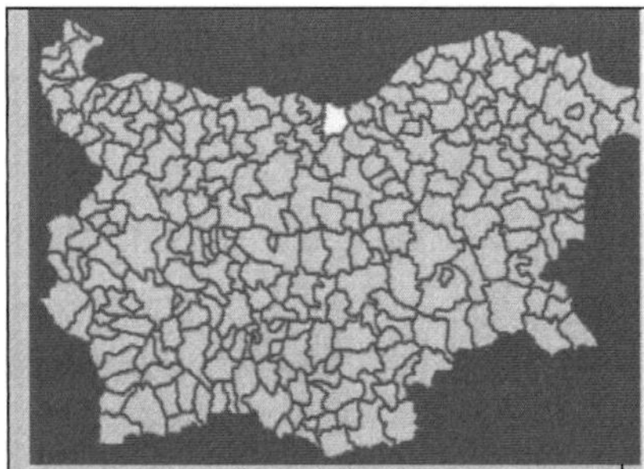		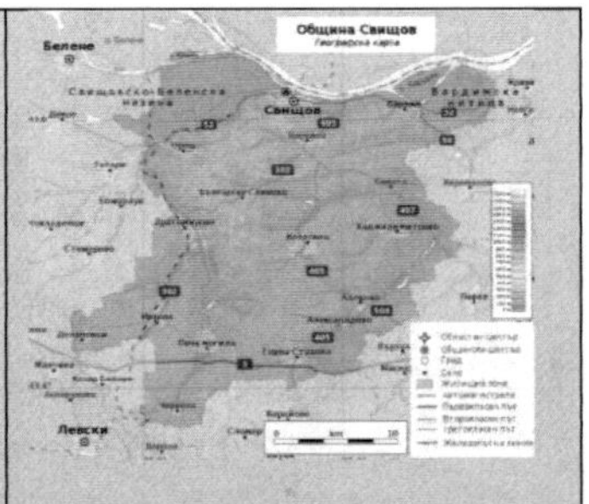
FIGURE 1. Map of Bulgaria, Svishtov municipality – in yellow.	**FIGURE 2.** Map of Svishtov municipality	**FIGURE 3.** Relief map of Svishtov region

Air pollution is not an isolated problem of individual countries - it is a global problem. Pollutants that fall into the air over a country cannot be contained, they are spread in the atmosphere. Thus, the area of air pollution expands. Polluted air has a negative impact on the environment and poses a risk to human health and the environment. The regular impact of these pollutants manifests itself in a variety of ways - from disturbances of the respiratory system to premature death. Air pollution remains the most serious environmental threat to the health of Europeans [1, 2, 3, 4]. According to the European Environment Agency dirty air remains the biggest threat to human health and life.

There are various sources of air pollution, both anthropogenic and natural:

- combustion of fossil fuels in the production of electricity, in transport, industry and households;
- industrial processes and use of solvents, for example in the chemical and mining industries;
- agriculture;
- waste treatment;
- volcanic eruptions, windblown dust, sea salt spray, and emissions of volatile organic compounds from plants are examples of sources of natural emissions.

For Bulgaria, the main air pollutants are fossil fuels, composed of old cars and solid fuel stoves for domestic heating during winter periods. The simultaneous impact of these factors leads to the deterioration of the ecological situation, and hence serious damage to human health and ecosystems [9]. Some meteorological and climatic conditions such as air temperature, humidity, atmospheric pressure, etc. also have an influence on increasing the concentration levels of dust and gaseous pollutants [10 - 26]. Polluted air in urbanized conditions mainly contains sulfur dioxide SO_2, nitrogen dioxide NO_2 and ground-level ozone O_3.

In this article, the relationship between the concentration levels of the mentioned gaseous pollutants, measured for the period 20.02.2021 - 30.04.2023 from a manual measurement station in Svishtov, Bulgaria, and some of the characteristics of the atmospheric air for the same period is statistically analyzed [5].

Material and methods

The period selected for the purposes of this study covers the last 2 years (from 20.02.2021 to 30.04.2023 (800 days)). Daily data for this period is available on the official website of the Ministry of Environment and Water, Bulgaria. We use official data of measured concentrations of 3 gaseous pollutants - sulfur dioxide (SO_2), nitrogen dioxide (NO_2) and ozone (O_3) for the city of Svishtov, Bulgaria, published by Regional Inspectorate of Environment and Water [8]. To analyze the above data, we use specialized software SPSS [5, 6].

Results and comments

Concentration levels of 3 gaseous pollutants were investigated: sulfur dioxide SO_2, nitrogen dioxide NO_2 and ozone O_3 for the city of Svishtov, Bulgaria. Table 1 shows several statistic characteristics of the 3 gaseous pollutants. There are some missing measurements for the period under consideration. Their numbers for each pollutant are also given in the Table 1. The missing data were linearly interpolated.

TABLE 1. Statistics of the 3 gaseous pollutants for the city of Svishtov, Bulgaria for the period 20.02.2021 – 30.04.2023

		SO_2	NO_2	O_3	temp	rains	clouds	pa
N	Valid	755	753	754	315	315	315	315
	Missing	45	47	46	485	485	485	485
Mean		3.4346	18.6651	70.6299	14.9429	1.5933	4.9584	1010.8740
Median		2.7900	15.3200	67.7800	14.3000	0.0000	4.7000	1009.5000
Mode		1.90	6.58[a]	48.96[a]	7.40	0.00	10.00	1008.60[a]
Std. Deviation		2.33032	11.33845	20.86845	8.50060	4.29273	3.26776	7.16923
Variance		5.430	128.560	435.492	72.260	18.428	10.678	51.398
Skewness		1.781	1.162	0.186	0.119	4.032	0.130	0.318
Std. Error of Skewness		0.089	0.089	0.089	0.137	0.137	0.137	0.137
Kurtosis		4.485	0.998	-0.657	-1.141	20.477	-1.257	-0.095
Std. Error of Kurtosis		0.178	0.178	0.178	0.274	0.274	0.274	0.274
Range		16.16	60.92	110.12	34.20	35.80	10.00	41.60
Minimum		0.50	0.00	15.24	-2.20	0.00	0.00	990.30
Maximum		16.66	60.92	125.36	32.00	35.80	10.00	1031.90

a. Multiple modes exist. The smallest value is shown

The permissible concentration limits of gaseous pollutants according to EU legislation are shown in the following table 2

TABLE 2. Limits of gaseous pollutant concentrations according EU law [1].

LIMITS of concentration, according EU law	**Sulphur dioxide (SO_2) concentration $\mu g/m^3$**	**Nitrogen dioxide (NO_2) concentration $\mu g/m^3$**	**Ozone O_3 concentration $\mu g/m^3$**
For 1 hour mean	350	200	
Maximum daily 8 hour mean			120
24 hours mean	125		
1 year mean		40	

Figure 4 shows histograms for each measured gaseous pollutant for the whole period 20.02.2021 – 30.04.2023 for the city of Svishtov, Bulgaria.

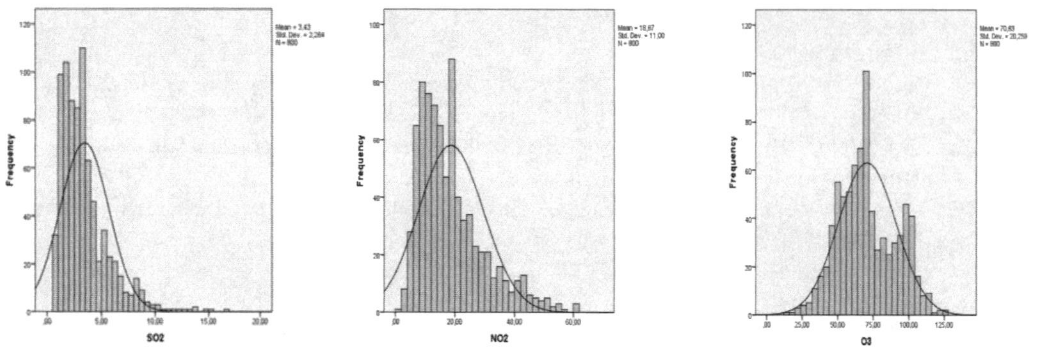

FIGURE 4. Histograms for SO_2, NO_2 and O_3 for the whole period 20.02.2021 – 30.04.2023

Figure 5 shows graphs for each measured gaseous pollutant for the whole period 20.02.2021 – 30.04.2023. It is obvious that all concentrations of the gaseous pollutants for the period examined for the city of Svishtov, Bulgaria are quite variable. The range for O_3 is 110.12µg/m^3 (minimum is 3.44 µg/m^3 and maximum is 103.99 µg/m^3), the range for NO_2 is 63.79 µg/m^3 (minimum is 3.99 µg/m^3 and maximum is 77.82 µg/m^3) and the range for SO_2 is 16.16 µg/m^3 (minimum is 4.19 µg/m^3 and maximum is 65.57 µg/m^3).

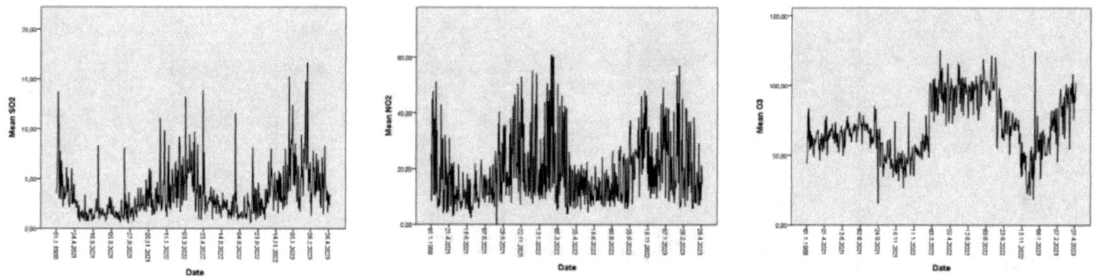

FIGURE 5. Graphs for SO_2, NO_2 and O_3 for the whole period 20.02.2021 – 30.04.2023

Among the most common methods for statistical processing of numerical data is Pearson's Correlation Analysis - a statistical method for measuring the strength of correlation dependence between two or more quantities. This is done through correlation coefficients. The Pearson coefficient is a number in the interval from -1 to 1, and the meaning of these values is as follows: 0 means that there is no correlation between the studied quantities, -1 corresponds to a complete negative correlation, and 1 is a sign of a complete positive dependence. The essence of these coefficients is as follows: a positive correlation between two quantities means that if one quantity increases its value, the second quantity will also increase its value; in the case of a negative correlation, the following is observed - if one quantity increases its value, the other quantity decreases in value. At a correlation coefficient of 0, there is no linear relationship between the quantities, but there may be a quadratic or other higher order relationship [5, 6]. The closer Peason's correlation coefficient is to 0, the weaker it is. The closer it is to +/-1, the stronger the dependence of the variables.

For the considered gaseous pollutants and for some atmospheric characteristics such as temperature, pressure, rains and clouds - we calculate the Pearson correlation coefficients (see Table 3). It can be seen from Table 3 that almost every gaseous pollutant has significant correlations mostly at the 0.01 level (two-tailed) with the other pollutants and with some of the atmospheric characteristics studied. This shows that there are linear relationships between these gaseous pollutants. Next, we will create linear regression models to examine them.

TABLE 3. Pearson Correlation Coefficients

		SO_2	NO_2	O_3	temp	rains	clouds	pa
SO_2	Pearson Correlation	1	0.544**	0.063	-0.450**	-0.074	-0.078	0.286**
	Sig. (2-tailed)		0.000	0.268	0.000	0.188	0.167	0.000
	N	315	315	315	315	315	315	315
NO_2	Pearson Correlation	0.544**	1	-0.121*	-0.370**	-0.103	-0.153**	0.385**
	Sig. (2-tailed)	0.000		0.032	0.000	0.068	0.006	0.000
	N	315	315	315	315	315	315	315
O_3	Pearson Correlation	0.063	-0.121*	1	0.580**	-0.217**	-0.550**	-0.159**
	Sig. (2-tailed)	0.268	0.032		0.000	0.000	0.000	0.005
	N	315	315	315	315	315	315	315
temp	Pearson Correlation	-0.450**	-0.370**	0.580**	1	-0.190**	-0.486**	-0.397**
	Sig. (2-tailed)	0.000	0.000	0.000		0.001	0.000	0.000
	N	315	315	315	315	315	315	315
rains	Pearson Correlation	-0.074	-0.103	-0.217**	-0.190**	1	0.312**	-0.098
	Sig. (2-tailed)	0.188	0.068	0.000	0.001		0.000	0.083
	N	315	315	315	315	315	315	315
clouds	Pearson Correlation	-0.078	-0.153**	-0.550**	-0.486**	0.312**	1	-0.057
	Sig. (2-tailed)	0.167	0.006	0.000	0.000	0.000		0.312
	N	315	315	315	315	315	315	315
pa	Pearson Correlation	0.286**	0.385**	-0.159**	-0.397**	-0.098	-0.057	1
	Sig. (2-tailed)	0.000	0.000	0.005	0.000	0.083	0.312	
	N	315	315	315	315	315	315	315

**. Correlation is significant at the 0.01 level (2-tailed).
*. Correlation is significant at the 0.05 level (2-tailed).

Model for the SO_2 (sulphur dioxide) air pollutant

Let Y (dependent variable) denote the level of SO_2 concentration, and let the independent variables be X_1 – NO_2 concentration, X_2 – temperature, for the entire considered period. Using SPSS software we get the following.

Table 4. Model Summary for sulphur dioxide SO_2

Model	R	R Square	Adjusted R Square	Std. Error of the Estimate
1	0606[a]	0.367	0.363	1.37351

a. Predictors: (Constant), temp, NO_2

Table 5. ANOVA[a] for the model of SO_2

Model		Sum of Squares	df	Mean Square	F	Sig.
1	Regression	341.699	2	170.850	90.562	0.000[b]
	Residual	588.600	312	1.887		
	Total	930.299	314			

a. Dependent Variable: SO_2
b. Predictors: (Constant), temp, NO_2, O_3

Table6. Coefficients[a] for the model (1)

Model		Unstandardized Coefficients		Standardized Coefficients	t	Sig.
		B	Std. Error	Beta		
1	(Constant)	2.242	0.248		9.025	0.000
	NO_2	0.071	0.008	0.437	9.012	0.000
	temp	-0.058	0.010	-0.288	-5.949	0.000

The model of sulphur dioxide SO_2 (according to Tables 4, 5, 6) is

$$Y = 2.242 + 0.071*X_1 - 0.058*X_2 . \quad (1)$$

All coefficients in (1) are statistically significant and the model is adequate. 36.7% of all measurements of SO_2 for the whole period are described by this linear regression model. We use this model to predict the sulphur dioxide concentration levels for the period 1.05.2023 – 15.05.2023. The results coud be seen on Figures 6, 7 and 8. The measured values are less than the model value. The average relative error for the 5-day forecast is 17.55%, for the 10-day forecast is 15.91% and for the 15-day forecast is 6.71%. It can be seen that the forecasting is more accurate for the longer time.

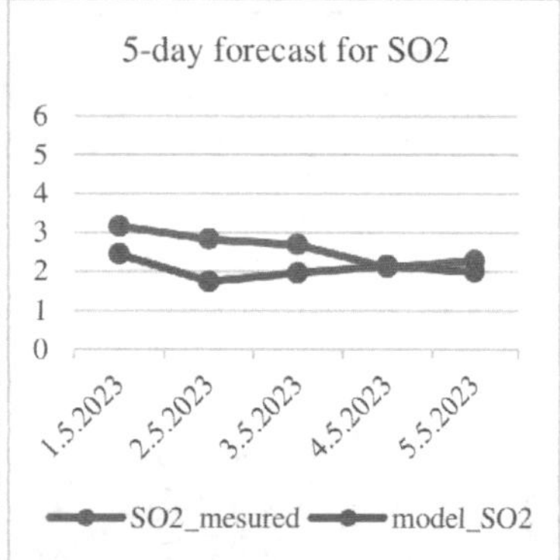

FIGURE 6. Forecast for sulphur dioxide SO_2 concentration levels for the city of Svishtov, Bulgaria, for the period 1.05.2023 – 5.05.2023. Blue line – measured values, red line – model (1).

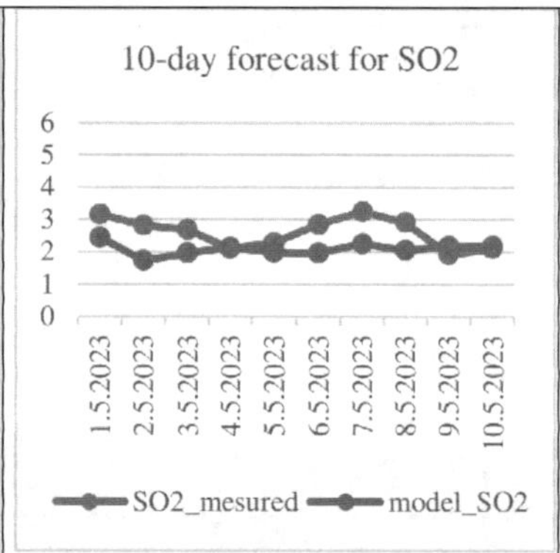

FIGURE 7. Forecast for sulphur dioxide SO_2 concentration levels for the city of Svishtov, Bulgaria, for the period 1.05.2023 – 10.05.2023. Blue line – measured values, red line – model (1).

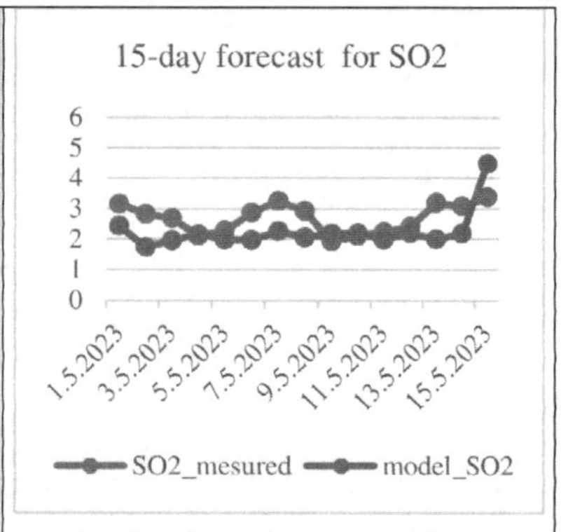

FIGURE 8. Forecast for sulphur dioxide SO_2 concentration levels for the city of Svishtov, Bulgaria, for the period 1.05.2023 – 15.05.2023. Blue line – measured values, red line – model (1).

Model for nitrogen dioxide NO_2 air pollutant

Let Y (dependent variable) denote the level of NO_2 concentration, and let the independent variables be X_1 – SO_2 concentration, X_2 – temperature, X_3 – clouds for the entire considered period. Using SPSS software we receive the following.

Table 7. Model Summary for nitrogen dioxide NO2

Model	R	R Square	Adjusted R Square	Std. Error of the Estimate
1	0.629a	0.395	0.389	8.54659

a. Predictors: (Constant), SO2, clouds, temp

Table 8. ANOVA for the model of NO2

Model		Sum of Squares	df	Mean Square	F	Sig.
1	Regression	13708.013	3	4569.338	62.556	0.000b
	Residual	20963.673	287	73.044		
	Total	34671.686	290			

a. Dependent Variable: NO2
b. Predictors: (Constant), O3, SO2, rains, clouds, temp

Table 9. Coefficients[a] for the model (2)

Model		Unstandardized Coefficients		Standardized Coefficients	t	Sig.
		B	Std. Error	Beta		
1	(Constant)	24.651	2.578		9.563	0.000
	temp	-0.485	0.081	-0.383	-6.000	0.000
	SO_2	2.203	0.343	0.357	6.430	0.000
	clouds	-1.063	0.190	-0.320	-5.607	0.000

a. Dependent Variable: NO_2

The model of nitrogen dioxide NO_2 (according to Tables 7, 8, 9) is

$$Y = 24.651 + 2.203*X_1 - 0.485*X_2 - 1.063*X_3 . \quad (2)$$

All coefficients in (2) are statistically significant and the model is adequate. This linear regression model describes 39.5% of all nitrogen dioxide measurements for the entire period.

Using this model, we can predict nitrogen dioxide concentration levels for the period 1.05.2023 – 15.05.2023. The results coud be seen on Figures 9, 10 and 11. Most of measured values are less than the model values. The average relative error for the 5-day forecast is 18.88%, for the 10-day forecast is 18.76% and for the 15-day forecast is 18.16%. As can be seen for longer period the forecasting is more accurate. Neverthelest this model has to be improved.

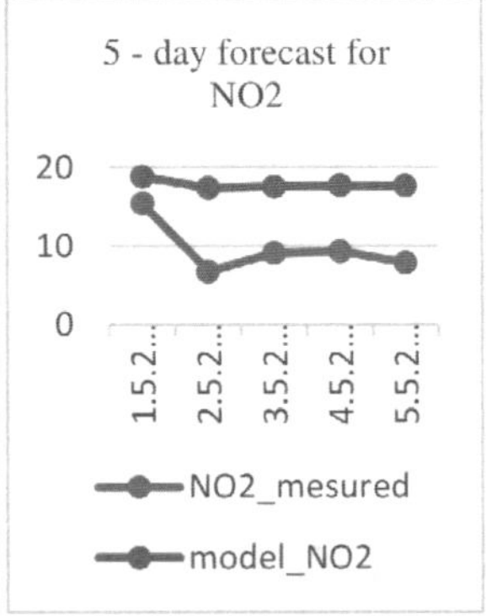

FIGURE 9. Forecast for nitrogen dioxide SO_2 concentration levels for the city of Svishtov, Bulgaria, for the period 1.05.2023 – 5.05.2023. Blue line – measured values, red line – model (1).

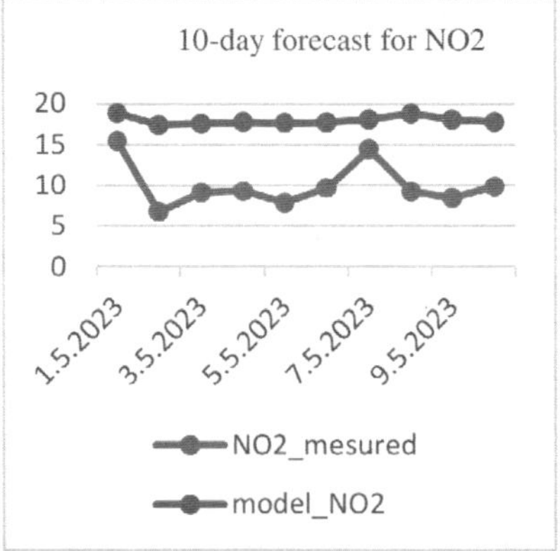

FIGURE 10. Forecast for nitrogen dioxide SO_2 concentration levels for the city of Svishtov, Bulgaria, for the period 1.05.2023 – 10.05.2023. Blue line – measured values, red line – model (1).

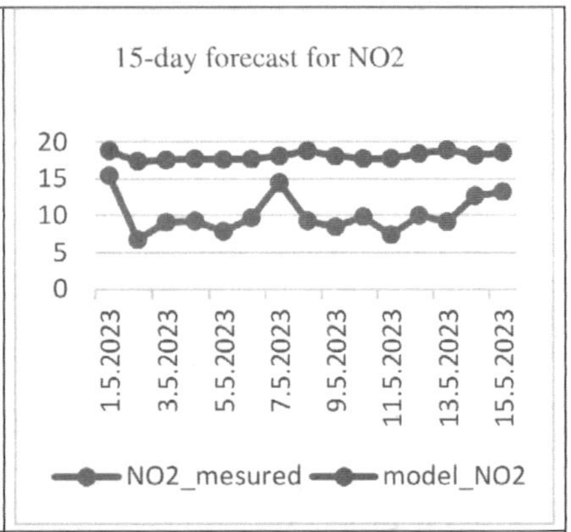

FIGURE 11 Forecast for nitrogen dioxide SO_2 concentration levels for the city of Svishtov, Bulgaria, for the period 1.05.2023 – 15.05.2023. Blue line – measured values, red line – model (1).

Model for ozone O_3 air pollutant

Let Y (dependent variable) denote the level of O_3 concentration, and let the independent variables be X_1 – temperature, X_2 – clouds for the entire considered period. Using SPSS software we receive the following.

Table10. Model Summary for O_3

Model	R	R Square	Adjusted R Square	Std. Error of the Estimate
1	0.656[a]	0.430	0.426	8.85762

a. Predictors: (Constant), temp, clouds

Table11. ANOVA[a] for the model of O_3

Model		Sum of Squares	df	Mean Square	F	Sig.
1	Regression	18466.432	2	9233.216	117.684	0.000[b]
	Residual	24478.740	312	78.458		
	Total	42945.172	314			

a. Dependent Variable: O_3
b. Predictors: (Constant), temp, clouds

Table12. Coefficients[a] for the model (3)

Model		Unstandardized Coefficients		Standardized Coefficients	t	Sig.
		B	Std. Error	Beta		
1	(Constant)	57.951	1.692		34.254	0.000
	temp	0.563	0.067	0.409	8.364	0.000
	clouds	-1.256	0.175	-0.351	-7.172	0.000

a. Dependent Variable: O_3

The model of ozone O_3 (according to Tables 10, 11, 12) is

$$Y = 57.951 + 0.563*X_1 - 1.256*X_2. \quad (3)$$

All coefficients in (3) are statistically significant and the model is adequate. 43.0% of all measurements of ozone for the whole period are described by this linear regression model. We use this model to predict the ozone concentration levels for the period 1.05.2023 – 15.05.2023. The result coud be seen on Figures 12, 13 and 14. The measured values exceed the model values. The average relative error for the 5-day forecast is 20.41%, for the 10-day forecast is 16.77% and for the 15-day forecast is 18.15%. These persentiges are not very small. Hence the model has to be improved.

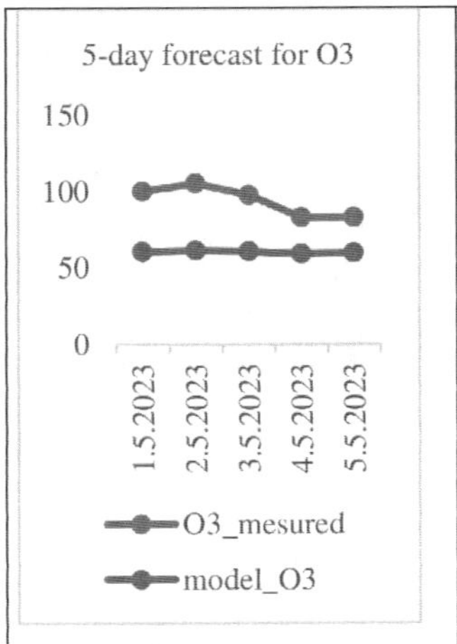

FIGURE 12. Forecast for nitrogen dioxide O_3 concentration levels for the city of Svishtov, Bulgaria, for the period 1.05.2023 – 5.05.2023. Blue line – measured values, red line – model (1).

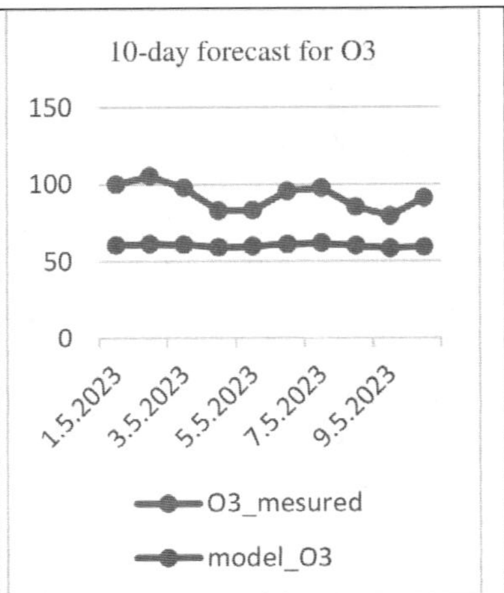

FIGURE 13. Forecast for nitrogen dioxide O_3 concentration levels for the city of Svishtov, Bulgaria, for the period 1.05.2023 – 10.05.2023. Blue line – measured values, red line – model (1).

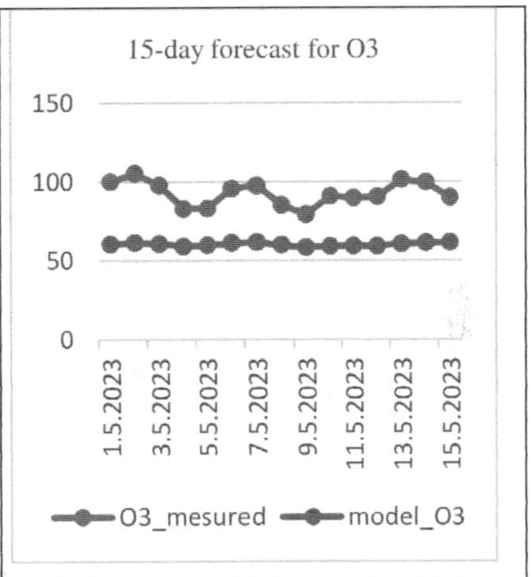

FIGURE 14 Forecast for nitrogen dioxide O_3 concentration levels for the city of Svishtov, Bulgaria, for the period 1.05.2023 – 15.05.2023. Blue line – measured values, red line – model (1).

Conclusions

This study is devoted to the dependencies between some of the main gaseous pollutants: sulfur dioxide SO_2, nitrogen dioxide NO_2, and ozone O_3 for the city of Svishtov, Bulgaria, for the period 20.02.2021–30.04.2023. We collected official data for daily measured concentration levels for these pollutants, published by the Regional Inspectorate of Environment and Water, Svishtov, Bulgaria. We use SPSS software to investigate the relationship between gaseous air pollutants and some meteorological factors through regression analysis.

As a result, we got that that all air pollutants are correlated and also they are correlated with some of the atmospheric characteristics. This means that there are linear regression relationships between them. Using SPSS we create linear regression models which are good enough to predict air contamination for the city of Svishtov, Bulgaria, for the period 1.05.2023 – 15.05.2023. The compiled models of air pollutants are of acceptable accuracy. In order to make the models more accurate, other dependances (not only linear regression) between the dependent (pollutants) and independent variables (meteo factors) have to be uesed, also more factors that affect the environment should also be taken into account in the modeling of air pollution future.

Acknowledgments

This paper contains results of the work on project No 2023–AIF–03, financed by Scientific Research Fund of Ruse University and Project BG05M2OP001 2.016 0008 C02 "Innovations, Science and Education for high quality and compliance with the labor market at the Technical University of Sofia and Partnership (INNOTEH PRO)", financed under the Operational Program "Science and Education for Smart Growth". Priority axis 2 "Education and lifelong learning", co-financed by the European Union through the European Social Fund.

References

1. https://ec.europa.eu/environment/topics/air/air-quality/eu-air-quality-standards_en
2. Pavlova S, Nenova L (2008), Peculiarities in climate variations during last 30-ty years at the region of "Obraztsov chiflik" – Rousse, Scientific works, University of Ruse, issue 47, series 1.2, pp. 75-80.
3. Alexandrov V et al. (2010), Climate change, BAS, Meteorology and Hydrology Institute, Sofia, Bulgaria
4. Takuchev N, Kostadinova P, Naydenova G, I. Stoilova I, (2018), Combined ground level and satellite monitoring of urban air pollution, estimation of the number of patients treated in hospitals due to air pollution and the cost of their treatment - case study of Ruse, Bulgaria Trakia Journal of Sciences, Vol. 16, Suppl. 1, pp 158-168, Available online at http://www.uni-sz.bg ISSN 1313-3551 (online) doi:10.15547/tjs.2018.s.01.03
5. Pavlov V, Applied Statistics, (2013), PREPRES, Sofia, Bulgaria, pp 160, ISBN 978-954-90443-6-2 (in Bulgarian)
6. Pavlov V, Mihova V, Applied Statistics with SPSS, (2016), Avangard Print, Bulgaria, pp 192, ISBN 978-954-337-308-6 (in Bulgarian)
7. https://www.moew.government.bg/bg/vuzduh/kachestvo-na-atmosferniya-vuzduh/normativni-aktove/
8. https://www.moew.government.bg/bg/prescentur/byuletini-na-mosv/vuzduh/
9. https://www.britannica.com/search?query=criteria+air+pollutants+
10. https://ec.europa.eu/environment/topics/air/air-quality/eu-air-quality-standards_en
11. Pavlova S, Nenova L, (2008), Peculiarities in climate variations during last 30-ty years at the region of "Obraztsov chiflik" – Rousse, Scientific works, University of Ruse, issue 47, series 1.2, pp. 75-80
12. Alexandrov V et al, (2010), Climate change, BAS, Meteorology and Hydrology Institute, October Sofia, Bulgaria

13. Zheleva I, Veleva E, Filipova M, (2017) Analysis and modeling of daily air pollutants in the city of Svishtov, Bulgaria, in AMITANS'17, AIP Conference Proceedings 1895, 030007; DOI https://doi.org/10.1063/1.5007366
14. Veleva E, Zheleva I, Statistical modeling of particle mater air pollutants in the city of Ruse, Bulgaria, (2018) MATEC Web Conf. Volume 145, NCTAM 2017 – 13th National Congress on Theoretical and Applied Mechanic DOI: https://doi.org/10.1051/matecconf/201814501010
15. Filipova M, Zheleva I, Roussev P, (2013), Characteristics of PM air pollution along Bulgaria - Romania Danube region, ECOLOGICA, 71, pp. 215-217.
16. Ivanov I, Gocheva-Ilieva S, Stoimenova M, (2020), Hybrid Boosted Trees and Regularized Regression for tudying Ground Ozone and PM10 concentration, in AMiTaNS'20, AIP Conf. Proc, 2302, 060005, 9p, https://doi.org10.1063/5.0033591
17. Ivanov A, Voynikova D, Gocheva-Ilieva S, Kulina H, Iliev I, (2015), Using principal component analysis and general path seeker regression for investigation of air pollution and CO modeling, 7th Conference of the Euro-American Consortium for Promoting the Application of Mathematics in Technical and Natural Sciences, Albena, Bulgaria, June 28-July 3, AIP Conf. Proc. 1684, ed. M. Todorov, 100004, pp. 1-11, 2015. http://dx.doi.org/10.1063/1.4934341
18. Voynikova D S, Gocheva-Ilieva S G, Ivanov A V, Iliev I P, (2015), Studying the effect of meteorological factors on the SO2 and PM10 pollution levels with refined versions of the SARIMA model, 7th Conference of the Euro-American Consortium for Promoting the Application of Mathematics in Technical and Natural Sciences, Albena, Bulgaria, June 28-July 3, AIP Conf. Proc. 1684, ed. M. Todorov, 100005, pp. 1-12, 2015. http://dx.doi.org/10.1063/1.4934342
19. Zheleva I, Filipova M, Atmospheric characteristics statistic study of Ruse region, Bulgaria, (2016), AIP Conference Proceedings 1773, 110019 doi: 10.1063/1.4965023, http://scitation.aip.org/content/aip/proceeding/aipcp/1773
20. Filipova M, Zheleva I, Tsvetanova I, Stefanova A, Svishtov P, (2016), Analysis of the state of ambient air in the border region Bulgaria-Romania, International Symposium "The Environmental and The Industry" SIMI, Bucharest, Symposium Proceedings, pp. 440-450 http://www.simiecoind.ro/doi10-21698simi-2016-0062-analysis-of-the-state-of-ambient-air-in-the-border-region-bulgaria-romania/
21. Tsvetanova I, Zheleva I, Filipova M, Stefanova A (2018) Statistical analysis of ambient air PM10 contamination during winter periods for Ruse region, Bulgaria. MATEC Web Conf. Volume 145, NCTAM 2017 – 13th National Congress on Theoretical and Applied Mechanics DOI: https://doi.org/10.1051/matecconf/201814501007
22. Stefanova A, Zheleva I, Filipova M, Tsvetanova I, (2018) Examination of the possibility of transborder pollution in the days with registered exceedances of pollutant dust in the city of Ruse, Bululgaria, MATEC Web of Conferences 145, 01004, NCTAM 2017, https://doi.org/10.1051/matecconf/201814501004
23. Tsvetanova I, Zheleva I, Filipova M (2018) Statistical study of the influence of some atmospheric characteristics upon the particulate matter (PM10) air pollutant in the city of Ruse, Bulgaria, AIP Conference Proc. 2025, 110006; https://doi.org/10.1063/1.5064949
24. Gocheva-Ilieva G., Ivanov A V, (2019), Assaying stochastic SARIMA and generalized regularized regression for particulate matter PM10 modeling and forecasting – Intern J of Environment and Pollution, pp. 41-62.
25. Veleva E, Georgiev I R, Zheleva I, Filipova M (2020) Markov Chains Modelling of Particulate Matter (PM10) Air Contamination in the City of Ruse, Bulgaria. Application of Mathematics in Technical and Natural Sciences, No 2222 (SJR rank: 0.182 /2018, www.scimagojr.com). doi: 10.1504/IJEP.2019.104520

26. Gocheva-Ilieva S., Ivanov A., Iliev I., (2019) Exploring key air pollutants and forecasting particulate matter PM10 by a two-step SARIMA approach. AIP Conf. Proc. Vol. 2164, 020004-1–020004-10 https://doi.org/10.1063/1.5109327.

Intuitionistic fuzzy seasonal variation analysis of marine biotoxin distribution in the Black Sea

Velichka Traneva[1], **Stoyan Tranev**[2]
[1] Department of Natural Science, "Prof. Asen Zlatarov" University, "Prof. Yakimov" Blvd, Bourgas 8000, Bulgaria
[2] Department of Social Science, "Prof. Asen Zlatarov" University, "Prof. Yakimov" Blvd, Bourgas 8000, Bulgaria
E-mail: `veleka13@gmail.com`, `tranev@abv.bg`

Abstract. It is crucial to establish a seasonal dependence on the levels of marine biotoxins in the Black Sea to impose seasonal prohibitions on the gathering of marine species in this sea and safeguard human health. Fisher's analysis of variance (ANOVA) is a statistical technique for assessing how various variables affect a collection of data. In some cases, research must be carried out in an uncertain environment, with missing or insufficient data. An approach for solving this imprecision when a data set occurs in an ambiguous context is intuitionistic fuzzy logic, which Atanassov introduced in 1983. Compared to fuzzy data, intuitionistic fuzzy (IF) data also has a hesitation degree. In 2020, we first suggested the one-way (1-D) IFANOVA and the software utility for its performance, combining traditional variational analysis with modeling opportunities of Index Matrices (IMs) and Intuitionistic Fuzzy Sets (IFSs). The current study focuses on 1-D IFANOVA of the distribution of marine biotoxins in the Black Sea by the "season" factor on the IF dataset of the number of marine biotoxins for the period from February 1 to July 31, 2021. We will also compare the findings of 1-D IFANOVA and conventional ANOVA performed on the identical data set.

1. Introduction

Negative consequences of Black Sea pollution include reduced biological variety and toxicity to aquatic life. It is necessary to develop decision-making algorithms so they can be used with ambiguous or false data in the context of a pandemic and growing inflation. Since bivalve mollusks concentrate contaminants to quantities greater than those found in seawater, they are frequently used in monitoring programs as bioindicators. Because clams are aquatic organisms that lead a sedentary lifestyle, the amount of pollutants in their bodies can be used as an indicator of the level of pollution in a particular area. Mussels are a great choice for monitoring because of their vast geographic range, sedentary lifestyle, ease of sampling, resilience to stress, and relatively quick buildup of harmful compounds as a result of their eating patterns. The bivalve mollusks (mussels, oysters), which are the subject of marine aquaculture research in Bulgaria, filter the seawater during the feeding process and so collect the microscopic plankton species. Marine biotoxins accumulate, for example in mussels, without changing the vitality of the organisms, as well as their taste qualities. Thus, these pollutants of biogenic origin can reach the final consumer - man. Once in the human body, biotoxins can cause serious health problems.

The analysis of variance (ANOVA) was used to assess the significance of interseasonal variations in the levels of biotoxins found in mussels in the Black Sea from 1 February to 31 July 2021. Fisher is recognized for developing the ANOVA technique for determining statistical differences between group means [12]. The input data used in an analysis like an ANOVA may frequently have some irregularities, such as gaps in the data or unpredictable changes in successive number values, caused by factors like incomplete or delayed reporting of statistics. Uncertainty in the data cannot be analyzed by the traditional ANOVA method. To use ANOVA for this kind of data, it must be updated. A concept that can be utilized to express these abnormalities is fuzzy logic [33]. In fuzzy sets (FSs), each element has an acceptance and non-acceptance degree that sum to 1.

Fuzzy ANOVA (FANOVA) is a modified variation of ANOVA that has already been developed using fuzzy logic. FANOVA is used in [24], and is based on fuzzily generated random variables [19]. The moment adjustment is utilized in [17] to reduce the impact of the ambiguity on the sum of squares in an ANOVA with a fuzzy data approach. A bootstrap suggestion to FANOVA has reportedly been made, according to [13]. The study [9] claims that variance confidence intervals are used in the creation of FANOVA. The works [15, 22] provide a one-way triangular FANOVA. A FANOVA using Zadeh's extension notion is proposed in [20] A FANOVA approach based on least squares is developed in [14]. The traditional ANOVA is extended in [21] by the use of a unique approach for FANOVA over typical fuzzy numbers.

Fuzzy logic is extended by intuitionistic fuzzy sets (IFSs) [2, 4], which add a degree of falsity that, when subtracted from the degree of truth, yields the element's degree of hesitation. These sets also allow the degree of hesitancy in their elements to be accounted for. An improved ANOVA that makes use of intuitionistic fuzzy values is known as an Intuitionistic Fuzzy ANOVA (IFANOVA). The IFANOVA method is looked at in [16] to assess student homogeneity. The one-way IFANOVA approach is used in [1] to produce accurate results from intuitionistic fuzzy data.

The index-matrix interpretation of the traditional ANOVA in the form of the IFANOVA is produced in [31, 32] by applying the index-matrix perspective (IM, [3]) and IF logic [4]. This is done to ascertain the influence of the factor over IF data that arose in an uncertain environment. One-way IFANOVA is conducted in [30] utilizing the created software package. IFANOVA is a development of the traditional ANOVA that takes into account data that is fuzzy on the subject of intuition. For IFANOVA, the same criterion for data normality as for traditional ANOVA application applies. The ability to use IFANOVA with both precise and ambiguous data is one of its benefits. It is simply possible to extend the method to multidimensional data stored in multidimensional index matrices [7].

In this study, we apply one-way IFANOVA to the IF dataset of the number of marine biotoxins over the period from February 1 to July 31, 2021 [34], to examine the dispersion of marine biotoxins in the Black Sea by the "season" factor. In a later section of this research, we will present the one-way IFANOVA results we obtained utilizing the utility and contrast them with those using conventional ANOVA. The described IFANOVA and its application to determine the relationship between marine biotoxins in the Black Sea and the "season" factor are the sources of the work's scientific contribution.

The following are the sections of the remaining paper: In Sect. 2, basic justifications based on the theories of IMs and IF logic are given. A description of the ANOVA and an examination of how it was used to study marine biotoxins in the Black Sea based on a season factor are provided in Sect. 3. We define the 1-D IFANOVA procedure in Sect. 4 and present the findings after it has been used on the identical set of data. In Sect. 5, it is proposed what the conclusion and future study directions might be.

2. Fundamental Definitions of the IMs and IF Logic Concepts

The **IFP** is an object of the form $\langle a, b\rangle = \langle \mu(p), \nu(p)\rangle$, where $a, b \in [0, 1]$ and $a + b \leq 1$, that is used as an evaluation of a proposition p [8]. The degrees of membership and non-membership are determined by the variables $mu(p)$ and $nu(p)$, respectively. Let us have two IFPs $x = \langle a, b\rangle$ and $y = \langle c, d\rangle$. Basic operations over two IFPs are described in [4, 6, 8, 10, 25] as follows:

$$
\begin{aligned}
\neg x = \langle b, a\rangle; x \wedge_1 y = \langle \min(a,c), \max(b,d)\rangle;\\
x \vee_1 y = \langle \max(a,c)), \min(b,d)\rangle; x \wedge_2 y = x + y = \langle a + c - a.c, b.d\rangle;\\
x \vee_2 y = x.y = \langle a.c, b + d - b.d\rangle; \alpha.x = \langle 1-(1-a)^{\alpha}, b^{\alpha}\rangle (\alpha = n \text{ or } 1/n\ (n \in N));\\
x - y = \langle \max(0, a-c), \min(1, b+d, 1-a+c)\rangle;\\
x : y = \begin{cases} \langle \min(1, a/c), \min(\max(0, 1-a/c), \\ \max(0, (b-d)/(1-d))) \text{ if } c \neq 0 \,\&d \neq 1 \\ \langle 0, 1\rangle \text{ otherwise} \end{cases}
\end{aligned} \tag{1}
$$

Let $R_{\langle a,b\rangle} = 0.5(2-a-b)(1-a)$ [26] is the distance of the IFP to the truth. Next, according to [4, 31], we can compare IFPs using the following relation:

$$x \geq_{\diamond} y \quad \text{iff} \quad b \leq d; \quad x \geq_R y \quad \text{iff} \quad R_{\langle a,b\rangle} \leq R_{\langle c,d\rangle}. \tag{2}$$

Let $\mathcal{I}$ be a set. By two-dimensional IF index matrix (**2-D IFIM**) $A = [K, L, \{\langle \mu_{k_i,l_j}, \nu_{k_i,l_j}\rangle\}]$ with index sets K and L ($K, L \subset \mathcal{I}$), we denote [5]:

$$A \equiv \begin{array}{c|ccccc} & l_1 & \ldots & l_j & \ldots & l_n \\ \hline k_1 & \langle \mu_{k_1,l_1}, \nu_{k_1,l_1}\rangle & \ldots & \langle \mu_{k_1,l_j}, \nu_{k_1,l_j}\rangle & \ldots & \langle \mu_{k_1,l_n}, \nu_{k_1,l_n}\rangle \\ \vdots & \vdots & \ddots & \vdots & \ddots & \vdots \\ k_m & \langle \mu_{k_m,l_1}, \nu_{k_m,l_1}\rangle & \ldots & \langle \mu_{k_m,l_j}, \nu_{k_m,l_j}\rangle & \ldots & \langle \mu_{k_m,l_n}, \nu_{k_m,l_n}\rangle \end{array},$$

where for every $1 \leq i \leq m, 1 \leq j \leq n$: $0 \leq \mu_{k_i,l_j}, \nu_{k_i,l_j}, \mu_{k_i,l_j} + \nu_{k_i,l_j} \leq 1$.

There are specific operations with 2-D IFIMs, such as addition, transposition, multiplication, projection, substitution, aggregation operations, internal subtraction of IMs' components, termwise multiplication, and subtraction, that are also defined in [5, 27, 28, 29] in an approach similar to that of classical matrices.

3. Applying ANOVA to the Black Sea Biotoxins by the "Season" Factor

3.1. Classical one-way ANOVA approach

Fisher created the original ANOVA procedure for examining the variance in the means of multiple independent populations [12]. One-way ANOVA is the most fundamental instance and is carried out as follows [11]:

If $i = 1, 2, ..., m$ and $j = i_1, i_2, ...i_I (1 \leq i_I \leq n)$, then the data from the k_i–th level and the l_j–th observation are denoted as y_{k_i,l_j}. Suppose there are N observations. The ANOVA has been considered to confirm or deny the hypothesis.

$$H_0 : M_{k_1} = M_{k_2} = ... = M_{k_m}, \text{ against } H_1 : \text{ not all } M_{k_i} \text{ are equal},$$

where M_{k_i} are the factor level means. According to [11], the mean sum of squares (MST), the mean sum of squares for error (MSE), and the mean sum of squares for treatment (MSC) are all calculated.

If the test statistic

$$F = \frac{MSC}{MSE} \geq F_{(\alpha,m-1,N-m)} \text{ or } \frac{1}{F} \leq \frac{1}{F_{(\alpha,m-1,N-m)}} = F_{(1-\alpha,N-m,m-1)}, \tag{3}$$

where $F_{(\alpha,m-1,N-m)}$ is the α–quantile of the F–distribution with the $m-1$ and $N-m$ degrees of freedom, then H_0 is rejected at the significance level α [11, 12].

3.2. Implementation of one-way ANOVA to the biotoxins of the Black Sea by the "season" factor in 2021
To assess the condition of the marine environment, it is crucial to look into the seasonal dynamics of marine biotoxin levels. 17 samples of both wild and cultured mussels collected from the Bulgarian coast between February 2021 and July 2021 were tested for the presence of the phycotoxins Domoic acid (DA), Yesotoxin (YTX), and its derivatives OH-YTX and PTX2-sa [34]. The wild and cultured mussel sample is selected to coincide with the seasons when mussels are taken for commerce and private consumption. It was done along the northern and southern Black Sea coasts throughout the winter, spring, and summer of 2021.

Let us use Excel to apply one-way ANOVA on the data containing marine biotoxin values [34] for the period from February to July 2021. The Kolmogorov-Smirnov test [11] at the 5% significance level indicates that the data are regularly distributed. Winter, spring, and summer are the three different values for the "season" factor for the values of the marine biotoxins.

According to the number of observed biotoxins in the Black Sea from February 1, 2021, to July 31, 2021, the following table 1 shows the results of a one-way ANOVA by the factor "season" with $\alpha = 0.05$:

Table 1. ANOVA table for the marine biotoxins by the variable "season".

Source	SS	df	MS	F	p-value	F crit
Between groups	0.0017	2	0.000857	23.43	3.41E-05	3.74
Within groups	0.0005	14	3.66E-05			
Total	0.0022	16				

The ANOVA results lead us to the conclusion that the value of the "season" component does have a statistically significant impact on the values of the marine biotoxins. The winter (0.031 mg/g) and spring (0.014 mg/g) had the highest average levels of marine biotoxins in the Black Sea for 2021. The summer months have the lowest levels (0.0031 mg/g) (see table 2 and fig. 1). The average marine biotoxins (mg/g) in the Black Sea for the time frame under study are shown in Fig. 2 by elements and seasons. Different tendencies for the five determined toxins are seen in the analysis of the seasonal dynamics in the levels of the determined toxins in the shellfish samples for the period of February to July 2021(Fig. 2).

4. Seasonal Dependence of Biotoxins in the Black Sea using One-Way Intuitionistic Fuzzy ANOVA

In [31], we described a technique for combining intuitionist fuzzy logic and IMs with the advantages of conventional variational analysis. To enable the efficient application of the 1-D IFANOVA method on actual data, command-line software built on C++ has been developed [30]. The basic IM operations are implemented using the index matrix template class (*IndexMatrix*$\langle T \rangle$) from [18]. The software manipulates IM objects made up of IFIMs.

One IFIM's contents are loaded by the software into a tab-separated text file that the user specifies. The original data must be converted into IFPs using the method from [4]. The MSE, MSC, and $\frac{1}{F}$ values are then reported by the software. The interim matrices are displayed when the "verbose" option is selected.

To determine whether the levels of the "season" factor has a significant impact on the average value of marine biotoxins in the Black Sea using the same dataset [34], which we used for classical ANOVA,

Table 2. Average marine biotoxins (mg/g) in the Black Sea by "season" from February to July 2021

Winter	Spring	Summer
0.031	0.014	0.0031

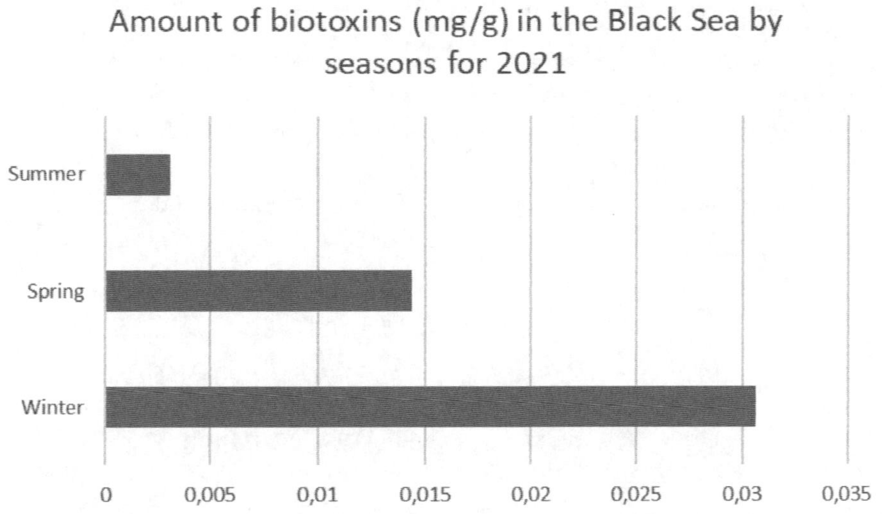

Figure 1. The average value of marine biotoxins in the Black Sea in mg/g from February to July 2021, depending on the "season" factor

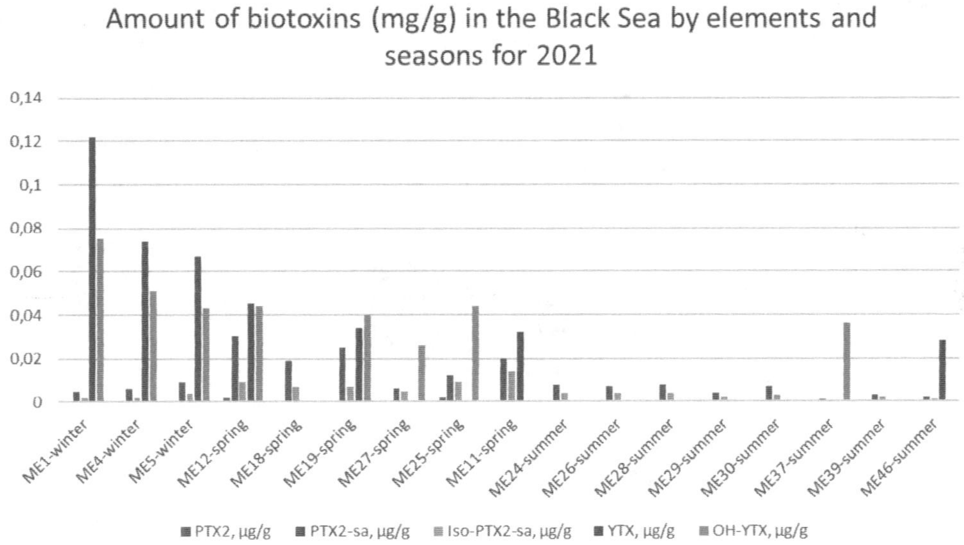

Figure 2. The average value of marine biotoxins in the Black Sea in mg/g by elements and seasons from February to July 2021

the next application clarifies the proposed 1-D IFANOVA in this section. First, using the methodology from previous research [4, 31], we must convert the data with values for marine biotoxins into IFPs. It is possible to convert the data to IFPs using the expert's method described in full in [4].

The values measured by the various levels of the factor under study are the components of the IFIM $Y[K, L]$, which is built. The matrix Y is originally configured without the final two rows and columns as follows:

```
IndexMatrix<IFPair> S5S = S5.Substitute({},{},{"Sri"},{"Dif"});
IndexMatrix<IFPair> S6S = S6.Substitute({},{},{"Sr"},{"Dif"});

IndexMatrix<IFPair> S7 = S5S.Add(S6S,[](IFPair A, IFPair B){return A-B;});

IndexMatrix<IFPair> S7xS7 = S7.TermwiseMultiply(S7, [](IFPair A, IFPair B){return A*B;});

IFPair AGIO2 = S7xS7.AggregateSummary([](IFPair A, IFPair B){return A.operOr1(B);});

IFPair MSC = (1.0/km1)*AGIO2;
```

Figure 3. An extract from the IFANOVA utility's code

$$Y[K/\{Sr_2, Sr\}, L/\{Sr_1, Sr\}] =$$

	$ME1$	$ME2$	$ME3$	$ME4$	$ME5$	$ME6$	$ME7$	$ME8$
Winter	$\langle 0.81, 0.18\rangle$	$\langle 0.52, 0.47\rangle$	$\langle 0.48, 0.51\rangle$	$\langle \bot, \bot\rangle$	$\langle \bot, \bot\rangle$	$\langle \bot, \bot\rangle$	$\langle \bot, \bot\rangle$	$\langle \bot, \bot\rangle$
Spring	$\langle 0.51, 0.48\rangle$	$\langle 0.08, 0.91\rangle$	$\langle 0.41, 0.58\rangle$	$\langle 0.13, 0.86\rangle$	$\langle 0.25, 0.74\rangle$	$\langle 0.24, 0.75\rangle$	$\langle \bot, \bot\rangle$	$\langle \bot, \bot\rangle$
Summer	$\langle 0.02, 0.97\rangle$	$\langle 0.02, 0.97\rangle$	$\langle 0.028, 0.97\rangle$	$\langle 0.004, 0.99\rangle$	$\langle 0.02, 0.97\rangle$	$\langle 0.13, 0.86\rangle$	$\langle 0.0, 1.0\rangle$	$\langle 0.1, 0.89\rangle$

The following results were obtained after we ran IFANOVA on Y:

$$MSC = \langle 0.076, 0.92\rangle, MSE = \langle 0.004, 0.98\rangle.$$

The Fig. 4 and Fig. 5 present a snippet of an IFP input file and the utility's results after processing the file:

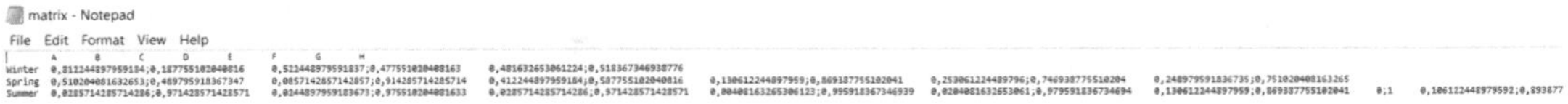

Figure 4. A snippet from an input file of IFPs

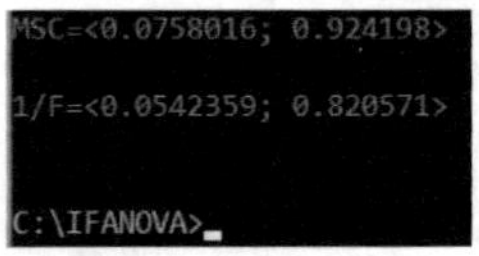

Figure 5. A snippet of an IFP input file and the utility's results after processing the file

Pietraszek's approach ([23], 2016) must be applied to generate the fuzzy estimate of the ANOVA key statistics F value. Figure Fig. 6 displays the fuzzy estimators for the bootstrapped F statistics. The fuzzy assessment $F_{fuzzy(0.95,16,2)} = \langle 0.97, 0\rangle$ is equivalent to the classic value $F(0.95, 16, 2) = 0.28$. Consequently,

$$\langle 0.05, 0.82\rangle = \frac{MSE}{MSC} = \frac{1}{F} \leq F_{fuzzy(0.95,55,4)} = \langle 0.97, 0\rangle.$$

This means that we can conclude from IFANOVA that the factor under study is dependent.

The results of the 1-D IFANOVA and ANOVA for a set of values for marine biotoxins show the influence of the "season" component.

This suggests that it is important to regularly check these contaminants' concentrations, especially during the winter. It is important to investigate the causes of the differences in their levels. Thus, policies to preserve the areas where aquaculture is practiced and to stop their levels from rising above health standards could be advocated.

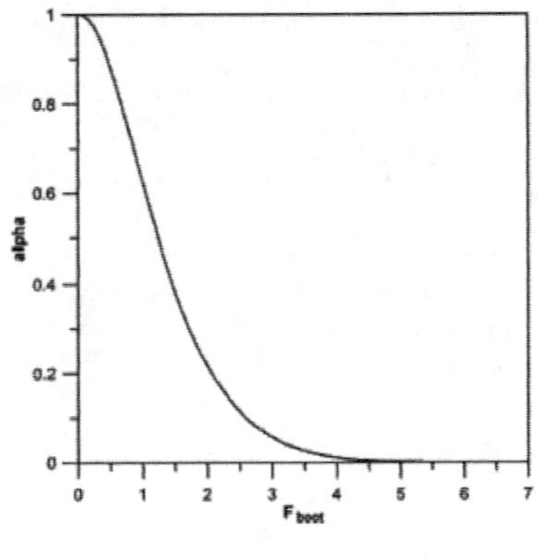

Figure 6. Fuzzy estimator of the bootstrapped *F* statistics

5. Conclusion

In this work we have applied one-way IFANOVA to analyze the distribution of marine biotoxins in the Black Sea by the "season" factor on the IF dataset of their values for the period from February 1 to July 31, 2021. In this study, we also compared the one-way IFANOVA results we acquired using the software tool for its performance to those of conventional ANOVA. After examining data on the levels of marine biotoxins in the Black Sea during the study period, we can say that the "season" factor affects the levels of marine biotoxins in a statistically significant way. As a result, we can see that the influence of the "season" factor on marine biotoxins was determined by both traditional ANOVA and IFANOVA. Future work on the suggested IFANOVA approach will involve using it with additional real-world datasets.

5.1. Acknowledgments

Work is supported by the Assen Zlatarov University through project Ref. No. NIX-482/2023 "Application of Intuitionistic Fuzzy Logic in Decision Making".

References

[1] Anuradha D and Kalpanapriya D 2018 Intuitionistic fuzzy ANOVA and its application in medical diagnosis *Research Journal of Pharmacy and Technology* **11** (2), 653-656
[2] Atanassov K 1983 Intuitionistic fuzzy sets *In Proceedings of VII ITKR's Session* (Sofia) (in Bulgarian)
[3] Atanassov K 1987 Generalized index matrices *Comptes rendus de l'Academie Bulgare des Sciences* **40**(11) pp. 15-18
[4] Atanassov K 2012 On Intuitionistic Fuzzy Sets Theory *STUDFUZZ* **283** (Heidelberg: Springer)
[5] Atanassov K 2014 *Index Matrices: Towards an Augmented Matrix Calculus* Studies in Computational Intelligence **573** (Cham: Springer)
[6] Atanassov K 2018-2019 Remark on an intuitionistic fuzzy operation "division" *Issues in IFS and GNs* **14** pp. 113-116
[7] Atanassov K 2018 n-Dimensional extended index matrices Part 1 *Advanced Studies in Contemporary Mathematics* **28** (2) pp. 245-259
[8] Atanassov K., Szmidt E, Kacprzyk J 2013 On intuitionistic fuzzy pairs *Notes on Intuitionistic Fuzzy Sets* **19**(3) pp. 1-13
[9] Buckley J J 2006*Fuzzy probability and statistics* (Berlin Heidelberg: Springer)
[10] De S K, Bisvas R, Roy R 2000 Some operations on IFSs *Fuzzy sets and systems* **114** (4) pp. 477-484
[11] Doane D and Seward L 2016 *Applied statistics in business and economics* (New York: USA McGraw-Hill Education)
[12] Fisher R 1925 *Statistical Methods for Research Workers* (London)
[13] Gil M A, Montenegro M, González-Rodríguez G, Colubi A, Casals M 2006 Bootstrap Approach to the Multi-sample Test of Means with Imprecise Data *Computer Statistics and Data Analysis* **51** pp. 148–162
[14] Jiryaei A., Parchami A, Mashinchi M 2013 One-way ANOVA and least squares method based on fuzzy random variables *Turk J Fuzzy Syst* **4** (1) pp. 18–33
[15] Kalpanapriya D and Pandian P 2012 Fuzzy hypothesis testing of ANOVA model with fuzzy data *Int. J. Mod. Eng. Res.* **2** (4) pp. 2951–2956
[16] Kalpanapriya D and Unnissa M 2017 Intuitionistic fuzzy ANOVA and its application using different techniques *In: Madhu*

V, Manimaran A, Easwaramoorthy D, Kalpanapriya D, Mubashir Unnissa M (eds) Advances in Algebra and Analysis. Trends in Mathematics pp. 457-468 (Birkhäuser: Cham)

[17] Konishi M, Okuda T, Asai 2006 Analysis of variance based on fuzzy interval data using moment correction method *International Journal of Innovative Computing, Information and Control* **2** (1), pp. 83-99

[18] Mavrov D 2019-2020 An Application for Performing Operations on Two-Dimensional Index Matrices *Annual of "Informatics" Section* Union of Scientists in Bulgaria **10** pp. 66–80

[19] Montenegro M., Gonzalez-Rodriguez G, Gil M A, Colubi A, Casals M R 2004 Introduction to ANOVA with fuzzy random variables *In: Lopez-Diaz, M., Gil, M.A., Grzegorzewski, P., Hryniewicz, O., Lawry, J. (Eds.) Soft Methodology and random information systems* (Berlin: Springer) p. 487–494

[20] Nourbakhsh M R., Parchami A, Mashinchi M 2013 Analysis of variance based on fuzzy observations *Int. J. Syst. Sci.* **44** (4) pp. 714-726

[21] Parchami A, Nourbakhsh M, Mashinchi M 2017 Analysis of variance in uncertain environments *Complex Intell. Syst.* **3** (3) pp. 189-196

[22] Parchami A, Mashinchi M, Kahraman C 2020 A case study on vehicle battery manufacturing using fuzzy analysis of variance *In: Kahraman, C. (eds.) INFUS 2020, Advances in Intelligent Systems and Computing* vol. 1197 (Cham: Springer) p. 916–923

[23] Pietraszek J, Ko lomycki M, etc. 2016 The Fuzzy Approach to Assessment of ANOVA Results *In: Nguyen NT., etc. (eds) Computational Collective Intelligence ICCCI 2016 LNCS* vol. 9875 (Cham: Springer) p. 260-268

[24] Puri M and Ralescu D Fuzzy random variables 1986 *J. Math. Anal. Appl.* **114** pp. 409-422

[25] Riecan B and Atanassov A 2010 Operation division by n over intuitionistic fuzzy sets *NIFS* **16** (4) pp. 1-4

[26] Szmidt E and Kacprzyk J 2009 Amount of information and its reliability in the ranking of Atanassov intuitionistic fuzzy alternatives *In: Rakus-Andersson (eds.) Recent Advances in Decision Making* vol. 222 (Heidelberg: Springer) p. 7-19

[27] Traneva V 2015 Internal operations over 3-dimensional extended index matrices *Proceedings of the Jangjeon Mathematical Society* vol. 18(4) p. 547-569

[28] Traneva V and Tranev S 2017 *Index Matrices as a Tool for Managerial Decision Making* (Bulgarian: Publ. House of the Union of Scientists Bulgaria) (in Bulgarian)

[29] Traneva V, Tranev S, Stoenchev M, Atanassov K 2018 Scaled aggregation operations over 2- and 3-dimensional IMs *Soft computing* **22**(15) p. 5115-5120

[30] Traneva V, Mavrov D, Tranev S 2022 Software Utility of One-Way Intuitionistic Fuzzy ANOVA *Lecture Notes in Networks and Systems* **504** pp. 681 - 689

[31] Traneva V and Tranev S 2021 Intuitionistic Fuzzy Analysis of Variance of Ticket Sales *In: Kahraman, C. (eds.) INFUS 2020 Advances in Intelligent Systems and Computing* vol. 1197 (Cham: Springer) p. 363-340

[32] Traneva V and Tranev S 2022 Digital Interpretation of Movie Sales Revenue Through Intuitionistic Fuzzy Analysis of Variance *Lecture Notes in Networks and Systems* **504** pp. 581 - 588

[33] Zadeh L 1965 *Fuzzy Sets* Information and Control **8** (3) pp. 338-353

[34] www.mu-varna.bg/BG/AboutUs/Pharmacy/Pages/deinosti-i-postignati-rezultati-morska-medicina.aspx" Last accessed 4 March 2023

Furniture market demand forecasting using machine learning approaches

P E Yaneva and H N Kulina

Department of Mathematical Analysis, Faculty of Mathematics and Informatics, University of Plovdiv Paisii Hilendarski 24 Tzar Asen Street, 4000 Plovdiv, Bulgaria
E-mails: polinayaneva@uni-plovdiv.bg, kulina@uni-plovdiv.bg

Corresponding author's e-mail address: polinayaneva@uni-plovdiv.bg

Abstract. Researching consumer demand in the market for goods is essential for any business. The purpose of this study is to forecast the demand for furniture from a large manufacturer in Bulgaria. Significant factors influencing customer flow, both in the company and online stores, are investigated. Daily observations for nearly two years are modelled using CART Ensemble with arcing. The constructed models describe the demand for furniture with high goodness-of-fit statistics: coefficient of determination up to 93% and determine the order of the factors influencing the demand.

1. Introduction

The worldwide furniture market is expected to rise at a compound annual growth rate of 5.2% from 2021 to 2028 [1] and is now an important part of the global economy. In recent years, this has given furniture companies access to a lot of information that can be used to improve their development decisions. For example, they can use machine learning (ML) to better expectations of accurate furniture demand. This allows corporations to optimize their production techniques, inventory management, and advertising strategies.

Business strategies now have increasingly involved demand forecasting. Studies have proven that ML algorithms including regression evaluation, time series analysis, and neural networks can enhance the accuracy of demand forecasting [2,3]. For instance, to analyse the demand for different products in the furniture market, researchers [4-6] have used machine learning approaches such as deep learning, application of regression, and artificial intelligence methods to predict demand in a construction machinery company and application of artificial neural networks using the Bayesian training rule in sales forecasting for the furniture industry. These studies reveal that, unlike conventional statistical patterns, machine learning algorithms can improve demand forecasting accuracy. Paper [7] describes how a long short-term memory (LSTM) network is used to forecast sales in the supply chain of the furniture industry. The authors note that successful marketing forecasting is a challenging task because of the complex nature of supply chain operations and the wide range of consumer demand. They point out that using LSTM networks can help improve the accuracy of sales forecasts by taking seasonality and trends into account. A similar proposal, but for artificial neural networks (ANN) is demonstrated in the paper [8]. This shows the potential of these machine learning techniques in the supply chain and planning processes of the furniture industry. The authors also emphasize the importance of choosing the right input variables and network architecture for ANN models to improve results. Paper [9] developed a better demand forecasting model using deep learning methods and also proposed a supply chain solution integration strategy. To improve the accuracy of sales forecasts, [10] proposed a demand-based tensor factorization approach. The paper [11] uses data mining techniques to forecast customer demand for remanufactured products. The process of discovering patterns and relationships from large data sets is called data mining. The authors use machine learning techniques to determine the requirements for refurbished products (RPDP) with the highest accuracy and adaptability. The purpose of the research is to create an appropriate and successful marketing plan. Article [12] proposes demand forecasting using machine learning and ordered summation methods. The method includes many learning algorithms such as linear regression, random forests, gradient boosting, and regression trees. The combination of these makes a final prediction based on the predictions of the original models.

The use of machine learning-based demand forecasting methods compared to traditional mathematical time series models, such as autoregressive integrated moving average (ARIMA), also improves forecasting accuracy [13]. Some of the most frequently used demand forecasting techniques are random forest regression (RFR), decision tree regression (DTR), and support vector regression (SVR) due to their ability to handle non-linear interactions and massive data sets [14]. Traditional forecasting methods have conditions for capturing nonlinear relationships and handling massive and complex datasets [15]. They usually presume a linear relationship between variables and are incapable of determining more complex patterns and features in the data. In comparison, machine learning methods are intended to handle non-linear relationships. They can handle large amounts of data and thus are well suited for demand forecasting in the furniture business [16, 17]. Machine learning has a large range of applications and can significantly improve demand forecasting. For example, it can be used to optimize manufacturing and store management, lower waste, and minimize costs [18, 19]. To improve customer contentment, we can use machine learning to provide that the right product is supplied at the right time [20, 21]. In the furniture industry, predicting demand has the future to change the way companies work and improve profitability and productivity.

In this paper, we research the application of machine learning methods in furniture market demand forecasting. Specifically, we focus on classification and regression trees (CART) ensemble methods that use lagged variables of independent factors to capture the lagged effect of advertising on business purchase characteristics. The objective of this article is to gain insight into the future of these technologies for accurate demand forecasting.

2. Data and methods

2.1. Data

This work used daily data for a big furniture producer in Bulgaria. The period is from 01 January 2020 to 30 October 2021, or N=648 recordings. The investigated data are as follows: the flow of people in the store network for the day (Store_Traffic); the flow of people who visited the online store (website) per day (Web_ Traffic); advertisement on Facebook (FB_Ads), in BGN; advertisement in Google (Google_Ads), in BGN; advertisement in two big private TV groups (TV1_Group), (TV2_Group), in target rating point (TRP). The dependent variables are the flow of people in the store network for the day (Store_Traffic) and those who visited the online store per day (Web_ Traffic). Visitors to the company store network are counted by cameras placed at each site. Website visitors are counted with a visit counter. We have analyzed customer traffic in the stores and website, summarizing the findings in Table 1. The data shows notable fluctuations in Store_Traffic and Web_Traffic, indicating varying levels of customer engagement.

2.2. Ensemble methods used in this study

Ensemble methods are machine learning techniques that combine several base models (such as CART, neural networks, etc.) to make the final decision, which is often better than individuals. These methods rely on "resampling" techniques to create different training sets for each of the models. According to how generated the base learners there are two popular methods for creating ensembles: sequential ensemble methods – boosting, where the base learners are generated sequentially, and parallel ensemble methods where the base learners are generated in parallel – bagging, represented by Breiman [22]. The parallel ensemble methods exploit the independence between the base learners and show that the error can be reduced. Ensemble methods are widely used in both classification and regression. In classification, the final decision is made by voting, and in regression, by averaging.

Table 1. Descriptive statistics of the used variables.

Variables	Minimum	Maximum	Range	Std. Deviation	Mean	
					Statistics	Std.Error
Store_Traffic (number)	720	5997	5277	974.008	2992.07	38.263
Web_Traffic (number)	14408	47390	32982	6000.337	28632.27	235.716

The arcing (arc-x4) framework is a sequential ensemble method introduced by Breiman in [23]. In the case of classification, arcing uses a simple mechanism for determining the probabilities of including observation in the training set. After the first K classifiers the next formula expresses the chance p_i that observation i will be picked for the training set of classifier K + 1 [23]:

$$p_i = (1+m_i^4)(\sum_{j=1}^{N}(1+m_j^4))^{-1}, \tag{1}$$

where the value m_i represents the number of times that the ith observation was incorrectly included by the previous K classifiers. In regression, the probability is used as a weight for the ith observation.

The advantage of arcing ensemble methods is that they can help to reduce bias and variance, especially when individual models. In general, ensemble methods are less sensitive to noise and outliers in the data. The combination of several models helps to smooth the errors of individual models. This increases the overall robustness of the predictions. The disadvantages of ensemble methods are that they often require more computational resources and time to train and predict than single models and can be challenging to interpret and explain.

In this paper, we use CART to build the base models in a regression model. Therefore, the predictions of the ensemble model are the arithmetic mean of the predictions of the trees of the ensemble. The hyperparameters set at the beginning of the analyses are the number of trees in the ensemble, the minimum number of cases in the parent node, and the minimum number of instances in a child node. Another main parameter that determines the algorithm is the cross-validation of the k-fold type.

This paper uses the Arcing ensemble algorithm for regression from the software engine CART ensembles and bagger of the Salford predictive modeler [24]. It also uses IBM SPSS Statistics [25].

2.3. Model evaluation metrics

When evaluating the performance of arcing, we utilize several predictive model evaluation measures. We chose: root mean squared error (RMSE), mean absolute percentage error (MAPE), and R-squared (R^2), using the formulas:

$$RMSE = \left(\frac{1}{N}\sum_{i=1}^{N}(Y_i - P_i)^2\right)^{1/2}, \; MAPE = \frac{100}{N}\sum_{i=1}^{N}\left|\frac{Y_i - P_i}{Y_i}\right|, \; R^2 = \frac{\left(\sum_{i=1}^{N}(P_i - \overline{P})(Y_i - \overline{Y})\right)^2}{\sum_{i=1}^{N}(P_i - \overline{P})^2 . \sum_{i=1}^{N}(Y_i - \overline{Y})^2} \tag{2}$$

where Y is the observed dependent variable (target), P is the model prediction, Y_i and P_i (i=1,2, …, N) are their values, respectively; $\overline{Y}$ and $\overline{P}$ are mean values, N is the sample size. A good predictive model should exhibit low errors, indicating accurate predictions with minimal deviation. Additionally, a high R^2 value suggests a strong explanatory power of the model, indicating a better fit for the observed data.

3. Results with discussion

3.1. Building ensemble models of Store_Traffic

Figure 1 illustrates the plot of the dependent variable Store_Traffic, revealing a noticeable upward linear trend. The plots also highlight numerous peaks in the variables, indicating potential factors such as the growing popularity of the store, an increasing number of customers, or successful marketing and advertising campaigns.

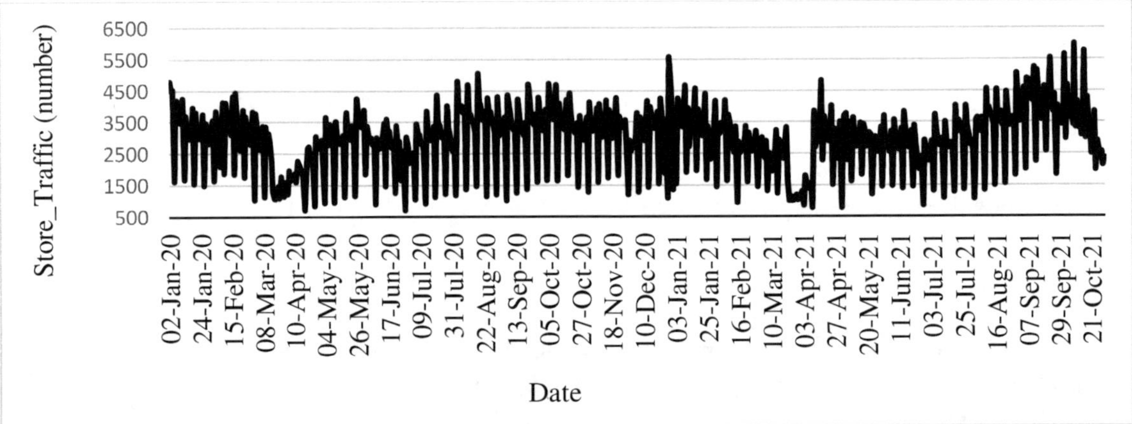

Figure 1. Observed values of Store_traffic.

The conducted Dickey-Fuller and Phillips-Perron tests showed a p-value < 0.01 rejecting the null hypothesis of non-stationarity, i. e. the time series is stationary. The study continues with the construction of the graph of the partial autocorrelation function (PACF) function in Figure 2.

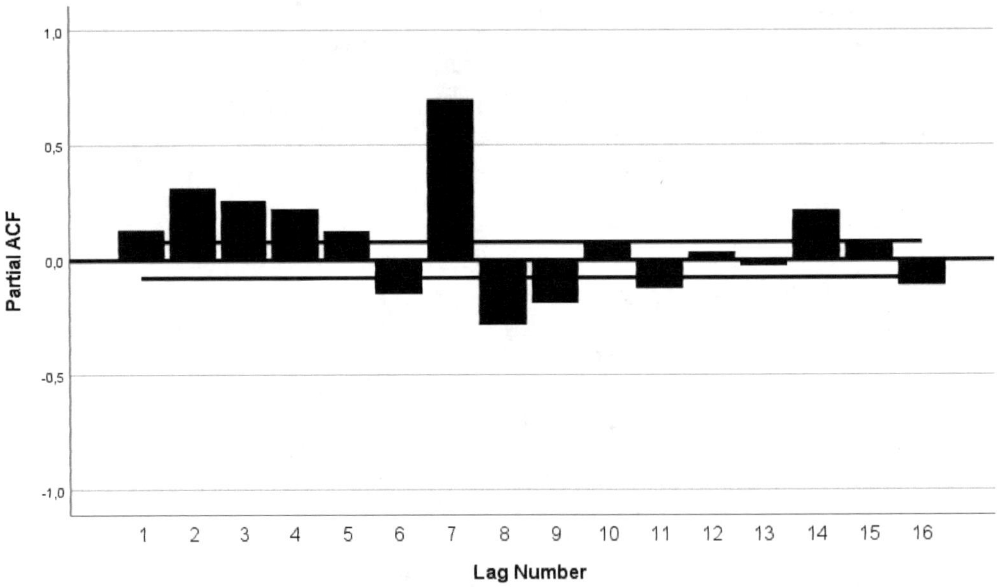

Figure 2. Partial autocorrelation function (PACF) of Store_Traffic.

From the PACF plots of the dependent variable Store_Traffic, we can observe that Store_Traffic lag 7 stands out with a relatively high partial autocorrelation coefficient. PACF shows us that we can use the lagged values of the variable for further analyses.

To validate the models, we constructed a new dependent variable called Store_Traffic_7. This variable was created by removing the last seven values from the initial time series, Store_Traffic.

For the model predictors, we conducted tests using various variables and identified the following lag values as the most influential: Web_ Traffic: lag 2, 4, and 7; Store Traffic: lag 1 and 7; Facebook (FB): lag 2 and 4; Google: lag 7; TV1_Group: lag 7; TV2_Group: lag 7. These lag values indicate the time intervals at which the predictors are most relevant and impactful in forecasting Store_Traffic_7. The hyperparameters of the ensemble models include a minimum number of cases in parent and child nodes of each tree – 5 and validation, which is performed by the 10-fold cross-validation algorithm. For each different choice of predictor, we compare models with different numbers of trees in the ensemble based on evaluation measures. We find that as the number of trees in the ensemble increases, R^2 increases and the corresponding MAPE decreases. Table 2 shows the prediction statistics of the results of ensemble models with 10, 20, 30, 40, 50, 60, and 70 trees.

Table 2. Prediction statistics of ensemble models of Store_Traffic_7

Model	Number of trees	RMSE	MAPE	R^2
Arc10	10	335.8009	11.6659	0.8929
Arc20	20	271.3181	9.3211	0.9337
Arc30	30	244.8294	8.2289	0.9462
Arc40	40	232.9647	7.7723	0.9514
Arc50	50	226.2525	7.5315	0.9542
Arc60	60	220.8948	7.2943	0.9564
Arc70	**70**	**218.3189**	**7.1542**	**0.957**

We have determined that the best model consists of 70 trees. Figure 3 displays a comparison between the observed values of the Store_Traffic time series, (represented by the red line,) and the predictions made by the Arc70 model, (represented by the blue dashed line.) This comparison focuses on the most recent 50 days of the time series. Beyond the vertical line in the plot, we find the forecasted values generated by the Arc70 model.

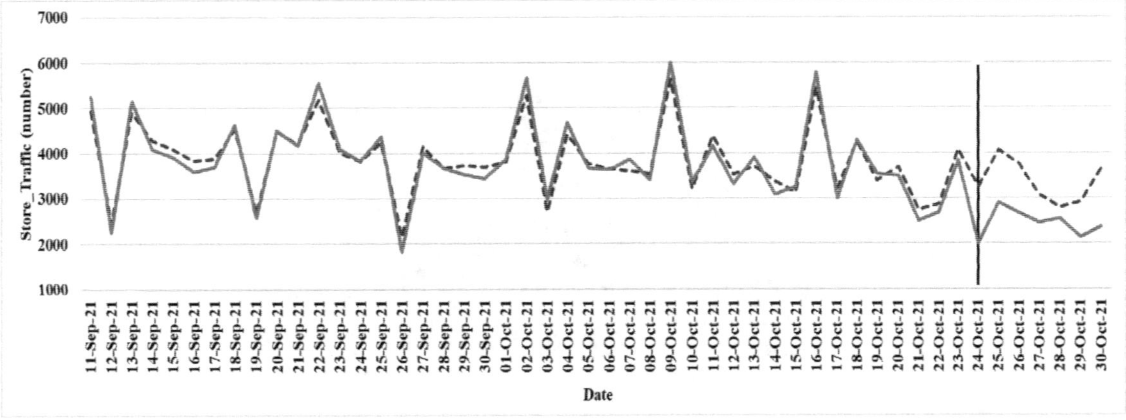

Figure 3. Comparison between observed values of the store traffic time series (red line) and predicted by the Arc70 model (blue dashed line). After the vertical line are the forecasted values.

The built model has a coefficient of determination R^2=0.957 (between Store_Traffic and predicted Store_Traffic _7) and explains around 96% of the change of the dependent variable Store_Traffic. The following performance measures are: RMSE = 218.3189, MAPE = 7.1542.

The importance of variables in the Arc70 model remains consistent regardless of the number of trees. The most significant influence on Store_ Traffic is the store traffic from 7 days ago with 100 points, as shown in Figure 4. The influence of the remaining variables is determined relative to it. The next highest importance is the website traffic from 2 days ago. FB advertising, website traffic, and Google advertising are subsequent variables in terms of their influence on store traffic. And finally, advertisements on television exhibit the least significant impact on store traffic according to the Arc70 model.

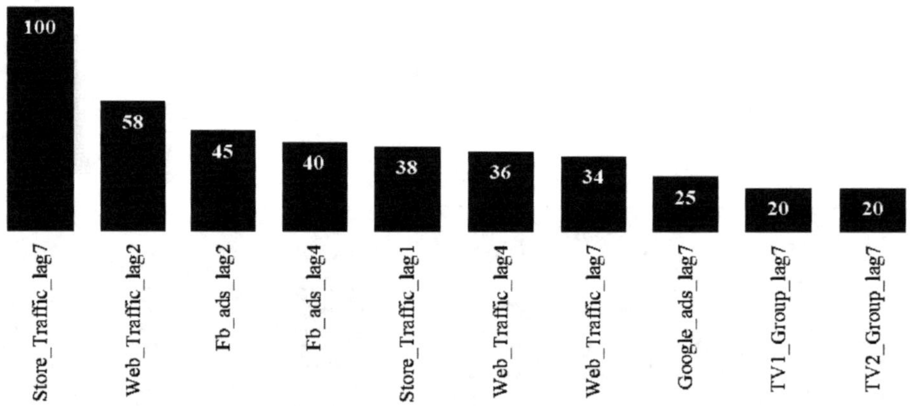

Figure 4. Relative variable importance of independent variables in the model Arc70.

Examination of the residuals for the selected model shows that they are close to normal distribution. This can be seen from the histogram of the residuals in Figure 5.

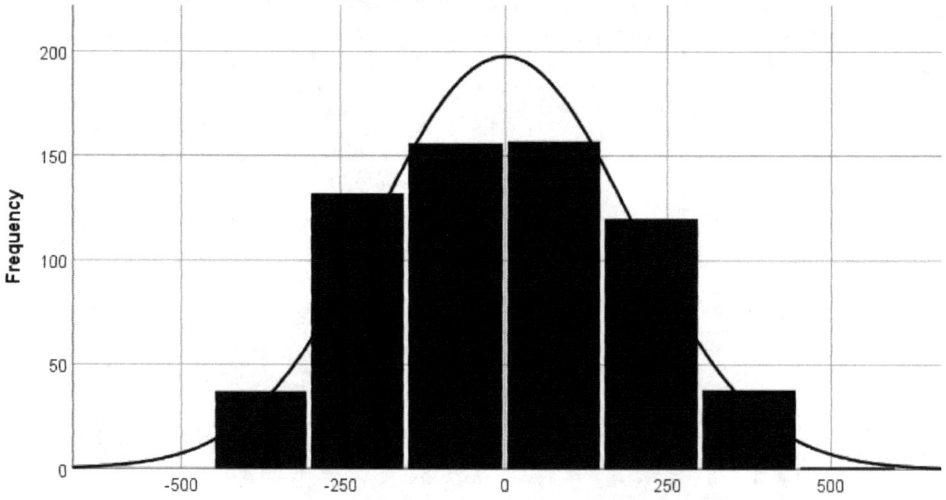

Figure 5. Histogram of residual scores of Store_Traffic for Arc70 model.

3.2. Building ensemble models of Web_Traffic

Like any large chain, the company offers its customers to view and purchase goods from the online store. Website visits are an indicator of customer interest and demand. Figure 6 illustrates the plot of the dependent variable Web_Traffic.

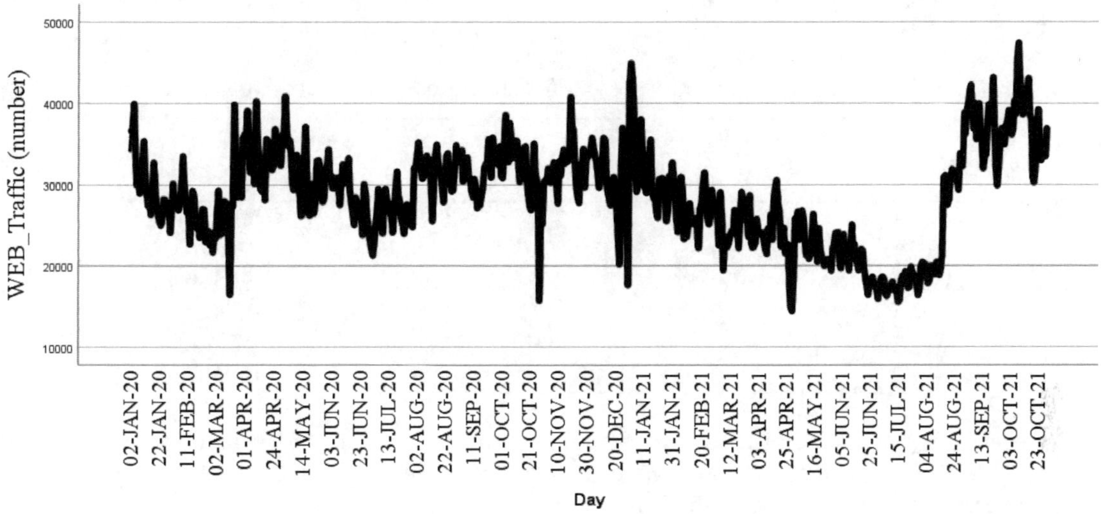

Figure 6. Observed values of the Web_Traffic.

The Dickey-Fuller, and Phillips-Perron tests performed showed a p-value > 0.05, indicating that the time series is non-stationary. Figure 7 showcases the PACF plot of the dependent variable Web_Traffic. Lag 1 has a high positive partial autocorrelation coefficient. Lags 4, 5, and 6 also exhibit significant partial autocorrelation, although with smaller coefficients, suggesting some correlation between the current value and the values at those lags. This gives us reason to include the corresponding lagged variables in the construction of the arcing models.

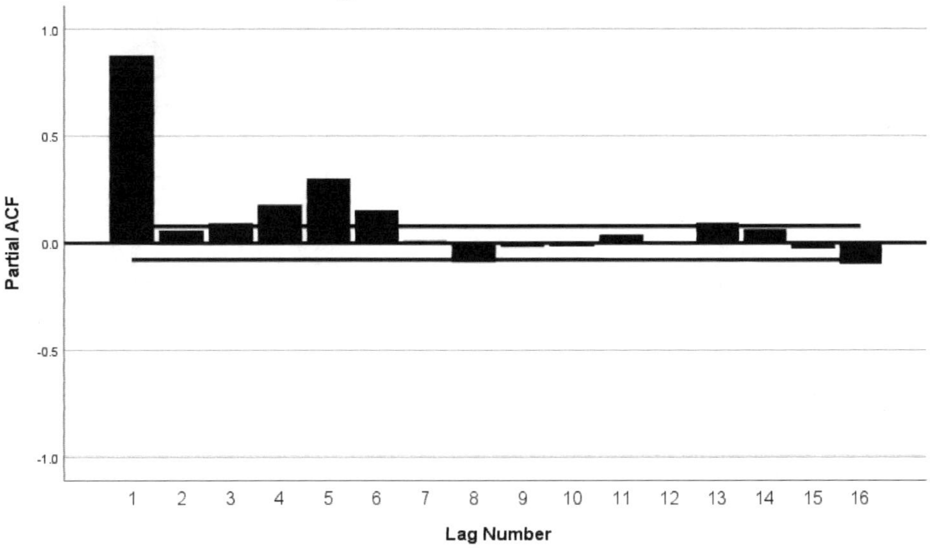

Figure 7. Partial autocorrelation function (PACF) of Web_Traffic.

To validate the models for the Web_Traffic variable, we created a new dependent variable called Web_Traffic_7. This variable was derived by removing the last 7 values from the initial time series, Web_Traffic. We tested various predictors and selected the following variables: Web Traffic: lag 2, 4, and 7; Store Traffic; Facebook (FB): lag 2 and 4; Google; TV1_Group; TV2_Group. The hyperparameters used in this evaluation process are the same as those employed for the arcing model of Store_Traffic. Table 3 shows the prediction statistics of ensemble models with 10, 20, 30, 40, 50, and 60 trees.

Table 3. Prediction statistics of ensemble models of Web_Traffic_7.

Model	Number of trees	RMSE	MAPE	R^2
Arc10	10	1672.211	5.2553	0.930516
Arc20	20	1359.629	4.2876	0.95676
Arc30	30	1239.262	3.9311	0.96539
Arc40	40	1163.306	3.6997	0.970065
Arc50	50	1128.799	3.5686	0.971628
Arc60	**60**	**1108.679**	**3.513**	**0.960**

For each different choice of predictor, we compare models with different numbers of trees in the ensemble based on evaluation measures. We can see that after the Arc10 model, the corresponding mean absolute percentage error decreases and is under 5%.

The best model for forecasting Web_Traffic_7 is the one with 60 trees. This model demonstrates the following performance measures: RMSE = 1108.679, MAPE = 3.513, R^2 = 0.960. They highlight the accuracy and precision of the chosen model in forecasting Web_Traffic_7.

Figure 8 shows the comparison between observed values of the Web_Traffic time series (red line) and predicted by the Arc60 model (blue dashed line) for the last 50 days. The forecasted values are after the vertical line.

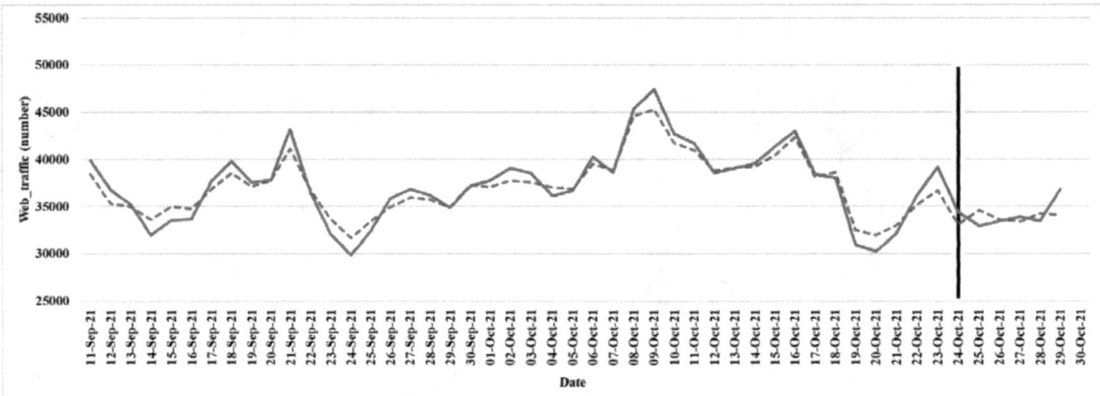

Figure 8. Comparison between observed values of the web traffic time series (red line) and predicted by the Arc60 model (blue dashed line). After the vertical line are the forecasted values.

The analysis of variable importance within the selected ensemble model reveals that the influence of variables remains consistent across different numbers of trees – 10, 20, 30, 40, 50, and 60. For the chosen model, a significant influence on the Web_Traffic variable has website traffic from 2 days ago. Following are the website traffic from 7 and 4 days ago, indicating that recent web traffic patterns strongly impact the current web traffic levels. Advertisements on television have the least significant impact on the Web_traffic variable.

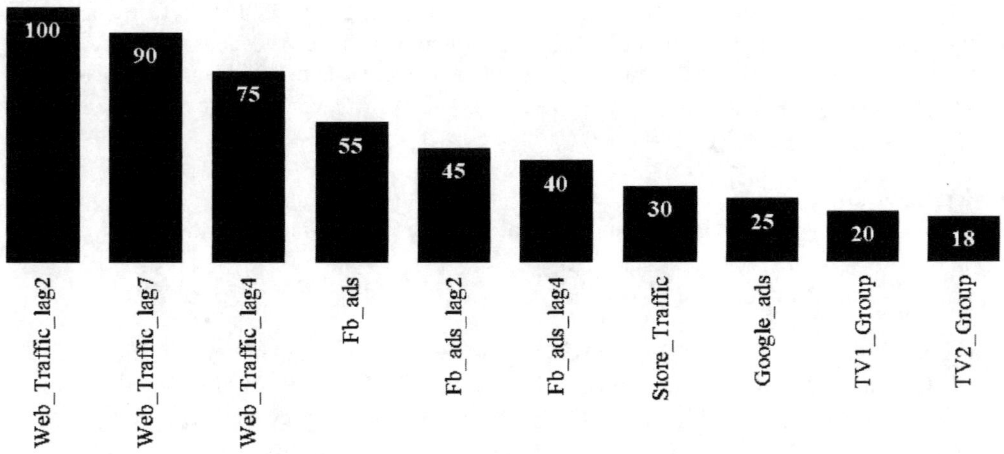

Figure 9. Relative variable importance of independent variables in the model Arc60.

Examination of the residuals for the selected model shows that they are close to normal distribution. This can be seen from the histogram of the residuals in Figure 10.

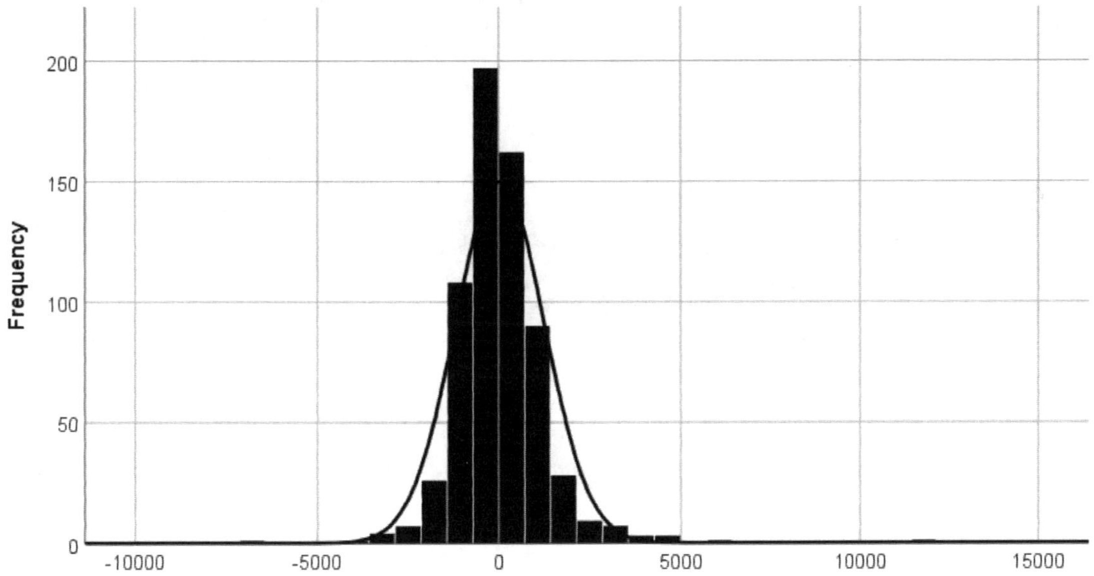

Figure 10. Histogram of residual scores of Web_Traffic for Arc60 model.

4. Conclusion

In conclusion, our study focused on utilizing machine learning techniques for furniture market demand forecasting. We evaluated the performance of these models on the Store_Traffic and Web_Traffic variables, providing valuable insights into the accuracy of our predictions.

Based on the analysis and interpretation of the results, it can be concluded that website traffic has the greatest influence on the demand for furniture, followed by online advertisements. This suggests that the online presence and marketing efforts of the furniture store significantly impact the demand for their products. It is likely that customers are increasingly relying on online channels to explore and make purchasing decisions. On the other hand, TV commercials have the least impact on demand. Furthermore, it was previously discussed that store traffic has the greatest impact on store turnover. This finding suggests that customers have a preference for physically visiting the store before making a purchase.

Acknowledgments
This study was supported by the Bulgarian National Science Fund, Grant KP-06-N52/9.

References

[1] Furniture Market Size, Share & Trends Analysis Report By Product (Beds, Tables & Desks, Sofa & Couch, Chairs & Stools, Cabinets & Shelves), By Material (Metal, Wood, Plastic, Glass), By Application, By Region, And Segment Forecasts, 2022 – 2030 https://www.grandviewresearch.com/industry-analysis/furniture-market

[2] Alsharari N M, Al-Shboul M and Alteneiji S 2020 Implementation of cloud ERP in the SME: evidence from UAE *Journal of Small Business and Enterprise Development* **27**(2) 299–327

[3] Al-Haidari R, Al-Rawashdeh S, Zeidan A, Omambala J and Nagarur N 2023 Impact of Covid-19 pandemic on demand and demand forecasting in a furniture wholesale company *Preprints* **2023040144**, https://doi.org/10.20944/preprints202304.0144.v1

[4] Deng W, Patil R, Liu F, Daji E and Shi Y 2021 Exploring freight loading management by deep learning: a case study in home furnishing industry *Annals of Data Science* **9**(2) 213–228

[5] Aktepe A, Yanik E and Ersöz S 2021 Demand forecasting application with regression and artificial intelligence methods in a construction machinery company *Journal of Intelligent Manufacturing* **32(6)**1587–1604

[6] Yucesan M, Gul M and Erkan E 2017 Application of artificial neural networks using bayesian training rule in sales forecasting for furniture industry *Drvna Industrija* **68**(3) 219–228

[7] Pliszczuk D, Lesiak P, Żuk K and Cieplak T 2021 Forecasting sales in the supply chain based on the LSTM network: the case of furniture industry *European Research Studies Journal* XXIV(Special Issue **2**) 627–636

[8] Bibaud-Alves J, Thomas P and Haouzi H B E 2019 Demand forecasting using artificial neuronal networks and time series: application to a French furniture manufacturer case study, *Conf. Proc. International Joint Conference on Computational Intelligence* **2019**

[9] Kilimci Z H, Akyuz A O, Uysal M, Akyokus S, Uysal M O, Bulbul B A and Ekmis M A 2019 An improved demand forecasting model using deep learning approach and proposed decision integration strategy for supply chain *Complexity* **2019** Article ID 9067367

[10] Bi X, Adomavicius G, Li W and Qu A 2022 Improving sales forecasting accuracy: a tensor factorization approach with demand awareness *Informs Journal on Computing* **34**(3) 1644–1660

[11] Van Nguyen T, Zhou L, Loong Chong Y A, Li B and Xiaodie P. 2020 Predicting customer demand for remanufactured products: A data-mining approach *European Journal of Operational Research* **281**(3) 543–558

[12] Tugay R and Oguducu S G 2020 Demand prediction using machine learning methods and stacked generalization *arXiv:2009.09756 (Cornell University)*

[13] Fattah J, Ezzine L, Aman Z, El Moussami H and Lachhab A 2018 Forecasting of demand using ARIMA model *International Journal of Engineering Business Management* **10**

[14] Birim S, Kazancoglu I, Mangla S K, Kahraman A and Kazancoglu Y 2022 The derived demand for advertising expenses and implications on sustainability: a comparative study using deep

learning and traditional machine learning methods *Annals of Operations Research* **Jan 7**, 1-31

[15] Younis H M, Sundarakani B and Alsharairi M 2021 Applications of artificial intelligence and machine learning within supply chains: systematic review and future research directions *Journal of Modelling in Management* **17**(3) 916–940

[16] Rosienkiewicz M, Kowalski A, Helman J and Zbieć M 2018 Development of lean hybrid furniture production control system based on glenday sieve, artificial neural networks and simulation modeling *Drvna Industrija* **69**(2) 163–173

[17] Abbasimehr H, Shabani M and Yousefi M 2020 An optimized model using LSTM network for demand forecasting *Computers & Industrial Engineering* **143** 106435

[18] Makkar S, Devi G N R and Solanki V K 2020 Applications of machine learning techniques in supply chain optimization *ICICCT 2019 - System Reliability, Quality Control, Safety, Maintenance and Management* ed V K Gunjan, V Garcia Diaz *et al* (Singapore: Springer Singapore) pp 861–869

[19] Song Y, Cao Q and Zhang C 2018 Towards a new approach to predict business performance using machine learning *Cognitive Systems Research* **52** 1004–1012

[20] Ersöz T, Güven İ and Ersöz F 2022 Defective products management in a furniture production company: A data mining approach *Applied Stochastic Models in Business and Industry* **38**(5) 901– 914

[21] Hernández N N M, Cano-Olivos P, Sánchez-Partida D, Martínez-Flores J L and Caballero-Morales S O 2020 Demand forecasting: a challenge for furniture SMEs *Handbook of Research on Developments and Trends in Industrial and Materials Engineering* ed P Sahoo (IGI Global) pp 1-21

[22] Breiman L 1996 Bagging predictors *Machine Learning* **24**(2) 123–140

[23] Breiman L 1996 Bias, variance, and arcing classifiers *Technical Report* **460**, Statistics Department, University of California, Berkeley, CA

[24] *Salford Predictive Modeler* www.minitab.com/en-us/products/spm/

[25] *IBM SPSS Statistics*, https://www.ibm.com/products/spss-statistics

Journal of Physics: Conference Series **2675** (2023) 012009 doi:10.1088/1742-6596/2675/1/012009

Development of an imaging method for in vivo single-cell tracking under high-noise conditions: a proof of principle

Yu-Fen Chang[1,4], Pavel Struchalin[1], Bodil Næss[2], Tom Christian Holm Adamsen[2,3] and Boris V. Balakin[1]

[1] Dept. of Mechanical and Marine Engineering, Western Norway University of Applied Sciences, Inndalsveien 28, 5063 Bergen, Norway
[2] Center for Nuclear Medicine and PET, Haukeland University Hospital, Haukelandsveien 22, 5021 Bergen, Norway
[3] Dept. of Chemistry, University of Bergen, Allégaten 41, 5007 Bergen, Norway
[4] UiT The Arctic University of Norway, Hansine Hansens veg 18, 9019 Tromsø Norway

E-mail: yu-fen.chang@hvl.no, yu-fen.chang@uit.no

Abstract. This paper reports the development of a positron imaging method for in vivo single-cell tracking under high-noise conditions. Following biological processes spatially and temporally at a single-cell level in a living organism is desirable for inquiring into the relationships between the behaviours and properties of cells. Positron-emitting radionuclides enable detecting and following radioactivity-labelled substances deep inside living organisms. However, positron imaging has several challenges, such as the distribution of high noise in other areas close to the cell under investigation. In this work, an algorithm for locating a cell with millisecond resolution to combat the strong interference of nearby noise is developed. The feasibility of the method is verified by the demonstration of particle tracking and detection of behavioural changes in an environment with the signal-to-noise ratio of 1:9.

1. Introduction

The relationship between single-cell molecular characteristics and in vivo behaviour can play an important role in many critical processes, such as cancer cell metastases, as well as stem cell migration and engraftment. A deepened understanding of the migration process and the factors that affect delivery is needed to detect latent obstacles and develop strategies to improve efficiency and functional outcomes of stem cell implantation. Given that cell populations are heterogeneous and the molecular dynamics amongst cells is inhomogeneous, relating the cell behaviour, biological processes and cell properties at the single-cell level is necessary.

The delivery and fate of implanted stem cells can be assessed by using ex vivo processing at multiple time points. In vitro studies can also shed light on the migration, interactions, proliferation, death and differentiation of cells. However, the properties of cells taken out of their native environment and studied in vitro or ex vivo may be considerably different. Following the biological processes of a single cell spatially and temporally in a living organism is desired to inquire into the behaviour and molecular characteristics of a single cell. Although in vivo single-cell detection has been attempted by using photo-acoustic imaging, bioluminescent imaging and magnetic resonance imaging, these techniques suffer from inadequate depths and/or limited temporal and spatial resolution. Jung et al. [1] demonstrated a positron emission tomography (PET)-based technique for following a single cell over time in a living mouse. In their study [1], the cells were labelled with nanoparticles carrying ^{68}Ga. A conventional PET image reconstruction technique and a new method requiring a small amount of radioactivity in combination with a spline method were used to derive the cell's trajectory. Cell movement was tracked at the velocity of approximately 50 mm/s. However, the method suffered from the reduction in cell-to-background image contrast within a few hours due to radionuclide efflux from nanoparticles [2].

In addition to radionuclide efflux from cells, radionuclide accumulation in other areas of the organism poses challenges to single-cell tracking by using PET-based techniques. In this study, we aim to develop and demonstrate a positron imaging method for the in-depth tracking of a single cell in vivo under high-noise conditions. Our method can be performed by employing a contemporary PET scanner or a modular PET system as the detector array. This technique intends to contribute to detailing cell biological processes, in vivo cell behaviour and their relationships with molecular characteristics.

2. Background and review of positron imaging methods

The technique developed in this study is based on the idea of positron emission particle tracking (PEPT) [3]. In PET and PEPT, tracers are labelled with positron-emitting radionuclides, such as ^{18}F, ^{68}Ga, and ^{64}Cu. As the radionuclide undergoes beta-plus decay, a positron is emitted and, after travelling a short distance [4,5], undergoes annihilation with an electron. This annihilation event results in two photons of 511 keV each that travel in back-to-back directions with angles close to 180°. By detecting these two photons, the radionuclide can be estimated to be located somewhere along the line connecting the two detection sites. This line is called the line of response (LOR).

In PET, the distribution and concentration of the tracer are 3D-mapped. By contrast, in PEPT, the tracer substances are deliberately confined in a very small volume. Therefore, one of the applications of PEPT can be the acquisition of information on the environment and processes by observing how this tiny amount of radioactive substance (hereafter called 'particle') reacts to its surroundings and environmental conditions. If placed in a process with a complex flow field, such as stirring [6], mixing [7], fluidisation [8] and separation [3], and if the particle properties are representative of the processed matter, flow patterns can be revealed in detail by following the radioactive particle with adequate spatial and temporal resolution.

In conventional PET, 3D tracer maps are reconstructed by processing LORs by using projection-based methods [4]. PEPT methods based on different principles and unlimited by projection have been developed and are undergoing innovation [9–12]. In this work, we developed methods that compare LORs with each other to find the most probable location of the particle centroid. In principle, if the radioactivity distribution is confined in a tiny volume, such as a cell, this volume should be located in the area of the closest distance between two LORs. Thus, the location of the cell can be approximated with a reasonably small amount of LORs because in an ideal case, the location can almost be determined by two LORs. In projection-based PET image reconstruction without time-of-flight information, the whole LOR is taken into account in data processing. However, most sections of the LOR are not the annihilation site. Therefore, the efficient use of LORs is emphasised in the single-cell tracking method presented here. The section of the radioactive decay event is estimated by the LOR's relationships with others at the outset of data processing. Consequently, the positron imaging method for single-cell tracking presented in this article requires a smaller amount of LORs than PET. This condition then allows the enhancement of temporal resolution.

3. Experimental

3.1. Tracer particle and setup for particle flow in a high-noise environment

As a proof of principle, an ice particle composed of ^{18}F aqueous solution was prepared to model a cell labelled with positron-emitting isotopes. This approach is a straightforward method for creating radioactive particles, and the radioactivity and size of the particle can be tuned easily. An ice particle of approximately 5.5 mm was used as the particle to be tracked.

Figure 1 shows the experimental setup for particle flow in a high-noise environment. The ice particle was released to fall freely in cold liquid with temperatures below 0 °C. An inclined ice surface was placed under the release site of the ice particle to model the behavioural changes of the cell in a biological process. The particle was tracked during its free fall in liquid, collision with the ice surface and movement along the ice surface in the environment with a massive amount of positron emission events originating from other sources.

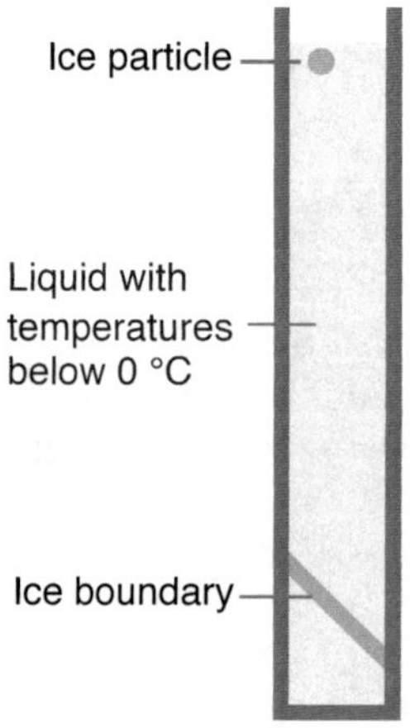

Figure 1. Setup to create the ice particle flow in a high-noise environment. The ice particle was released to fall freely and an inclined ice surface was placed under the release site. Several similar ice particles were placed in the same container to create intensive noise.

An environment with radioactivity much higher than that of a single particle was created to verify the algorithm that was developed to track a single cell under high-noise conditions. Several similar ice particles were placed in the same container where the target particle would move through. Figure 2 (a) shows the LORs generated by a tracer particle and a background source with considerably higher radioactivity. The LORs were drawn between the two detector locations of each photon pair.

3.2. Detector array

A PET scanner (Siemens Biograph Vision 600) was used as the detector array for photon-pair detection. A total of 60800 lutetium oxyorthosilicate (LSO) crystals were arranged to enclose a cylindrical detection space with the radius of 410 mm and axial length of 263 mm, where 760 crystals were in the tangential direction and 80 crystals were in the axial direction. Each crystal had the dimensions of $3.2\ mm \times 3.2\ mm \times 20\ mm$. Each 5×5 array of crystal was optically coupled to a 4×4 array of silicon photomultipliers (SiPMs). The coincidence processing unit compared the times of photon arrivals. If the arrival times were within 4.73 ns, the unit determined that the two events formed a coincidence. Additional characteristics of the scanner, such as energy window and electronics details, can be found in the references (see, e.g., [13]).

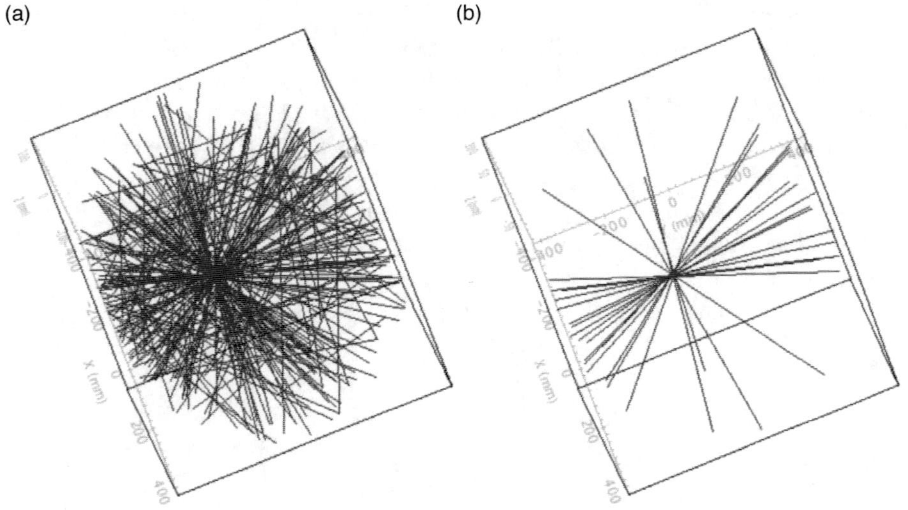

Figure 2. Setup to test the algorithm with an ice particle passing through an environment with high radioactivity. In the instance of the data shown, the edge-to-edge distance between the ice particle and a strong noise source is approximately 13 mm, and the signal-to-noise ratio in the field of view is approximately 1:8. (a) LORs within a period of 0.06 ms. (b) Remaining LORs indicating the location of the ice particle after the data were processed by the developed algorithm.

3.3. Single-cell tracking algorithm

The goal of this work is to develop a single-cell tracking algorithm that is in principle different from the projection-based tomography image reconstruction. While the fundamental aim of these algorithms is to find the most probable particle location by evaluating the relations between LORs, their implementation can be of great variety. For instance, the relationships between two LORs can be evaluated on the basis of the existence of a cutpoint (cross point) in each 2D plane and further evaluated by comparing their cutpoint with others' cutpoints. If this cutpoint falls very far from most of the other cutpoints, the likelihood of an annihilation event occurring around this cutpoint is determined as low. By running this kind of evaluation iteratively, the most likely location can be determined [3, 14]. The evolutions of the algorithms, such as fitting the density of cutpoints with peak shape functions for single and multiple particle tracking and weighting the likelihood of a cutpoint on the basis of the particle location found in the previous time step, have been demonstrated to be useful under different conditions [12, 15].

In this work, the algorithm was redeveloped to evaluate the relationships between LORs directly in 3D. The algorithm was designed to allow users to define region(s) of interest/noninterest with a relevant time point or period. A filter was then applied in accordance with time- and space-specific prior information, if available, to reduce the influences of LORs that were likely generated by the nonrelevant background. The filter was designed to evolve in accordance with the result of every time step. In this work, an ice particle was set to pass by an area with considerably higher radioactivity to test the capability of the algorithm. Figure 2 (a) shows all the LORs within a period of 0.06 ms and (b) shows the remaining LORs, indicating the location of the ice particle, after processing by the algorithm. In the situation shown by Figure 2, the edge-to-edge distance between the ice particle and an intensive noise source was approximately 13 mm, and the signal-to-noise ratio was approximately 1:8.

4. Results and discussion

Figure 3 shows the pathline of an ice particle freely falling in liquid with a temperature between 0 °C and -20 °C, colliding with an inclined ice surface and continuing to move along the ice surface. The temporal resolution of 4 ms was chosen for processing the LORs (i.e. the location of the particle was obtained every 4 ms) because the particle was moving at a speed below 0.3 mm/ms. Properties, such as velocity component, speed (as shown in Figure 3 (b)), acceleration and kinetic energy can be derived from the ms positions of the particle. The colour scale bar in Figure 3 (a) indicates the elapsed time in ms and that in (b) indicates the speed in m/s.

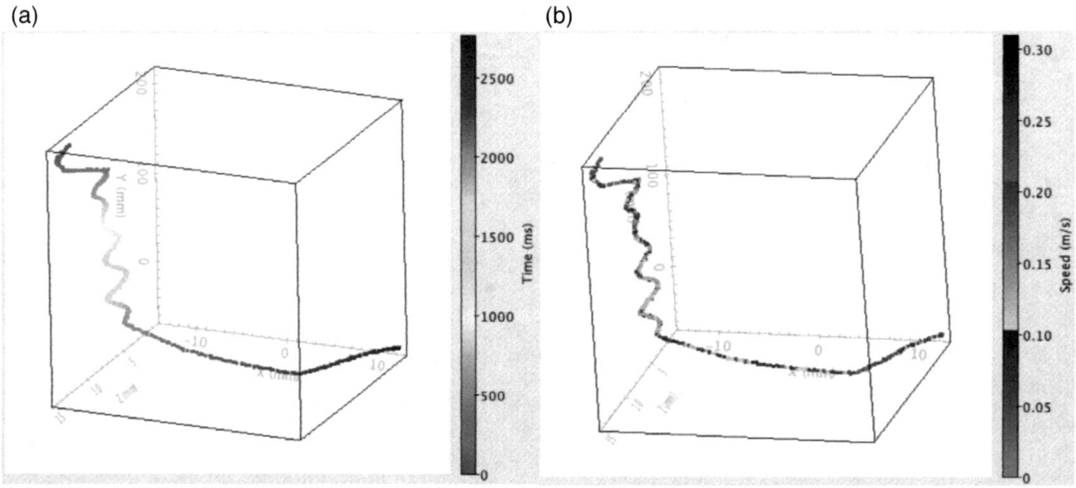

Figure 3. Pathline of an ice particle freely falling in liquid with a temperature between 0 °C and -20 °C, colliding with an inclined ice surface and continuing to move along the ice surface. The colour scale bar in (a) indicates the elapsed time in millisecond and that in (b) indicates the speed in m/s.

During this tracking period, the average signal-to-background ratio was approximately 1:9 to demonstrate the capability of the single-cell tracking method under high-noise conditions. Figure 3 (b) illustrates that the behavioural changes (here the speed changes) in time and space due to interactions with the environment can be clearly seen by using the single-cell tracking method. The particle speed dropped abruptly as the particle hit the ice surface.

5. Conclusions

In this work, a positron imaging method for tracking a single cell in vivo under high-noise conditions was developed. A setup of ice particle flow and ice-ice collision in a high-noise environment was utilised to test the method. The capability of the method to follow a particle and unveil its behavioural changes at the signal-to-noise ratio of 1:9 was successfully demonstrated.

Acknowledgments

The authors gratefully acknowledge funding for FLOWCHART project (300286) from the Research Council of Norway as well as funding from the University and College Network in Western Norway (UH-nett Vest). The authors are appreciative of support of Siemens Healthineers.

References

[1] Jung K O, Kim T J, Yu J H, Rhee S, Zhao W, Ha B, Red-Horse K, Gambhir S S and Pratx G 2020 *Nature Biomedical Engineering* **4** 835–844

[2] van Dongen G A M S, Boellaard R and Vugts D J 2020 *Nature Biomedical Engineering* **4** 765–766

[3] Chang Y F, Ilea C G, Aasen Ø L and Hoffmann A C 2011 *Chemical Engineering Science* **66** 4203–4211

[4] Bailey D L, Townsend D W, Valk P E and Maisey M N 2006 *Positron Emission Tomography: Basic Sciences* (Springer London) ISBN 9781846280078

[5] Jødal L, Le Loirec C and Champion C 2014 *Physics in Medicine and Biology* **59** 7419–7434

[6] Stewart R L, Bridgwater J and Parker D J 2001 *Chemical Engineering Science* **56** 4257–4271

[7] Broadbent C J, Bridgwater J, Parker D J, Keningley S T and Knight P 1993 *Powder Technology* **76** 317–329

[8] Leadbeater T W, Seville J P K and Parker D J 2022 *The Canadian Journal of Chemical Engineering* 1–14

[9] Blakemore D M, Govender I, McBride A T and Mainza A N 2019 *Chemical Engineering Science* **207** 780–789

[10] Nicusan A and Windows-Yule C 2020 *Review of Scientific Instruments* **91** 013329

[11] Wiggins C, Santos R and Ruggles A 2017 *Nuclear Instruments and Methods in Physics Research Section A: Accelerators, Spectrometers, Detectors and Associated Equipment* **843** 22–28

[12] Chang Y F and Hoffmann A C 2015 *Experiments in Fluids* **56** 4

[13] Casey M E and Osborne D R 2020 *Advances in PET* 71–91

[14] Chang Y F, Adamsen T C H and Hoffmann A C 2012 *IEEE International Instrumentation and Measurement Technology Conference Proceedings* 2326–2330

[15] Chang Y F 2016 Method for tracking multiple positron-emitting particles Australian Provisional Patent Application No. 2016901634

Asymptotic behaviour of the non-oscillating solutions of first order nonlinear impulsive functional-differential equations of neutral type

V I Donev

Department of Electrical Engineering, Automation and Information technology, Technical University of Sofia, branch Sliven, Bulgaria

* Corresponding Author E-mail: vanyo_ivanov@tu-sofia.bg

Abstract. This paper deals with nonlinear first-order impulsive functional-differential equations of neutral type with variable coefficients. The asymptotic behaviour of the non-oscillating solutions of such equations is investigated as a function of the values of the variable coefficient in the neutral term to serve as a basis for further studies of the oscillation of the solutions of such equations.

1. Introduction

Impulsive Ordinary Differential Equations (IODE) are a natural generalization of Impulsive Functional Differential Equations (IFDE). Generally speaking, IFDE are very interesting mixture between the Impulsive Differential Equations (IDE) (see [2], [3] and [11]) and the Functional-Differential Equations (FDE) (see [6], [7] and [10]). In their composition, IFDE possess impulses that reflect first-order discontinuities of their solutions. These impulses are very convenient for mathematical modeling of some processes in nature, subject to momentary disturbances of almost zero duration. The differential equation itself in the composition of the IFDE has deviating arguments (retarded or advanced), which is a convenient tool for simulating the dependence of the modeled processes on their history or future. Such processes mentioned above are found in optimal control theory (manipulators with controlled constraints), physics (the movement of emitting electrons), population dynamics (population growth, spread of epidemics), neural networks (pattern recognition, associative memory), biotechnology, industrial robotics, medicine, etc.

We note here that [8] is the first work where IFDE were considered.

In IFDE theory, the following classification is adopted – retarded-type equations, advanced-type equations and neutral-type equations. The most interesting and complex object of study is the impulsive functional-differential equations of neutral type, both linear (see [1], [4], [5] and [9]) and non-linear (see [12]). The neutral character of these equations is due to the highest order derivative of the unknown solution appearing in the differential equation with the argument t (the current state of the system) as well as one or more lagged and/or forward arguments (the past and/or future state of the system). However, the nonlinear IFDE of neutral type are still not widely investigated, so we pay special attention to some of them in this work.

2. Preliminary notes

In this paper we consider the nonlinear first order functional-differential equation of neutral type with variable coefficients, of the form:

$$\begin{cases} [y(t)-c(t)y(h_1(t))]' + \varepsilon q(t) f(y(t), y(h_1(t)), y(h_2(t))) = 0, \ t \neq \tau_k, \ k \in N \\ \Delta[y(\tau_k) - c_{\tau_k} y(h_1(\tau_k))] + \varepsilon q_{\tau_k} I(y(\tau_k), y(h_1(\tau_k)), y(h_2(\tau_k))) = 0, \ k \in N. \end{cases} \quad (1)$$

In what follows, we denote $J=[\tau_0,+\infty) \subseteq R^+$ and assume the following hypotheses are satisfied:

$$0 \le \tau_0 < \tau_1 < \tau_2 < \ldots < \tau_k < \ldots, \ k \in N, \ \lim_{k\to\infty} \tau_k = \infty, \ \min_{k\in N}\{\tau_{k+1}-\tau_k\} \le \max_{k\in N}\{\tau_{k+1}-\tau_k\} < \infty \quad \text{(H)}$$

$$\begin{aligned} &h_1, \ h_2 \in C^1(R^+,R^+), && h_1'(t)>0, \ h_2'(t)>0, \\ &h_1(t)<t, \quad h_2(t)<t, && \lim_{t\to\infty} h_1(t) = +\infty, \quad \lim_{t\to\infty} h_2(t) = +\infty; \end{aligned} \quad \text{(H1)}$$

$$c \in C^1(J,R), \quad c_{\tau_k} = c(\tau_k); \quad \text{(H2)}$$

$$q \in PC_\tau(J,R^+), \quad q_{\tau_k} \in R_0^+; \quad \text{(H3)}$$

$$\sum_{k=1}^{\infty} q_{\tau_k}^2 \neq 0, \quad \int_J q(s)ds + \sum_{k=1}^{\infty} q_{\tau_k} = +\infty, \quad \text{(H4)}$$

which are valid alongside one of the following options:

$$c \in C^1(J,(-\infty,0)), \quad c_{\tau_k} \in (-\infty,0]; \quad \text{(H5)}$$

$$\exists\, c_0 \in (0,1): \ c \in C^1(J,(0,c_0]), \quad 0 < c_{\tau_k} \le c_0, \ k \in N; \quad \text{(H6)}$$

$$\exists\, C_0 > 1: \ c \in C^1(J,[C_0,+\infty)), \quad c_{\tau_k} \in (C_0,+\infty), \ k \in N. \quad \text{(H7)}$$

Regarding the nonlinear nature of (1), we also introduce the following hypotheses:

$$\begin{aligned} &f \in C(J\times J\times J,R), \ u_i f(u_1,u_2,u_3)>0, \ (u_1,u_2,u_3)\neq(0,0,0), \\ &f_{u_i}'(u_1,u_2,u_3)>0, \quad \exists\, C_f \in R^+: \ 1 \le \frac{|f(u_1,u_2,u_3)|}{|u_i|} \le C_f, \ i=1,2,3; \end{aligned} \quad \text{(H8)}$$

$$\begin{aligned} &I \in C(J\times J\times J,R), \ u_i I(u_1,u_2,u_3)>0, \ (u_1,u_2,u_3)\neq(0,0,0), \\ &\exists\, C_I \in R^+: \ 1 \le \frac{|I(u_1,u_2,u_3)|}{|u_i|} \le C_I, \ i=1,2,3. \end{aligned} \quad \text{(H9)}$$

The division of points $\tau_Y = \{\tau_k\}_{k=1}^{\infty}$, $\tau_k \in (0,+\infty)$, $k \in N$ consists of the moments of impulsive effect (jump points), where the respective function reveals its discontinuities of first kind as jumps. In order to manifest these jumps of the unknown function $y(t)$, we use the notation:

$$\Delta y(\tau_k) = y(\tau_k+) - y(\tau_k-),$$

where

$$y(\tau_k-) = y(\tau_k-0) = \lim_{t\to\tau_k-0} y(t) = y(\tau_k), \quad y(\tau_k+) = y(\tau_k+0) = \lim_{t\to\tau_k+0} y(t).$$

Denote by $PC_{\tau_Y}(J,R)$ the set of all functions $u(t),\ u:J\to R$, which satisfy the following conditions:

1. u is piece-wise continuous on $[\tau_0,\tau_1)\bigcup\{(\tau_k,\tau_{k+1})\}_1^\infty$;
2. u is continuous from the left at $\tau_k\in\tau_Y$, i.e. $u(\tau_k-0)=\lim\limits_{t\to\tau_k-0}u(t)=u(\tau_k)$;
3. there exists a sequence of real numbers $\{u(\tau_k+0)\}_{k=1}^\infty$: $u(\tau_k+0)=\lim\limits_{t\to\tau_k+0}u(t)$;
4. u may have discontinuities of first kind at $\tau_k\in\tau_Y$, i.e. $\Delta u(\tau_k)=u(\tau_k+)-u(\tau_k-)\neq 0$, $k\in N$, that we characterize as down-jumps ($\Delta u(\tau_k)<0$) or as up-jumps ($\Delta u(\tau_k)>0$).

Further, we denote by $PC_{\tau_Y}{}^k(J,R),\ k\in N,$ the set of all functions $u:J\to R$, for which $\dfrac{d^k u}{dt^k}\in PC_{\tau_Y}(J,R)$ and by $i[\tau_0,t)$ – the number of points τ_k, situated in $[\tau_0,t)$.

We say, that the real valued function $y(t)$ is a <u>solution</u> of (1), if there exists a number $T_0>0$, such that $y\in PC_{\tau_Y}{}^1([h_1(T_0),+\infty),R)$, the function $z(t)=y(t)-c(t)y(h_1(t))$ is continuously differentiable for $t\geq T_0,\ t\neq\tau_Y$ and $y(t)$ satisfies (1) for all $t,\tau_k\geq T_0$.

The solution $y(t)$ of (1) is said to be <u>regular</u>, if it is defined on $[T_y,+\infty)\subseteq[T_0,+\infty)$ and $\sup\limits_{T\geq T_y}\{|y(t)|:t\geq T\}>0$.

The regular solution $y(t)$ of (1) is said to be finally positive (finally negative), if there exists a number $T_1>0$, such that $y(t)>0$ ($y(t)<0$) for $t\geq T_1$. The regular solution is said to be oscillating if it is not of constant sign, for $t\geq T_1$. Otherwise, it is said to be non-oscillating.

For our needs, we also define an auxiliary function $z(t)$, considering $y(t)$ to be a solution of (1):

$$z(t)=y(t)-c(t)y(h_1(t)),\ \Delta z(\tau_k)=\Delta y(\tau_k)-c_{\tau_k}\Delta y(h_1(\tau_k)). \qquad (*)$$

In our investigations we use the following lemma, which is a generalization of Lemma 1.5.2. in [6].

<u>Lemma 1:</u> Let $J=[t_0,+\infty)\subset R^+$ and $c,z,y:J\to R$ be such that

$$z(t)=y(t)+c(t)y(h(t)),\quad t>h(t)\geq t_0.$$

Assume there exist numbers $C_1, C_2, C_3, C_4\in R$, such that one of the following statements is true:

(A) $C_1\leq c(t)\leq 0$; (B) $0\leq c(t)\leq C_2<1$; (C) $1<C_3\leq c(t)\leq C_4$.

Suppose $h\in C^1(J,R^+),\ h'(t)>0,\ y(t)>0,\ t\in J$ and there exists $L=\lim\limits_{t\to\infty}z(t)$. Then $L=0$.

<u>Proof:</u> We shall prove the lemma when (A) holds. The proofs of cases (B) and (C) are similar and will be omitted. So, from the representation of the function $z(t)$ we have:

$$z(h^{-1}(t))-z(t)=y(h^{-1}(t))+[c(h^{-1}(t))-1]y(t)-c(t).y(h(t)).$$

Obviously $\lim\limits_{t\to\infty}\{z(h^{-1}(t))-z(t)\}=0=\lim\limits_{t\to\infty}\{y(h^{-1}(t))+[c(h^{-1}(t))-1]y(t)-c(t).y(h(t))\}$.

Therefore, because $c(t)$ is bounded, if we choose a sequence of real numbers $\{t_n\}$, such that $\lim_{n\to\infty} t_n = \infty$ and $\lim_{n\to\infty} y(t_n) = 0$, we may obtain $\lim_{t\to\infty} \{y(h^{-1}(t_n)) - c(t_n).y(h(t_n))\} = 0$.

Observe that $y(h^{-1}(t_n)) > 0$ and $c(t_n).y(h(t_n)) < 0$. This implies $\lim_{t\to\infty} \{c(t_n).y(h(t_n))\} = 0$ and therefore $L = \lim_{n\to\infty} z(t_n) = \lim_{n\to\infty} [y(t_n) + c(t_n).y(h(t_n))] = 0$. **The proof is complete.**

The next two lemmas explore the asymptotic behavior of the function $z(t)$, defined on (*), in cases, when it is constructed from non-oscillating solutions of (1).

Lemma 2: Let the hypotheses (H), (H1)-(H4), (H8) and (H9) hold and $y(t)$ is a finally positive solution of (1) when $\varepsilon = +1$. Therefore, the function $z(t)$, defined on (*), is decreasing and:

(A) If (H5) holds, then $z(t)$ is finally positive and $\lim_{t\to+\infty} z(t) = L$, $0 \le L < +\infty$, $\lim_{\tau_k\to+\infty} |\Delta z(\tau_k)| = 0$;

(B) If (H6) holds, then $z(t)$ is finally positive and $\lim_{t\to+\infty} z(t) = 0$, $\lim_{\tau_k\to+\infty} |\Delta z(\tau_k)| = 0$;

(C) If (H7) holds, then $z(t)$ is finally negative and $\lim_{t\to+\infty} z(t) = -\infty$.

Proof: Let $y(t)$ is a solution of (1) and there exists such a real number $t_0 > 0$, that $y(t)$ is defined and positive for every $t \ge t_0$. Therefore, for $t \ge t_0$, $\tau_k \ge t_0$, $k \in N$ from (1), we have:

$$\begin{cases} z'(t) = -q(t)f(y(t), y(h_1(t)), y(h_2(t))), \ t \ne \tau_k, \\ \Delta z(\tau_k) = -q_{\tau_k} I(y(\tau_k), y(h_1(\tau_k)), y(h_2(\tau_k))), \ k \in N. \end{cases} \tag{2}$$

From (2), (H2), (H3), (H8) and (H9) we conclude $z'(t) < 0$, $\Delta z(\tau_k) \le 0$. Therefore $z(t)$ is a decreasing function in every interval $(\tau_{k-1}, \tau_k]$, $k \in N$, with down-jumps (non-strict) in the impulse points τ_κ.

(A) In this case it is obvious that $z(t)$ is positive. Therefore $L = \lim_{t\to+\infty} z(t)$ exists, $L \ge 0$, and this implies $\lim_{\tau_k\to+\infty} |\Delta z(\tau_k)| = 0$.

(B) Assume for the sake of contradiction that $z(t)$ is finally negative. Therefore, $\exists\, t_1 \ge t_0$, for which $\exists\, \varepsilon(t_1) > 0$: $z(t) < -\varepsilon(t_1)$, $t \ge t_1$, i.e. $y(t) < -\varepsilon(t_1) + c_0 y(h_1(t))$, $t \ge t_1$. The last inequality is developed iteratively to $y(t) < -\varepsilon(t_1)\left(c_0 + c_0^2 + \ldots + c_0^{n-1}\right) + c_0^n y(h_1^n(t))$, $h_1^n(t) \ge t_1$, which finally leads us to the contradiction $y(t) < -\varepsilon(t_1)\dfrac{1}{1-c_0}$.

Therefore, $z(t)$ is finally positive, $L = \lim_{t\to+\infty} z(t)$ exists and $L \ge 0$. Let further integrate (2) from t_0 to t. We have:

$$\begin{cases} z(t) - z(t_0) - \sum_{t_0 \le \tau_k < t} \Delta z(\tau_k) = -\int_{t_0}^{t} q(s) f(y(s), y(h_1(s)), y(h_2(s))) ds \\ \Delta z(\tau_k) = -q_{\tau_k} I(y(\tau_k), y(h_1(\tau_k)), y(h_2(\tau_k))), \ k \in N, \end{cases}$$

which after passing to limit as $t \to \infty$ leads us to

$$L+\sum_{t_0\le\tau_k<\infty} q_{\tau_k} I(y(\tau_k), y(h_1(\tau_k)), y(h_2(\tau_k)))+\int_{t_0}^{\infty} q(s)f(y(s), y(h_1(s)), y(h_2(s)))ds = z(t_0). \quad (3)$$

The result in (3) shows that $\sum_{t_0\le\tau_k<\infty} q_{\tau_k} I(y(\tau_k), y(h_1(\tau_k)), y(h_2(\tau_k)))$ is summable and $\int_{t_0}^{\infty} q(s)f(y(s), y(h_1(s)), y(h_2(s)))ds < \infty$. Considering hypotheses (H4), (H8) and (H9), we deduce $\liminf_{t\to\infty}|y(t)| = 0$. However, then condition (B) from Lemma 1 holds, whence it follows $\lim_{t\to+\infty} z(t) = 0, \ \lim_{\tau_k\to+\infty} |\Delta z(\tau_k)| = 0$.

(C) Assume for the sake of contradiction that $z(t)$ never becomes negative by sign in the intervals $(\tau_k, \tau_{k+1}]$, $k \in N$, i.e. $y(t)-c(t)y(h_1(t)) \ge 0$, $L = \lim_{t\to+\infty} z(t)$ exists and $L \ge 0$. But $y(t) \ge c(t)y(h_1(t)) \ge C_0 y(h_1(t))$, $t \ne \tau_k$, $k \in N$, because $c(t) \ge C_0 > 1$. This iteratively leads us to $y(h_1^{\,n}(t)) \ge C_0^{n+1} y(h_1(t))$, $t \ne \tau_k$, $k \in N$, $n \in N$, whence it follows that $y(t)$ increases to infinity, regardless of the jumps in the impulse points, i.e. $\lim_{t\to+\infty} y(t) = \infty$. But, when we integrate (2) from t_0 to t, we obtain (3), which is a contradiction, considering hypothesis (H4), (H8) and (H9) and taking into account that $\liminf_{t\to\infty}|y(t)| > 0$. Therefore, the function $z(t)$ is finally negative. Thus, if $\lim_{t\to+\infty} z(t) = L$, then we have $-\infty < L < 0$ or $L = -\infty$. Assume that $-\infty < L < 0$. Then $0 < \liminf_{t\to\infty} y(t)$ and $\lim_{\tau_k\to+\infty} |\Delta z(\tau_k)| = 0$. However, when we integrate (2) from t_0 to t, we have:

$$L = z(t_0) - \sum_{t_0\le\tau_k<\infty} q_{\tau_k} I(y(\tau_k), y(h_1(\tau_k)), y(h_2(\tau_k))) - \int_{t_0}^{\infty} q(s)f(y(s), y(h_1(s)), y(h_2(s)))ds.$$

Here we see, considering monotonicity of f and hypotheses (H8) and (H9), that we have a contradiction with hypothesis (H4), as follows:

$$L \le z(t_0) - \liminf_{t\to\infty} y(t)\left(\sum_{t_0\le\tau_k<\infty} q_{\tau_k} + \int_{t_0}^{\infty} q(s)ds\right).$$

Therefore, $\lim_{t\to+\infty} z(t) = -\infty, \ \Delta z(\tau_k) \le 0$. **The proof is complete.**

Lemma 3: Assume that hypotheses (H), (H1)-(H4), (H8), (H9) are satisfied and $y(t)$ is finally positive solution of (1) when $\varepsilon = -1$. Therefore, the function $z(t)$, defined on (*), is increasing and:

(A) If (H5) holds, then $z(t)$ is finally negative and $\lim_{t\to+\infty} z(t) = L, \ 0 \ge L > -\infty, \ \lim_{\tau_k\to+\infty} |\Delta z(\tau_k)| = 0$;

(B) If (H6) holds, then $z(t)$ is finally negative and $\lim_{t\to+\infty} z(t) = 0, \ \lim_{\tau_k\to+\infty} |\Delta z(\tau_k)| = 0$;

(C) If (H7) holds, then $z(t)$ is finally positive and $\lim_{t\to+\infty} z(t) = +\infty$.

The proof is analogous and reciprocal to the proof of Lemma 2, respectively.

3. Main results

The following results are related to the asymptotic behavior of the non-oscillatory solutions of (1).

THEOREM 1: Assume that hypotheses (H), (H1)-(H4) and (H5) are satisfied and $y(t)$ is finally non-oscillating solution of (1) when $\varepsilon=+1$. Therefore:

(A) $y(t)$ is finally bounded function, i.e. $0<\liminf_{t\to+\infty}|y(t)|\le \limsup_{t\to+\infty}|y(t)|<+\infty$ and $\exists\ L_1, L_2\in R^+ : L_1\le|\Delta y(\tau_k)|\le L_2,\ k\in N$;

(B) If (H8) and (H9) hold and $\lim_{t\to+\infty} c(t)=0$, then $\lim_{t\to+\infty} y(t)=0,\ \lim_{\tau_k\to+\infty}|\Delta y(\tau_k)|=0$;

(C) If (H8) and (H9) hold and $\lim_{t\to+\infty} c(t)=-\infty$, then $\lim_{t\to+\infty} y(t)=0,\ \lim_{\tau_k\to+\infty}|\Delta y(\tau_k)|=0$.

Proof: Assume that $y(t)$ is a solution of (1) and there exists such a real number $t_0\ge\tau_0$, that $y(t)$ is defined and positive for every $t\ge t_0$. The case with a finally negative solution can be proved analogously. From Lemma 2 we know that $z(t)$, defined on (*), is finally positive decreasing function, with down-jumps (non-strict) in the impulse points $\tau_k\in\{\tau_i\}_1^\infty$, as $L=\lim_{t\to+\infty} z(t)$ exists, assuming $\lim_{\tau_k\to+\infty}|\Delta z(\tau_k)|=0$. Thus, $L\ge 0$.

(A) It is obvious that $z(t)\ge y(t)$. This means that regardless of the jumps of $y(t)$, we have $0\le\liminf_{t\to+\infty}|y(t)|\le \limsup_{t\to+\infty}|y(t)|<+\infty$ and $\exists\ M_1, M_2\in R^+ : M_1\le|\Delta y(\tau_k)|\le M_2,\ k\in N$.

(B) If we integrate (2) from t_0 to t and pass to the limit as $t\to\infty$, we obtain (3). The result in (3) shows that $\sum_{t_0\le\tau_k<\infty} q_{\tau_k} I(y(\tau_k), y(h_1(\tau_k)), y(h_2(\tau_k)))$ is summable and with respect to the integral we have $\int_{t_0}^{\infty} q(s)f(y(s), y(h_1(s)), y(h_2(s)))ds<\infty$. This implies $\liminf_{t\to\infty}|y(t)|=0$. Let us suppose that:

$$\sum_{t_0\le\tau_k<t} q_{\tau_k} I(y(\tau_k), y(h_1(\tau_k)), y(h_2(\tau_k)))=L_I>0,\quad \int_{t_0}^{t} q(s)f(y(s), y(h_1(s)), y(h_2(s)))ds=L_f>0$$

and consider the left side of (3) as a function of its lower limit, i.e.

$$z(\tilde t)=L+\sum_{\tilde t\le\tau_k<\infty} q_{\tau_k} I(y(\tau_k), y(h_1(\tau_k)), y(h_2(\tau_k)))+\int_{\tilde t}^{\infty} q(s)f(y(s), y(h_1(s)), y(h_2(s)))ds,\ \tilde t\ge t_0. \quad (4)$$

Assume that $y(t)$ does not converge to zero, i.e. $0=\liminf_{t\to+\infty}|y(t)|<\limsup_{t\to+\infty}|y(t)|<+\infty$ and $\exists\ M_1, M_2\in R^+ : M_1\le|\Delta y(\tau_k)|\le M_2,\ k\in N$. Therefore, there exist $M_3>0$ and an increasing sequence $\{t_n\}_1^\infty\to\infty$, such that

$$y(h_1(t_n))>M_3,\ n\ge 1. \quad (5)$$

However, if $\lim_{t\to+\infty} c(t)=0$, then $\{c(t_n)y(h_1(t_n))\}_1^\infty\to 0$, because $y(t)$ is bounded function. Hence, considering (4), we have $L>z(t_n)\ge y(t_n)\ge L+L_I+L_f+c(t_n)y(h_1(t_n))$, which leads us to the contradiction $0>L_I+L_f$.

Therefore, $0=\liminf_{t\to+\infty}|y(t)|=\limsup_{t\to+\infty}|y(t)|$, i.e. $\lim_{t\to+\infty} y(t)=0$, $\lim_{\tau_k\to+\infty}|\Delta y(\tau_k)|=0$.

(C) Following the logic in (B), we have (5). But if $\lim_{t\to+\infty} c(t)=-\infty$, then $\exists\, \nu\in N:\ -c(t_n)y(h_1(t_n))>L+L_I+L_f$, $n\ge\nu$, which contradicts to (4). Therefore, $0=\liminf_{t\to+\infty}|y(t)|=\limsup_{t\to+\infty}|y(t)|$, i.e. $\lim_{t\to+\infty} y(t)=0$, $\lim_{\tau_k\to+\infty}|\Delta y(\tau_k)|=0$. **The proof is complete.**

THEOREM 2: Assume that hypotheses (H), (H1)-(H4) and (H6) are satisfied and $y(t)$ is finally non-oscillating solution of (1) when $\varepsilon=+1$. Therefore:

(A) $y(t)$ is finally bounded function, i.e. $0<\liminf_{t\to+\infty}|y(t)|\le \limsup_{t\to+\infty}|y(t)|<+\infty$ and $\exists\ L_1,\, L_2\in R^+ : L_1\le|\Delta y(\tau_k)|\le L_2,\ k\in N$;

(B) If $\lim_{t\to+\infty} c(t)=\bar{c}_0\le c_0$, then $\lim_{t\to+\infty} y(t)\ge 0$, $\lim_{\tau_k\to+\infty}|\Delta y(\tau_k)|=0$;

(C) If (H8) and (H9) hold, then $\liminf_{t\to+\infty}|y(t)|=0$;

(D) If (H8) and (H9) hold and $\lim_{t\to+\infty} c(t)=\bar{c}_0\le c_0$, then $\lim_{t\to+\infty} y(t)=0$, $\lim_{\tau_k\to+\infty}|\Delta y(\tau_k)|=0$.

Proof: Assume that $y(t)$ is a solution of (1) and there exists such a real number $t_0\ge\tau_0$, that $y(t)$ is defined and positive for every $t\ge t_0$. The case with finally negative solution can be proved analogously. From Lemma 2 we know that $z(t)$ is finally positive decreasing function, with down-jumps (non-strict) in the impulse points $\tau_\kappa\in\{\tau_i\}_1^\infty$, as $L=\lim_{t\to+\infty} z(t)$ exists, assuming $\lim_{\tau_k\to+\infty}|\Delta z(\tau_k)|=0$. Thus, $L=0$.

(A) From $\lim_{t\to+\infty} z(t)=L=0$, $\lim_{\tau_k\to+\infty}|\Delta z(\tau_k)|=0$ follows that $\exists\ L_1>0$ and $\exists\ t_1\ge t_0$: $0<z(t)<L_1$, $t\ge t_1$. Then we have:

$$y(t)<c_0 y(h_1(t))+L_1\ ,\ t\ge t_1. \tag{6}$$

Assume that $y(t)$ is not bounded. Then there exists such an increasing sequence $\{t_i\}_{i=1}^\infty\to+\infty$, for which the corresponding sequence $\{y(t_i)\}_{i=1}^\infty\to+\infty$, where $y(t_i):=\max\{y(t): t_1\le t\le t_i\}$, $i\in N$. However, using (6), we see that $y(t_i)<c_0 y(h_1(t_i))+L_1<c_0 y(t_i)+L_1$, $i\in N$, which is a contradiction. Therefore, $y(t)$ is finally bounded function, i.e. $0\le\liminf_{t\to+\infty}|y(t)|\le\limsup_{t\to+\infty}|y(t)|<+\infty$ and $\exists\ M_1,\, M_2\in R^+ : M_1\le|\Delta y(\tau_k)|\le M_2,\ k\in N$.

(B) Because $y(t)$ is bounded, let $m=\liminf_{t\to+\infty} y(t)$, $M=\limsup_{t\to+\infty} y(t)$, where $m\le M$. It is obvious that $m\ge\dfrac{L-m}{-\bar{c}_0}$ and $M\le\dfrac{L-M}{-\bar{c}_0}$. So, we conclude that $m(1-\bar{c}_0)\ge L\ge M(1-\bar{c}_0)$, i.e. $m\ge M$, which establishes that there exists $\lim_{t\to+\infty} y(t)\ge 0$, whence it follows $\lim_{\tau_k\to+\infty}|\Delta y(\tau_k)|=0$.

(C) If we integrate (2) from t_0 to t and pass to the limit as $t \to \infty$, we obtain (3). The result in (3) shows that $\sum_{t_0 \le \tau_k < \infty} q_{\tau_k} I(y(\tau_k), y(h_1(\tau_k)), y(h_2(\tau_k)))$ is summable, and with respect to the integral we have $\int_{t_0}^{\infty} q(s) f(y(s), y(h_1(s)), y(h_2(s)))ds < \infty$. This implies $\liminf_{t\to\infty} |y(t)| = 0$.

(D) Combining the proofs of (2) and (3), we prove (4). **The proof is complete.**

THEOREM 3: Assume that hypotheses (H), (H1)-(H4) and (H7) are satisfied and $y(t)$ is finally non-oscillating solution of (1) when $\varepsilon = +1$. Therefore, if $\exists\, C_1 > 1 : c(t) \le C_1$, then $\limsup_{t\to+\infty} y(t) = +\infty$.

Proof: Assume that $y(t)$ is a solution of (1) and there exists such a real number $t_0 \ge \tau_0$, that $y(t)$ is defined and positive for every $t \ge t_0$. The case with finally negative solution can be proved analogously. From Lemma 2 we know that $z(t)$ is finally negative decreasing function, with down-jumps (non-strict) in the impulse points $\tau_\kappa \in \{\tau_i\}_1^\infty$, as $z(t) \xrightarrow[t\to\infty]{} -\infty$.

Thus, $z(t) = y(t) - c(t)y(h_1(t))$, $t \ge t_0$, i.e. $-C_1 y(h_1(t)) \le -c(t)y(h_1(t)) \le z(t)$, $t \ge t_0$.

After passing to the limit as $t \to +\infty$, we have $C_1 y(h_1(t)) > -z(t) \xrightarrow[t\to\infty]{} +\infty$, that proves $\limsup_{t\to+\infty} y(t) = +\infty$. **The proof is complete.**

Conclusion

The results in this paper reveal the basic asymptotic behavior of the non-oscillating solutions of nonlinear first-order impulsive functional-differential equations of neutral type with variable coefficients in order to serve as a basis for further studies of the oscillation of such solutions.

Acknowledgement

The author would like to thank the Research and Development Sector at the Technical University of Sofia for the financial support.

References

[1] Yeniçerioğlu A F 2019 *J. Math. Anal. Appl.* **479**(2) 2196–2213
[2] Bainov D D, Lakshmikantham V and Simeonov P S 1989 *Theory of Impulsive Differential Equations* (Singapore: World Scientific)
[3] Bainov D D and Simeonov P S 1998 *Oscillation Theory of Impulsive Differential Equations: Periodic solutions and Applications* (Orlando: International Publications)
[4] Guo S, Shen Y and Shi B 2014 *Abstr. Appl. Anal.* **2014** 1–5
[5] Abasiekwere U A and Moffat I U 2017 *IJMTT* **48**(2) 128–132
[6] Gyori I and Ladas G 1991 *Oscillation Theory of Delay Differential Equations with Applications* (Oxford: Clarendon Press)
[7] Erbe L H, Kong Q and Zhang B G 1995 *Oscillation Theory for Functional Differential Equations* (New York: Marcel Decker Inc.)
[8] Gopalsamy K and Zhang B G 1989 *J. Math. Anal. Appl.* **139**(1) 110–122
[9] Grammatikopoulos M K, Dimitrova M B and Donev V I 2007 *Technical Report* (Greece: University of Ioannina) 16 183–193
[10] Hale J K 1977 *Theory of Functional Differential Equations* (New York: Springer Verlag)
[11] Samoilenko A M and Perestyuk N A 1987 *Differential Equations with Impulse Effect* (Kiev: Vishcha Shkola) (in Russian)
[12] Santra S S 2019 *Eng. Math. Lett.* **2019** Article ID 6

Approximate Analytical Solution for Turbulent Flow in Channel

Alex Fedoseyev
Ultra Quantum Inc, Huntsville, Alabama, USA
E-mail: af@ultraquantum.com

Abstract. In this work, the approximate analytical solutions of the Generalized Hydrodynamic Equations (GHE) for turbulent flow in channel is presented. The general solution is a superposition of the laminar (parabolic) and turbulent (superexponential) solutions. The analytical solution compares well with the experimental data by Van Doorne (2007) and Nikuradse (1933).

It is proposed to explain the nature of turbulence as oscillations between the laminar (parabolic) and turbulent (superexponential) solutions. The suggestion is confirmed by comparing the results for turbulent velocity fluctuations of the analytical formula with the experiment by Van Doorne (2007). The Navier-Stokes equations do not have the superexponential solution.

The obtained analytical solution provides a complete structure of the turbulent boundary layer that compares well with the experiments by Wei and Willmarth (1989).

1. Introduction

Generalized Hydrodynamic Equations (GHE) have been proposed by Boris Alexeev in 1994 [1]. They were used for the simulations of incompressible viscous flows for a wide range of problems and flow parameters, including high Reynolds number turbulent flows with thin boundary layers in 3D driven cavity flow at Re = 3200 and 10,000, 2D backward facing step flow at Re = 132,000, flow in channels for Reynolds number up to $Re = 10^6$, magnetohydrodyamic flows [7, 8, 9, 10, 12], resulting in good agreement with experiments [20], [19].

GHE model has been applied to compressible hypersonic flows that exhibit both continuum and non-continuum flow regimes. The shock wave (bow shock) can be detached from the vehicle at high altitude, and near boundary slip-flow is typical for such regimes. Results for hypersonic GHE model have been reported in [13, 14, 15] are in close agreement with the experiments by [3, 17, 18] and other works even for highly rarefied flows.

In this work the approximate analytical solution of GHE for turbulent flow in channel is presented and compared with the experimental data.

Generalized Hydrodynamic Equations (GHE) are based on a new set of conservation equations obtained from a generalized version of the Boltzmann equation (GBE) by Alexeev that includes more details of the molecular collision processes [1], [2]. The model combines continuum-to-free molecular flow physics in one consistent formulation by accounting for the kinetic effects (intermediate Knudsen number, fluctuations and turbulence) in the continuum approximation. A brief outline of the basic ideas of GBE and derivation of governing Generalized Hydrodynamic Equations is presented in Section 1.1.

1.1. Generalized Boltzmann Transport Equation (GBE)

Physical derivation of the standard Navier-Stokes equations (NS) can be obtained from the kinetic theory of gases, which is based on the solution to the Boltzmann transport equation for space-time evolution of particle velocity distribution function, f, written in the form

$$\frac{Df}{Dt} = J, \tag{1}$$

where

$$\frac{D}{Dt} = \frac{\partial}{\partial t} + \mathbf{v} \cdot \frac{\partial}{\partial \mathbf{r}} + \mathbf{F} \cdot \frac{\partial}{\partial \mathbf{v}}$$

represents material derivative in space, velocity space and time and J is the collision integral ([2], p.11), where $\mathbf{v}$ and $\mathbf{r}$ are the velocity and the radius-vector of the particle, respectively, and $\mathbf{F}$ is the force acting on the particle. The standard Boltzmann transport equation takes into account the changes in distribution function f on hydrodynamic and mean time between collision scales of infinitesimal particles. Accounting for a third time scale associated with finite dimensions of interacting particles gives rise to an additional term in the Boltzmann transport equation resulting in a generalization of the form as

$$\frac{Df}{Dt} - \frac{D}{Dt}(\tau^* \frac{Df}{Dt}) = J, \tag{2}$$

where τ^* is the mean time between particle collisions. The new term is thermodynamically consistent and is proportional to the Knudsen number, $Kn = l/L$ (l is the mean free path between collisions, and L is the hydrodynamic scale), and therefore in the hydrodynamic limit, to viscosity. More details on the GBE are provided in Alexeev's book [2].

2. Generalized Hydrodynamic Equations

Hydrodynamic equations can be obtained from Eq. (2) by multiplying the latter by the standard collision invariants (mass, momentum, energy) and integrating the result in the velocity space. These equations are valid for incompressible viscous flow. They have the following form and have originally been present in [12]:

$$\frac{\partial \mathbf{V}}{\partial \mathrm{t}} + (\mathbf{V}\nabla)\mathbf{V} - Re^{-1}\nabla^2\mathbf{V} + \nabla \mathrm{p} - \mathbf{F} = \tau\left\{2\frac{\partial}{\partial \mathrm{t}}(\nabla \mathrm{p}) + \nabla^2(\mathrm{p}\mathbf{V}) + \nabla(\nabla \cdot (\mathrm{p}\mathbf{V}))\right\} \tag{3}$$

while continuity equation is

$$\nabla \cdot \mathbf{V} = \tau\left\{\mathbf{2}\frac{\partial}{\partial \mathbf{t}}(\nabla \cdot \mathbf{V}) + \nabla \cdot (\mathbf{V}\nabla)\mathbf{V} + \nabla^{\mathbf{2}}\mathrm{p} - \nabla \cdot \mathbf{F}\right\} \tag{4}$$

where $\mathbf{V}$ and p are nondimensional velocity and pressure, $Re = V_0 L/\nu$ - the Reynolds number, V_0 - velocity scale, L - hydrodynamic length scale, ν - kinematic viscosity, $\mathbf{F}$ is a body force and a nondimensional $\tau = \tau^* L^{-1} V_0$. Note that the right-hand side of (3) is the divergence of the fluctuation part of flow velocity, expressed explicitly through original primitive variables according to [1].

We made the following assumptions deriving Eq. (3, 4):

- τ is assumed to be constant.
- the nonlinear terms of the third order in the fluctuations, and the terms of the order τ/Re, are neglected. The focus is on large Re numbers.

- Assumed slow flow variation, so second derivatives in time are neglected.

Additional boundary conditions on walls are for fluctuations to be zero. The following boundary condition for pressure on walls is

$$(\nabla \mathrm{p} - \mathbf{F}) \cdot \mathbf{n} = 0, \tag{5}$$

where $\mathbf{n}$ is a wall normal.

Note, that the dimension of a product $\tau^*\nu$ is a square of length. We introduce fluctuation length scale l, $l^2 = \tau^*\nu$ and rewrite nondimensional τ as $\tau = l^2 L^{-2} Re = K \cdot Re$, where $K = l^2/L^2$. The value of τ^* is a material property and not known in advance, but we provide speculation on a choice of τ^* value at the end of Section 5 and in the discussion of results, Section 7.

The obtained GHE model is not a turbulence model, and no additional equations are introduced. Kinetic effects (small flow scales) have been successfully captured with the GHE. The obtained small scale of turbulence compared well with observations in experiments by Koseff and Street [20]. 2D and 3D Naver-Stokes solutions and k-ε turbulent model solutions have been outperformed by GHE results [12].

In this paper, to obtain the analytical solution, we further simplify the GHE equations: (a) temporal derivatives are neglected in the fluctuations, (b) the nonlinear terms are neglected in the fluctuations.

2.1. Governing Equations for 2D Stationary Incompressible Flow

The case of 2D statinary incompressible fluid flow is considered. The equations are taken from [12] where they are presented explicitly, and further simplification is done by dropping all the terms (with coefficient τ) in momentum equations, and keeping only Laplacian term in the continuity equation.

The resulting continuity equation of GHE model is the following:

$$u_x + v_y = \tau \nabla^2 p \tag{6}$$

while the momentum equations are :

$$u_t + u\, u_x + v\, u_y + p_x = Re^{-1} \nabla^2 u \tag{7}$$

$$v_t + u\, v_x + v\, v_y + p_y = Re^{-1} \nabla^2 v \tag{8}$$

where $Re = LV_0/\nu$ is Reynolds number, L is the length scale, V_0 is velocity scale, ν is kinematic viscosity, $\tau = \tau^* V_0/L$ is the nondimensional time scale from GHE (τ^* is the dimensional time).

One may point out that the continuity Eq. (6) is not correct, and the mass is not conserved for incompressible fluid (may have sources and sinks). Let us note, that

- Eq.(6), as well as Eq.(7, 8) are obtained by a standard procedure, by multiplying Eq. (2) by standard collision invariants (mass, momentum, energy) and integrating the result in the velocity space. Then the simplification was done, and only the highest derivative was left in Eq. (6).
- Recall that the particles now are not the material points, but finitely-sized particles, that may be partially in and partially out of any control volume. "The characteristic length of this control volume is much bigger than mean free path between collisions but much smaller than hydrody-namic scale" ([2], p.99). The pressure Laplacian term in Eq.(6) is the consequence of GBE.

- When solving Eq.(6, 7, 8) numerically, the value of div(**V**) for the problems with Laplacian term and without that term have been verified [11]. The values of div(**V**) have been small and of the same order of magnitude in both cases, so no sources and sinks have been observed for wide range of flows. Still the results using the Navier-Stokes equation with div(**V**)=0 were far from the experimental data while the GHE results with $div(\mathbf{V}) = \tau \nabla^2 p$ fit the experimental data well [11].

3. Laminar Flow Solution of GHE

The channel flow problem is a stationary flow in x direction in a 2D (horizontal) channel of width L with a center line velocity V_0. So the time derivatives are dropped in Eq.(7, 8). We consider the flow where everything is the same in any x cross section, so all the partial derivatives in x are zero, except for the applied pressure p, and the pressure gradient is p_x (given). We assume a symmetry about a center line $y = L/2$, and will solve the problem in the half of the domain. The boundary conditions are: $u = 0$, $v = 0$ and the normal derivative of the pressure $p_n = 0$ at the wall $y = 0$; $u = V_0$, and $v = 0$, $\frac{\partial p}{\partial y} = 0$ (the symmetry conditions) at $y = L/2$.

A flow with $p_x =$ const has a parabolic velocity u profile:

$$U_L = u(y) = 4V_o y(L - y)/L^2, \tag{9}$$

where $V_0 = -\frac{1}{8} Re\; p_x L^2$, and $v = 0$ everywhere.
This solution of Eq.(6),(7),(8) is also a solution of the Navier-Stokes equations, as $\nabla^2 p = 0$ in Eq.(6) in this case.

The equations (6),(7),(8) have a second solution, that was initially discovered numerically for a flow in 2D channel [11], Figure 1(a): both turbulent and laminar numerical solutions are shown for $Re = 5000$.

In this paper the second solution is obtained analytically, it is called a turbulent solution, and it is a super exponential function:

$$U_T = V_0 \left(1 - e^{1-e^{y/\delta}}\right). \tag{10}$$

Figure 1(b) shows example of both U_L (Eq.(9), parabolic, green line) and U_T (Eq.(10), super exponential, blue line) solutions for laminar and turbulent ($\delta = 0.025$) flows respectively. It is quite interesting that U_T for $\delta = 0.280$ (red dashed line) coincides very well with U_L (maximum deviation is 1.4%).

In Section 4 it is explained how to obtain the exact and approximate analytical solution (second solution). The construction of general solution of GHE is proposed in Section 5, and a comparison of this analytical solution with the experimental data is provided in Section 6.

4. Analytical Solution of GHE

A general solution can be constructed from a particular solution of a problem with non-zero right hand side (pressure gradient) and general solution with zero pressure gradient. Below, it is described how to find a second solution of GHE and then obtain a general solution.
We consider stationary flow regime

$$u\, u_x + v\, u_y + p_x = Re^{-1}(u_{xx} + u_{yy}) \tag{11}$$
$$u\, v_x + v\, v_y + p_y = Re^{-1}(v_{xx} + v_{yy}) \tag{12}$$
$$u_x + v_y = \tau(p_{xx} + p_{yy}) \tag{13}$$

Let us designate $\alpha = Re^{-1}$ and assume $u_x = v_x = 0, p_x =$cont. The boundary conditions are at $y = 0$: $u, v = 0$, and at $y = L/2$: $u = V_0, v = 0$ (Section 3).

In the Section 4.1 we obtain explicit analytical expressions for the approximate solution.

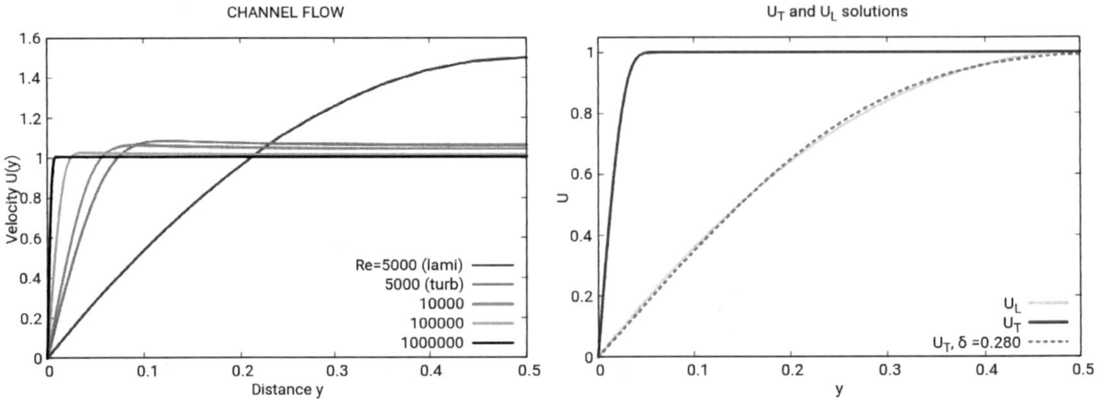

(a) (b)

Figure 1. (a) Simulation of 2D channel flow where the turbulent solution has been discovered, Fedoseyev and Alexeev (2010) : horizontal velocity U(x), for $Re = 5 \cdot 10^3$ both laminar and turbulent solutions are shown (the brown and red lines), $10^4, 10^5$ and 10^6 (each solution has the same total flux). (b) Example of solutions for laminar and turbulent flows, U_L, Eq.(9) (parabolic solution, green line), and U_T, Eq.(10) (super exponential solution, blue line, $\delta = 0.025$). It is interesting to note that U_T for $\delta = 0.280$ (red dashed line) coincides very well with U_L.

4.1. Approximate Analythical Solution of GHE

Let us designate $\Psi = v_y$, $\alpha = Re^{-1}$ and assume $u_x = v_x = 0, p_x =$cont, and neglect small $v\, v_y$ term. Then

$$p_x = \alpha u_{yy} - v u_y \tag{14}$$

$$p_y = \alpha \Psi_y \tag{15}$$

$$\tau p_{yy} = \Psi \tag{16}$$

Substituting Eq.(15) into Eq.(16) obtain

$$\alpha \tau \Psi_{yy} - \Psi = 0.$$

The solution is $\Psi = Ae^{\pm y/\sqrt{\alpha\tau}}$, and recalling what is $\Psi = v_y$, then we get $v_y = Ae^{\pm y/\sqrt{\alpha\tau}}$. Integrating that one obtains $v = A_1 e^{\pm y/\sqrt{\alpha\tau}} + B_1$, and then applying boundary condition $v = 0$ at y=0, obtain the solution v:

$$v = A(1 - e^{\pm y/\sqrt{\alpha\tau}}), \tag{17}$$

Here

$$\alpha\tau = \tau Re^{-1} = \frac{\tau^* V}{L} \cdot \frac{\nu}{VL} = \frac{\tau^* \nu}{L^2} = \delta^2,$$

where we introduced δ,

$$\delta = \frac{\sqrt{\tau^* \nu}}{L} \tag{18}$$

4.2. Case of the Positive Sign in the Exponent

In this case the solution for $v = A(1 - e^{y/\delta})$ grows exponentially, still for $y \leq L/2$ it is limited. So we try to get the solution u using obtained solution for v. Let us recall Eq.(14) and put the solution for v there, we have:

$\alpha u_{yy} - A(1 - e^{y/\delta})u_y - p_x = 0$, where $p_x = 0$, as we search for a solution of a homogeneous equation

$$\alpha u_{yy} - A(1 - e^{y/\delta})u_y = 0, \tag{19}$$

To integrate once in y we rewrite Eq.(19) as:

$$\frac{u_{yy}}{u_y} = \frac{A}{\alpha}(1 - e^{y/\delta}),$$

and obtain integral as

$$\ln u_y = \frac{A}{\alpha}(y - \delta e^{y/\delta}) + C_1$$

or

$$\ln u_y = \frac{A\delta}{\alpha}(y/\delta - e^{y/\delta}) + C_1$$

or

$$u_y = \exp(\frac{A\delta}{\alpha}(y/\delta - e^{y/\delta}) + C_1) \tag{20}$$

The explicit expression for the solution of $u(y)$ (integral of u_y, Eq.(20)) is available if

$$A = \frac{\alpha}{\delta}. \tag{21}$$

Since A is an arbitrary constant, we may choose A as in (21). With this choice, the general solution is

$$u(y) = -\delta e^{C_1} e^{-e^{y/\delta}} + C_2.$$

If $u(0) = 0$, and $u(\infty) = V_0$, then the constants C_1 and C_2 are

$$C_2 = V_0,$$

and

$$\delta e^{C_1} = V_0 e^1.$$

We get the solution for $u(y)$ that we call a turbulent solution U_T as:

$$U_T = V_0 \left(1 - e^{1-e^{y/\delta}}\right),$$

that is limited, $u(0) = 0$, $u(L/2) \approxeq V_0$, and is acceptable.

4.3. Case of the Negative Sign in the Exponent

Again recall Eq.(14) and putting the solution for v there, we have:

$\alpha u_{yy} + A(1 - e^{-y/\delta})u_y - p_x = 0$, where $p_x = 0$, as we seek a solution of a homogeneous equation

$$\alpha u_{yy}/u_y = -A(1 - e^{-y/\delta}).$$

Solution for u_y :

$\ln u_y = -Ay - A\delta e^{-y/\delta} + C$

or

$\ln u_y = -A\delta(y/\delta + e^{-y/\delta}) + C,$

and

$u_y = De^{-A\delta(y/\delta+e^{-y/\delta})}$, where $D = e^C$.

Integrating u_y we get

$$u = D\int e^{-A\delta(y/\delta+e^{-y/\delta}}dx = A_1\left[e^{y/\delta+e^{-y/\delta}} - Ei(e^{-y/\delta})\right] + A_2,$$

where $A_1 = \delta e^{-A\delta+C}$.

As $u(0) = 0$, and $Ei(0) = e^0 = 1$, then $A_2 = 0$ and the solution u is the following:

$$u = A_1\left[e^{y/\delta+e^{-y/\delta}} - Ei(e^{-y/\delta})\right], \tag{22}$$

where Ei(x) is the integral exponential function. This solution for u grows indefinitely, so it is not physical, and will be disregarded.

5. General Solution for the Turbulent Flow

The general solution is proposed as a linear superposition of two solutions, laminar and turbulent:

$$U_{GHE} = \gamma U_T + (1 - \gamma)U_L, \tag{23}$$

where the coefficients γ and $(1 - \gamma)$ are introduced, and $U_{GHE} = V_0$ at the center line. The first term containing U_T grows superexponentially in the boundary layer, while the second term containing U_L is nearly zero. Outside the boundary layer, the first term is constant, and the second term starts to grow.

Substituting U_T Eq.(10) and U_L Eq. (9) we get

$$U_{GHE} = V_0\left[\gamma\left(1 - e^{1-e^{y/\delta}}\right) + (1 - \gamma)4y(L - y)/L^2\right]. \tag{24}$$

5.1. Calculation of parameter γ from experimental data

The value of γ can be calculated from the Eq.(14) at some point for small y, for example, $y = 0$, then $v = 0$.

Substituting the solution U_{GHE} into Eq.(14) we get

$$\alpha U_{GHE},_{yy} = p_x \tag{25}$$

and substituting the expression of U_{GHE} from Eq.(23):

$$\alpha(\gamma U_{T},_{yy} + (1-\gamma)U_{L},_{yy}) = p_x,$$

or

$$\gamma(\alpha(U_{T},_{yy} - U_{L},_{yy}) + \alpha U_{L},_{yy} = p_x,$$

and

$$\gamma = (p_x - \alpha U_{L},_{yy})/(\alpha(U_{T},_{yy} - U_{L},_{yy})) \tag{26}$$

evaluated at $y = 0$. Here $U_{T},_{yy}$ and $U_{L},_{yy}$ are the 2nd partial derivatives of U_T and U_L in y, $U_{T},_{y}$ in y, and the experimental data for p_x is used. As $U_{T},_{yy} = 0$ at $y = 0$, and recalling that $\alpha = 1/Re$, we obtain

$$\gamma = \frac{U_{L},_{yy} - p_x Re}{U_{L},_{yy}} = 1 - \frac{p_x Re}{U_{L},_{yy}}. \tag{27}$$

We can see from Eq.(27) that the parameter $\gamma \to 0$, if $Re \to 0$, as $p_x \to \alpha U_{L},_{yy}$ (laminar solution). The parameter γ grows for $Re \to \infty$ as we can see from the experimental data in Section 6.

5.2. Finding parameters by fitting the experimental data

The parameter δ and the value of γ in the comparisons with the experiments below will be chosen to fit the experimental data, as we do not have the material property τ^* for particular liquids used in the experiments, as $\delta = \sqrt{\tau^*\nu}/L$. The value of γ can be calculated from the parameters of experiments using Eq.(27) if the values of p_x and L are known.

Each of the two laminar and turbulent solutions has different similarity parameters. The laminar parabolic solution depends on the Re number, while the turbulent solution depends on the parameter $\delta = \sqrt{\tau^*\nu}/L$.

The parameter τ^* (dimension of time) is a mean time between particle collisions. For gases this mean time between collisions can be calculated using hard-sphere model of particles at pressure p, and kinematic viscosity μ as $\tau^* = \Pi\mu/p$, where Π is a constant, $\Pi = 0.786$ [4]. For liquids this is more complicated (see [2], p.324 discussion about Frenkel work [16]), there is no explicit formula, and the experimental measurements are needed. It can be found indirectly by fitting the experimental velocity data for δ as $\tau^* = \delta^2 L^2/\nu$.

6. Comparison of Analytical Solution with Experiments

In this Section we will compare the analytical solution with the experimental data from several high quality experiments. All the experiments represent the turbulent flows, and our analytical solution for the flow velocity will correspond to the mean flow velocity from the experiment.

6.1. Pipe Flow Experiment by Van Doorne at Re=7200

Here a comparison will be provided for the approximate GHE and other analytical solutions with the pipe flow data Van Doorne 2004, 2007 [5, 6] (Prof. Bruno Eckhardt group). The analytical solution was obtained for 2D flow in Cartesian coordinates, but still we can use it for pipes (cylindrical coordinates) as $\delta \ll 1$. Figure 2 shows the experimental data digitized from [5], and a number of different plots from (a) laminar (parabolic) flow profile (green line); (b) turbulent solution (blue line) and (c) GHE solution. The figure demonstrates that neither the laminar nor turbulent solution fit the data, but the superposition U_{GHE} provides an excellent fit to the experimental mean velocity profile for γ=0.66 and δ=0.047. The fluid velocity is a sum of $\gamma = 0.66$, 66% of turbulent solution and $(1-\gamma) = 0.34$, 34% of laminar solution, and δ is assumed to be a Kolmogorov length scale: in [12] the GHE solution compared well with the experiment for the driven cavity flow [20] for $\delta = 0.0036$ (hydrodynamic scale $L = 15$ cm), and the dimensional value of $\delta = 0.58$ mm was a good approximation to the experimentally observed "Kolmogorov microscale" $\delta_{exp} = 0.5$ mm ([20], p. 398).

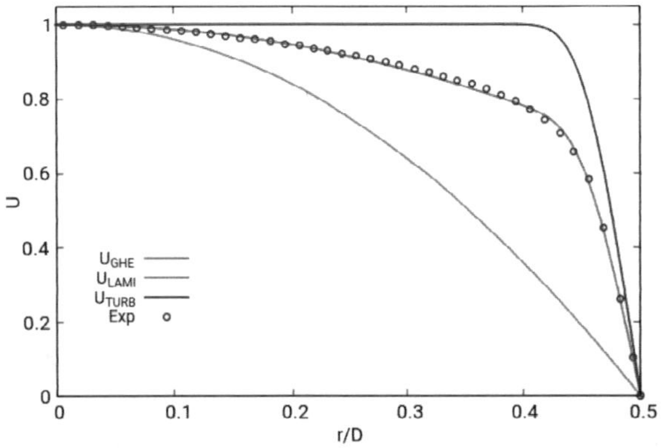

Figure 2. A comparison of experimental data for horizontal velocity by Van Doorne [5], water at Re=7200, with the GHE model. The GHE solution which is the linear superposition of the turbulent and laminar profiles (U_{GHE} on the plot) fits the experimental velocity profile.

The next Figure 3(a) is a comparison of the values of experimental turbulent intensity for horizontal velocity rms(u') with the corresponding analytical turbulent intensity. As the solution of the unsteady equations is not provided (not possible analytically), it is suggested that turbulence is the oscillation between different stationary solutions, two solutions in our case, the laminar (parabolic) velocity profile U_L and the turbulent one given by U_T. Therefore, the rms(u') at every point is

$$rms(u^{'}) = \sqrt{\frac{\sum (U_m - u(t_n))^2}{N}}$$

where $U_m(y)$ is a mean velocity, $u(y, t_n)$ is a velocity at time t_n, and N is the number of measurements.

The analytical rms(u') is presented as a difference between two solutions U_T and U_L:

$$rms(u') = B|\ U_T - \beta U_L\ | \tag{28}$$

and shown in Figure 3(a). As one can see, the data and analytical curve are in concordance. The parameters of the Eq.(28) are the following: $B = 1.57, \beta = 0.77$. Figure 3(b) shows a comparison of the experimental normalized shear stress from [5] with the analytical solution. The comparisons in Figure 3 are quite satisfactory.

6.2. Experiments by Nikuradse for High Reynolds Numbers

The solution superposition principle works well for higher values of Reynolds number, up to nearly 10^6, a range of nearly 3 orders of magnitude. Nikuradse (1932, 1933) (Prandtl group) experiments have been done for Re from 27,000 to Re=970000 [21, 22]. We present a comparison for Re=427,000 and Re=970,000 in Figure 4. The approximate analytical solution U_{GHE} (pink curve) compares well with the experimental data. The parameters for the analytical solution are the same for all the cases in [21], $\gamma = 0.71, \delta = 0.010$. The approximate analytical solution compares well with the experimental data.

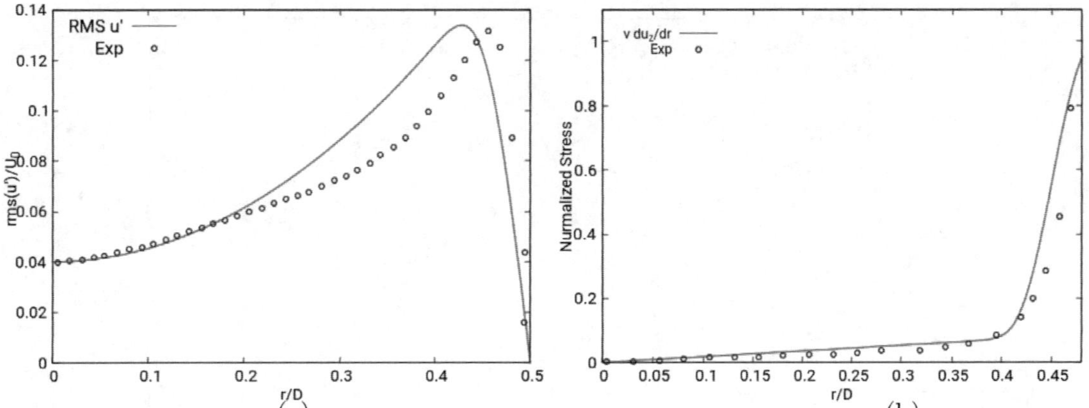

(a) (b)

Figure 3. (a) Comparison of experimental rms(u') data [5] (blue circles) for turbulent intensities and analytical solution from GHE, Eq.(28) (red curve). (b) Comparison of the experimental normalized shear stress $\nu d{<}u_z{>}/dr$ from [5] with the analytical solution. The comparisons are quite satisfactory.

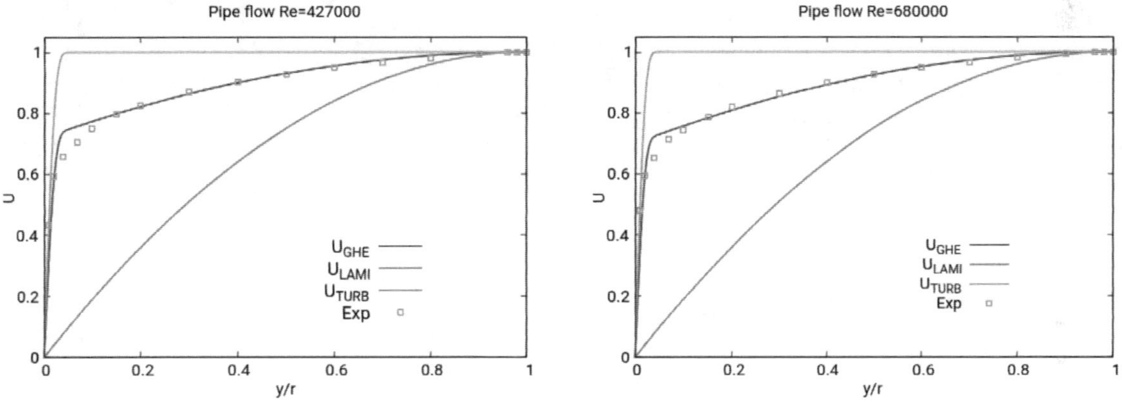

(a) (b)

Figure 4. Experimental data for mean velocity profiles by horizontal velocity Nikuradse [21] for Re=427,000 (a) and Re=970,000 (b) compared to laminar flow profile (green). Turbulent flow profile (blue) and analytical solution GHE flow profile (purple). Approximate analytical solution U_{GHE} (purple) compares well with the experimental data (circles).

6.3. Experiments by Wei and Willmarth

For the experiments with a turbulent boundary layer, the velocity U^+ plots versus y^+ does not depend on Re number, see e.g. data from Wei and Willmarth (1989) [23], Figure 5. The parameter $y^+ = yu_\tau/\nu$ where u_τ is so called friction velocity, y is the absolute distance from the wall, and ν is a kinematic viscosity. One can interpret y^+ as a local Reynolds number. The velocity scale u_τ is defined as $u_\tau = \sqrt{\tau_w/\rho}$, where wall shear stress τ_w, $\tau_w = \rho\nu\frac{dU}{dy}$ at y=0, and the dimensionless velocity is given by $U^+ = u/u_\tau$.

The analytical solution U_{GHE} is presented in Figure 5 as red line, and is in a good concordance with data (for $\delta = 0.052, \gamma = 0.65$).

Typically, Figure 5, the inner boundary layer (BL) region (viscous sublayer) is in the range $0 < y^+ < 5$, where $U^+ = y^+$. The near-middle (buffer) BL region is in the range $5 < y^+ < 30$, a strictly nonlinear region. The far-middle (inner) BL region is in the range $30 < y^+ < 200$.

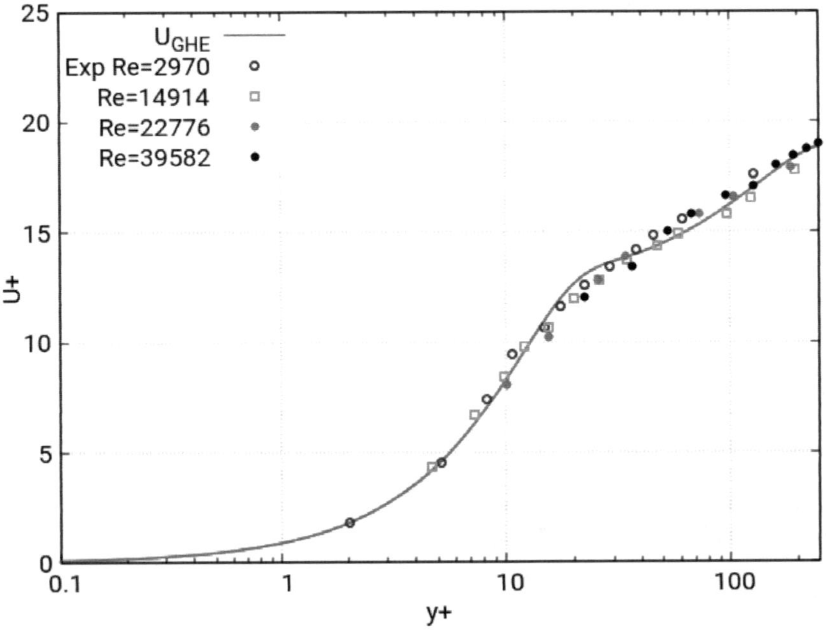

Figure 5. Mean velocity profiles in turbulent boundary layer from experiment [23] (distilled water), non-dimensionalized on inner variables, for the four Reynolds numbers examined: Re = 2970 (circles); Re = 14914 (squares); Re = 22776 (red dots); Re = 39582 (black dots). The data for different Re numbers merge well in coordinates U^+, y^+. The red line is an approximate analytical GHE solution that compares well with the experiments in all the boundary layer regions.

Velocity profile in this region can be expressed as the logarithmic "law-of-the-wall" von Karman law

$$U^+ = 1/k \; log \; y^+ + B, \tag{29}$$

where k = 0.41, B = 5.2.

The outer (non-linear, essentially inviscid)) region starts at $y^+ > 200$ and continues to the center line.

The analytical solution U_{GHE} fits well to the experimental data in all the regions.

7. Discussion

The approximate analythical GHE solution compares well with the experimental data. The GHE solution depends on two parameters: the Reynolds number Re, and $Al = \sqrt{\tau^* \nu}/L$, the introduced Alexeev number (Alexeev is the author of GHE and GBE). The solution of the Navier-Stokes equations depends on the Reynolds number Re only.

The time parameter τ^* is a material property. For gases this is a mean time between collisions according to Alexeev [1], and is $\tau^* = \Pi\mu/p$, for hard-sphere model of particles, where p is pressure, μ is dynamic viscosity, Π is a constant, $\Pi = 0.786$ [4].

There is no explicit formula for τ^* for liquids, and the experimental measurements are needed (see [2], p.324 about Frenkel work [16]). The parameter τ^*, if not known for a given fluid, can be found by fitting the experimental velocity data (as we did in the previous Sections). In this case we will obtain the value of δ which will be used to calculate τ^*: $\tau^* = \delta^2 L^2/\nu$.

8. Conclusions

The exact and approximate analytical solutions of Generalized Hydrodynamic Equations (GHE) have been obtained for channel flow (that are valid for the pipe flow if $\delta \ll 1$), that constructed from two solutions, the laminar (parabolic) and turbulent (superexponential) solutions, presented as a linear superposition of these solutions. The Navier-Stokes equations do not have the superexponential solution.

The approximate analytical solution for the GHE model has been compared with the experimental data for turbulent flows. The linear superposition of two solutions (turbulent and laminar) provides good agreement with the experimental axial mean velocity data in a wide range of Reynolds number from Re=7200 to Re = 960,000.

The experimental data for turbulent intensities for axial velocity are in concordance with the proposed analytical equation (weighted difference of the turbulent and laminar solutions). Apparently the turbulent oscillations are the oscillations between two solutions (two in considered case): the turbulent solution U_T and the laminar solution U_L. GHE intrinsically include a turbulence model.

Moreover, the obtained analytical solution was able to capture correct velocity behavior across the whole turbulent boundary layer (BL) and into the external flow: from the inner viscous sublayer to the outer layer of the turbulent BL. It compares well with the experiments by Wei and Willmarth (1989).

The analytical solution presents the explicit verifiable proof that the Alexeev generalized hydrodynamic theory (GHE and GBE) is in good agreement with experiments for turbulent flows.

Acknowledgments

Author would like to extend the sincere thanks to Dr. Yuri Bozhkov (University of Campinas, Brazil) for discussions of the text during manuscript preparation.

References

[1] Alexeev B.V., The generalized Boltzmann equation, generalized hydrodynamic equations and their applications. Phil. Trans. Roy. Soc. London, A. 349 (1994), 417-443.
[2] Alexeev B.V.,Generalized Boltzmann Physical Kinetics, Elsevier, 2004.
[3] Allegre J., D. Bisch, J.C. Lengrand, J. of Spacecraft and Rockets, 34 (6), 724-728 (1997).
[4] Cercignani C. Theory and Application of the Boltzmann Equation, Scottish Academic Press, Edinburgh and London, 1975.
[5] Van Doorne C.W.H., Jerry Westerweel, 2007, Measurement of laminar, transitional and turbulent pipe flow using Stereoscopic-PIV, February 2007, Experiments in Fluids 42(2), DOI: 10.1007/s00348-006-0235-5.
[6] Van Doorne C.W.H., Stereoscopic PIV on transition in pipe flow, Ph.D. Thesis, TU Delft, Netherlands, 2004.
[7] Fedoseyev A.I., Alexeev, B.V. Mathematical model for viscous incompressible fluid flow using Alexeev equations and comparison with experimental data. In: Dey, S.K., Ziebarth, G., Ferrandiz (Eds.), Proceedings of Advances in Scientific Computing Modelling (Special Proceedings of IMACS'98). Alicante, Spain, 1998, 158-163.
[8] Fedoseyev A.I., Kansa E.J., C. Marin, M. Volz, and A. G. Ostrogorsky, AIAA Paper 2000-0698.
[9] Fedoseyev A., A regularization approach to solving the Navier-Stokes Equations for Problems with Boundary Layer, Comput. Fluid Dynamics J., 9, (2001) 317-324.
[10] Fedoseyev A.I., Kansa E.J., C. Marin, and A.G. Ostrogorsky, Japanese Computational Fluid Dynamics Journal, 10(3), (2001), 325-333.
[11] Fedoseyev A.I., Alexeev, B.V., Simulation of viscous flows with boundary layers within multiscale model using generalized hydrodynamics equations , Procedia Computer Science, 1 (2010) 665-672.
[12] Fedoseyev A., Alexeev B.V., Generalized hydrodynamic equationsfor viscous flows-simulation versus experimental data,in AMiTaNS-12, American Institute of Physics AIP CP 1487, 2012, pp.241-247.
[13] Fedoseyev A., Finite element method stabilization for supersonic flows with flux correction transport method, AIP CP 2302, 120003 (2020), Ed. M.Todorov, https://doi.org/10.1063/5.003351
[14] Fedoseyev A., Griaznov V., Simulation of Rarefied Hypersonic Gas Flow and Comparison with Experimental Data, in AMiTaNS-2021, Conf. Proc., AIP CP 2522, 100003, 2021, Ed. M.Todorov.

[15] Fedoseyev A., Griaznov V., Ouazzani J., Simulation of rarefied hypersonic gas flow and comparison with experimental data II, Proc. AMITANS-2022 Conf., AIP CP 2953, 2023, Ed. M.Todorov.
[16] Frenkel', Ya.I. . Kineticheskaya Teoriya Zhidkostei, (Kinetic Theory of Liquids),1945, Izd. AN SSSR, Moscow–Leningrad.
[17] Harvey J.K., Holden M. S., and Wadhams, T. P., Code Validation Study of Laminar Shock/Boundary Layer and Shock/Shock Interactions in Hypersonic Flow, Part B: Experimental Measurements," AIAA Paper 2001-1031, January 2001.
[18] Holden, M. S., Wadhams, T. P., Candler, G. V., Harvey J. K., Measurements in Regions of Low Density Laminar Shock Wave/Boundary Layer Interaction in Hypervelocity Flows and Comparison with Navier-Stokes Predictions," AIAA Paper 2003-1131, January 2003.
[19] Kim J., S.J. Kline, and J. P. Johnston (1980) ASME J. Fluids Engng. 102, 302-308.
[20] Koseff R., R. L. Street , Trans. ASME/Journal of Fluids Engineering 106 (1984) 390-398.
[21] Nikuradse J., Laws of Flow In Rough Pipes, NASA Technical Memorandum 1292, 1950. Translation of Stromungsgesetze in rauhen Rohren., VDI-Forschungsheft 361. Beilage zu Forschung auf dem Gebiete des Ingenieurwesens Ausgabe B Band 4, July/August 1933.
[22] Nikuradse J., Laws of turbulent flow in smooth pipes (English translation), NASA (1932) TT F-10: p. 359 (1966). Available from https:// www.Princeton .edu/mae/people/ faculty/smits/homepage/data-1/nikuradse data/Nikuradse German_1932.pdf
[23] Wei T., Willmarth W.W., Reynolds-number effects on the structure of a turbulent channel flow, J. Fluid Mech. (1989), vol 204, pp. 57-95.

Radiation effects model for Ultra-Thin Silicon Solar Cells

Alex Fedoseyev[1], Yana Gurimskaya[1], Stanislau Herasimenka[1], Mikhail Reginevich[1], Charles Cooper[2], Slavica Grdanovska[2], Thomas Kroc[2]

[1]Solestial Inc., Tempe, Arizona, US
[2]Fermilab, Batavia, Illinois, US

E-mail: yg@solestial.com

Abstract. This work is focused on investigation of radiation effects in Ultra-Thin Silicon solar cells with a particular emphasis on electron irradiation. The study is motivated by the application of these solar cells, developed by Solestial, Inc., for powering space missions, including those led by NASA, the Department Of Defence (DOD), and commercial satellite missions. Our research encompasses both experimental and theoretical approaches, addressing unique challenges posed by ultra-thin solar cell technology that deviates from the traditional models. From the experimental point of view, the test structures were irradiated with 1 MeV electrons up to the fluence of 1E+15 e/cm^2. Radiation effects due to accumulated electron dose were noticed as a drop in open-circuit voltage (V_{oc}). An analytical expression for modeling the Voc characteristic of the UT-Si cell after exposure to electron irradiation was formulated and subsequently compared with experimental data. The theoretical foundation of the proposed approach builds upon the Non-Ionising Energy Loss (NIEL) concept, a fundamental parameter in addressing radiation damage modelling and survivability predictions.

1. Introduction

Degradation of solar cells in space environment mainly arises from their interaction with incident protons and electrons [1]. These charged particles can originate from two sources: either being trapped within the Earth's radiation belts, known as the Van Allen belts, or emitted during solar events. It becomes essential to devise a method to predict a degradation level cells undergo in such radiation harsh space environment. The degradation rate for a particular type of solar cell depends initially on the energies of the incident particles, but also on the shielding mechanisms. Typically a front side of the solar cell is shielded by a cover glass, and a rear side - by the substrate material and/or supporting array structure. These serve to attenuate the incident particles prior to they reach bare cell active region. Furthermore, the response to irradiation varies across different types of solar cell technologies [2]. This divergence arises from such factors as materials used, thickness of the active regions, the specific types and concentrations of dopants and impurities, solar cell architecture but also environmental factors. All these parameters should be taken into account to develop a comprehensive understanding of how different solar cell technologies will perform in such challenging radiation conditions.
Two main approaches are currently employed to model solar cells degradation in space - the first formulated at the US Jet Propulsion Laboratory (JPL) [3,4], and the second one developed

Journal of Physics: Conference Series **2675** (2023) 012012 doi:10.1088/1742-6596/2675/1/012012

at the US Naval Research Laboratory (NRL) [5]. Although both methods are similar, the key difference is that the NRL approach is based on the classical Non-Ionizing Energy Loss (NIEL) hypothesis built-on the assumption that all damage scales with the displacement energy, so-called direct hardness factor scaling. Unfortunately it doesn't take into account the fact that same imparted displacement energy results in different distribution of damage on the microscopic level (point-like defects to the clusters) for different particles and particle energies. The issue of NIEL violation has arisen within both the High-Energy Physics (HEP) and Space communities [6–8], indicating that the conventional model occurs incomplete describing the radiation damage effects in the given circumstances. Radiation-induced damage effects associated with the creation of point defects do not comply with the conventional NIEL scaling concept, which typically relates damage susceptibility to particle type, e.g. proton versus neutron damage, despite having the same NIEL value. Furthermore, these damage effects are often influenced by the presence of impurities within the silicon material, which significantly impact the kinetics of point defect formation, thereby altering the damage effects observed. On the other hand, damage effects that do scale with the classical NIEL concept typically originate from the intrinsic defects. These defects consist solely of silicon vacancies (V) and interstitials (I), such as e.g. clusters V_2, V_3 or I_2. A prominent example of this phenomenon is observed in leakage current, where the defect generation rate is predominantly driven by the creation of cluster defects [9]. Typically, clusters are large agglomerations of vacancies or interstitials in a volume of $\sim$20 nm^3 with 1E+5-1E+6 atoms. Driven by the Shockley2013Read2013Hall (SRH) mechanism, the defects might act like trapping, generation or recombination centers, depending on the energy level of the defect within the bandgap. Additionally, recombination centers have the effect of diminishing the diffusion length, whereas trap centers reduce the overall quantity of carriers, resulting in so-called carrier removal effect. Consequently there is a crucial need to develop a more sophisticated model but also rely on experimental data to better understand and predict the radiation damage within particular scenarios.

The purpose of this paper is to make a direct comparison of the experimental and theoretical approaches using a particular case of electrons incident on Ultra-Thin Silicon (UT-Si) solar cells as radiation response of the specific solar cell technology.

Figure 1. *Solestial* 20 μm thick Silicon Heterojunction solar cell.

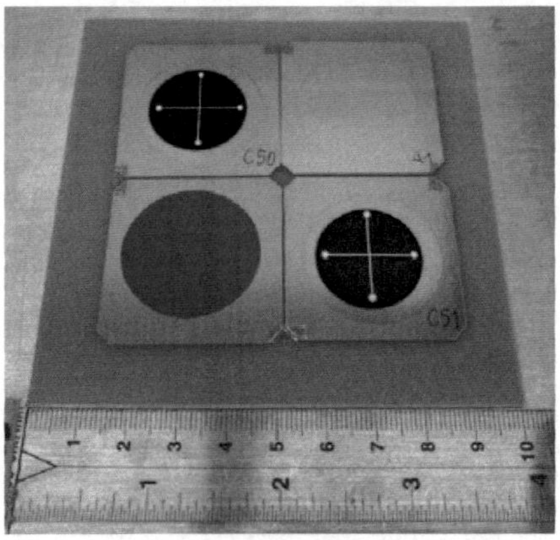

Figure 2. Packaging of the *Solestial* prototypes of selected dimensions assigned for 1 MeV electron irradiation at FNAL.

2. Experimental Approach

2.1. Test structures

Several Ultra-Thin (UT) silicon (Si) prototypes with an area of 3 cm^2, suitable for irradiation purposes, were manufactured by *Solestial, Inc.* The samples used in the described below irradiation experiment were Silicon Heterojunction (HJT) solar cells with thicknesses ranging from 20 to 80 μm, example of the thinnest cell is shown in Fig. 1. It is worth to mention that besides being radiation hard, the *Solestial* core technology - Ultra-Thin Silicon HTJ solar cell - has unique radiation damage self-curing capabilities [10], confirmed by a third-party verification as reported elsewhere [11].

2.2. Test facilities

The electron irradiation experiments were conducted at the Fermi National Accelerator Laboratory (FNAL), the Accelerator Applications Development and Demonstration (A2D2) research platform, led by Principal Investigator Charles A. Cooper. FNAL is equipped with a state-of-the-art linear accelerator, operated by a team with world-known expertise. The original electron beam parameters were as follows: a kinetic energy of 9 MeV, beam intensity of 8.33E+14 e/s and a Gaussian beam shape with a sigma value of 8.8 cm. To reduce the electron energy to the required 1 MeV, an aluminum plate, namely "Aluminum-6061", with dimensions of 20 by 20 cm^2 and thickness of 1.725 cm was designed and used as a beam degrader. Such aluminium block size was carefully selected in order to avoid undesired beam heating effects. The test prototypes were irradiated at nominal room temperature to fluences of 1E+12, 5E+12, 1E+13, 5E+13, 1E+14, 5E+14 and 1E+15 e/cm^2, two or more samples at each dose. The duration of the exposure required to achieve the highest fluence was approximately 82 s. In order to accommodate the 9 cm diameter of the electron beam and to ensure the safety of to the solar cells during irradiation, test structures were mounted by four onto a 1.5 mm thick fiberglass plate, as illustrated in Fig. 2.

3. Experimental results and discussion

In this paper we present the latest results on the electrical characterization of *Solestial* UT-Si solar cells. The gradual degradation of open-circuit voltage V_{oc} after electron irradiation is observed with a received fluence for all cell thicknesses. For the thinnest cell of 20 μm the results are shown in Fig. 3. On this plot blue empty symbols correspond to the experimental values whereas red curve represents the parametrization results. Values presented are normalized to those corresponding to non-irradiated devices.

We assessed the accuracy of the irradiation data by comparing obtained values for V_{oc} for a 20 μm thick solar cell irradiated at FNAL and previously at NIST. Notably, the first three digits of Voc measurements were identical, demonstrating an accuracy level of 0.2%.

It is worth to mention that *Solestial* UT-Si solar cells undergo significant annealing of radiation damage at temperatures below 80 °C [12]. To demonstrate this, three sets of the cells of different thicknesses were subjected to a 36-hours or less isothermal annealing after electron irradiation at the highest available - 1E+15 e/cm^2 - fluence under the light exposure. The data do not follow the trends predicted by any methodology described in the literature [[13–15]]. This is primarily due to the fact that neither GEANT4 [16] nor TRIM simulations [17], which are commonly used in HEP and Space communities, can individually resolve the NIEL violation puzzle. Instead, a combined approach is necessary to address this complex issue. The results of such annealing studies performed at 100 °C are shown in Fig. 4 showing major radiation resistance.

The degradation of device parameters, as estimated through a semi-empirical equation derived from the NRL model, can be expressed using the following formula:

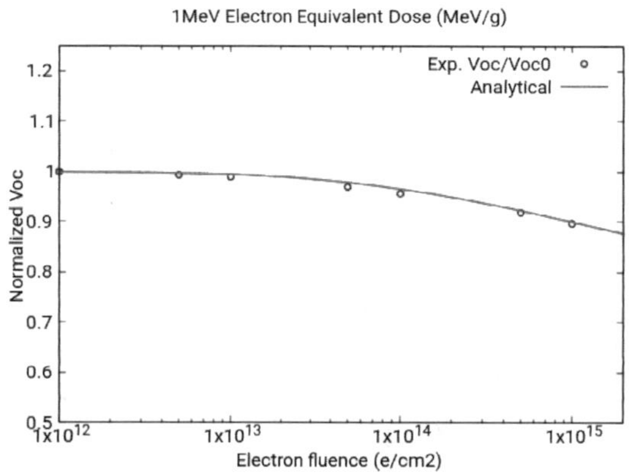

Figure 3. The comparison of V_{oc} from the analytical model to the experimental results for 20 μm thick *Solestial* solar cell.

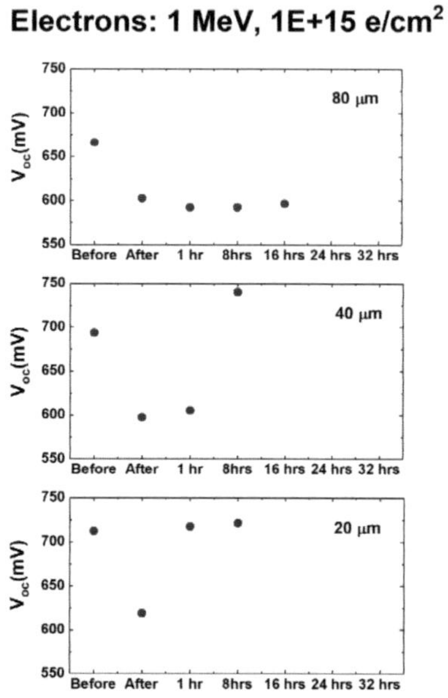

Figure 4. Curing of *Solestial* UT-Si solar cells of selected thicknesses at °C.

$$\frac{V_{oc}}{V_{oc0}} = 1 - C_x \log_{10}(1 + \frac{D_d}{D_x}). \tag{1}$$

Here, C_x and D_x represent the fitting parameters, D_d stands for the Displacement Damage Dose (DDD) in [MeV/g], V_{oc} corresponds to the measured parameter value after irradiation, and V_{oc0} is the value before irradiation. The interpolation parameters have been derived from the measured data to describe this relationship accurately. The DDD can be calculated by considering the flux of particles, their energy distribution, and the material's response to them:

$$DDD = \int \frac{\delta\phi(E)}{\delta E} S(E) dE, \tag{2}$$

where $\phi(E)$ is the irradiation fluence, $\delta\phi(E)$ stands for the differential fluence and $S(E)$ corresponds to the NIEL value for a given particle and energy. By analyzing the data in terms of DDD, data from a variety of different radiations can be easily extrapolated to a common curve.

The performance of UT-Si solar cells in the challenging space environment is a critical consideration for space missions. As mentioned above, these solar cells are exposed to a variety of radiation sources, including high-energy (1-10 MeV) electrons and protons. Understanding how these particles impact on the cells performance, particularly in terms of the production of Primary Knock-On Atoms (PKAs), is essential for designing reliable space-based power systems. When it comes to PKAs generation, the mechanisms differ significantly between electrons and protons. Electrons, as elementary charged particles, interact primarily through continuous scattering interactions with lattice electrons in the material. These interactions can lead to the creation of electron-hole pairs and atomic displacements. In this process, electrons lose energy and create PKAs more uniformly through the material due to their frequent electron-electron scattering and electron-phonon interactions. Protons, on the other hand, penetrate deeper into the material before losing most of their energy. As a result, PKAs generated by proton irradiation are concentrated in a specific area along the proton path, so-called displacement cascade. In conclusion, understanding how high-energy electrons and protons affect UT-Si solar cells is crucial for modelling and consequently mitigating radiation damage.

4. Conclusions

The assessment of solar cell performance in space relies on its key characteristics measurements under controlled conditions on the ground. This article outlines the procedure for processing and interpreting the data obtained from so-called ground-based testing. The central objective of this analysis is to predict the test structure response to radiation exposure using obtained test data acquired after subjecting them to reference particle irradiation. The proposed method benefits from its predictive capabilities in this context.

5. Acknowledgments

The material is based upon work supported by NASA under award No 80NSSC22CA123. We would like to acknowledge Arizona State University for providing access to their R&D pilot line and personally to Richard King for his valuable contribution. We extend our gratitude to the DOE Fermi National Laboratory for their support and collaboration in this research.

References

[1] NASA. (2005). *Solar Cell Degradation in Space Environments.* NASA Technical Report. Retrieved from `https://ntrs.nasa.gov/citations/20050071537`

[2] Bailey, Sheila J. (2022). *Photovoltaics for Space.* Elsevier. ISBN: 978-0-12-823300-9. URL: `https://shop.elsevier.com/books/photovoltaics-for-space/bailey/978-0-12-823300-9`

[3] Tada, H. Y., Carter, J. R., Anspaugh, B. E., & Downing, R. G. (1982). *The Solar Cell Radiation Handbook, 3rd edition.* JPL Publication, 69-82.

[4] Anspaugh, B. E. (1989). *The Solar Cell Radiation Handbook - Addendum 1: 1982-1988.* JPL Publication, Addendum 1, 69-82.

[5] Messenger, S. R., Summers, G. P., Burke, E. A., Walters, R. J., & Xapsos, M. A. (2001). *Modeling solar cell degradation in space: A comparison of the NRL displacement damage dose and the JPL equivalent fluence approaches. Progress in Photovoltaics: Research and Applications*, 9(4), 243-252. doi: 10.1002/pip.357 URL: `https://doi.org/10.1002/pip.357`

[6] Boschini, M. J., et al. (2011). *Nuclear and Non-Ionizing Energy-Loss for Coulomb Scattered Particles from Low Energy up to Relativistic Regime in Space Radiation Environment.* In: *Cosmic Rays for Particle and Astroparticle Physics*, Volume 6, WORLD SCIENTIFIC, pp. 9-23. doi: 10.1142/9789814329033_0002. URL: `https://doi.org/10.1142/9789814329033_0002`

[7] Bergmann, Benedikt, et al. (2016). *Ionizing Energy Depositions After Fast Neutron Interactions in Silicon. IEEE Transactions on Nuclear Science*, 63, 2372-2378. doi: 10.1109/TNS.2016.2574961. URL: `https://doi.org/10.1109/TNS.2016.2574961`

[8] Xapsos, M. A., Burke, E. A., Badavi, F. F., Townsend, L., Wilson, J., & Jun, I. (2005). *NIEL calculations for high-energy heavy ions. IEEE Transactions on Nuclear Science*, 51, 3250-3254. doi: 10.1109/TNS.2004.839136.

[9] Radu, Roxana, Fretwurst, Eckhart, Klanner, Robert, Lindstroem, Gunnar, & Pintilie, Ioana (2013). *Radiation damage in n-type silicon diodes after electron irradiation with energies between 1.5 MeV and 15 MeV. Nuclear Instruments and Methods in Physics Research Section A: Accelerators, Spectrometers, Detectors and Associated Equipment*, 730, 84-90. doi: 10.1016/j.nima.2013.04.080. URL: `https://doi.org/10.1016/j.nima.2013.04.080`

[10] S. Herasimenka, M. Reginevich, Y. Gurimskaya, and A. Fedoseyev (2023). *Full Recovery of Radiation Damage in Silicon Solar Cells at 80C Under Light.* Poster presentation at the Space Power Workshop 2023. URL: `https://spw.aerospace.org/files/2023/06/2023-05_08_Poster_S.-Herasimenka_Presentation.pdf`

[11] The Press release published on 28th of March, 2023, `http://bit.ly/3LLb7eN`.

[12] S. Herasimenka, M. Reginevich, A. Fedoseyev, S. Bowden, N. Honesty, M. Eyink and Y. Gurimskaya, "Recovery of electron and proton damage in Ultra-thin Silicon Solar Cells at 65°C and 80°C Under Light:

Wafer Thickness, Wafer Type and Cell Technology Dependence," 27th Space Photovoltaic Research And Technology Conference, Cleveland, OH, 2022.

[13] R. A. Weller, M. H. Mendenhall, and D. M. Fleetwood (2004). *A screened Coulomb scattering module for displacement damage computations in Geant4. IEEE Transactions on Nuclear Science*, 51(6), 3669-3678.

[14] Allison, J., Amako, K., Apostolakis, J., Araujo, H., et al. (2006). *Geant4 Developments and Applications. IEEE Transactions on Nuclear Science*, 53(1), 270-278. doi: 10.1109/TNS.2006.869826. URL: https://doi.org/10.1109/TNS.2006.869826

[15] Raine, Melanie, et al. (2016). *Simulation of Single Particle Displacement Damage in Silicon - Part I: Global Approach and Primary Interaction Simulation. IEEE Transactions on Nuclear Science*, 64(1), 133-140. doi: 10.1109/TNS.2016.2615133. URL: https://doi.org/10.1109/TNS.2016.2615133

[16] Agostinelli, S., et al. (2003). *Geant4Ž014a Simulation Toolkit. Nuclear Instruments and Methods in Physics Research Section A: Accelerators, Spectrometers, Detectors and Associated Equipment*, 506(3), 250-303. doi: 10.1016/S0168-9002(03)01368-8. URL: https://doi.org/10.1016/S0168-9002(03)01368-8

[17] SRIM (The Stopping and Range of Ions in Matter) [Online]. Available: http://www.srim.org

Improving 3-Junction and 2-Junction solar cell efficiency by down-converting solar UV light to visible

Alex Fedoseyev[1], Stan Herasimenka[1], Sergey Sarkisov[2], Brandon Bogner[3], Stephen Polly[3], and Seth Hubbard[3]
[1] Solestial Inc, Tempe, Arizona, USA
[2] SSS Optical Technologies LLC, Huntsville, Alabama, USA
[3] Rochester Institute of Technology, Rochester, New York, USA

E-mail: af@solestial.com

Abstract. The approach to efficiency improvement of space 3-Junction and 2-Junction solar cells, that involves a layer of down-converting material on the top of the cell, is presented. The layer consists of a polymer nanocomposite coating impregnated with the nanoparticles of inorganic compounds (fluorides or oxides) doped with rare earth ions down-converting solar UV light to visible matching the spectrum of the responsivity of the cell. Computational model of a solar cell is presented and used for processing the experimental data. Experimental results demonstrate an increase of the generated current and photoelectric efficiency.

1. Introduction

Solar cells are subject to bombardment by high-energy particles, which induces a degradation referred to as radiation damage. Additional damaging phenomenon in space is UV radiation. Cover glass is a typical choice to minimize solar cell degradation. Alternative to cover glass is a polymer layer that provides similar protection of the solar cell from UV and space radiation. We propose to use such layer to increase solar cell photoelectric efficiency by down-converting the UV part of solar spectrum to visible-NIR light that matches the spectrum of responsivity of the cell. The layer in this case can be designed as a polymer nanocomposite impregnated with the nanoparticles of the inorganic compounds (fluorides or oxides) doped with the ions of rare earth (RE) that down-convert solar UV to visible-NIR light.

2. Down-conversion principle and implementation

A nanocomposite layer used to convert UV light to visible, thus improving PV power conversion efficiency, is presented in Figure 1. A nanocomposite layer made of RE-doped nanoparticles embedded in a transparent polyimide is deposited on the top of the polyimide film that protects PV cell from UV radiation. Both UV and visible-NIR components of solar radiation reach the layer. UV component is partially down-converted into visible-NIR radiation. All the remaining UV radiation is blocked by the protective polyimide film. The down-converted radiation along with the major visible-NIR component of solar radiation passes unabsorbed through the protective polyimide film to the PV cell and get transformed into electricity. The

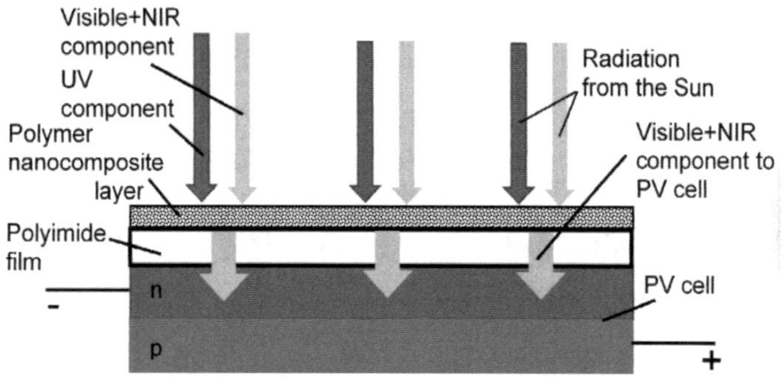

Figure 1. Schematic illustrating the principle of operation of a down converting layer bringing the PV conversion efficiency improvement.

down-converted radiation contributes to the production of extra electricity thus increasing PV efficiency.

Fluoride $NaYF_4$ has been selected as a host for RE ions. This is justified by its low phonon energy ($\sim$ 300 cm^{-1}) that makes multi-phonon assisted nonradiative relaxation of excited ions negligible [3]. The nanoparticles (NPs) made of such compound are environmentally stable and not toxic. The energy level diagram of Er^{3+} and Yb^{3+} ions in the compound is presented in Figure 2.

Fluorides doped with the ions of rare earth (RE) elements $NaYF_4$:Yb^{3+}, Er^{3+} [1, 2] have been used as a spectrum down-converting matter.

The erbium ion gets excited with a solar UV photon and emits down-shifted visible and NIR radiation (single photon) in one of the spectral bands around 520, 540, 650, 850, or 980 nm (marked with vertical arrows down). It can also perform energy exchange with the ytterbium ion through two quantum-cutting mechanisms (marked by slanted dash-dotted arrows down) when the energy of the UV solar photon is cut into the energies of two NIR photons emitted by two ytterbium ions.

We synthesized the RE-doped fluoride phosphor using the wet method followed by baking the obtained micro-powder in ambient air and reducing it with ball-milling to nano-powder [1]. Average size of the NPs as measured by the dynamic light scatterometry (DLS) with a Zetasizer instrument from Malvern Instruments was estimated as 172 to 228 nm (Figure 2b). The X-ray diffraction (XRD) spectrum of the ball-milled nanopowder (Figure 2c) showed that the NPs were the nanocrystals of hexagonal beta-phase of $NaYF_4$. Photoluminescence (PL) spectral bands of the phosphor excited with UV radiation included downshifting and quantum cutting components (Figure 2e) that corresponded to the optical absorption bands (Figure 2d). Photoluminescence quantum yield (PLQY) determined as the ratio of the emitted visible-NIR photons to the absorbed UV photons was ≤ 10 % .

We used polymers CP-1 and CORIN as hosts for the NPs. They are space qualified polymers.

We call the polymer nanocomposite layer made in such way the Layer of UV CONverting coating or LUVCON.

3. Computational model of solar cell

3.1. Key parameters of solar cell

The solar cell is characterized by the following key parameters: $V_{oc}, I_{sc}, V_{mp}, I_{mp}$ and FF, where V_{oc} is the open circuit voltage (no load), I_{sc} is a short circuit current (maximum current the the cell can produce), V_{mp} is the voltage at which the cell produces maximum power $P = I \cdot V$ (a product of current and voltage), and I_{mp} is a current at maximum power. The fill factor FF is defined as $FF = V_{mp} \cdot I_{mp}/(V_{oc} \cdot I_{sc})$. The approximate value of FF can be estimated by formula:

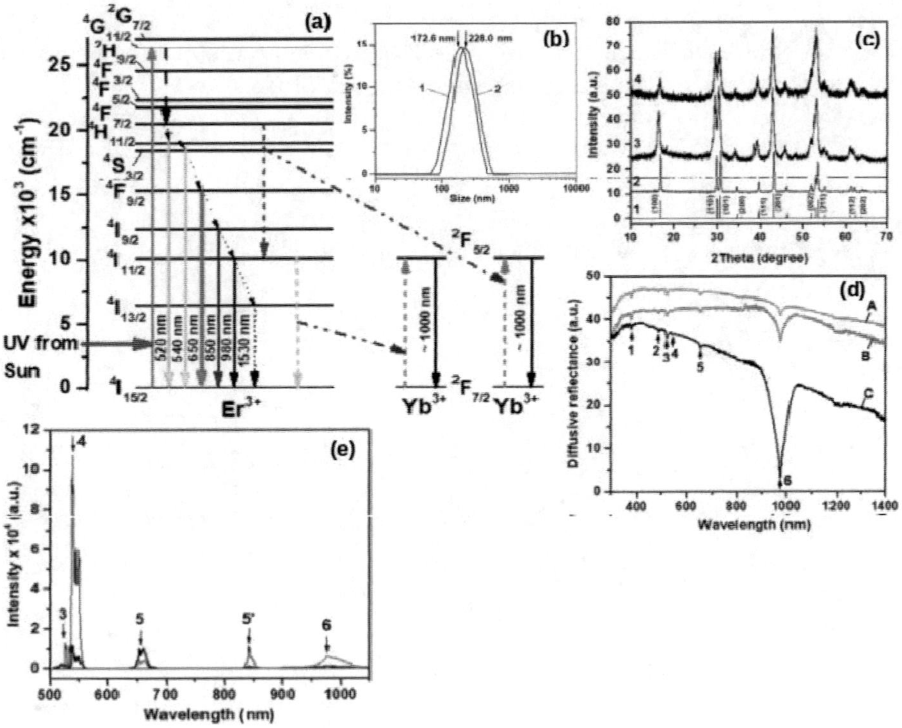

Figure 2. a) Energy level diagram of coupled ions Er^{3+} - Yb^{3+}. (b) Size distributions of two exemplary colloids of ball-milled NPs of phosphor $NaY_{0.83}F_4$: $Yb^{3+}_{0.14}Er^{3+}_{0.03}$ in 1-propanol obtained with dynamic light scattering (DLS) measurement. Vertical arrows down mark the average size of the NPs (assuming their spherical shape) on size scale (c) X-ray diffraction (XRD) spectra. Plot 1- XRD spectrum (computed, JCPDS card No. 28-1192) of $NaYF_4$ beta-phase (hexagonal). 2- synthesized $NaY_{0.83}F_4$: $Yb^{3+}_{0.14}Er^{3+}_{0.03}$ powder baked at 500^oC for 1h before ball milling; 3, 4 - two exemplary ball milled nanopowders of $NaY_{0.83}F_4$:$Yb^{3+}_{0.14}Er^{3+}_{0.03}$. (d) Diffuse reflectance spectra of the powder samples of RE-doped fluoride down-conversion phosphors. Curve A corresponds to $NaY_{0.94}F_4$: $Yb^{3+}_{0.03}Er^{3+}_{0.03}$; B - $NaY_{0.78}F_4$: $Yb^{3+}_{0.2}Er^{3+}_{0.02}$; C - $NaY_{0.69}F_4$: $Yb^{3+}_{0.3}Er^{3+}_{0.01}$. Absorption peak 1 corresponds to transition in ion of Er^{3+}: ${}^4I_{15/2} \to {}^2G_{11/2}$ (380 nm); 2 - Er^{3+}: ${}^4I_{15/2} \to {}^4F_{7/2}$ (480 nm); 3 - Er^{3+}: ${}^4I_{15/2} \to {}^2H_{11/2}$ (520 nm); 4 - Er^{3+}: ${}^4I_{15/2} \to {}^4S_{3/2}$ (540 nm); 5 - Er^{3+}: ${}^4I_{15/2} \to {}^4F_{9/2}$ (650 nm); 6 - Er^{3+}: ${}^4I_{15/2} \to {}^4I_{11/2}$ and Yb^{3+}: ${}^2F_{7/2} \to {}^2F_{5/2}$ ($\sim$ 980 - 1000 nm) as presented in the energy level diagram (a). All the spectra are normalized to peak 3: its depth is (-1). (e) Visible and NIR emission spectrum of the powder of phosphor $NaY_{0.83}F_4$: $Yb^{3+}_{0.14}Er^{3+}_{0.03}$ excited by a 372-nm UV laser (black curve) and by 488-nm line of an Ar-ion laser (blue). The peaks correspond to the absorption bands with the same numbers as in plot (d). New peak 5* corresponds to the band at 850 nm. This band does not exhibit prominent absorption in plot (d).

$$FF = \frac{v_{oc} - ln(v_{oc} + 0.72)}{v_{oc} + 1},$$

where v_{oc} is a normalized V_{oc} by thermal voltage, $v_{oc} = V_{oc} \cdot \frac{q}{nkT}$ (see below for explanation of thermal voltage).

3.2. Characteristic IV equation

Mathematical model of solar cell is used to access the solar parameters that are impossible to measure directly, for example, the shunt resistance and serial resistance.

The equivalent circuit of solar cell is shown in Figure 3. The current produced by the solar cell is given by the following equation Eq.(1),

$$I = I_L - I_D - I_{SH}, \tag{1}$$

here I_L is a load current, I_D is the current through diode (cell), and I_{SH} is a shunt current.

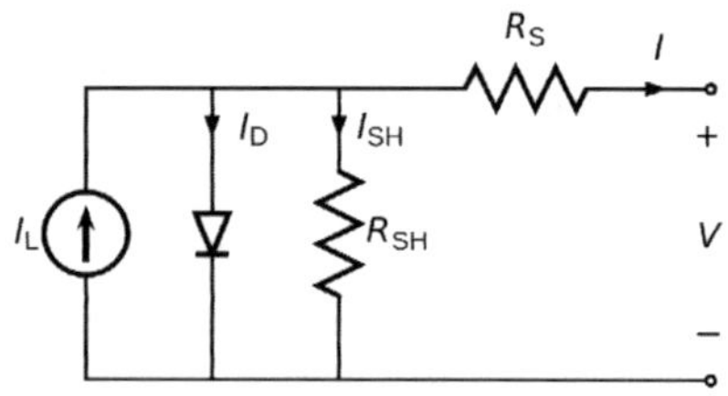

Figure 3. Schematic of solar cell equivalent circuit.

The voltage across the solar cell diode V_J is given by Eq.(2):

$$V_J = V + IR_S, \tag{2}$$

Shockley diode equation [6] describes the current driven through the diode:

$$I_D = I_0 \left[\exp(\frac{V_J}{nV_T}) - 1\right], \tag{3}$$

here $V_T = kT/q$ is the thermal voltage, k is the Boltzmann constant, T is a temperature in K, q is the electron charge, and n is the ideality factor, also known as the quality factor or emission coefficient($V_T = 25.85$mV at room temperature).

The current through the shunt resistor is $I_{SH} = V_J/R_{SH}$. Substituting Eq.(2) and Eq.(3) into Eq.(1) we obtain the characteristic equation of a solar cell:

$$I = I_L - I_0 \left[\exp\left(\frac{V + IR_S}{nV_T}\right) - 1\right] - \frac{V + IR_S}{R_{SH}}, \tag{4}$$

If R_S is not zero, there is no direct expression of the current versus voltage. Still the equation can be solved, and the solution of I versus V is expressed using the Lambert W function:

$$I = \frac{(I_L + I_0) - V/R_{SH}}{1 + R_S/R_{SH}} - \frac{nV_T}{R_S} \cdot$$

$$W\left[\frac{I_0 R_S}{nV_T(1 + R_S/R_{SH})} \exp\left(\frac{V}{nV_T}(1 - \frac{R_S}{R_S + R_{SH}}) + \frac{(I_L + I_0)R_S}{nV_T(1 + R_S/R_{SH})}\right)\right], \tag{5}$$

where $W(x)$ is a Lambert function.

3.3. Processing of experimental data

Since the parameters I_0, n, R_S, and R_{SH} can not be measured directly, the above equation is used to extract these parameters from the experimental IV curve in the next sections. In particular, the model is used in the paper to calculate the current-voltage IV curve for a given shunt and serial resistances, which are of true interests, as actual values of these resistances vary from one cell to another. The experimentally obtained IV-curve is approximated by Eq.(5) through adjusting the values of I_0, n, R_S, and R_{SH} to get the minimum difference (best fitting). Having that done we can get IV-curve for any of given R_S, and R_{SH}.

4. Experimental results on material stability

UV stability of LUVCON coating and its ability to protect PV cells was tested using Flood Lamp System Model 38125 from Dymax (Figure 4). UV irradiance on the top of the atmosphere is $\sim 8\%$ of the total Solar irradiance. UV irradiance in Dymax is equivalent to 37 $mW/cm^2/0.08 = 425$ mW/cm^2 or ~ 3 total Solar irradiances. One hour of exposure in Dymax is thus ~ 3 ESH.

We have identified the material that satisfy all the key requirements: stable coating under UV light for 2500+ hours, high optical transmittance for visible light, minimal optical darkening under UV light, good suppression of the UV light.

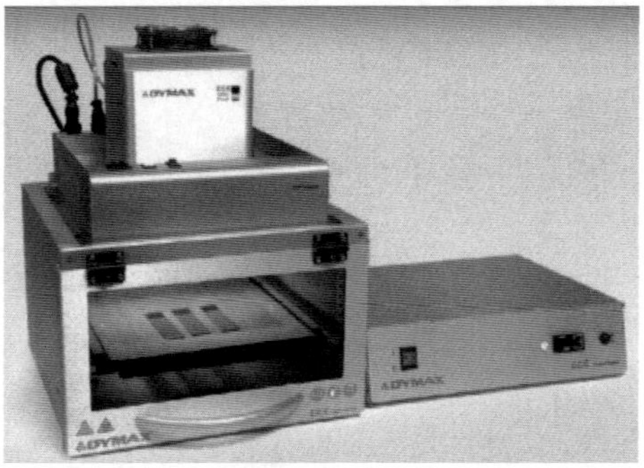

Figure 4. Dymax UV-A+visible light source (metal halide lamp) used in the experiments on UV stability.

5. AM0 conditions experimental results

We have performed the experiments with several types of solar cells. The results with UT-Si silicon solar cells [4] have been presented in [5]. UT-Si solar cells coated with LUVCON increased the current by 22.0% compared to cells coated with CORIN polymer only. That translates into an absolute 4.3% efficiency increase for the 22% efficient UT-Si silicon solar cells, resulting in 26.3% efficiency. As UT-Si solar cell degrades very little under space radiation, the EOL efficiency expected to be 26.0%.

In this work we present the results for 3-Junction and 2-Junction solar cells. A solar simulator with a Xenon arc lamp and AM0 filter has been used to measure the AM0 IV characteristics of our solar cells coated with LUVCON. For comparison, we also provide the results for the cells coated with pure polymer coating.

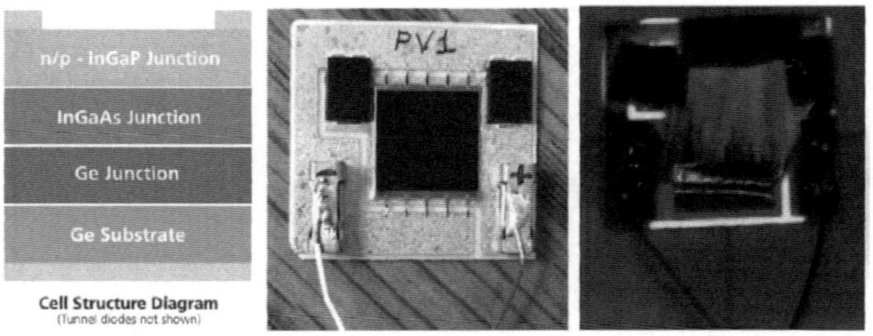

(a) (b) (c)

Figure 5. (a) Emcore 3J PV cell schematics, (b) Emcore 3J PV, cell area 1.1x1.1 cm = 1.21 cm^2, with wires, and (b) Emcore 3J PV cell covered with CP-1 film coated with LUVCON and illuminated with UV LED.

A 3-junction solar cell (made by EMCORE) received from long-time collaborator Dr. Seth Hubbard, Rochester Institute of Technology (RIT), is shown in Figure 5. It is the solar cell for concentrated solar power applications with 1-cm^2 area with a documented AM1.5 efficiency of 39% under concentrated illumination at 800 Suns. AM0 efficiency is estimated to be 30% at 1 Sun.

3J solar cell improvement with LUVCON coating (CORIN + NPs) is demonstrated in Figure 6. 3J solar cells covered with LUVCON increased the current by 5.1% compared to the cells covered with the polymer only, and absolute 2.4% efficiency increase for 33% efficient 3J solar cells, resulting in 35.4% BOL efficiency (FF=0.90). Mathematical model, Section 3, has been used to extract IV data from the experiment.

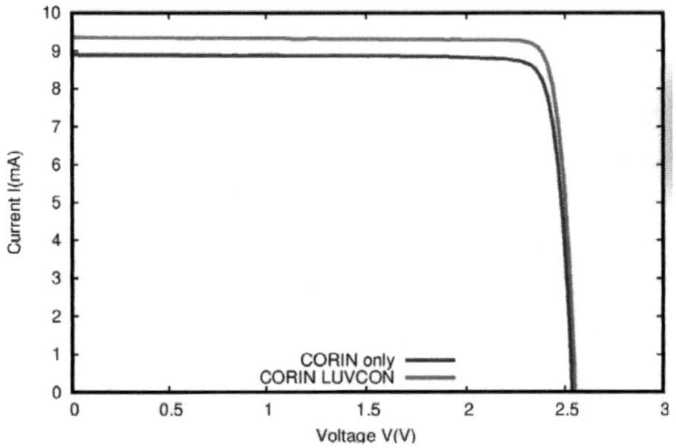

Figure 6. Improving conversion efficiency of the 3J solar cells for AM0 conditions: I-V curves for CORIN coating only (blue color) and LUVCON coating (red color). 3J solar cell coated with LUVCON increased the current by 5.1% compared to the cell coated with CORIN polymer only, and 2.4% efficiency increase for 33% efficient 3J solar cells, resulting in 35.4% BOL efficiency (FF=0.90).

2J solar cells. These cells were fabricated at RIT (Dr. Seth Hubbard Lab). RIT has a good established process for InGaP/GaAs 2J (with 50x quantum wells) resulting (theoretically) in 24% AM0 efficiency. InGaP and GaAs+QW cells were adjusted for current matching, Figure 7. The provided device did not have antireflective coating (ARC). The estimated efficiency was $\sim$ 18-19% AM0 without ARC (Anti Reflection Coating).

Similar improvements have been observed in the experiments with 2J solar cells covered with LUVCON: the current increase was 2.0% and efficiency increase 0.45%, compared to the same cells covered with CORIN polymer only (Figure 8) resulting in 24.5% BOL efficiency (FF=0.90). Mathematical model, Section 3, has been used to extract IV data from the experiment.

6. Discussion

Experimental results have shown improvements for two types of solar cells coated with LUVCON: (a) 3J solar cells, current increase by 5.1%, (b) 2J cells, current increase by 2.0%, compared to the same cells covered with CORIN polymer only. In those experiments (i) we did not coat these cells directly with LUVCON (we had just 2 and 3 of these cells), but used extra film with LUVCON coating instead. This film exhibits incident light reflection due to two additional interfaces, negatively affecting the current. (ii) LUVCON was not optimized for these cells yet (in terms of the down-conversion spectrum).

Mathematical model has been used to extract IV data from the experiments.

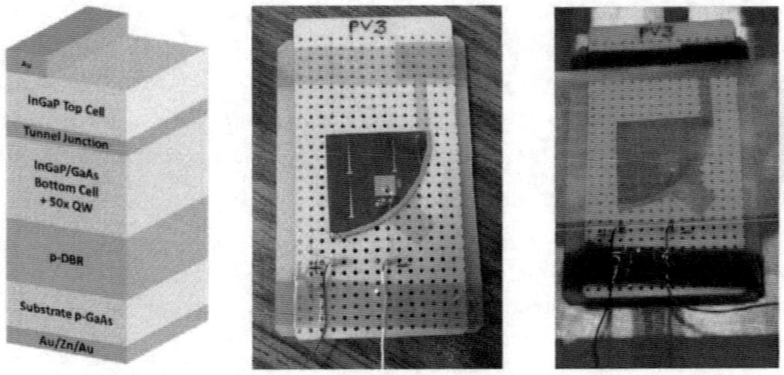

(a) (b) (c)

Figure 7. 2J solar cells (a) schematics, (b) InGaP-GaAs 1/4 wafer with three 2J PV cells ($1cm^2$ each) with wires without antireflective coating (ARC), and (c) InGaP-GaAs 1/4 wafer with three 2J PV cells covered with a glass slide coated with LUVCON illuminated with UV LED.

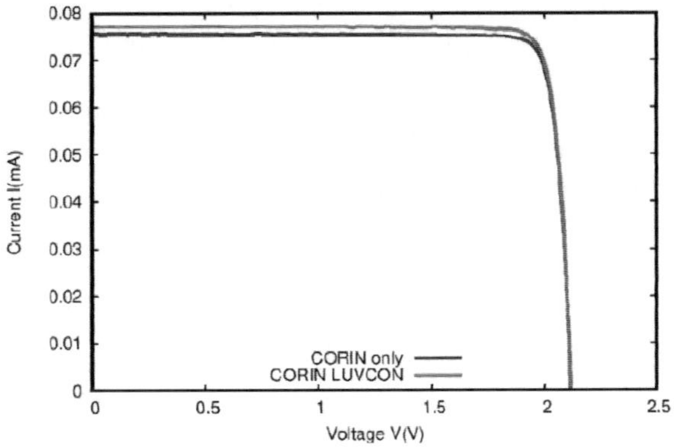

Figure 8. Improving conversion efficiency of 2J solar cells for AM0 conditions. The I-V curve for the cell covered with CORIN only (blue color curve) and LUVCON coating (red color curve). 2J solar cell coated with LUVCON increased the current by 2.0% compared to the cells coated with CORIN polymer only, and absolute 0.45% efficiency increase for 24% efficient 2J solar cells, resulting in 24.5% BOL efficiency (FF=0.90).

7. Conclusion

The experimental results have proved the feasibility of the proposed down converting concept. A significant improvement of solar cell efficiency by LUVCON coating, down-converting UV light to visible and near infra-red and therefore increasing the current generated has been observed. UT-Si solar cell coated with LUVCON increased the current by 22.0% compared to cells coated with CORIN polymer only [5]. 3J solar cells from EMCORE showed a current increase of 5.1% and absolute 2.4% efficiency increase, and 2J solar cell showed current increase by 2.0% and absolute 0.45% efficiency increase, while LUVCON coating was not optimized for those 3J and 2J solar cells yet.

Acknowledgments

The material is based upon work supported by NASA under award No 80NSSC22PB171.

References

[1] Sarkisov S.S., Patel D.N., et al. Polymer nanocomposite luminescent films for solar energy harvesting made by concurrent multi-beam multi-target pulsed laser deposition. Proc. SPIE 2018;10755,1075502.

[2] Sarkisov S.S., Patel D.N., et al. Polymer nanocomposite sunlight spectrum down-converters made by open-air PLD. Nanotech Rev 2020;9,1044-58.

[3] Baker C.C., Fontana J., et al. Nanoparticle doping for high power fiber lasers at eye-safer wavelengths. Opt Express 2017;25(12):3903-13915.

[4] S.Y. Herasimenka, M. Reginevich, Y. Gurimskaya and A. Fedoseyev, Full Recovery of Radiation Damage in Silicon Solar Cells at 80° C Under Light, poster at 40th Space Power Workshop, torrence, CA, 2023.

[5] A. Fedoseyev, S. Herasimenka, S. Sarkisov, Improving the Space Silicon Solar Cell Efficiency by Adding the Layer Down-converting UV Light to Visible, Proc. PVSC-50, 2023.

[6] Shockley diode equation, WikipediA, *https : //en.wikipedia.org/wiki/Shockley_diode_equation*

Assessing the Impact of Speed and Type of Road Surface on the Noise Level in a Passenger Car Cabin

N Y Kovachev[1], V A Mihova[2,*], I R Georgiev[2] and V T Pavlov[2]

[1]Department of Heat, Hydraulics & Environmental Engineering, University of Ruse, 8 Studentska Str, 7017 Ruse, Bulgaria
[2]Department of Applied Mathematics and Statistics, University of Ruse, 8 Studentska Str, 7017 Ruse, Bulgaria

[*]Corresponding author's e-mail address: vmicheva@uni-ruse.bg

Abstract. This paper presents a study of noise pollution in the cabin of a passenger car in the context of various types of road. The purpose of the study is to understand how the type of road surface can influence the levels of noise inside the car. Sound pressure levels were measured on the A-scale of a sound level meter for 3 types of road surface - crushed stone pavement, asphalt, and cobblestone. The study was conducted at varying speeds of the car within the range of 10-60 km/h. Due to limitations of homogeneous road conditions, a duration of 10 seconds was chosen for each measurement. The sound level meter was placed at a selected measuring point, close to the driver's head. On the data collected in the above-mentioned way, various regression models have been developed to assess the impact of speed and type of road surface on the noise level. It was found that the noise level increases with the roughness of the road surface, and the resulting risks were analysed.

1. Introduction

The major types of pollution threatening modern human life include air pollution [1], water pollution [2], soil pollution, and noise pollution [3]. Air pollution, primarily from industrial emissions and vehicle exhausts, poses significant respiratory health risks, while water and soil pollution, often due to chemical waste and agricultural runoff, can contaminate food and water supplies, leading to a range of health problems. Noise pollution, largely from urbanization and machinery, can lead to psychological and physiological issues, impacting overall well-being. In this paper, we primarily pay attention to the noise pollution. Nowadays, people spend up to several hours per day travelling in a car for diverse activities as commuting, shopping, recreationing, etc., which makes the question for the noise pollution inside the car of a paramount importance.

The permissible levels of sound pressure in the individual octave frequency bands are given in the standard BNS 14478:1982 [4]. These permissible values (see table 1) are defined for drivers and service personnel of transport, agricultural and construction machinery.

Table 1. Permissible levels of sound pressure in octave frequency bands.

Octave band, Hz	63	125	250	500	1000	2000	4000	8000
Sound pressure level, dB	99	92	86	83	80	78	76	74

The permissible sound pressure level according to A-weighted curve, determined by hygiene standards is LpA = 85 dBA [4].

Modern cars rely on suitable conditions created on the earth's surface that allow the car to move, defined as the road surface. Most often, these are road surfaces made of asphalt-concrete, which are made of several layers, in order to achieve the most rational transmission of the load. In general, the parameters of this type of flooring are standardized surfaces [5]. But very often they are built to a wide quality tolerance, and additional depreciation and deterioration changes their qualities. Apart from these pavements, there are also a number of others that can be used by cars for a certain period of time due to various reasons such as the lack of a built asphalt road, repairs, natural disasters, etc. Such are dirt roads, roads made of crushed stone pavement, pavement, etc [3].

The objective of the present paper is to model and evaluate the noise in the compartment of a car powered by an internal combustion engine, depending on the speed and the type of road surface of the traffic lane. To achieve it, the following tasks are solved:

1. Development of a methodology for conducting the measurements.
2. Carrying out the experimental studies of noise emissions in the passenger car compartment on different types of road surfaces and at different vehicle speeds
3. Statistical processing, modeling and analysis of the obtained data.

2. Methodology for experimental study of the noise imissions

The methodology for conducting the measurements includes the following basic steps:

• Determination of the measurement points in the compartment of a car;
• Determination of measurement parameters;
• Meter settings;
• Installation of the noise meter and carrying out noise measurements at different speeds of movement and different road surfaces [6];
• Transferring the data from the sound meter to a computer;
• Processing, regression analysis of the results and evaluation of the obtained data.

An important element in the methodology is related to determining the locations for performing the measurements. The measurements were made according to the requirements of the standard ISO 5128:1996 for measuring the internal noise in motor vehicles [7]. A measuring point was selected at a height of $0.8\,m \pm 0.05\,m$ above the intersection of the planes of the unoccupied seat next to the driver and the backrest in the plane of symmetry in the horizontal direction. The location of the microphone was at a distance from the walls and upholstery not less than $15\ cm$, and the direction of its highest sensitivity coincided with the direction of movement of the vehicle. The specified standard defines the conditions to which the technical condition of the car must meet, as well as the climatic conditions and the condition of the terrain under which the test takes place.

For the second task, the measurement parameters are determined. For the A-weighted sound pressure level used, the A scale of the level meter is applied and the unit measured is marked as $Lp(A)$ in relative units $dB(A)$. In spectral measurements, the frequency range of the octave filters is given to be in the range of $31.5\ Hz$ to $16000\ Hz$ [8].

3. Experimental study

When conducting the research, a SVANTEK 971 sound meter was used [9], located at the specified point. A tripod SDV 10 was used to mount the sound meter. Before the start of the measurement, calibration of the measurement system is performed. An acoustic calibrator is used. Measurements were made at speeds that were constant over the measurement range. The selected speed points are respectively $10, 20, 30, 40, 50$ and $60\ km/h$. The maximum speed limits are due to requirements in the Urban Road Traffic Act as well as the type of road surface which does not allow higher speeds in this measurement. The object of the study is the noise in a Citroen C3 car, manufactured in 2003, $1.4\ HDi\ 16V\ (90\ hp)$, with an engine displacement of $1398\ cm^3$.

In all measurements, the motor shaft rotation frequency is in the range of $2000 - 2500\ min^{-1}$. For each measurement, the noise was recorded for $10\ seconds$.

According to the requirements of the standard, tests were carried out at constant speed. An on-site test was also carried out, for $300\ seconds$, aiming to represent the main sources of engine noise without the influence of speed and road surface type [10].

Measurements are made on road sections indicated in figure 1, with a different composition of the mixture used to build the road surface. At figure 1a) it is a pavement with a thickness of the elements over $120\ mm$ and a distance between them of up to $20\ mm$, figure 1b) presents crushed stone pavement, with elements of different sizes and shapes, and figure 1c) - an asphalt surface with a thickness of mineral inclusions between $6 - 12\ mm$.

a) b) c)

Figure 1. Types of road surfaces: a) pavement with elements dimension up to 120 mm and gaps 20 mm; b) cobblestone with randomly used elements; c) asphalt with mineral ingredients dimensions 6-12 mm.

After the measurements are completed, the data is transferred to the SVAN PC++ software product. The program allows numerous analyzes and statistical processing of the recorded signal. EXCELL, MATLAB were used to process the obtained results.

3.1. Noise Measurements and Analysis

Table 2 shows the sound pressure levels in the individual octave frequency bands, as well as the A-weighted sound pressure level. Values represent averaged levels for each $10 - second$ measurement. The table shows the levels at individual speeds and road surfaces.

Table 2. Sound pressure level as a function of car's speed at different octave band and A-weighted corrections on 3 types of road surfaces – pavement, cobblestone and asphalt.

Pavement

V, km/h	31.5 Hz	63 Hz	125 Hz	250 Hz	500 Hz	1000 Hz	2000 Hz	4000 Hz	8000 Hz	16000 Hz	Total A
10	61.00	57.00	58.95	51.37	52.93	44.39	41.57	37.39	29.41	19.35	64.73
20	64.69	66.27	64.29	57.52	57.20	49.90	46.14	45.39	36.23	24.10	70.45
30	66.99	68.28	68.08	63.62	63.40	55.67	51.66	51.24	43.35	33.20	73.63
40	66.85	68.10	69.05	66.35	65.63	57.53	53.13	51.69	43.49	33.52	74.51
50	65.79	66.37	75.76	68.69	67.63	59.39	54.71	51.53	43.02	32.75	77.83
60	65.46	66.66	75.59	69.23	68.91	61.23	57.89	53.02	43.13	31.96	78.04

Cobblestone

V, km/h	31.5 Hz	63 Hz	125 Hz	250 Hz	500 Hz	1000 Hz	2000 Hz	4000 Hz	8000 Hz	16000 Hz	Total A
10	55.78	52.87	54.13	47.86	50.22	44.73	38.43	30.58	21.96	11.10	60.15
20	60.51	58.03	60.75	54.50	55.22	51.49	44.77	38.05	30.21	19.23	65.80
30	64.97	62.35	65.00	58.96	58.79	54.88	49.17	42.35	33.34	21.47	70.05
40	67.43	66.19	69.31	63.91	62.28	58.73	53.53	47.79	39.39	27.39	73.73
50	67.49	66.85	69.10	68.94	66.01	60.79	56.12	55.21	44.42	31.88	75.23

60	68.65	67.89	69.87	69.29	65.57	60.37	54.65	48.79	41.65	29.41	75.65

Asphalt

V, km/h	**31.5 Hz**	**63 Hz**	**125 Hz**	**250 Hz**	**500 Hz**	**1000 Hz**	**2000 Hz**	**4000 Hz**	**8000 Hz**	**16000 Hz**	**Total A**
10	50.21	45.29	54.83	48.96	47.67	42.71	35.13	25.35	12.21	3.03	57.91
20	47.45	52.03	53.71	52.69	51.71	43.72	36.81	26.72	15.04	4.58	59.49
30	53.91	49.15	54.35	55.21	53.49	45.31	37.47	27.13	14.53	4.33	60.85
40	49.35	55.64	52.83	55.07	54.37	46.39	41.55	30.05	18.10	7.18	61.18
50	51.77	54.10	56.67	57.63	55.95	49.27	44.43	33.30	21.37	8.59	62.95
60	54.07	54.67	56.61	60.15	57.55	52.91	46.97	36.14	24.00	11.35	64.54

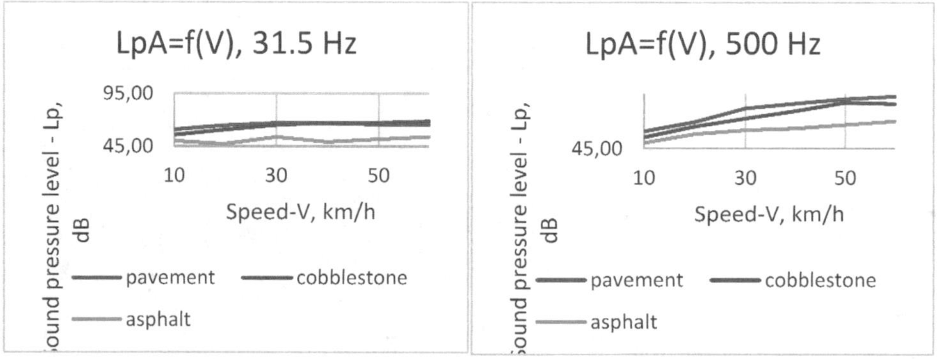

a) b)

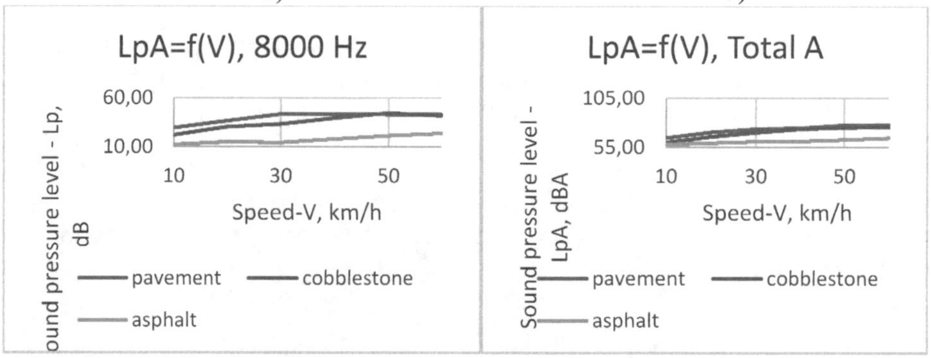

c) d)

Figure 2. Sound pressure level as a function of car's speed at different octave band and A-weighted corrections on 3 types of road surfaces – pavement, cobblestone and asphalt: a) in octave band with geometric mean frequency - 31.5 Hz;); b) at 500 Hz; c) at 8000 Hz; d) A-weighted sound pressure levels

Figure 2 shows 4 graphs in which the dynamics of sound pressure levels can be presented as a function of car's speed. This was done for 3 octave frequency bands, with frequencies, respectively: figure 2a) 31.5 Hz, figure 2b) 500 Hz, figure 2c) 8 kHz, and figure 2d) respectively shows the variation of levels as a function of speed when measuring the corrected sound level pressure. There are 3 lines on each diagram that represent a visualization of the functions of the different road surface types.

In the measured sections, the unevenness in the uniformity along the length of the coatings affects the dynamics of the sound pressure levels. However, similar trends of increasing sound pressure levels with increasing speed are observed.

Figure 3 shows graphs that present the dynamics of sound pressure levels over the time of measurement at different speeds and road surfaces. It can be seen that the levels are highest in the pavement noise measurement, but remain stable throughout the measurement. There are significant deviations from the mean on the crushed stone surface, especially at higher speeds. This is due to the bluer strokes in the chaotic construction of the stone pavement. Fine asphalt shows stability in results.

Analysis of table 2 and figure 2 leads to some interesting results. In general, the maximum noise values are when driving on a paved road, and they increase with speed. At low frequencies, however, this is true up to $40\ km/h$, where the noise levels when driving on pavement and cobblestone equalize, after which the noise in the passenger compartment above the crushed stone pavement is higher than that on the pavement and reaches values of $Lp = 68.65\ dB$ at $60\ km/h$. Briefly, at an octave bandwidth of $8000\ Hz$, road traffic noise on a paved road is lower than that on a gravel road, with a difference of $\Delta Lp = 1.4\ dB$.

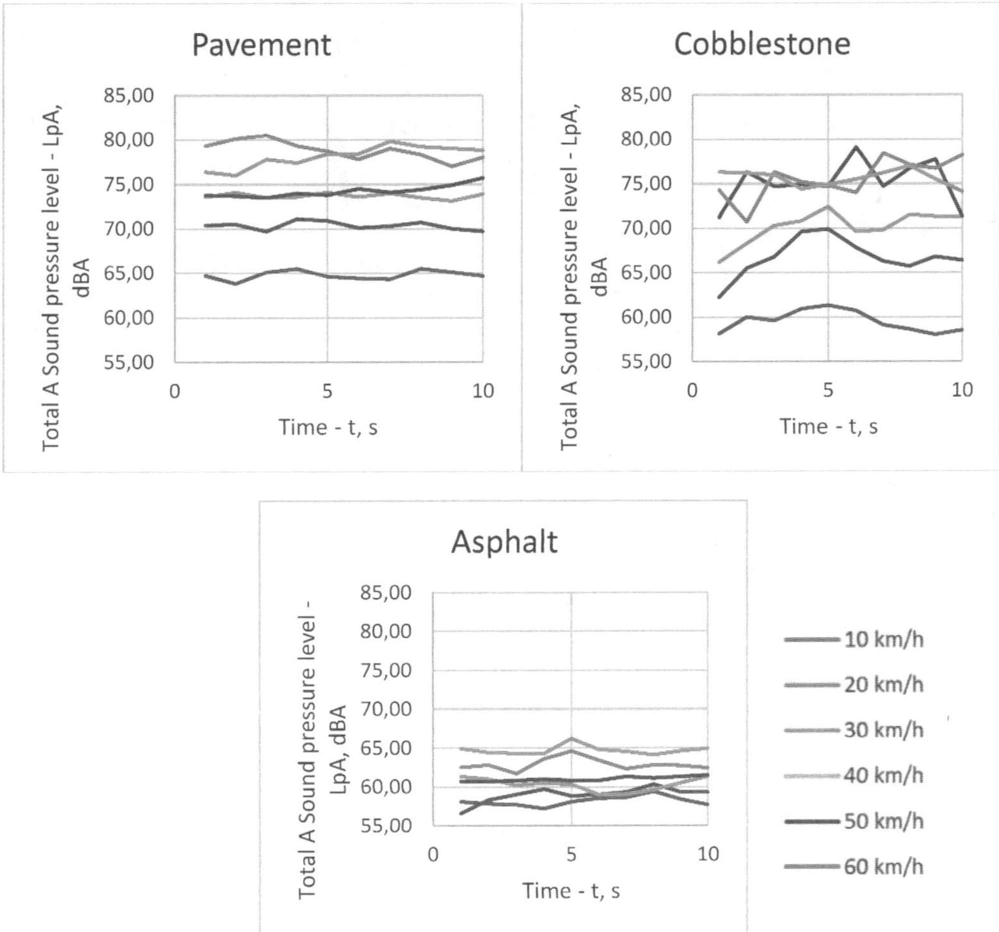

Figure 3. Dynamics over measurement time of A-weighted sound pressure level at different car's speed on 3 types of road surfaces – pavement, cobblestone and asphalt.

At low frequencies $31.5\ Hz$, the greatest is the increase in noise when driving on crushed stone pavement $\Delta Lp = 12.9\ dB$, while for pavement and asphalt these values are $4.5\ dB$ and $3.9\ dB$, respectively. At medium frequencies $500\ Hz$, the difference between the noise at the minimum measurement speed and the maximum is $\Delta Lp = 16\ dB$ for pavement, $\Delta Lp = 15.4\ dB$ for cobblestone and $\Delta Lp = 9.98\ dB$ when driving on asphalt. The difference in the levels at high frequencies is more significant, where for pavement $\Delta Lp = 13.7\ dB$, for cobblestone $19.7\ dB$ and for asphalt $11.8\ dB$.

At the corrected values on the A scale, the differences are $13.3\ dB$ respectively; $15.5\ dB$, and $6.63\ dB$ for paving, cobblestone and asphalt.

When analyzing the individual frequency bands, the following results are observed in figure 4. In all measurements in the individual octave frequency bands, low frequency noise predominates. When driving on a paved road, according to figure 4a), low-frequency noise predominates. It can be seen that at $31.5\ Hz$ the value is $Lp = 61\ dB$, while at $16\ kHz$ it is $10\ dB$. When driving on a gravel road, the values are lower, but the distribution is similar. Here at $31.5\ Hz$ the pressure level value is $Lp = 55.8\ dB$. Driving on asphalt generates noise in the car compartment with maximum values in an octave frequency band of $125\ Hz$ equal to $56.7\ dB$. Here, high frequencies show extremely low values.

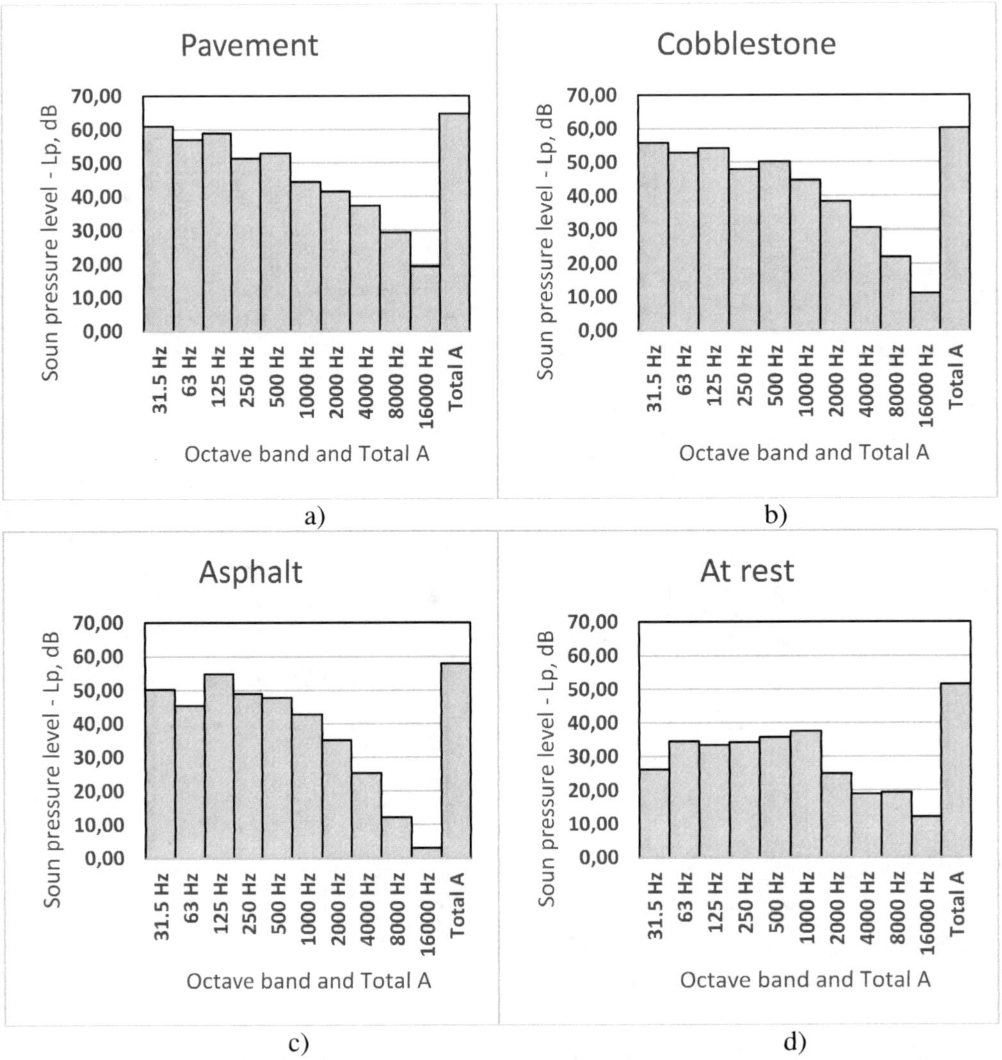

Figure 4. Spectral distribution of sound pressure levels with a width of 1/1 octave band at 3 types of road surface: a) pavement; b) cobblestone; c) asphalt; d) at rest.

To visualize the difference that occurs in the noise when the vehicle is in motion, the spectrum of the noise in a state of rest is shown - figure 4d). It can be seen that the low frequencies have significantly lower values and spectral distribution, indicating that the road surface type is a prerequisite for their generation during traffic. The differences are at the order of $20 - 30\ dB$, which is significant and should be taken into account when designing noise reduction measures.

3.2. Statistical modelling

Different types of statistical models were performed to study how the noise in the passenger car cabin depends on the octave frequency band, the speed and the different types of road surface. The nature of the data suggests starting from a full quadratic model with 10 coefficients of the form (1).

$$Z = \beta_0 + \beta_1 X + \beta_2 Y + \beta_3 U + \beta_4 X^2 + \beta_5 Y^2 + \beta_6 U^2 + \beta_7 XY + \beta_8 XU + \beta_9 YU + \varepsilon, \quad (1)$$

where
Z – noise (in dB)
X – octave frequency bands
Y – velocity (km/h)
U – type of road surface ("-1" - asphalt, "0" - cobblestone, "1" - pavement)
ε – stochastic error.

For the selection of a model in this article, the statistical significance of the coefficients, the adequacy of the model and the adjusted R-squared play a decisive role. From the full quadratic model, due to the insignificance of the coefficient in front of XY, the model (2) is reached.

$$\hat{Z} = 49.4211 - 0.0054X + 0.5876Y + 4.0326U + 0.0000X^2 - 0.0046Y^2 - 4.1150U^2 + 0.0004XU + 0.0462YU. \quad (2)$$

The confidence intervals for the parameters show that all of them are statistically significant (at confidence level $> 99\%$), except the coefficient in front of X^2. The adjusted R-squred is 0.9584, which shows that 95.84% of the variance in the noise could be explained by the factors in the model. The model is adequate (at confidence level $> 99\%$). The confidence intervals for the residuals from the model show that there are 5 observations (from total of 180) for which we have statistically significant errors.

A model excluding X^2 has also been tested. This led to insignificant coefficient in front of YU. Finally, the model (3) is obtained.

$$\hat{Z} = 46.8809 - 0.0027X + 0.5876Y + \widehat{5.6502}U - 0.0046Y^2 - 4.1150U^2 + 0.0004XU. \quad (3)$$

This model is slightly simplified. The adjusted R-squared is 0.9003, which shows slight decrease compared to the previous model. The confidence intervals for the parameters show that all of them are statistically significant (at confidence level $> 99\%$). The model is adequate (at confidence level $> 99\%$). The confidence intervals for the residuals from the model show that there are 8 observations (from total of 180) for which we have statistically significant errors.

The advantage of the last model is all the coefficients are statistically significant and the model is slightly simplified, while the disadvantages include the lower adjusted R-squared and biggest number of statistically significant errors.

Furthermore, different models have been tested depending on the road surface type, again starting from full quadratic models and simplifying the models due to insignificance of the coefficients.

For asphalt the linear model (4) is obtained.

$$Z = \alpha_0 + \alpha_1 X\ . \quad (4)$$

Using the estimates of the coefficients, obtained by the Least Squares Method, the equation (5) is examined.

$$\hat{Z} = 50.8498 - 0.0032X. \quad (5)$$

The confidence intervals for the parameters show that all of them are statistically significant (at confidence level > 99%). The model is adequate (at confidence level > 99%), the adjusted R-squared is 0.8674. The confidence intervals for the residuals from the model show that there are 3 observations (from total of 60) with statistically significant errors.

The regression model for the noise in the passenger car cabin when driving on asphalt in dependence of octave frequency bands is shown on figure 5.

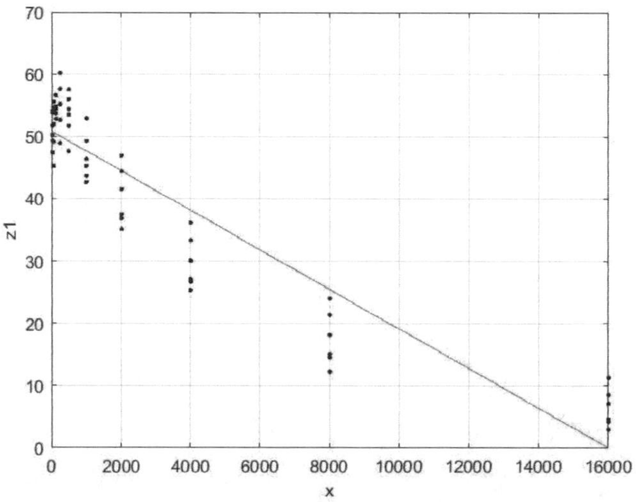

Figure 5. Regression model for asphalt.

For cobblestone and pavement, the models are from type (6).

$$Z = \gamma_0 + \gamma_1 X + \gamma_2 Y + \gamma_3 Y^2 \ . \tag{6}$$

The estimates of the coefficients in the case of cobblestone are as in equation (7).

$$\hat{Z} = 40.9249 - 0.0026X + 0.8631Y - 0.0073Y^2. \tag{7}$$

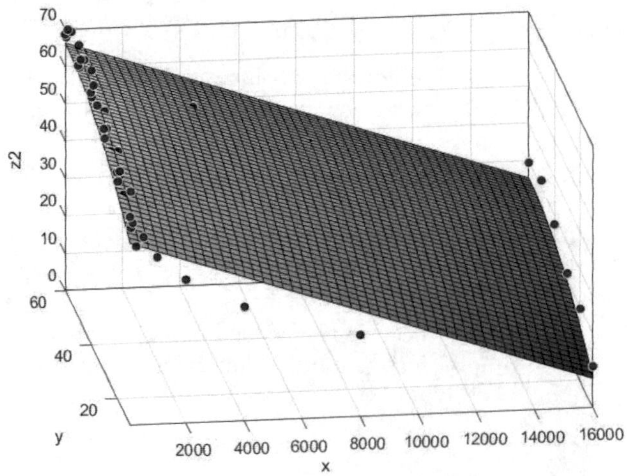

Figure 6. Regression model for cobblestone.

The confidence intervals for the parameters show that all of them are statistically significant (at confidence level > 99%). The model is adequate (at confidence level > 99%), the adjusted R-squaredis 0.9148. The confidence intervals for the residuals from the model show that there are no observations with statistically significant errors.

The regression model for the noise in the passenger car cabin when driving on cobblestone in dependence of octave frequency bands and the velocity is shown on figure 6.

The estimates of the coefficients in the case of pavement are as in equation (8).

$$\hat{Z} = 45.7068 - 0.0023X + 0.8014Y - 0.0077Y^2. \quad (8)$$

The confidence intervals for the parameters show that all of them are statistically significant (at confidence level > 99%). The model is adequate (at confidence level > 99%), the adjusted R-squared is 0.8685. The confidence intervals for the residuals from the model show that there is 1 observation (from a total of 60) with statistically significant error.

The regression model for the noise in the passenger car cabin when driving on pavement in dependence of octave frequency bands and the velocity is shown on figure 7.

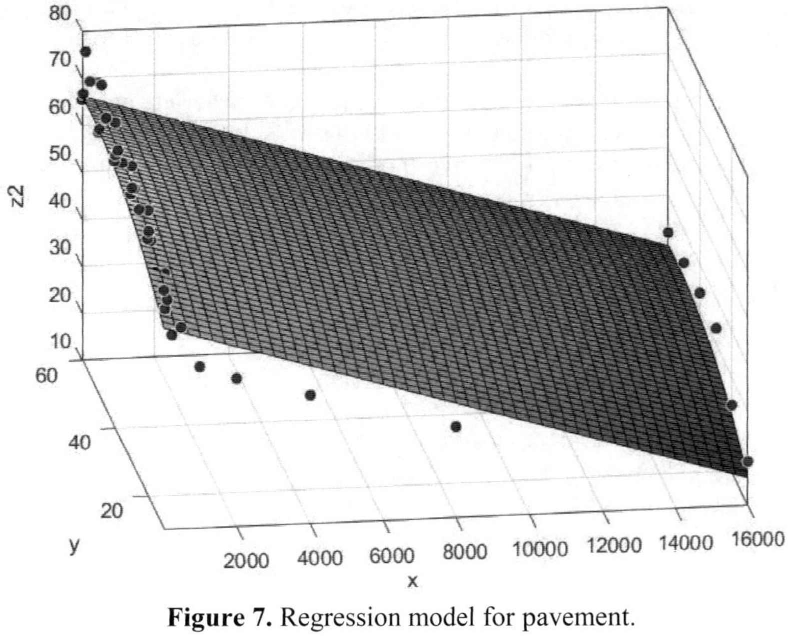

Figure 7. Regression model for pavement.

The models developed on different type of surface show that the noise generated on cobblestone and pavement has similar behavior and depends on both octave frequency bands and velocity, while the noise, generated on asphalt, depends only on the octave frequency bands. The reason for that is most probably because the asphalt is the smoothest surface.

4. Conclusions

The main findings and contributions of this study could be summarized as follows:

1. *Noise Measurements and Analysis:* The study systematically measured and analyzed the noise levels in the cabin of a Citroen C3 car, under different conditions of speed and road surface type (asphalt, cobblestone, and pavement). It sheds light on how these factors interact to affect the comfort and potentially the health of car passengers and drivers. In more details, the noise levels were found to be highest on the paved road and generally increased with speed. Furthermore, the low-frequency noise has been predominant.

2. *Statistical Models:* The study illustrates how such models can help in isolating the effects of individual factors and providing a quantitative understanding of their roles. Specifically, several statistical models are utilized to investigate the factors influencing noise levels. The study started with a full quadratic model, which was then simplified due to the insignificance of certain coefficients. The final model successfully explained a significant proportion of the variance in noise levels (95.84% in the more complex model and 90.03% in the simplified one).
3. *Comparisons Across Road Surfaces:* A notable finding is that the noise generated on asphalt depends only on the octave frequency bands, probably due to the smoother nature of asphalt surfaces. In contrast, noise on cobblestone and pavement depends on both the octave frequency bands and the velocity of the car.
4. *Implications for Noise Reduction Measures:* The findings of this study could be crucial in informing noise reduction measures in cars. For instance, knowing that low-frequency noise is predominant and that the type of the road surface is a significant factor could help in designing specific noise reduction strategies.
5. *Road Design and Planning:* The research also has implications for road design and urban planning. By revealing how different types of road surfaces contribute to noise levels, it suggests that the choice of road surface could be an important consideration in managing urban noise pollution.

The study therefore offers valuable insights for car manufacturers, urban planners, and policymakers interested in improving the noise conditions inside vehicles and in urban environments. Further research may extend this work by investigating other factors affecting car cabin noise and by exploring more complex statistical models.

Acknowledgments
This paper contains results of the work on project No 2023 - FNSE – 04, financed by "Scientific Research" Fund of Ruse University.

References
[1] Todorov V and Georgiev S 2023 *Highly accurate scrambled stochastic approaches for multidimensional sensitivity analysis in air pollution modelling* (Springer Proceedings in Mathematics & Statistics vol 412) ed A Slavova (USA: Springer) pp 389-399
[2] Georgiev S and Vulkov V 2022 *Numerical solving of parabolic Cauchy problems by reduction on bounded domain and application to solute water pollution* (AIP Conf Proc vol 2505) ed G Venkov, N Popivanov and V Pasheva (USA: AIP) 080027
[3] Kovachev N , Manev P and Vladimirov L 2012 *Protection against noise and vibration. Guide for practical exercises* (Pleven: "Mediatech" Publishing house) p. 96
[4] BNS 14478:1982 Noise. Permissible noise levels at work. General information for noise measurements (Sofia)
[5] Ordinance No. 02-20-19 from 12.11.2012 For the Maintenance And Current Repair Of The Roads. Issued by the Minister of Regional Development And Public Works, Promulgation, State Gazette, no. 91 from 20.11.2012 (Sofia)
[6] Ordinance No. 02-20-2 from August 28, 2018 for Design of Roads (Sofia)
[7] BNS ISO 5128:1996 Measurement of noise inside motor vehicles (Sofia)
[8] Mirchev P and Ovtcharov V 1981 *Noise from machines and the macine manufacture* (Sofia)
[9] Hassall J R and Zaveri K 1988 *Acoustic noise measurements* (Sweden: Bruel & Kaer)
[10] EN 13043:2005+AC:2005/NA:2012 Aggregates for bituminous mixtures and surface treatments for roads, airfields and other trafficked areas

Raster image processing using 2D Padé-type approximations

V I Olevskyi[1], Yu B Olevska[2], O V Olevskyi[3] and V V Hnatushenko[1]

[1] Department of Information Technology and Computer Engineering, Dnipro University of Technology, 19 Dmytro Yavornytskyi Avenue, Dnipro, 49005, Ukraine
[2] Department of Applied Mathematics, Dnipro University of Technology, 19 Dmytro Yavornytskyi Avenue, Dnipro, 49005, Ukraine
[3] Department of Computer Technologies, Oles Honchar Dnipro National University, 72 Gagarina Avenue, Dnipro, 49010, Ukraine

Olevskyi.V.I@nmu.one

Abstract. We have developed a method called the two-dimensional Padé-type approximants method, which can be used to reduce the Gibbs phenomenon in the harmonic two-dimensional Fourier series. This method can be applied to both monochrome and color raster images. To do this, we implement the generalized two-dimensional Padé approximation proposed by Chisholm. In this approach, we select the range of frequency values on the integer grid according to the Vavilov method. We propose a definition of a Padé-type functional and provide examples of its application to simple discontinuous templates represented as raster images. Through this study, we are able to draw conclusions about the practical usage and advantages of the Padé-type approximation. We demonstrate that the Padé-type approximant effectively eliminates distortions associated with the Gibbs phenomenon, and it is visually more appropriate than the Fourier approximant. Additionally, the application of the Padé-type approximation reduces the number of parameters without sacrificing precision.

1. Introduction

The primary method of displaying graphic files on a computer screen is through raster images. These images are created by compiling pixels, each containing unique color and tonal information, to form the overall image. Raster image files consist of a rectangular array of pixel parameter values, typically stored in a compressed format. Depending on the type of compression used, it may or may not be possible to fully restore the original image.

Expanding a raster image signal into a two-dimensional Fourier series is a common approach to reducing the size of the graphic file. However, this often results in significant distortions, particularly due to the Gibbs effect. The Gibbs phenomenon is a property of one-dimensional Fourier series that occurs when a discontinuous periodic function is truncated, and only a finite number of terms are used [1-3]. This truncation causes distortions near the points of discontinuity, which cannot be eliminated by increasing the number of series terms.

In the case of two-dimensional images, the Gibbs phenomenon significantly degrades the quality of processed images for most popular graphic standards that utilize a finite sum of harmonics. Distortions occur along the borders of sharp contrast changes, resulting in the appearance of false optical shadows. This negative impact on image quality also affects the analysis of x-ray and sonar studies when processing their results.

The presence of the Gibbs phenomenon is not limited to one-dimensional signals but also applies to the two-dimensional case. This phenomenon has a significant negative impact on the quality of image processing in widely-used graphic standards that rely on a finite sum of harmonics [2]. Distortions occur along the borders where there are sharp changes in contrast, resulting in the false appearance of optical shadows [3, 4]. This issue is particularly problematic when analyzing the results of x-ray and sonar studies. In the context of magnetic resonance imaging (MRI), the Gibbs phenomenon manifests as a type of artifact. In MRI images, parallel lines can be observed that are associated with abrupt and intense changes in the object being imaged, such as the CSF-spinal cord and skull-brain interface [4, 5].

The field of mathematical physics approximation is experiencing rapid development, with the theory of approximation being the most prominent area of progress [1, 6, 7]. Traditionally, the focus has been on approximation techniques that utilize polynomials [8, 9] and trigonometric functions [6, 7, 10]. One particularly successful type of approximation technique is the use of fractional rational functions [8, 11, 12]. The interest in this type of approximation has been steadily growing due to its wide range of applications in various fields, including theoretical physics, applied mechanics, geophysics, and more [8, 11, 12]. These techniques enable the generalized summation of series and the extension of the function being approximated into the meromorphic domain.

Lately, there has been a significant focus on extending the classical theory of approximation using fractional-rational functions. This includes exploring different types of basis functions and various methods of constructing approximants, such as Padé-type approximations [12, 14-19]. By carefully selecting the construction method, significant enhancements in the desirable properties of the approximants can be achieved for specific classes of functions [12-14].

The objective of this study is to address two main goals: eliminating the Gibbs phenomenon in image processing and reducing the size of resulting image files. To accomplish this, we propose utilizing the two-dimensional Padé-type approximants method that we have previously developed. This method has shown promise in achieving these objectives in our previous work [12].

2. Technic of Padé-type approximants usage

2.1. Connection establishing between power series and Fourier series for construction of Padé-type approximants

In our approach, we aim to construct Padé-type approximants for the harmonic two-dimensional Fourier series as a subset of power series [12]. In order to achieve this, we will present some basic principles of construction. We consider the separable space $L_2\left[(a_2,b_2)\times(a_2,b_2)\right]$ of two-dimensional complex functions $f\left(x_1,x_2\right)$, which are integrable on this finite or infinite rectangle. We can choose a countable set of functions

$$B_i=\{e_{ij}, j=\overline{1,\infty}\}, i=\overline{1,2}, \tag{1}$$

as a basis of space with respect to individual coordinates.

In the situation involving trigonometric functions, the fundamental functions can be expressed in the following manner:

$$e_{nk}=\mathrm{e}^{\mathrm{i}kx_n}=\left(\mathrm{e}^{\mathrm{i}x_n}\right)^k=\left(e_{n1}\right)^k, n=1,2. \tag{2}$$

In the given context, the representation of an arbitrary function in the considered space based on the given basis as in equation (2) can be viewed as a two-dimensional generalized power series in the following form:

$$f=\sum_{k,p=1}^{\infty} a_{kp}(e_{11})^k(e_{21})^p. \tag{3}$$

In [12] we have proposed the definition of the functional of the Padé-type in following form.

Definition. Suppose a two-dimensional power series $S=\sum_{k,p=1}^{\infty} a_{kp}(x_1)^k(x_2)^p$ of complex variables x_1, x_2 and the associated Padé approximant $P[m_1,n_1/m_2,n_2](x_1,x_2)$ in the proper sense are given. Then the Padé-type functional $GP=\sum_{k,p=1}^{\infty} a_{kp}(x_1)^k(x_2)^p$ associated with the given generalized power series $GS=\sum_{k,p=1}^{\infty} a_{kp}(f_1)^k(f_2)^p$ for the complex functions of these variables is defined as

$$GP_{GS}[m_1,n_1/m_2,n_2](f_1,f_2)=P[m_1,n_1/m_2,n_2](x_1,x_2)\big|_{x_1=f_1,\,x_2=f_2}\,. \tag{4}$$

The construction process of Padé approximant is outlined as follows:

1. We first select the types of bases for the individual variables as in equation (1).
2. The function *f* that we want to approximate is then represented in the form as in equation (3).
3. We construct a Padé approximant $P[m_1,n_1/m_2,n_2](x_1,x_2)$ in the proper sense for the power series of two complex variables x_1, x_2 using coefficients which correspond to the representation as in equation (3).
4. Finally, we apply a substitution of the basis functions into a functional of Padé-type as shown in equation (4).

The suggested method enables the determination of the essential and complete set of coefficients for constructing a Padé-type approximant with a predefined numerator and denominator structure.

2.2. Technic of Padé-type approximants construction

Suppose we have a function $f(x_1,x_2)$ of monochrome image point brightness in the range of [0,1] on the rectangle $(a_2,b_2)\times(a_2,b_2)$. It can be approximately represented in form of truncated Fourier series

$$f(x_1,x_2)\approx\sum_{m=0}^{M}\sum_{n=0}^{N} f_{mn}\cos(m\lambda_1 x_1)\cos(n\lambda_2 x_2)\,.$$

Let's denote it's Padé approximant $P[m_1,n_1/m_2,n_2](x_1,x_2)$ as $P(x_1,x_2)$.Then to obtain cosine part of two-dimensional exponent, we use the following equality [18]:

$$f(x_1,x_2)\approx\frac{1}{2}\mathrm{Re}\left[P(x_1,x_2)+P(x_1,-x_2)\right]. \tag{5}$$

In cases where the image is stored as a two-dimensional array of points, such as a bmp file, the Padé approximation can be implemented using discrete Fourier transform (DFT) procedures [2]. DFT serves as the foundation for various image and video compression algorithms, including the widely used jpeg and mpeg standards for compressing both still and video images.

To perform the Padé approximation, the input image is first decomposed into spectral components using a two-dimensional discrete cosine transform (2D DCT). The 2D DCT can be computed by applying a one-dimensional DCT algorithm to each row or column of the two-dimensional matrix representing the input signal. This is possible because the DCT is a separable function. The direct calculation of the 2D DCT for an $M\times N$ matrix of a two-dimensional signal $f(x,y)$ can be expressed as:

$$F(u,v)=\frac{2}{\sqrt{MN}}C(u)C(v)\sum_{n=0}^{N-1}\sum_{m=0}^{M-1} f(n,m)\cos\left(\frac{\pi(2n+1)u}{2N}\right)\cos\left(\frac{\pi(2m+1)v}{2M}\right) \tag{6}$$

where $C(x)=2^{-0,5}$ for $x=0$ and $C(x)=1$ for $x\neq 0$.

Coefficients of a truncated cosine Fourier series for $f\left(x_1, x_2\right)$ can be obtained as the values of $F(u,v)$ in equation (6), divided by the step.

Afterwards, the Vavilov method is employed to select the range of frequency values on the integer grid [13]. The size of this range directly influences the number of equations that need to be generated.

To achieve this objective, we employ Chisholm's suggested method for the generalized two-dimensional Padé approximation when $N=M$ [11]. In the case where $f(x,y)$ represents a function of two variables, it can be represented as a power series in two dimensions of the form

$$f(x,y)=\sum_{k,p=1}^{\infty} a_{kp} x^k y^p$$

expressing the N-th Chisholm approximation can be done as follows:

$$f_{N,N}(x,y)=\left(\sum_{k,p=1}^{N} p_{kp} x^k y^p\right)\left(\sum_{k,p=1}^{N} q_{kp} x^k y^p\right)^{-1}$$

If we utilize a collection of power values that is confined within a right triangle where the legs correspond to the axes, then we can determine the coefficients p_{kp} and q_{kp} by employing the subsequent equations:

$$\sum_{\sigma=0}^{\gamma}\sum_{r=0}^{\delta} q_{\sigma r} a_{\gamma-\sigma,\delta-r} = p_{\gamma\delta}, \left(\gamma,\delta=0,1,\ldots,2N, 1\le\delta+\gamma\le 2N\right), \tag{7}$$

$$\sum_{\sigma=0}^{\gamma}\sum_{r=0}^{\delta}\left(q_{\sigma r} a_{\gamma-\sigma,\delta-r} + q_{r\sigma} a_{\delta-r,\gamma-\sigma}\right)=0, \left(\gamma=1,2,\ldots,2N, \delta+\gamma=2N\right),\ p_{00}=1.$$

The solution of this system of equations is a matrix of coefficients p and q.

In our investigations we use the Vavilov method for obtaining approximants coefficients [12]. In this case we use only equation (7) but for a set of power values bounded by two rectangles as it shown in figure 1 and supposing $p_{\gamma\delta}=0,\left(\gamma,\delta>N\right)$.

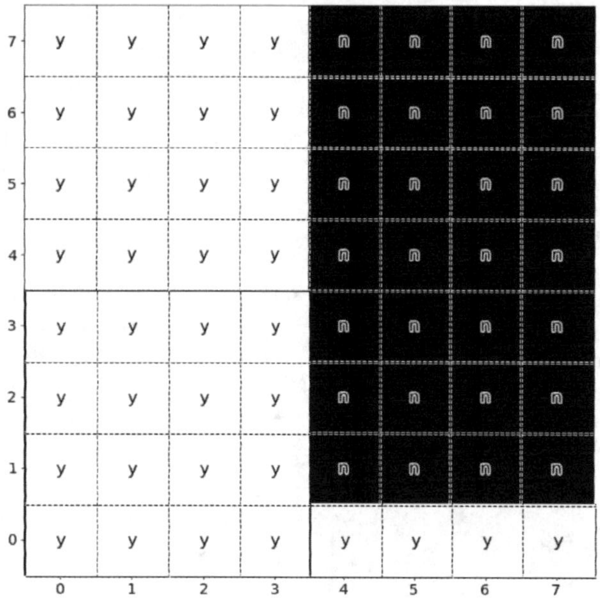

Figure 1. An integer grid of power values for the Vavilov method when N=4: white points with "y" are used for calculations, and black points with "n" aren't.

The process of compressing two-dimensional signals using two-dimensional discrete cosine transformation description can be found in reference [12].

3. Examples of Padé-type approximants application

3.1. Examples for monochrome image case

Here we examine the utilization of Padé-type approximations to basic, non-continuous template functions in both symbolic and discrete forms [12].

3.1.1. Example for formulaic representation of raster. In the general case, such a representation of a raster image assumes the presence of Fourier series already calculated analytically for the brightness function [2].

Let's examine a template function in a periodic form with a period of 2π

$$f(x_1,x_2)=\begin{cases}1, x_1^2+x_2^2\le\pi,\\0, x_1^2+x_2^2>\pi.\end{cases}$$

This function is symmetrical about both x_1 and x_2 axes. Therefore, the truncated Fourier series is considered to be the most appropriate method for its approximation in form

$$f(x_1,x_2)\approx\sum_{m=0}^{4}\sum_{n=0}^{4}f_{mn}\cos(mx_1)\cos(nx_2)$$

We utilized the Padé-type approximation with a structure of [2,2 / 2,2] to equation (6), taking a subset of previously derived Fourier coefficients without any supplementary information. To extract the cosine component of the two-dimensional exponent, we applied the inverse transformation as in equation (5) to the Padé-type approximant. The initial raster image, the desired real approximation and image Fourier series representation is shown in column *No* 1 in table 1.

Table 1. Initial and restored monochrome images.

No	1	2	3	4	5	6	7
Raster image	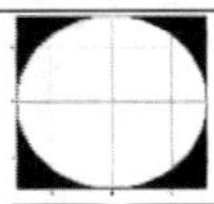	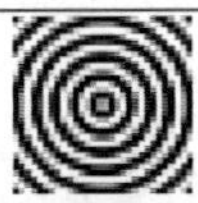		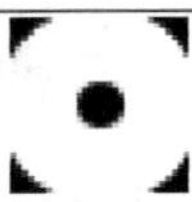	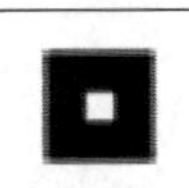		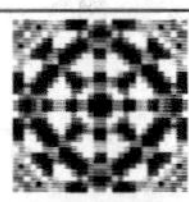
Padé-type approximant	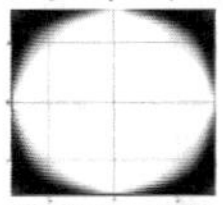	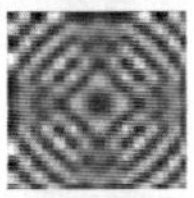					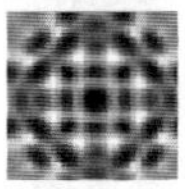
Fourier series with compression	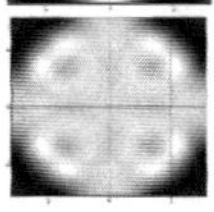						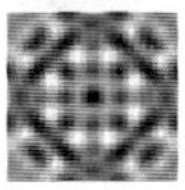
Number of harmonics for minimum error	8	7	6	4	4	5	8

The transformation used for the cosine series in this context is identical to the one employed in the field of radio physics [10].

The image shows distortions that are inherent to the Gibbs phenomenon in the Fourier series, but these distortions are effectively absent in the Padé-type approximant. It is clear that the Padé-type approximant is visually more appropriate than the Fourier series.

In order to assess the accuracy of both the Fourier series method and the Padé-type approximation, we have included a comparison of their respective one-dimensional sections with the template function in figure 2. This comparison highlights the advantage of using the Padé approximation.

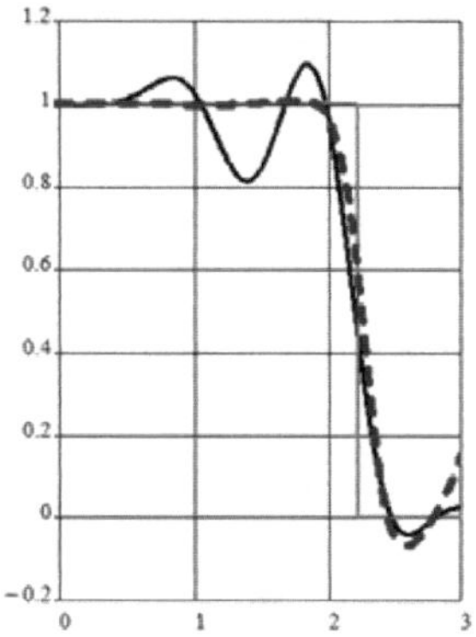

Figure 2. The profiles of the approximations and the template function along line $x_1 = x_2$: red – the template, black – Fourier series, dashed blue – Padé-type approximant.

3.1.2. Examples for discrete representation of raster. In our study, we selected six monochrome symmetrical bitmap test images from the digital library [20], which are listed in columns No 2-7 of table 1. For asymmetrical input signals, we artificially expanded the images to make them symmetrical.

To reduce the file size according to the jpeg standard, insignificant decomposition coefficients were excluded. We applied this method by processing the results of 2D DCP on the input images, considering the rapid decline of harmonic amplitudes. The resulting image was used for comparison with the quality of the image compressed by the proposed method. The standardized root mean square error and the normalized mean absolute error were used as criteria for comparison. The size of the area on the integer grid ranged between 2 and 8, and the number of coefficients used for reconstructing the compressed image was gradually increased. For the compressed image, this number (and thus the file size) was approximately half of the original value.

3.2. Example for color image case

Let's consider processing of color images in RGB format [2]. The main objective of the RGB color model is to facilitate the sensing, representation, and display of images in electronic systems like computers and televisions. This model employs the additive color approach, where the primary colors of red, green, and blue light are combined with varying intensities to reproduce a wide range of colors in a pixel. So, instead one matrix of brightness of monochrome image point, we have three matrixes coinciding every color. and proposed technic can be used for processing every color matrix separately.

The jpeg format, which is most used for storing bitmap files, divides the original image into blocks of 8x8 pixels, followed by processing each block using DCT separately (figure 3).

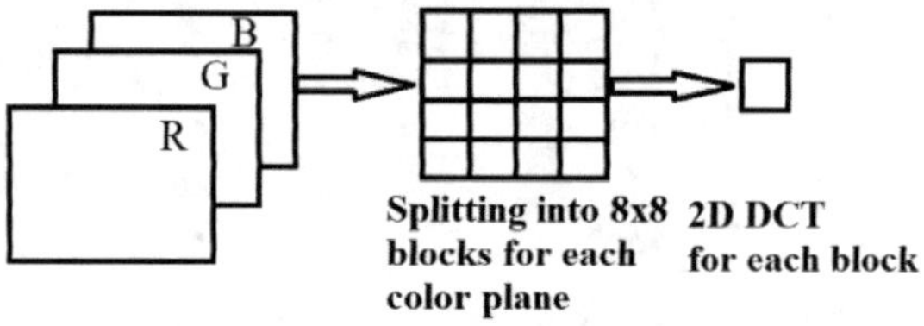

Figure 3. Part of an algorithm for processing of 2D color raster images in jpeg format.

The proposed technic was used for color bitmap "Test card Philips PM5544" (GNU Free Documentation License) [21]. Consecutive image processing steps are presented in figure 4.

4-th Chisholm approximation is used for each 8x8 block of every color matrix. Results of comparison are represented in figure 5. In the approximate calculation of the parameters of the Padé-type approximation, the appearance of conjugate false poles and zeros is possible [11, 22]. This can lead to image scaling failure. It should be noted that similar scaling violations also occur when deploying jpeg files [23]. This phenomenon is eliminated either by cutting off the values of the brightness function, or by using an averaging filter. For this purpose, other methods developed by us for increasing the spatial resolution of multispectral images with automation of the choice of the best pansharpening method can be effectively applied [24, 25].

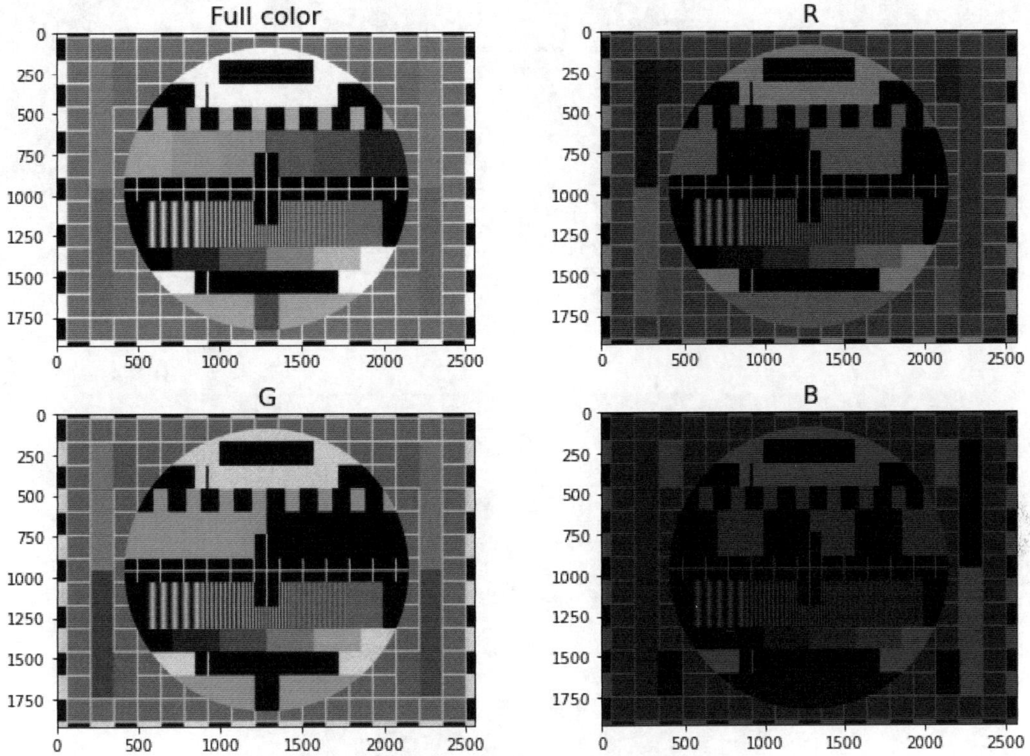

(a)

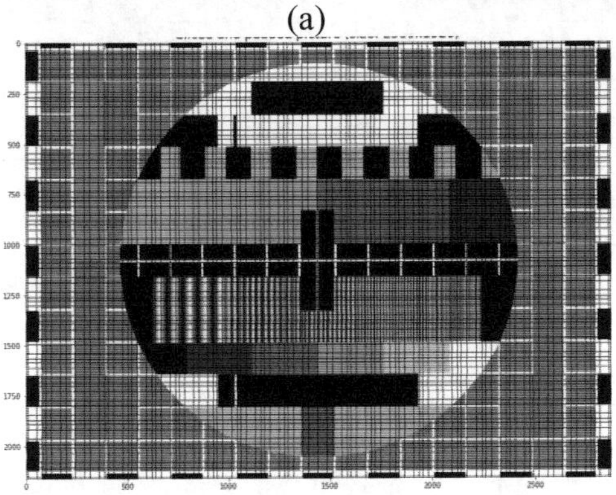

(b)

Figure 4. The algorithm steps for processing of 2D color raster images: (a) – color division, (b) – 8x8 blocks division.

The results shown in figure 5 illustrate that the suggested approach successfully approximates the original image in terms of both clarity and color reproduction. Although there are minor clarity distortions found only in areas with very high spatial frequencies, they do not significantly impact the overall perception of the image. Therefore, the proposed method is suitable for storing and displaying color raster images on electronic devices effectively.

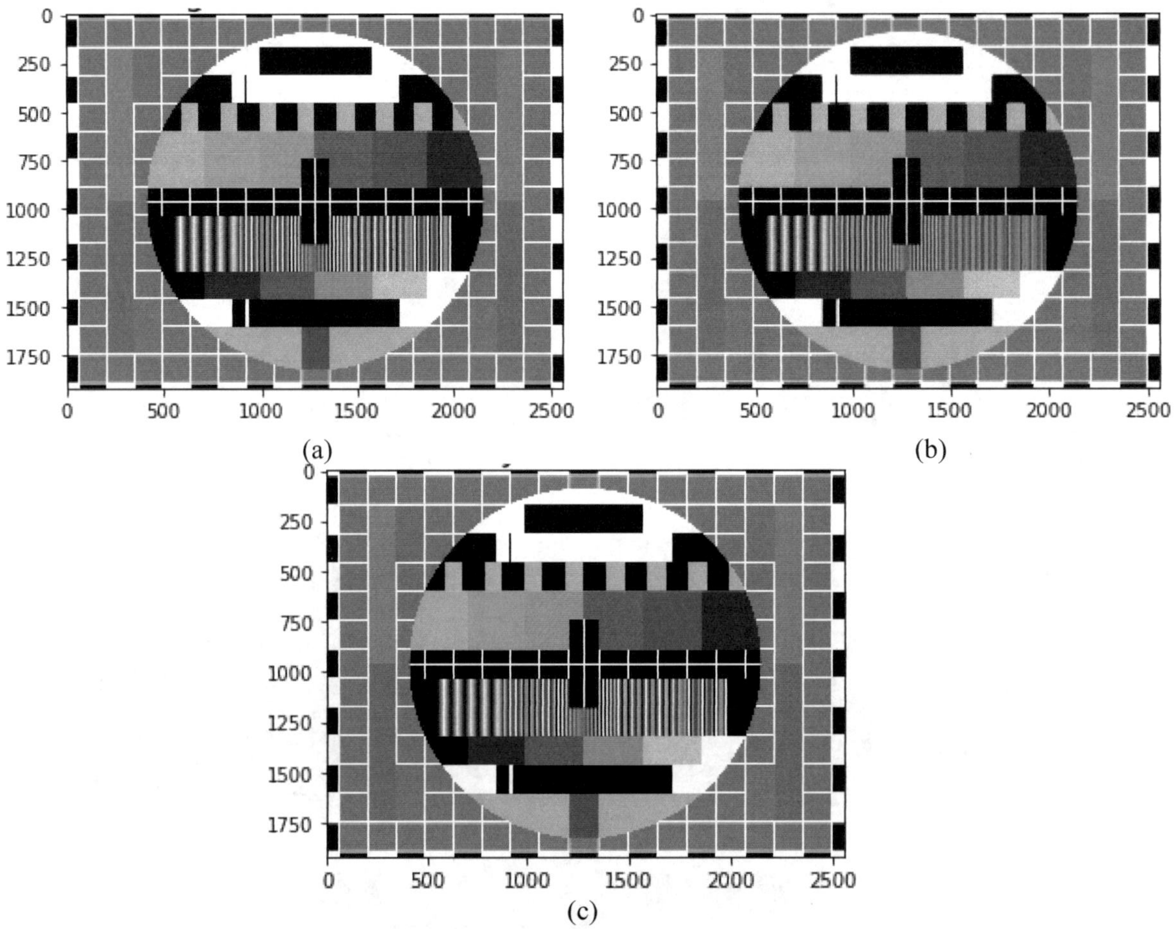

Figure 5. Comparison of processing results: (a) – original raster image, (b) – proposed technics result, (c) – jpeg format image.

4. Analysis of Padé-type approximants usage

Upon studying the mean square error graphs of the images enumerated in table 1, it is noticeable that as the visual similarity between the image and the original improves, there is a substantial reduction in the mean square error. The findings obtained from evaluating the criteria demonstrate that each image type possesses its own distinct threshold, at which point the reconstructed image achieves visual resemblance to the original (see table 1).

The analysis of the Padé-type approximation method and its benefits allows us to draw certain conclusions.

Firstly, Figure 6b illustrates that the Padé approximation exhibits a lower level of noise compared to the Fourier series for the cosine (Figure 6a).

Secondly, the utilization of the Padé-type approximation considerably reduces the number of approximant parameters without compromising accuracy (and sometimes even improving it). Specifically, the Fourier series with N harmonics in both directions requires N^2 parameters. However, employing the Padé-type approximation with the numerator and denominator powers equal to $N/2$ results in $N^2/2-1$ parameters, reducing the number of parameters by more than half.

The results obtained demonstrate that the Padé-type approximant effectively eliminates the distortions inherent to the Gibbs phenomenon. It is clear that the Padé-type approximant is visually more appropriate than the Fourier series for both monochrome and color images.

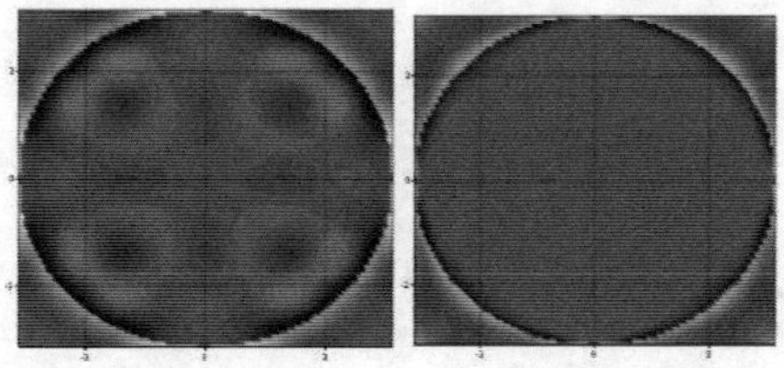

Figure 6. The discrepancies between the template functions and their corresponding approximations for (a) – Fourier cosine series, (b) – Padé approximation

Additionally, using the Padé-type approximation reduces the number of parameters without sacrificing precision. The software that implements this method is suitable for practical use and, furthermore, it can serve as a theoretical basis for developing a new effective color image format in addition to the jpeg format [2, 12].

Acknowledgments
The work is supported by the state budget scientific research project of Dnipro University of Technology "Models and information technologies of data processing and analysis in complex computer systems and networks" (state registration number 0121U114523).

References

[1] Olevskyi V and Olevska Y 2018 Geometric aspects of multiple Fourier series convergence on the system of correctly counted sets *Geom. Integr. Quantization* **19** 159–167
[2] Mitra S K 2010 *Digital Signal Processing: A Computer-Based Approach* (New York: McGraw-Hill)
[3] Helmberg G 2002 Localization of a Corner-Point Gibbs Phenomenon for Fourier Series in Two Dimensions *J. Fourier Anal. Appl.* **8** 29–42
[4] Archibald R and Gelb R 2002 A method to reduce the Gibbs ringing artifact in MRI scans while keeping tissue boundary integrity *IEEE Trans. Med. Imaging* **21** 305–319
[5] Veraart J, Fieremans E, Jelescu I O, Knoll F and Novikov D S 2016 Gibbs ringing in diffusion MRI *Magn. Reson. Med.* **76** 301–314
[6] Olevskyi V and Olevska Y 2018 Mathematical model of elastic closed flexible shells with nonlocal shape deviations *J. Geom. Symmetry Phys.* **50** 57–69
[7] Maggioli F, Melzi S, Ovsjanikov M, Bronstein M M and Rodolà E 2021 Orthogonalized fourier polynomials for signal approximation and transfer *Comput. Graph. Forum* **40** 435–447
[8] Andrianov I, Olevskyi V I and Olevska Y B 2017 *AIP Conf. Proc.* **1895** 080001 https://pubs.aip.org/aip/acp/article-abstract/1895/1/080001/696845/Asymptotic-estimation-of-free-vibrations-of?redirectedFrom=fulltext
[9] Mossakovskii V I, Mil'tsyn A M, Selivanov Yu M and Olevskii V I 1994 Automating the analysis of results of a holographic experiment *Strength Mater.* **26**(5) 385–391
[10] Drobakhin O O, Olevskyi O V and Olevskyi V I 2017 *AIP Conf. Proc.* **1895** 060001 https://pubs.aip.org/aip/acp/article-abstract/1895/1/060001/696883/Study-of-eigenfrequencies-with-the-help-of-Prony-s?redirectedFrom=fulltext
[11] Baker J A Jr and Graves-Morris P 1996 *Padé approximants* (New York: Cambridge University Press)
[12] Olevskyi V I, Hnatushenko V V, Korotenko G M, Olevska Yu B and Obydennyi Y O 2023 Application of two-dimensional Padé-type approximations for image processing *Radio Electronics, Computer Science, Control* **1** 99-106
[13] Olevskyi V I, Smetanin V T and Olevska Yu B 2017 *AIP Conf. Proc.* **1895** 070003 https://pubs.aip.org/aip/acp/article-abstract/1895/1/070003/696753/Fuzzy-method-of-recognition-of-high-molecular?redirectedFrom=fulltext
[14] Labych Yu A and Starovoitov A P 2009 Trigonometric Padé approximants for functions with regularly decreasing Fourier coefficients *Sb. Math.* **200** 1051–1074

[15] Buslaev V I and Suetin S P 2016 On the existence of compacta of minimal capacity in the theory of rational approximation of multi-valued analytic functions *J. Approx. Theory.* **206** 48–67
[16] Holub A P and Lysenko L O 2017 Padé approximants for some classes of multivariate functions *Ukrains'kyi Matematychnyi Zhurnal* **69** 631-40
[17] Holub A P 2014 Many-dimensional generalized moment representations and Padé-type approximants for functions of many variables *Ukrains'kyi Matematychnyi Zhurnal* **66** 1166–1174
[18] Bosuwan N and López Lagomasino G 2015 Inverse theorem on row sequences of linear Padé-orthogonal approximation *Comput. Methods Funct. Theory* **15** 529–554
[19] Daras, N.J. 2003 Generalized Padé-type approximation and integral representations *Adv. Comput. Math.* **18** 1–39
[20] TESTIMAGES free collection of digital images for testing *Access mode:* https://testimages.org/Received 00.00.2023.
[21] Ebnz 2008, Jun. 13 Test card Philips PM5544 *Access mode:* https://en.wikipedia.org/wiki/File:Philips_PM5544.svg
[22] Yamada H and Ikeda K 2014 A Numerical Test of Padé approximation for some functions with singularity *International Journal of Computational Mathematics* **2014**
[23] Khalid Y Understanding and Decoding a JPEG Image using Python *Access mode:* https://yasoob.me/posts/understanding-and-writing-jpeg-decoder-in-python/
[24] Hnatushenko V and Kashtan V 2021 Automated pansharpening information technology of satellite images *Radio Electronics, Computer Science, Control* **2** 123–132
[25] Gnatushenko V 2003 The use of geometrical methods in multispectral image processing *J. Autom. Inf. Sci.* **35** 1–8

Orthonormal eigenfunction expansions for sixth-order boundary value problems

N C Papanicolaou[1] and I C Christov[1,2]

[1] Department of Computer Science, University of Nicosia, 46 Makedonitissas Avenue, CY-2417 Nicosia, Cyprus

[2] School of Mechanical Engineering, Purdue University, West Lafayette, Indiana 47907, USA

E-mail: papanicolaou.n@unic.ac.cy, christov@purdue.edu

Abstract. Sixth-order boundary value problems (BVPs) arise in thin-film flows with a surface that has elastic bending resistance. To solve such problems, we first derive a complete set of odd and even orthonormal eigenfunctions — resembling trigonometric sines and cosines, as well as the so-called "beam" functions. These functions intrinsically satisfy boundary conditions (BCs) of relevance to thin-film flows, since they are the solutions of a self-adjoint sixth-order Sturm–Liouville BVP with the same BCs. Next, we propose a Galerkin spectral approach for sixth-order problems; namely the sought function as well as all its derivatives and terms appearing in the differential equation are expanded into an infinite series with respect to the derived complete orthonormal (CON) set of eigenfunctions. The unknown coefficients in the series expansion are determined by solving the algebraic system derived by taking successive inner products with each member of the CON set of eigenfunctions. The proposed method and its convergence are demonstrated by solving two model sixth-order BVPs.

1. Introduction

Sixth-order parabolic equations arise in a number of physical contexts. For example, early work by King [1] on isolation oxidation of silicon led to a model featuring a sixth-order parabolic equation due to the effect of diffusion and reaction on substrate deformation during the oxidation process. Importantly, King [1] observed that the resulting partial differential equation is a higher-order version of the well-known nonlinear, degenerate parabolic equations of second and fourth order that describe the height of a thin liquid film during gravity- and surface-tension-driven spreading on a rigid surface. At vastly different scales, the flow of magma under earthen layers leads to the formation of domes called *laccoliths*. The model used the describe the formation and spread of these geological features [2, 3, 4, 5] also led to sixth-order parabolic equations, similar to the one derived by King [1], but usually in axisymmetric polar coordinates.

In the last two decades, there has been significant interest in fundamental questions about the behavior of solutions of nonlinear sixth-order parabolic equations, motivated by problems of thin liquid films spreading underneath an elastic membrane [6, 7, 8, 9, 10, 11]. With the advent of cheap and rapid fabrication methods and improved imaging modalities, a growing area of application for these models and studies has been actuation in microfluidics [12, 13], including impact mitigation [14]. In these contexts, it is often of interest to understand infinitesimal perturbations of the film and their stability, leading to sixth-order eigenvalue problems.

In the mathematics literature, general properties of sixth-order eigenvalue problems have been discussed by Greenberg and Marletta [15, 16]. However, the eigenfunctions are seldom explicitly constructed, the studies focusing instead on calculation of eigenvalues for problems arising in hydrodynamic stability [17] or vibrations of sandwich beams [18, 19]. We surmise that constructing the eigenfunctions of sixth-order eigenvalue problems would be enlightening, and that these functions may resemble the well-known "beam" functions [17, 20, 21] of the fourth-order eigenvalue problem.

Thus, in the present work, motivated by recent interest in sixth-order parabolic equations, we study a model initial-boundary-value problem for a sixth-order equation (section 2), construct the eigenfunctions of its associated eigenvalue problem (section 3), use the latter to propose a Galerkin spectral method [22, 23] for the solution of the former (section 4), and demonstrate the approach on two elementary model problems (section 5).

2. Problem formulation: elastic-plated thin film

2.1. Problem definition and flow geometry

Here, we study a thin fluid film of equilibrium height h_0, underneath an elastic interface in a closed trough of width 2ℓ, as shown in figure 1. The fluid is considered incompressible and Newtonian with (constant) density ρ_f and (constant) dynamic viscosity μ_f. The elastic sheet has (constant) in-plane tension T and (constant) bending resistance B. The sheet is idealized as an interface with no mass (thus, no inertia during its motion). Gravity is the only body force considered to act on the fluid, and it acts in the $-y$ direction. The elastic interface is free to move vertically along the lateral walls at $x = \pm\ell$ while maintaining a 90° contact angle (non-wetting).

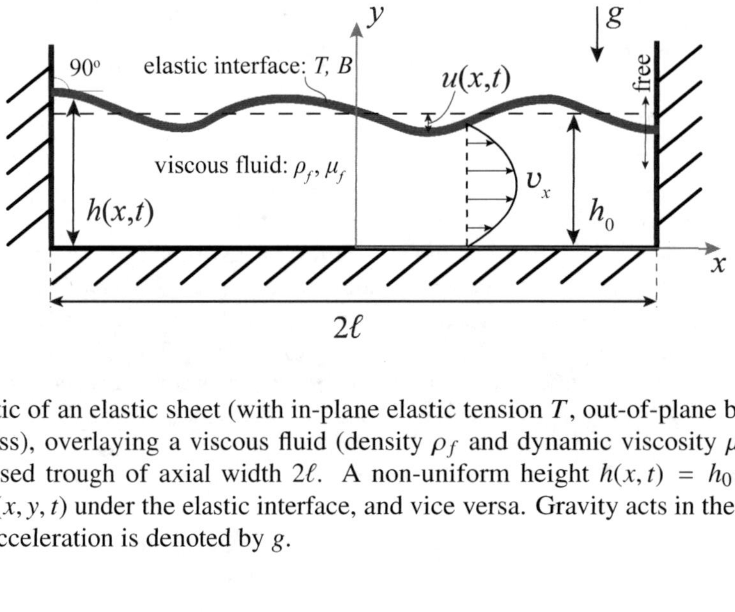

Figure 1: Schematic of an elastic sheet (with in-plane elastic tension T, out-of-plane bending rigidity B, and negligible mass), overlaying a viscous fluid (density ρ_f and dynamic viscosity μ_f) of equilibrium height h_0 in a closed trough of axial width 2ℓ. A non-uniform height $h(x,t) = h_0 + u(x,t)$ drives a horizontal flow $v_x(x,y,t)$ under the elastic interface, and vice versa. Gravity acts in the $-y$-direction, and the gravitational acceleration is denoted by g.

2.2. Derivation of the governing thin-film equation

We study the dynamics of the thin film in the long-wave limit [24, 25], also known as the lubrication approximation [26, 27], such that $h_0 \ll \ell$ as well as $\max_{x,t} h(x,t) \ll \ell$ during the film's evolution. Under this assumption, it can be shown [7, 8, 9, 10] that the dimensional governing equations for the viscous fluid flow under the elastic sheet reduce to:

$$\text{conservation of mass:} \qquad 0 = \frac{\partial v_x}{\partial x} + \frac{\partial v_y}{\partial y}, \tag{2.1a}$$

$$x\text{-momentum balance:} \qquad 0 \approx -\frac{\partial p}{\partial x} + \mu_f \frac{\partial^2 v_x}{\partial y^2}, \tag{2.1b}$$

y-momentum balance: $$0 \approx -\frac{\partial p}{\partial y} - \rho_f g, \tag{2.1c}$$

dynamic interface condition: $$-p|_{y=h} \approx T\frac{\partial^2 h}{\partial x^2} - B\frac{\partial^4 h}{\partial x^4}, \tag{2.1d}$$

kinematic interface condition: $$\frac{\partial h}{\partial t} = v_y|_{y=h}. \tag{2.1e}$$

See also [6, 28] for more in-depth discussions of the dynamic boundary condition at a fluid–elastic solid interface and its linearization in the long-wave limit.

Under the lubrication approximation, the flow is approximately unidirectional and $v_y \ll v_x$. The dominant velocity component can be obtained from (2.1b) as

$$v_x(x, y, t) = -\frac{1}{2\mu_f}\frac{\partial p}{\partial x}y[h(x, t) - y], \tag{2.2}$$

having enforced the no-slip conditions $v_x(x, 0, t) = v_x(x, h, t) = 0$ at the bottom of the trough ($y = 0$) and along the elastic interface ($y = h$). Based on (2.2), the volumetric flux *per unit width* is

$$q := \int_0^{h(x,t)} v_x(x, y, t)\,\mathrm{d}y = -\frac{h(x, t)^3}{12\mu_f}\frac{\partial p}{\partial x}. \tag{2.3}$$

The vertical velocity can be obtained from the conservation of mass equation (2.1a) using v_x from (2.2) as

$$v_y(x, y, t) = -\frac{\partial}{\partial x}\int_0^y v_x(x, \tilde{y}, t)\,\mathrm{d}\tilde{y}, \tag{2.4}$$

having enforce the no-penetration condition $v_y(x, 0, t) = 0$ at the bottom of the trough ($y = 0$). Using (2.3) and (2.4), the kinematic condition (2.1e) becomes

$$\frac{\partial h}{\partial t} = -\frac{\partial q}{\partial x}. \tag{2.5}$$

Next, fixing the pressure at the interface $y = h(x, t)$ to be the "ambient pressure" $p_a = \text{const.}$, equations (2.1c) and (2.1d) allow us to solve for the pressure distribution in the thin fluid film:

$$p(x, y, t) = p_a + \rho_f g[h(x, t) - y] - T\frac{\partial^2 h}{\partial x^2} + B\frac{\partial^4 h}{\partial x^4}. \tag{2.6}$$

Finally, using the pressure (2.6) to eliminate $\partial p/\partial x$ from the flow rate expression (2.3), and substituting the result into (2.5), we obtain the (long-wave) *thin film equation*:

$$\frac{\partial h}{\partial t} = \frac{\partial}{\partial x}\left[\frac{h^3}{12\mu_f}\left(\rho_f g\frac{\partial h}{\partial x} - T\frac{\partial^3 h}{\partial x^3} + B\frac{\partial^5 h}{\partial x^5}\right)\right]. \tag{2.7}$$

Observe that (2.7) is a nonlinear partial differential equation (PDE), nominally of parabolic type, but with possible degeneracy at a "contact line" where $h \to 0$, which leads to a wealth of possible contact-line behaviors [7, 29] beyond the scope of this work.

2.3. Nondimensionalization

We introduce the dimensionless ("hat") variables via the transformations:

$$x = \ell\hat{x}, \qquad t = \frac{12\mu_f \ell^6}{B h_0^3}\hat{t}, \qquad h = h_0\hat{h}. \tag{2.8}$$

We focus on the case in which the dominant force on the interface is the bending resistance, thus the choice of time scale is $t_c = 12\mu_f \ell^6 / Bh_0^3$, as in [14, 30]. Substituting the variables from (2.8) into (2.7) and dropping the hats, we arrive at:

$$\frac{\partial h}{\partial t} = \frac{\partial}{\partial x}\left[h^3\left(\mathcal{B}\frac{\partial h}{\partial x} - \mathcal{T}\frac{\partial^3 h}{\partial x^3} + \frac{\partial^5 h}{\partial x^5}\right)\right]. \tag{2.9}$$

The following dimensionless numbers have arisen:

$$\text{Elastic Bond number:} \qquad \mathcal{B} = \frac{\rho_f g \ell^4}{B}, \tag{2.10a}$$

$$\text{Tension number:} \qquad \mathcal{T} = \frac{T\ell^2}{B}. \tag{2.10b}$$

The elastic Bond number $\mathcal{B}$ quantifies the relative importance of gravity's role in deforming the elastic interface relative to its bending resistance (see also discussion in [31]). The tension number $\mathcal{T}$ quantifies the relative importance of in-plane tension's role in deforming the elastic interface relative to its bending resistance (see also the discussions in [28, 32] and [12, 13]).

We are interested in small perturbations to a flat film, $h = 1$, so we introduce $h(x,t) = 1 + \varepsilon u(x,t)$ with $\varepsilon \ll 1$ into (2.9) and keep only terms at $O(\varepsilon)$, to obtain

$$\frac{\partial u}{\partial t} = \mathcal{B}\frac{\partial^2 u}{\partial x^2} - \mathcal{T}\frac{\partial^4 u}{\partial x^4} + \frac{\partial^6 u}{\partial x^6}. \tag{2.11}$$

Equation (2.11) is the *linear* sixth-order PDE of interest in the remainder of this work.

2.4. Boundary conditions

Equation (2.11) requires six boundary conditions (BCs). Some are discussed in [8] for a thin film problem and in [28, section 4.4.2] in the context of beams. According to the problem posed in section 2.1 and shown in figure 1, we are interested in a nonwetting, non-pinned film in a closed trough. A non-wetting film will maintain a 90° contact at the wall, thus

$$\left.\frac{\partial u}{\partial x}\right|_{x=\pm 1} = 0. \tag{2.12a}$$

Since the film is not pinned (i.e., it is free), $u(\pm 1, t)$ is free and not necessarily $= 0$. Similarly, since the film is confined in a trough with walls, the walls provide shear force/resistance to the motion, so that $(\partial^3 u/\partial x^3)|_{x=\pm 1}$ is not necessarily $= 0$, but this in turn requires that there be no moment at the wall:

$$\left.\frac{\partial^2 u}{\partial x^2}\right|_{x=\pm 1} = 0. \tag{2.12b}$$

Finally, we observe from the derivation of (2.7), that if the fluid flux q is to vanish at $x = \pm 1$ (closed trough) then, taking into account (2.12a) and (2.12b), we must additionally require that

$$\left.\left[-\mathscr{T}\frac{\partial^3 u}{\partial x^3} + \frac{\partial^5 u}{\partial x^5}\right]\right|_{x=\pm 1} = 0.$$

If we restrict ourselves to cases in which tension in negligible compared to bending ($\mathscr{T} \to 0$), then

$$\left.\frac{\partial^5 u}{\partial x^5}\right|_{x=\pm 1} = 0. \tag{2.12c}$$

3. The sixth-order Sturm–Liouville boundary value problem

In the context of the PDE (2.11) subject to the BCs (2.12), we consider the following sixth-order initial boundary value problem (IBVP) for $u = u(x,t)$:

$$\begin{cases} \dfrac{\partial u}{\partial t} - \mathcal{B}\dfrac{\partial^2 u}{\partial x^2} + \mathcal{T}\dfrac{\partial^4 u}{\partial x^4} - \dfrac{\partial^6 u}{\partial x^6} = f(x,t), & (x,t) \in (-1,1)\times \in (0,\infty), \quad (3.1a) \\ \left.\dfrac{\partial u}{\partial x}\right|_{x=\pm 1} = \left.\dfrac{\partial^2 u}{\partial x^2}\right|_{x=\pm 1} = \left.\dfrac{\partial^5 u}{\partial x^5}\right|_{x=\pm 1} = 0, & t \in (0,\infty), \quad (3.1b) \\ u(x,0) = u^0(x), & x \in (-1,1), \quad (3.1c) \end{cases}$$

where $u^0(x)$ and $f(x,t)$ are given smooth functions. Although, on physical grounds, we took $\mathscr{T} \to 0$ to derive BC (2.12c), we have kept $\mathscr{T}$ in (3.1a) for the purposes of the mathematical discussion that follow.

To solve IBVP (3.1), we will employ a Galerkin method (section 4). To this end, we must construct a suitable set of eigenfunctions, and determine their properties, which is the subject of the remainder of this section.

3.1. Self-adjointness, orthogonality and completeness

The Sturm–Liouville eigenvalue problem (EVP) associated with IBVP (3.1) reads:

$$-\frac{d^6\psi}{dx^6} = \lambda^6\psi, \tag{3.2a}$$

$$\left.\frac{d\psi}{dx}\right|_{x=\pm 1} = \left.\frac{d^2\psi}{dx^2}\right|_{x=\pm 1} = \left.\frac{d^5\psi}{dx^5}\right|_{x=\pm 1} = 0. \tag{3.2b}$$

Before we proceed with the derivation of the solutions (eigenfunctions) of EVP (3.2), we first discuss why they possess certain properties required for the development of our Galerkin method in section 4. The desired results follow as direct applications of the comprehensive general theory on eigenvalue problems found in the seminal book of Coddington and Levinson [33]. To facilitate this discussion for the reader, we quote or paraphrase the relevant definitions and theorems from [33] accordingly. We begin with the definitions.

Definition 3.1. [Adjoint of a linear operator (from [33], chapter 3, section 6)] Consider the linear differential operator

$$\mathscr{L}_n(\,\cdot\,) := a_0\frac{d^n}{dx^n}(\,\cdot\,) + a_1\frac{d^{n-1}}{dx^{n-1}}(\,\cdot\,) + \cdots + (a_n\,\cdot), \tag{3.3}$$

where $a_0, a_1, \ldots, a_n$ are continuous (possibly complex-valued) functions of x on some interval $[a,b] \subset \mathbb{R}$, with $a_0(x) \neq 0\ \forall x \in [a,b]$. Then, the adjoint of $\mathscr{L}_n$ is another linear operator, denoted $\mathscr{L}_n^+$, also of order n given by

$$\mathscr{L}_n^+(\,\cdot\,) := (-1)^n\frac{d^n}{dx^n}(\bar{a}_0\,\cdot\,) + (-1)^{n-1}\frac{d^{n-1}}{dx^{n-1}}(\bar{a}_1\,\cdot\,) + \cdots + (\bar{a}_n\,\cdot), \tag{3.4}$$

where the bar ($\bar{}$) denotes the complex conjugate.

Throughout this work we will use the notation $\langle\cdot,\cdot\rangle$ for the L^2 inner product on the interval $[a,b]$, that is, for any $\phi, \psi \in L^2[a,b]$,

$$\langle\phi,\psi\rangle := \int_a^b \phi(x)\psi(x)\,dx. \tag{3.5}$$

Also, $\|\cdot\|$ will denote the associated L^2 norm unless explicitly specified otherwise.

Definition 3.2. [Self-adjoint eigenvalue problem (from [33], chapter 7, section 2)]
Let $\mathscr{L}_n$ be the n-th order operator given by

$$\mathscr{L}_n\phi = p_0\frac{\mathrm{d}^n\phi}{\mathrm{d}x^n} + p_1\frac{\mathrm{d}^{n-1}\phi}{\mathrm{d}x^{n-1}} + \cdots + p_n\phi, \tag{3.6a}$$

where the $p_i(x)$ are (possibly complex-valued) functions of class C^{n-j}, $j = 0, 1, \ldots, n$ on the closed interval $[a, b]$ and $p_0(x) \neq 0$ on $[a, b]$. Let

$$U_j\phi = \sum_{k=1}^{n}\left(M_{jk}\phi^{(k-1)}(a) + N_{jk}\phi^{(k-1)}(b)\right) \qquad (j = 1, \ldots, n), \tag{3.6b}$$

where the M_{jk} and N_{jk} are constants. Denote the boundary conditions $U_j\phi = 0$, $j = 1, \ldots, n$ by $\mathscr{U}\phi = 0$. The eigenvalue problem

$$\mathscr{L}_n\phi = \lambda\phi\,, \qquad \mathscr{U}\phi = 0, \tag{3.6c}$$

is *self-adjoint* if

$$\langle\mathscr{L}_n\phi, \psi\rangle = \langle\phi, \mathscr{L}_n\psi\rangle \tag{3.6d}$$

for all $\phi, \psi \in C^n[a, b]$ which satisfy the boundary conditions $\mathscr{U}\phi = \mathscr{U}\psi = 0$.

The first important consequence of self-adjointness is reflected in theorem 3.1.

Theorem 3.1. [From [33], theorem 2.1 of chapter 7, section 2] A self-adjoint EVP has a real, enumerable set of eigenvalues with corresponding distinct and mutually-orthogonal eigenfunctions.

We now show the following proposition.

Proposition 3.1. The sixth-order EVP (3.2) with BCs (3.2b) is *self-adjoint*.

Proof. The operator $\mathscr{L} := -\mathrm{d}^6/\mathrm{d}x^6$ from (3.2a) is symmetric ($\mathscr{L} = \mathscr{L}^+$). Let $\psi(x)$ be a solution of EVP (3.2) and $\phi(x) \in C^6[-1, 1]$ be any function satisfying BCs (3.2b). The $L^2[-1, 1]$ inner product between equation (3.2a) and ϕ is:

$$\int_{-1}^{+1} -\frac{\mathrm{d}^6\psi}{\mathrm{d}x^6}\phi\,\mathrm{d}x = \lambda^6\int_{-1}^{+1}\psi\phi\,\mathrm{d}x. \tag{3.7}$$

Integrating the left-hand side by parts yields

$$-\left[\frac{\mathrm{d}^5\psi}{\mathrm{d}x^5}\phi\right]\Bigg|_{x=-1}^{x=+1} + \int_{-1}^{+1}\frac{\mathrm{d}^5\psi}{\mathrm{d}x^5}\frac{\mathrm{d}\phi}{\mathrm{d}x}\,\mathrm{d}x = \lambda^6\int_{-1}^{+1}\psi\phi\,\mathrm{d}x. \tag{3.8}$$

Repeating five more times, we find

$$\langle\mathscr{L}\psi, \phi\rangle = J(\psi, \phi) + \langle\psi, \mathscr{L}\phi\rangle, \tag{3.9}$$

where

$$J(\psi, \phi) := \left[\frac{\mathrm{d}^5\psi}{\mathrm{d}x^5}\phi - \frac{\mathrm{d}^4\psi}{\mathrm{d}x^4}\frac{\mathrm{d}\phi}{\mathrm{d}x} + \frac{\mathrm{d}^3\psi}{\mathrm{d}x^3}\frac{\mathrm{d}^2\phi}{\mathrm{d}x^2} - \frac{\mathrm{d}^2\psi}{\mathrm{d}x^2}\frac{\mathrm{d}^3\phi}{\mathrm{d}x^3} + \frac{\mathrm{d}\psi}{\mathrm{d}x}\frac{\mathrm{d}^4\phi}{\mathrm{d}x^4} - \psi\frac{\mathrm{d}^5\phi}{\mathrm{d}x^5}\right]\Bigg|_{x=-1}^{x=+1}. \tag{3.10}$$

For the BCs (3.2b), we find $J(\psi, \phi) = 0$. Hence,

$$\langle\mathscr{L}\psi, \phi\rangle = \langle\psi, \mathscr{L}\phi\rangle, \tag{3.11}$$

for any solution $\psi(x)$ of EVP (3.2) and any function $\phi(x) \in C^6[-1, 1]$ satisfying the BCs (3.2b). □

The second important result regarding self-adjoint EVPs was shown in the expansion and completeness theorems, namely, theorem 4.1 and theorem 4.2 of chapter 7, section 4, of [33]. We paraphrase this result as follows.

Proposition 3.2. [Expansion and completeness] Let $\{\chi_m\}_{m=1}^{\infty}$ be an orthonormal sequence of solutions of the self-adjoint EVP (3.6) on $[a,b]$. Then, for any $f \in L^2[a,b]$,

$$f = \sum_{m=1}^{\infty} \langle f, \chi_m \rangle \chi_m, \tag{3.12}$$

where the equality implies convergence in the L^2 norm.

3.2. Constructing the orthonormal eigenfunctions

We now solve EVP (3.2) and derive our complete orthonormal (CON) set of solutions. The characteristic equation corresponding to (3.2a) has roots

$$\left\{+\lambda \mathrm{i}, -\lambda \mathrm{i}, \left(\tfrac{\mathrm{i}}{2} + \tfrac{\sqrt{3}}{2}\right)\lambda, \left(\tfrac{\mathrm{i}}{2} - \tfrac{\sqrt{3}}{2}\right)\lambda, \left(-\tfrac{\mathrm{i}}{2} + \tfrac{\sqrt{3}}{2}\right)\lambda, \left(-\tfrac{\mathrm{i}}{2} - \tfrac{\sqrt{3}}{2}\right)\lambda\right\}. \tag{3.13}$$

Therefore, the general solution of the homogeneous version of (3.2a) can be written in the form

$$\begin{aligned}\psi(x) = {} & \overbrace{c_1 \sin(\lambda x) + c_5 \sin\left(\tfrac{1}{2}\lambda x\right)\cosh\left(\tfrac{\sqrt{3}}{2}\lambda x\right) + c_6 \cos\left(\tfrac{1}{2}\lambda x\right)\sinh\left(\tfrac{\sqrt{3}}{2}\lambda x\right)}^{\psi^s(x)\ \text{(odd)}} \\ & + \underbrace{c_2 \cos(\lambda x) + c_3 \sin\left(\tfrac{1}{2}\lambda x\right)\sinh\left(\tfrac{\sqrt{3}}{2}\lambda x\right) + c_4 \cos\left(\tfrac{1}{2}\lambda x\right)\cosh\left(\tfrac{\sqrt{3}}{2}\lambda x\right)}_{\psi^c(x)\ \text{(even)}}. \end{aligned} \tag{3.14}$$

We have chosen to express the solution in this way, following the example of the so-called "beam" functions [17, 20, 21], in order to have odd and even sets of eigenfunctions resembling trigonometric sines and cosines.

Imposing BCs (3.2b) on each of the $\psi^s(x)$ and $\psi^c(x)$ separately yields homogeneous linear systems for the vectors $\mathbf{c}_{\text{odd}} = (c_1, c_5, c_6)^\top$ and $\mathbf{c}_{\text{even}} = (c_2, c_3, c_4)^\top$, respectively. To find nontrivial solutions we set the determinants of the coefficient matrices equal to zero and arrive at the eigenvalue relations:

$$\text{even}: \quad \cos(2\lambda^c) + \sqrt{3}\sin\lambda^c \sinh(\sqrt{3}\,\lambda^c) - \cos\lambda^c \cosh(\sqrt{3}\,\lambda^c) = 0, \tag{3.15a}$$

$$\text{odd}: \quad \sin(2\lambda^s) + \sqrt{3}\cos\lambda^s \sinh(\sqrt{3}\,\lambda^s) + \sin\lambda^s \cosh(\sqrt{3}\,\lambda^s) = 0. \tag{3.15b}$$

Each solution λ_m^c of (3.15a) is an eigenvalue corresponding to an even eigenfunction ψ_m^c of EVP (3.2), whereas each solution λ_m^s of (3.15b) corresponds to an odd eigenfunction ψ_m^s. Eigenvalue relations (3.15) were solved numerically using a highly accurate numerical solver—Mathematica's `FindRoot` [34] with 16 digits of working precision. We also derived large-λ asymptotic formulas for the eigenvalues. Using (3.15a), we found that $\lambda_m^c \sim (m + 1/6)\pi$ as $\lambda \to \infty$, whereas (3.15b) yields $\lambda_m^s \sim (m - 1/3)\pi$ as $\lambda \to \infty$. The results are presented in table 1. As we can see, the asymptotic formulas are extremely accurate even for $m = 6$, with the formulas for λ_m^c and λ_m^s agreeing with the results obtained with `FindRoot` for 12 and 11 digits, respectively. Thus, in practice when implementing a Galerkin method, there will be no need to use numerical root-finding to determine λ_m^c and λ_m^s for $m \geq 7$.

What remains is to determine the constants c_2, c_3 and c_4 for the set of even functions $\{\psi_m^c\}$ and c_1, c_5 and c_6 for the set of odd functions $\{\psi_m^s\}$, which were introduced in (3.14). These constants are found by imposing BCs (3.2b), utilizing the eigenvalue relations (3.15), and normalizing the eigenfunctions

Table 1: "Even" and "odd" eigenvalues, evaluated by solving the transcendental equations (3.15) with MATHEMATICA's [34] `FindRoot`, as well as the proposed asymptotic formulas.

	even		odd	
m	λ^c_m	$(m+1/6)\pi$	λ^s_m	$(m-1/3)\pi$
0	0	–	–	–
1	3.66606496814	3.66519142919	2.07175679767	2.09439510239
2	6.80678029161	6.80678408278	5.23608751229	5.23598775598
3	9.94837675280	9.94837673637	8.37757997731	8.37758040957
4	13.0899693899	13.0899693900	11.5191730650	11.5191730632
5	16.2315620436	16.2315620435	14.6607657167	14.6607657168
6	19.3731546971	19.3731546971	17.8023583704	17.8023583703

with respect to the L^2 norm. After a lengthy calculation, using MATHEMATICA's algebraic manipulation capabilities, we arrive at the expressions:

$$\psi^c_m(x) = c^c_m\Bigg\{\frac{4\sin\lambda^c_m}{\cos\lambda^c_m - \cosh(\sqrt{3}\,\lambda^c_m)}\Bigg[-\cos\tfrac{\lambda^c_m}{2}\sinh\tfrac{\sqrt{3}\,\lambda^c_m}{2}\sin\tfrac{\lambda^c_m}{2}x\sinh\tfrac{\sqrt{3}\,\lambda^c_m}{2}x + \sin\tfrac{\lambda^c_m}{2}\cosh\tfrac{\sqrt{3}\,\lambda^c_m}{2}\cos\tfrac{\lambda^c_m}{2}x\cosh\tfrac{\sqrt{3}\,\lambda^c_m}{2}x\Bigg] + \cos\lambda^c_m x\Bigg\}, \tag{3.16a}$$

$$\psi^s_m(x) = c^s_m\Bigg\{\frac{4\cos\lambda^s_m}{\cos\lambda^s_m + \cosh(\sqrt{3}\,\lambda^s_m)}\Bigg[-\cos\tfrac{\lambda^s_m}{2}\cosh\tfrac{\sqrt{3}\,\lambda^s_m}{2}\sin\tfrac{\lambda^s_m}{2}x\cosh\tfrac{\sqrt{3}\,\lambda^s_m}{2}x + \sin\tfrac{\lambda^s_m}{2}\sinh\tfrac{\sqrt{3}\,\lambda^s_m}{2}\cos\tfrac{\lambda^s_m}{2}x\sinh\tfrac{\sqrt{3}\,\lambda^s_m}{2}x\Bigg] + \sin(\lambda^s_m x)\Bigg\}, \tag{3.16b}$$

where

$$c^c_m = 2\sqrt{\tfrac{\lambda^c_m}{d^c_m}}\left[\cos\lambda^c_m - \cosh(\sqrt{3}\,\lambda^c_m)\right], \tag{3.16c}$$

$$c^s_m = 2\sqrt{\tfrac{\lambda^s_m}{d^s_m}}\left[\cos\lambda^s_m + \cosh(\sqrt{3}\,\lambda^s_m)\right], \tag{3.16d}$$

$$\begin{aligned} d^c_m &= \sin 4\lambda^c_m - 6\lambda^c_m(\cos 2\lambda^c_m - 2) + 2\lambda^c_m\cosh(2\sqrt{3}\lambda^c_m) + 2\sin 2\lambda^c_m\cosh^2\sqrt{3}\,\lambda^c_m \\ &\quad + \cosh\sqrt{3}\,\lambda^c_m\,[\sin\lambda^c_m - 3\sin 3\lambda^c_m + 4\lambda^c_m(\cos 3\lambda^c_m - 3\cos\lambda^c_m)] \\ &\quad + 4\sqrt{3}\sin^2\lambda^c_m\sinh\sqrt{3}\,\lambda^c_m(\cos\lambda^c_m - \cosh\sqrt{3}\lambda^c_m), \end{aligned} \tag{3.16e}$$

$$\begin{aligned} d^s_m &= 12\lambda^s_m - 3\sin(2\lambda^s_m) - \sin(4\lambda^s_m) + 10\lambda^s_m\cos(2\lambda^s_m) \\ &\quad - [\sin(2\lambda^s_m) - 2\lambda^s_m]\cosh(2\sqrt{3}\lambda^s_m) - 4\sqrt{3}\cos^2\lambda^s_m\sinh(\sqrt{3}\lambda^s_m)[\cos\lambda^s_m + \cosh(\sqrt{3}\lambda^s_m)] \\ &\quad + 2\cos\lambda^s_m\cosh(\sqrt{3}\lambda^s_m)\,(4\lambda^s_m[\cos(2\lambda^s_m) + 2] - 3\sin(2\lambda^s_m)), \end{aligned} \tag{3.16f}$$

for $n, m \in \mathbb{N}$.

It is important to note here that $\lambda^c_0 = 0$ is an eigenvalue of EVP (3.2) with corresponding eigenfunction $\psi^c_0(x) = 1$. The notation is due to the fact that $\psi(x) = 1$ is an even function. The profiles of the first members of the two sets of eigenfunctions are presented in figure 2.

We conclude the this section by stating the following proposition, which forms the foundation of our spectral method.

Proposition 3.3. [Complete orthonormal set] The two sets of solutions ψ^c_m and ψ^s_m of Sturm–Liouville problem (3.2) given by (3.16) and supplemented by $\psi^c_0(x) = 1$ form a *complete orthonormal set of*

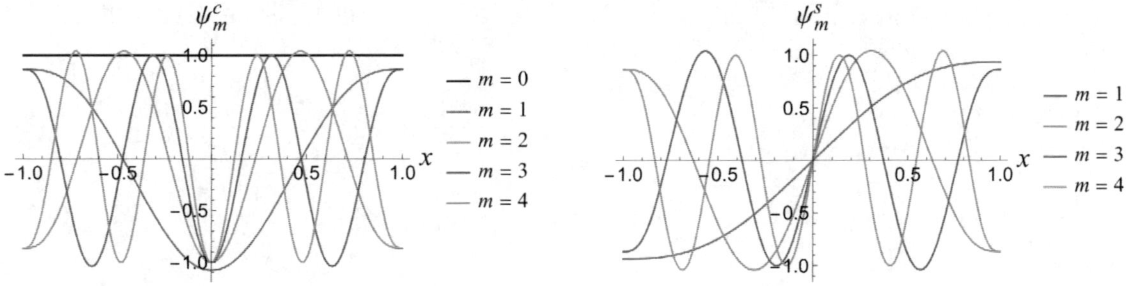

(a) Even eigenfunctions. (b) Odd eigenfunctions.

Figure 2: The profiles of the (a) even and (b) odd eigenfunctions for $m = 1, 2, 3, 4$, along with them the eigenfunction $\psi_0^c(x) = 1$ for $m = 0$.

functions (CON) on $L^2[-1, 1]$. Therefore, any function $u(x) \in L^2[-1, 1]$ can be expanded as

$$u(x) = \frac{1}{2} u_0^c \psi_0^c(x) + \sum_{m=1}^{\infty} u_m^c \psi_m^c(x) + u_m^s \psi_m^s(x), \tag{3.17a}$$

$$u_0^c := \langle u(x), \psi_0^c(x) \rangle = \int_{-1}^{+1} u(x)\, \mathrm{d}x, \tag{3.17b}$$

$$u_m^c := \langle u(x), \psi_m^c(x) \rangle = \int_{-1}^{+1} u(x) \psi_m^c(x)\, \mathrm{d}x, \quad m \geq 1, \tag{3.17c}$$

$$u_m^s := \langle u(x), \psi_m^s(x) \rangle = \int_{-1}^{+1} u(x) \psi_m^s(x)\, \mathrm{d}x, \quad m \geq 1, \tag{3.17d}$$

with the series (3.17) converging to $u(x)$ in $L^2[-1, 1]$.

The proposition follows from proposition 3.1, theorem 3.1 and proposition 3.2. It is clear that the only nonzero coefficients in the spectral expansion (3.17) of any even (or odd) function $f \in L^2[-1, 1]$ are the u_m^c (or u_m^s).

In this section n indicated the order of the EVP, and m was the index used for the countable infinity of eigenvalues and eigenfunctions, we have now restricted ourselves to a specific sixth-order EVP. Thus, in the following sections, we reuse n as the index for the countable infinity of eigenvalues and eigenfunctions to simplify the notation.

4. The Galerkin spectral method

4.1. Method formulation

Having established the necessary theoretical foundation and derived our CON basis functions, we proceed with the development of the proposed Galerkin spectral method [22, 23] for IBVP (3.1). To solve the sixth-order PDE (3.1a) using a Galerkin method based on the results above, we expand the sought function (i.e., the solution of the IBVP) $u(x, t)$ as in (3.17), allowing the expansion coefficients to be functions of time, and truncate the series at M terms:

$$u(x, t) \approx u_{\text{spectral}}(x, t) = \frac{1}{2} u_0^c(t) \psi_0^c(x) + \sum_{n=1}^{M} u_n^c(t) \psi_n^c(x) + u_n^s(t) \psi_n^s(x). \tag{4.1}$$

Note that the spectral expansion (4.1) will intrinsically satisfy BCs (3.1b), which is an important advantage of the classical Galerkin approach.

Next, we must express all the x-derivatives that appear in the PDE (3.1a) as linear combinations of the same basis functions:

$$\frac{d^2\psi_n^{c,s}}{dx^2} = \frac{1}{2}\beta_{n0}^{c,s} + \sum_{m=1}^{\infty}\beta_{nm}^{c,s}\psi_m^{c,s}(x), \qquad \text{where} \quad \beta_{nm}^{c,s} := \left\langle \frac{d^2\psi_n^{c,s}}{dx^2}, \psi_m^{c,s}\right\rangle = \int_{-1}^{+1}\frac{d^2\psi_n^{c,s}}{dx^2}\psi_m^{c,s}\,dx, \tag{4.2a}$$

$$\frac{d^4\psi_n^{c,s}}{dx^4} = \frac{1}{2}\gamma_{n0}^{c,s} + \sum_{m=1}^{\infty}\gamma_{nm}^{c,s}\psi_m^{c,s}(x), \qquad \text{where} \quad \gamma_{nm}^{c,s} := \left\langle \frac{d^4\psi_n^{c,s}}{dx^4}, \psi_m^{c,s}\right\rangle = \int_{-1}^{+1}\frac{d^4\psi_n^{c,s}}{dx^4}\psi_m^{c,s}\,dx. \tag{4.2b}$$

Note that, while $\beta_{n0}^s = \gamma_{n0}^s = 0$ by definition, it can be shown that $\beta_{n0}^c = 0$ but $\gamma_{n0}^c \neq 0$. The expressions for $\beta_{nm}^{c,s}$ and $\gamma_{nm}^{c,s}$ will be given in section 4.2 below. It follows that

$$\frac{\partial^2 u_{\text{spectral}}}{\partial x^2} = \sum_{n=1}^{M}\sum_{m=1}^{M} u_n^c(t)\beta_{nm}^c\psi_m^c(x) + \sum_{n=1}^{M}\sum_{m=1}^{M} u_n^s(t)\beta_{nm}^s\psi_m^s(x), \tag{4.3a}$$

$$\frac{\partial^4 u_{\text{spectral}}}{\partial x^4} = \sum_{n=1}^{M} u_n^c(t)\left[\frac{1}{2}\gamma_{n0}^c + \sum_{m=1}^{M}\gamma_{nm}^c\psi_m^c(x)\right] + \sum_{n=1}^{M}\sum_{m=1}^{M} u_n^s(t)\gamma_{nm}^s\psi_m^s(x), \tag{4.3b}$$

$$\frac{\partial^6 u_{\text{spectral}}}{\partial x^6} = -\sum_{n=1}^{M}(\lambda_n^c)^6 u_n^c(t)\psi_n^c(x) - \sum_{n=1}^{M}(\lambda_n^s)^6 u_n^s(t)\psi_n^s(x). \tag{4.3c}$$

Next, substituting (4.1) and (4.3) into (3.1a), taking successive inner products with $\psi_0^c(x) = 1$, $\{\psi_\ell^c(x)\}_{\ell=1}^M$ and $\{\psi_\ell^s(x)\}_{\ell=1}^M$, and using the orthogonality of the eigenfunctions, we obtain a semi-discrete (dynamical) system:

$$\frac{du_0^c}{dt} = \sum_{n=1}^{M}\left[-\mathcal{T}\gamma_{n0}^c\right]u_n^c(t) + f_0^c, \qquad (\ell = 0), \tag{4.4a}$$

$$\frac{du_\ell^c}{dt} = \sum_{n=1}^{M}\left[-\mathcal{T}\gamma_{n\ell}^c + \mathcal{B}\beta_{n\ell}^c - (\lambda_n^c)^6\delta_{nl}\right]u_n^c(t) + f_\ell^c, \qquad \ell \geq 1, \tag{4.4b}$$

$$\frac{du_\ell^s}{dt} = \sum_{n=1}^{M}\left[-\mathcal{T}\gamma_{n\ell}^s + \mathcal{B}\beta_{n\ell}^s - (\lambda_n^s)^6\delta_{nl}\right]u_n^s(t) + f_\ell^s, \qquad \ell \geq 1. \tag{4.4c}$$

where $\{f_0^c(t), f_m^c(t), f_m^s(t)\}$ are the expansion coefficients of the known right-hand side $f(x,t)$ of (3.1a), according to proposition 3.3. System (4.4) can be solved using a Crank–Nicolson-type approach for the time discretization, but its solution is the object of future work.

4.2. Expansion formulas

We first present the formula for expanding the second derivative of an even eigenfunction into a series of the even basis functions. After a lengthy calculation, using MATHEMATICA's algebraic manipulation capabilities, from (4.2a) we find that for a fixed $n \in \mathbb{N}$, and any $m = 1, 2, 3, \dots$

$$\beta^c_{nm} = \begin{cases} \frac{6c^c_n c^c_m (\lambda^c_m)^3 (\lambda^c_n)^3}{(\lambda^c_m)^6 - (\lambda^c_n)^6} \left[\frac{\lambda^c_m \sin \lambda^c_n \left(\cos 2\lambda^c_m - \sqrt{3} \sin \lambda^c_m \sinh \sqrt{3}\lambda^c_m - \cos \lambda^c_m \cosh \sqrt{3}\lambda^c_m \right)}{\cos \lambda^c_m - \cosh \sqrt{3}\lambda^c_m} \right. \\ \qquad \left. + \frac{\lambda^c_n \sin \lambda^c_m \left(-\cos 2\lambda^c_n + \sqrt{3} \sin \lambda^c_n \sinh \sqrt{3}\lambda^c_n + \cos \lambda^c_n \cosh \sqrt{3}\lambda^c_n \right)}{\cos \lambda^c_n - \cosh \sqrt{3}\lambda^c_n} \right] & \text{for } n \neq m, \\ -\frac{\lambda^c_n (c^c_n)^2}{8(\cos \lambda^c_n - \cosh \sqrt{3}\lambda^c_n)^2} \left[\lambda^c_n \left(3\cos 2\lambda^c_n + \sinh^2 \sqrt{3}\lambda^c_n + \cosh^2 \sqrt{3}\lambda^c_n \right. \right. \\ \qquad \left. + 4\sqrt{3} \sin^3 \lambda^c_n \sinh \sqrt{3}\lambda^c_n - 4\cos^3 \lambda^c_n \cosh \sqrt{3}\lambda^c_n \right) \\ \qquad + 2 \sin \lambda^c_n (\cos \lambda^c_n - \cosh \sqrt{3}\lambda^c_n) \\ \qquad \left. \times \left(\sqrt{3} \sin \lambda^c_n \sinh \sqrt{3}\lambda^c_n + \cos \lambda^c_n \cosh \sqrt{3}\lambda^c_n - \cos 2\lambda^c_n \right) \right] & \text{for } n = m, \end{cases} \tag{4.5}$$

where c^c_n and c^c_m are given by (3.16c).

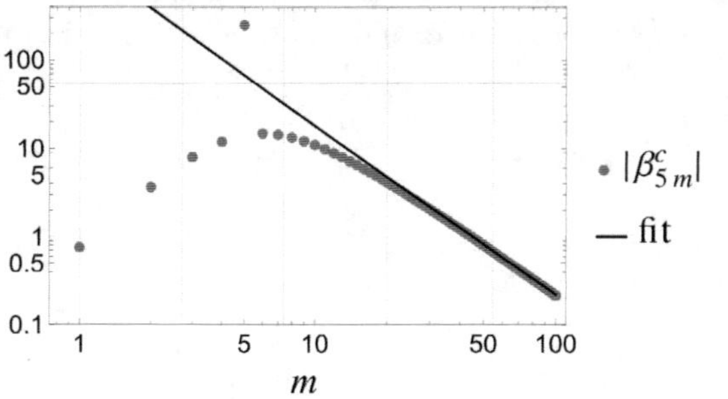

(a)

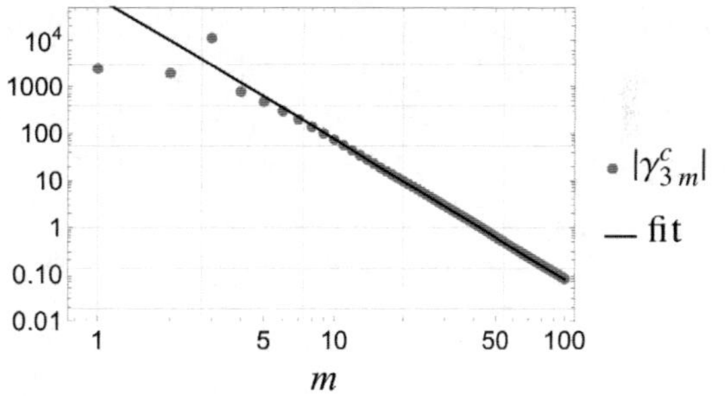

(b)

Figure 3: Log-log plots demonstrating the convergence rate of the spectral expansions of (a) the second derivative of the fifth even eigenfunction and (b) the fourth derivative of the third even eigenfunction. Symbols correspond to (a) $|\beta^c_{5m}|$ and (b) $|\gamma^c_{3m}|$. Solid lines are the best fits based on values for $m \geq 50$; specifically: (a) $1\,455\, m^{-1.91}$ and (b) $78\,160\, m^{-3}$. The series are truncated at $m = M = 100$.

The 'c' case of the expansion (4.2a) and (4.5) for the second derivative of even eigenfunctions was verified numerically for different values of n. In figure 3(a) the convergence rate of the spectral series (4.2a) for $n = 5$ is demonstrated. We observe that $|\beta^c_{5m}|$ decays algebraically for large m as $|\beta^c_{5m}| \sim m^{-1.91}$. This decay was confirmed for all the examined values of n.

This convergence rate could be very roughly estimated by inspecting the derived expression for β_{nm} and noting that, in this context, m is increasing with n being fixed. Looking at the upper branch of (4.5), i.e., the case $n \neq m$, and noting that the dominant terms as $m \to \infty$ are the powers of λ^c_m, the hyperbolic functions of λ^c_m and the term c^c_m, we find $|\beta^c_{nm}| \sim c^c_m \cdot m^{-2}$ for m large. A similar analysis of the lower branch of (4.5) explains why terms of the form β_{nn} do not follow the overall convergence rate. This observation is confirmed for β^c_{55} in figure 3(a).

For completeness, we also present the corresponding formula for expanding the second derivative of an odd eigenfunction into a series of the odd basis functions. From (4.2a), we find that

$$\beta^s_{nm} = \begin{cases} \frac{6c^s_n c^s_m (\lambda^s_m)^3 (\lambda^s_n)^3}{(\lambda^s_m)^6 - (\lambda^s_n)^6} \Bigg[-\frac{\lambda^s_m \cos\lambda^s_n \left(\sin 2\lambda^s_m - \sqrt{3}\cos\lambda^s_m \sinh\sqrt{3}\,\lambda^s_m + \sin\lambda^s_m \cosh\sqrt{3}\lambda^s_m\right)}{\cos\lambda^s_m + \cosh\sqrt{3}\lambda^s_m} \\ \qquad + \frac{\lambda^s_n \cos\lambda^s_m \left(\sin 2\lambda^s_n - \sqrt{3}\cos\lambda^s_n \sinh\sqrt{3}\,\lambda^s_n + \cos\lambda^s_n \cosh\sqrt{3}\lambda^c_n\right)}{\cos\lambda^c_n - \cosh\sqrt{3}\lambda^c_n} \Bigg] & \text{for } n \neq m, \\ \frac{\lambda^s_n (c^s_n)^2}{2(\cos\lambda^s_n + \cosh\sqrt{3}\lambda^s_n)^2} \Big[\lambda^s_n \Big(-\cos 2\lambda^s_n + \sinh^2\sqrt{3}\lambda^s_n + \cosh^2\sqrt{3}\,\lambda^s_n \\ \qquad -4\sqrt{3}\cos^2\lambda^s_n \sin\lambda^s_n \sinh\sqrt{3}\,\lambda^s_n + 4\sin^2\lambda^s_n \cos\lambda^s_n \cosh\sqrt{3}\,\lambda^s_n \Big) \\ \qquad + 2\cos\lambda^s_n (\cos\lambda^s_n + \cosh\sqrt{3}\lambda^s_n) \\ \qquad \times \Big(\sqrt{3}\cos\lambda^s_n \sinh\sqrt{3}\lambda^s_n - \sin\lambda^s_n \cosh\sqrt{3}\lambda^s_n - \sin 2\lambda^s_n \Big) \Big] & \text{for } n = m, \end{cases} \tag{4.6}$$

where c^s_n and c^s_m are given by (3.16d). The 's' case of the expansion (4.2a) and (4.6) for the second derivative of odd eigenfunctions was also verified numerically for different values of n. Although, we do not show it in a separate plot, we checked the convergence rate of the spectral series (4.2a) and (4.6) for $n = 5$, observing that $|\beta^s_{5m}| \sim m^{-1.94}$. This decay was confirmed for all the examined values of n, and can again be justified by examining the derived expression (4.6) for β^s_{nm}.

We also derived the corresponding expressions for expanding the fourth derivatives of our basis functions (see (4.2b)). We first present the formulas for the even case. For a fixed $n \in \mathbb{N}$ and $m = 1, 2, 3, \ldots$, we found

$$\gamma^c_{nm} = \begin{cases} \frac{3c^c_n c^c_m (\lambda^c_n)^6}{(\lambda^c_m)^6 - (\lambda^c_n)^6} \Bigg[-\frac{(\lambda^c_m)^3 \sin\lambda^c_m \left(-3 + \cos 2\lambda^c_n + 2\cos\lambda^c_n \cosh\sqrt{3}\lambda^c_n\right)}{\cos\lambda^c_n - \cosh\sqrt{3}\lambda^c_n} \\ \qquad + \frac{(\lambda^c_n)^3 \sin\lambda^c_n \left(-3 + \cos 2\lambda^c_m + 2\cos\lambda^c_m \cosh\sqrt{3}\lambda^c_m\right)}{\cos\lambda^c_m - \cosh\sqrt{3}\lambda^c_m} \Bigg] & \text{for } n \neq m, \\ \frac{(\lambda^c_n)^3 (c^c_n)^2}{8\left(\cos\lambda^c_n - \cosh\sqrt{3}\lambda^c_n\right)^2} \Big[6\sin 2\lambda^c_n \left(\cosh 2\sqrt{3}\lambda^c_n + 4\right) - 3\sin 4\lambda^c_n - 48\sin\lambda^c_n \cosh\sqrt{3}\,\lambda^c_n \\ \qquad +4\lambda^c_n \Big(-4\sqrt{3}\sin^3\lambda^c_n \sinh\sqrt{3}\lambda^c_n - 4\cos^3\lambda^c_n \cosh\sqrt{3}\lambda^c_n \\ \qquad +3\cos 2\lambda^c_n + \cosh 2\sqrt{3}\,\lambda^c_n \Big) \Big] & \text{for } n = m. \end{cases} \tag{4.7a}$$

As already mentioned in subsection 4.1, the coefficients $\gamma^c_{n0} \neq 0$ and are given by

$$\gamma^c_{n0} = 6c^c_n (\lambda^c_n)^3 \sin\lambda^c_n \qquad \text{for } n \in \mathbb{N}. \tag{4.7b}$$

In figure 3(b) the convergence rate of the spectral series (4.2b) and (4.7) for $n = 3$ is demonstrated, observing that $|\gamma^c_{3m}| \sim m^{-3}$. This decay was confirmed for all the examined values of n, and can again be justified by examining the derived expression (4.7a) for γ^s_{nm}.

The corresponding expression for the fourth derivative of the odd eigenfunctions reads

$$\gamma^s_{nm} = \begin{cases} \frac{6c^s_n c^s_m (\lambda^s_n)^6}{(\lambda^s_n)^6 - (\lambda^s_m)^6} \Bigg[\frac{(\lambda^s_m)^3 \cos\lambda^s_m \sin\lambda^s_n \left(\cosh\sqrt{3}\lambda^s_n - \cos\lambda^s_n\right)}{\cos\lambda^s_n + \cosh\sqrt{3}\lambda^s_n} \\ \qquad + \frac{(\lambda^s_n)^3 \sin\lambda^s_m \cos\lambda^s_n \left(\cos\lambda^s_m - \cosh\sqrt{3}\lambda^s_m\right)}{\cos\lambda^s_m + \cosh\sqrt{3}\lambda^s_m} \Bigg] & \text{for } n \neq m, \\ \frac{(\lambda^s_n)^3 (c^s_n)^2}{8(\cos\lambda^s_n + \cosh\sqrt{3}\lambda^s_n)^2} \Big[6\sin 2\lambda^s_n (\cos 2\lambda^s_n - \cosh 2\sqrt{3}\lambda^s_n) \\ \qquad + 4\lambda^s_n \times \Big(-\cos 2\lambda^s_n + \cosh^2\sqrt{3}\,\lambda^s_n + \sinh^2\sqrt{3}\lambda^s_n \\ \qquad +2\sin 2\lambda^s_n \left(\sin\lambda^s_n \cosh\sqrt{3}\,\lambda^s_n + \sqrt{3}\cos\lambda^s_n \sinh\sqrt{3}\,\lambda^s_n \right) \Big) \Big] & \text{for } n = m. \end{cases} \tag{4.8}$$

The decay rate of the odd coefficients (not plotted here) is $|\gamma^s_{3m}| \sim m^{-3}$, matching that of its even counterparts. This was confirmed for all the examined values of n and is expected from the expression (4.8) for γ^s_{nm}.

For the model problems considered in section 5, the following expansion formulas for even powers of x are also required:

$$x^p = \sum_{m=0}^{\infty} \chi_m^{\{p\}} \psi^c_m(x), \qquad \chi_m^{\{p\}} = \left\langle x^p, \psi^c_m(x) \right\rangle, \tag{4.9}$$

where $p = 2, 4, \ldots, 12$. The expressions for the coefficients $\chi_m^{\{p\}}$ for $m \geq 1$ are given in the Appendix. The case $m = 0$ is not needed, as will become apparent through (5.4) and (5.7) below.

5. Model problems: Results and discussion

In this first foray into Galerkin methods for sixth-order-in-space parabolic equations, and the construction of the associated higher-order beam eigenfunctions, we would like to restrict to the steady case, such that $\partial u/\partial t = 0$. In this context, we would like to demonstrate the spectral accuracy of the eigenfunction expansion on two simple *model* boundary-value problems (BVPs), which have simple exact solutions.

5.1. Model problem I

Model problem I is the BVP:

$$\frac{\mathrm{d}^6 u}{\mathrm{d}x^6} + 14\,400\, u = f(x), \tag{5.1a}$$

$$\left.\frac{\mathrm{d}u}{\mathrm{d}x}\right|_{x=\pm 1} = \left.\frac{\mathrm{d}^2 u}{\mathrm{d}x^2}\right|_{x=\pm 1} = \left.\frac{\mathrm{d}^5 u}{\mathrm{d}x^5}\right|_{x=\pm 1} = 0, \tag{5.1b}$$

where

$$f(x) = 216\,000x^2 - 691\,200x^4 + 377\,280x^6 + 216\,000x^8 - 86\,400x^{10} + 14\,400x^{12}. \tag{5.1c}$$

The problem (5.1) admits the exact solution:

$$u_{\text{exact}}(x) = (x-1)^6(x+1)^6. \tag{5.2}$$

The exact solution (5.2) as well as $f(x)$ in (5.1a) are even functions, and BCs (5.1b) are symmetric. Thus, we need *only* use the *even* eigenfunctions $\psi^c_0(x) = 1$ and $\{\psi^c_n(x)\}_{n=1}^{\infty}$ for the spectral expansion.

Applying the Galerkin spectral method introduced in section 4 leads us to introduce the truncated series

$$u(x) \approx u_{\text{spectral}}(x) = \frac{1}{2} u^c_0 \psi^c_0(x) + \sum_{n=1}^{M} u^c_n \psi^c_n(x)\,. \tag{5.3}$$

Now, we substitute (5.3) into (5.1a), take inner products with $\psi^c_0(x) = 1$ and $\{\psi^c_m(x)\}_{m=1}^{M}$, and employ the orthogonality relations, as in the derivation of (4.4). First, the coefficient u^c_0 is found as

$$14\,400 u^c_0 = \int_{-1}^{+1} f(x)\psi_0(x)\,\mathrm{d}x \quad \Rightarrow \quad u^c_0 = \frac{2\,048}{3\,003}\,. \tag{5.4}$$

Next, the remaining coefficients $\{u^c_n\}_{n=1}^{M}$ are obtained as the solution of the *diagonal* system of equations

$$\begin{aligned}[14\,400 - (\lambda^c_n)^6]\delta_{nm} u^c_n = {} & 216\,000\chi_m^{\{2\}} - 691\,200\chi_m^{\{4\}} + 377\,280\chi_m^{\{6\}} \\ & + 216\,000\chi_m^{\{8\}} - 86\,400\chi_m^{\{10\}} + 14\,400\chi_m^{\{12\}}, \qquad m = 1, 2, \ldots, M,\end{aligned} \tag{5.5}$$

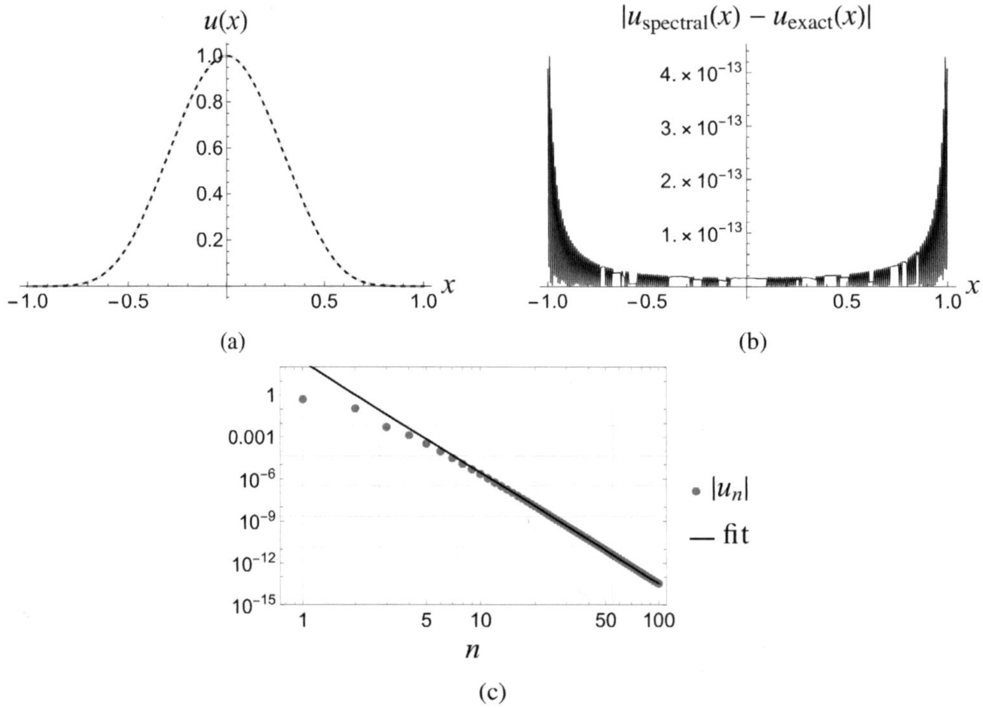

Figure 4: (a) Comparison of the exact (5.2) (solid) and spectral expansion (5.3)–(5.5) (dashed) solution profiles for model problem I with $M = 100$. (b) Absolute error between the exact solution and the spectral expansion. (c) Convergence rate of the spectral expansion with symbols showing $|u_n^c|$ and the solid line being the best fit $259n^{-7.94}$ based on values for $n \geq 50$.

where δ_{nm} is Kronecker's delta.

The exact solution (5.2) and the spectral approximation (5.3)–(5.5) using the sixth-order eigenfunctions are compared in figure 4(a), showing they agree identically (at least visually). As seen in figure 4(b), the spectral expansion is *highly* accurate with the maximum absolute error being $\leq 5 \times 10^{-13}$. Despite the Gibbs effect near $x = \pm 1$, figure 4(c) shows that the overall convergence rate of the spectral expansion is $\mathrm{O}(n^{-8})$. This extremely rapid convergence is due to the very definition of our eigenfunctions, namely that they satisfy a sixth-order BVP. Put differently: when solving the algebraic system (5.5), we are essentially dividing with quantities involving $(\lambda_n^c)^6$, and it is thanks to this fact that we have the impressive convergence rate, overcoming the more moderate decay rate of the coefficients appearing in the right-hide side of (5.5).

5.2. Model problem II

Model problem II is the BVP:

$$\frac{\mathrm{d}^6 u}{\mathrm{d}x^6} - 5544\frac{\mathrm{d}^2 u}{\mathrm{d}x^2} - 199\,584\,u = f(x), \tag{5.6a}$$

$$\left.\frac{\mathrm{d}u}{\mathrm{d}x}\right|_{x=\pm 1} = \left.\frac{\mathrm{d}^2 u}{\mathrm{d}x^2}\right|_{x=\pm 1} = \left.\frac{\mathrm{d}^5 u}{\mathrm{d}x^5}\right|_{x=\pm 1} = 0, \tag{5.6b}$$

where

$$f(x) = -199\,584x^{12} + 465\,696x^{10} - 574\,560x^4 + 50\,1984x^2 - 147\,456. \tag{5.6c}$$

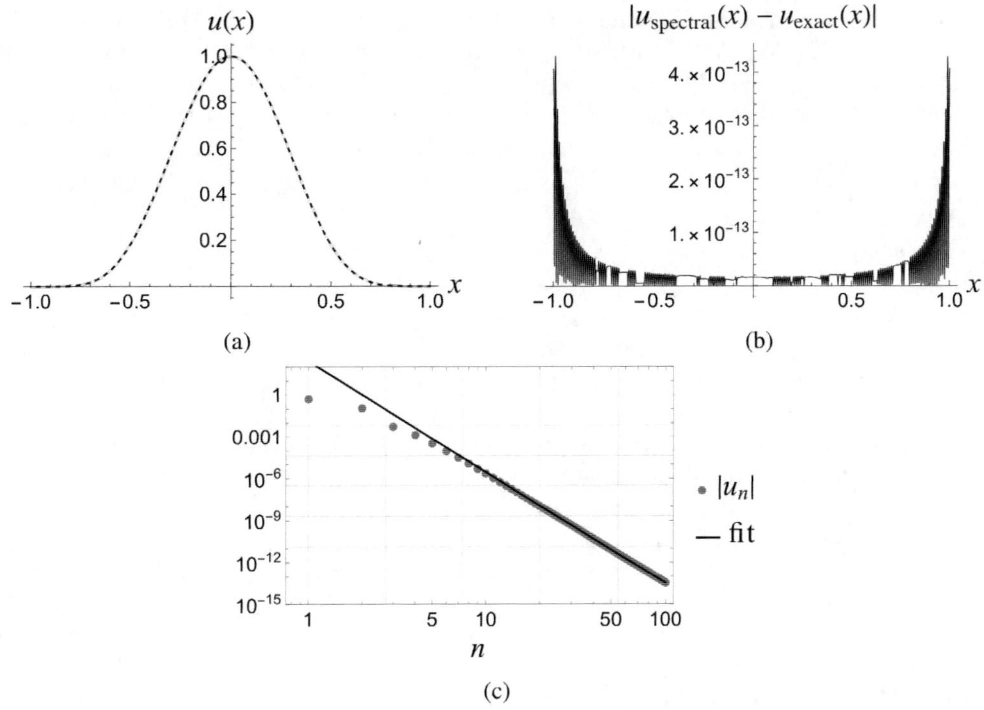

Figure 5: (a) Comparison of the exact (5.2) (solid) and spectral expansion (5.3), (5.7), (5.8) (dashed) solution profiles for model problem II with $M = 100$. (b) Absolute error between the exact solution and the spectral expansion. (c) Convergence rate of the spectral expansion with symbols showing $|u_n^c|$ and the solid line being the best fit $259n^{-7.94}$ based on values for $n \geq 50$.

Problem (5.6) admits the same exact solution (5.2) as problem (5.1). For the same reasons as above, we only use the even eigenfunctions in the spectral expansion for this problem.

We again use expansion (5.3), substitute it into (5.6a), take successive inner products with $\psi_0(x) = 1$ and $\{\psi_m^c(x)\}_{m=1}^M$, and employ the orthogonality relations, as in the derivation of (4.4). First, the coefficient u_0^c is found to be

$$-199\,584 u_0^c = \int_{-1}^{+1} f(x)\psi_0(x)\,\mathrm{d}x \quad \Rightarrow \quad u_0^c = \frac{2\,048}{3\,003}. \tag{5.7}$$

Then, the remaining coefficients $\{u_n^c\}_{n=1}^M$ are obtained as the solution of the *linear* system of equations:

$$\sum_{n=1}^{M} \left\{ \left[-199\,584 - (\lambda_n^c)^6 \right] \delta_{nm} - 5\,544 \beta_{nm}^c \right\} u_n^c$$
$$= -199\,584\chi_m^{\{12\}} + 465\,696\chi_m^{\{10\}} - 574\,560\chi_m^{\{4\}} + 50\,1984\chi_m^{\{2\}}, \qquad m = 1, 2, \ldots, M. \tag{5.8}$$

The coefficient matrix

$$\mathbf{A} = [a_{nm}], \quad \text{where} \quad a_{nm} = \left[-199\,584 - (\lambda_n^c)^6 \right] \delta_{nm} - 5\,544 \beta_{mn}^c, \qquad n, m = 1, 2, \ldots, M, \tag{5.9}$$

of the linear system (5.8) is now full, but *symmetric negative definite* (with $\|\mathbf{A}^{-1}\| < 1$). Thus, system (5.8) can be easily and efficiently solved using the $\mathbf{LDL}^\top$ factorization.

The exact solution (5.2) and the spectral approximation (5.3), (5.7), (5.8) using the sixth-order eigenfunctions are compared in figure 5(a). Again, as seen in figure 5(b), the spectral expansion is *highly* accurate with the maximum absolute error being again $\leq 5 \times 10^{-13}$. Model problem I and II have the same exact solution (5.2), so the spectral expansions in this and the previous subsection are of the same function, even if the BVPs are different. Thus, it should be no surprise that figures 4(b) and 5(b) look almost identical. Note that, as shown in figure 5(c), the spectral expansion has rapid convergence, with the coefficients decaying in magnitude as $O(n^{-8})$, *despite* the moderate convergence rate of β_{nm} demonstrated in figure 3(a). This observation further emphasizes the motivation for the proposed approach.

6. Conclusions

We proposed an efficient and highly accurate Galerkin spectral method for the solution of sixth-order boundary-value problems arising in the study of elastic-plated thin films. We explicitly constructed the sixth-order eigenfunctions and appealing to classical results from Sturm–Liouville theory, we showed that these functions form complete orthonormal bases of even and odd functions for $L^2[-1, 1]$. The considered boundary conditions were dictated by a specific physical problem under consideration, and they led to a self-adjoint eigenvalue problem. Accurate asymptotic formulas for the corresponding eigenvalues were found. Further, we developed exact expansion formulas for the second and fourth derivatives of the eigenfunctions and for even powers of the independent spatial variable, x.

The proposed spectral method was tested by solving two model problems using 101 terms ($M = 100$) in the spectral expansion. In both cases, the absolute error was less than 5×10^{-13} and the convergence rate of the spectral series was found to be eighth-order algebraic ($O(n^{-8})$, where n is the number of terms), which is even greater than the order (sixth) of the EVP.

In future work, we would like to consider the unsteady IBVP and develop time-stepping schemes (say, of Crank–Nicolson type) for the proposed Galerkin method. In addition, we would like to consider non-self-adjoint versions of EVP (3.2), which can arise when both bending and tension are important ($\mathscr{T} \neq 0$) and/or due to the importance of gravity (when the elastic Bond number $\mathscr{B} \neq 0$, see [35]).

Acknowledgments

ICC would like to acknowledge the hospitality of the University of Nicosia, Cyprus, where this work was completed thanks to a Fulbright US Scholar award from the US Department of State, and the US National Science Foundation, which supports his research on interfacial dynamics under grant CMMI-2029540.

Appendix

The expressions for the coefficients with $m \geq 1$ in the expansion formulas (4.9) for the even powers of x are:

$$\chi_m^{\{2\}} = -\frac{4\,c_m^c}{(\lambda_m^c)^3(\cos\lambda_m^c - \cosh\sqrt{3}\lambda_m^c)} \times \Big[-\lambda_m^c \cos 2\lambda_m^c + \sqrt{3}\lambda_m^c \sin\lambda_m^c \sinh\sqrt{3}\lambda_m^c + 3\sin\lambda_m^c\cos\lambda_m^c + (\lambda_m^c\cos\lambda_m^c - 3\sin\lambda_m^c)\cosh\sqrt{3}\lambda_m^c\Big], \tag{A.1a}$$

$$\chi_m^{\{4\}} = \frac{8\,c_m^c}{(\lambda_m^c)^4(\cos\lambda_m^c - \cosh\sqrt{3}\lambda_m^c)} \times \Big[((\lambda_m^c)^2 - 6)(\cos 2\lambda_m^c - \cos\lambda_m^c\cosh\sqrt{3}\lambda_m^c) - \Big(\sqrt{3}((\lambda_m^c)^2 + 6)\sinh\sqrt{3}\lambda_m^c + 9\lambda_m^c(\cos\lambda_m^c - \cosh\sqrt{3}\lambda_m^c)\Big)\sin\lambda_m^c\Big], \tag{A.1b}$$

$$\chi_m^{\{6\}} = \frac{6\,c_m^c}{(\lambda_m^c)^6(\cos\lambda_m^c - \cosh\sqrt{3}\lambda_m^c)} \times \Big[360 + 2((\lambda_m^c)^4 - 20(\lambda_m^c)^2 - 60)\cos 2\lambda_m^c - 15(\lambda_m^c)^3\sin 2\lambda_m^c + \Big(30(\lambda_m^c)^3\sin\lambda_m^c - 2((\lambda_m^c)^4 - 20(\lambda_m^c)^2 + 120)\cos\lambda_m^c\Big)\cosh\sqrt{3}\lambda_m^c - 2\sqrt{3}(\lambda_m^c)^2((\lambda_m^c)^2 + 20)\sin\lambda_m^c\sinh\sqrt{3}\lambda_m^c\Big], \tag{A.1c}$$

$$\chi_m^{\{8\}} = \frac{8\,c_m^c}{(\lambda_m^c)^9(\cos\lambda_m^c - \cosh\sqrt{3}\lambda_m^c)} \times \Big[2520(\lambda_m^c)^3 - 21((\lambda_m^c)^6 - 720)\sin 2\lambda_m^c + 2((\lambda_m^c)^6 - 42(\lambda_m^c)^4 - 420(\lambda_m^c)^2 - 5040)\lambda_m^c\cos 2\lambda_m^c \tag{A.1d}$$

$$- 2\sqrt{3}\lambda_m^c((\lambda_m^c)^6 + 42(\lambda_m^c)^4 - 5040)\sin\lambda_m^c \sinh\sqrt{3}\lambda_m^c$$
$$+ 2\Big(21((\lambda_m^c)^6 - 720)\sin\lambda_m^c - \lambda_m^c((\lambda_m^c)^6 - 42(\lambda_m^c)^4 + 840(\lambda_m^c)^2 - 5040)\cos\lambda_m^c\Big)\cosh\sqrt{3}\lambda_m^c\Big],$$

$$\chi_m^{\{10\}} = \frac{20\,c_m^c}{(\lambda_m^c)^{10}(\cos\lambda_m^c - \cosh\sqrt{3}\lambda_m^c)} \times \Big[4536(\lambda_m^c)^4 - \tfrac{27}{2}((\lambda_m^c)^6 - 20160)\lambda_m^c \sin 2\lambda_m^c \tag{A.1e}$$
$$+ ((\lambda_m^c)^8 - 72(\lambda_m^c)^6 - 1512(\lambda_m^c)^4 - 60480(\lambda_m^c)^2 + 362880)\cos 2\lambda_m^c$$
$$- \sqrt{3}((\lambda_m^c)^8 + 72(\lambda_m^c)^6 - 60480(\lambda_m^c)^2 - 362880)\sin\lambda_m^c \sinh\sqrt{3}\lambda_m^c$$
$$+ \cosh\sqrt{3}\lambda_m^c \cdot \Big(27\lambda_m^c((\lambda_m^c)^6 - 20160)\sin\lambda_m^c$$
$$- ((\lambda_m^c)^8 - 72(\lambda_m^c)^6 + 3024(\lambda_m^c)^4 - 60480(\lambda_m^c)^2 + 362880)\cos\lambda_m^c\Big)\Big],$$

$$\chi_m^{\{12\}} = \frac{12\,c_m^c}{(\lambda_m^c)^{10}(\cos\lambda_m^c - \cosh\sqrt{3}\lambda_m^c)} \times \Big[23760((\lambda_m^c)^6 - 5040) - 33(\lambda_m^c)^3((\lambda_m^c)^6 - 151200)\sin 2\lambda_m^c \tag{A.1f}$$
$$+ 2((\lambda_m^c)^{10} - 110(\lambda_m^c)^8 - 3960(\lambda_m^c)^6 - 332640(\lambda_m^c)^4 + 6652800(\lambda_m^c)^2 + 19958400)\cos 2\lambda_m^c$$
$$- 2\sqrt{3}(\lambda_m^c)^2((\lambda_m^c)^8 + 110(\lambda_m^c)^6 - 332640(\lambda_m^c)^2 - 6652800)\sin\lambda_m^c \sinh\sqrt{3}\lambda_m^c$$
$$+ 2\cosh\sqrt{3}\lambda_m^c \cdot \Big(33(\lambda_m^c)^3((\lambda_m^c)^6 - 151200)\sin\lambda_m^c$$
$$- \Big((\lambda_m^c)^{10} - 110(\lambda_m^c)^8 + 7920(\lambda_m^c)^6 - 332640(\lambda_m^c)^4 + 6652800(\lambda_m^c)^2 - 39916800\Big)\cos\lambda_m^c\Big)\Big].$$

References

[1] King J R 1989 *SIAM. J. Appl. Math.* **49** 1064–1080
[2] Michaut C 2011 *J. Geophys. Res.: Solid Earth* **116** B05205
[3] Bunger A P and Cruden A R 2011 *J. Geophys. Res.: Solid Earth* **116** B02203
[4] Bunger A P and Cruden A R 2011 *J. Geophys. Res.: Solid Earth* **116** B08211
[5] Thorey C and Michaut C 2014 *J. Geophys. Res.: Planets* **119** 286–312
[6] Huang R and Suo Z 2002 *J. Appl. Phys.* **91** 1135–1142
[7] Flitton J C and King J R 2004 *Eur. J. Appl. Math.* **15** 713–754
[8] Hosoi A E and Mahadevan L 2004 *Phys. Rev. Lett.* **93** 137802
[9] Hewitt I J, Balmforth N J and De Bruyn J R 2015 *Eur. J. Appl. Math.* **26** 1–31
[10] Pedersen C, Niven J F, Salez T, Dalnoki-Veress K and Carlson A 2019 *Phys. Rev. Fluids* **4** 124003 (*Preprint* `arXiv:1902.10470`)
[11] Peng G G and Lister J R 2020 *J. Fluid Mech.* **905** A30
[12] Boyko E, Eshel R, Gommed K, Gat A D and Bercovici M 2019 *J. Fluid Mech.* **862** 732–752 (*Preprint* `arXiv:1703.06820`)
[13] Boyko E, Ilssar D, Bercovici M and Gat A D 2020 *Phys. Rev. Fluids* **5** 104201
[14] Tulchinsky A and Gat A D 2016 *J. Fluid Mech.* **800** 517–530 (*Preprint* `arXiv:1512.00730`)
[15] Greenberg L and Marletta M 1998 *SIAM J. Numer. Anal.* **35** 2070–2098
[16] Greenberg L and Marletta M 2000 *J. Comput. Appl. Math.* **125** 367–383
[17] Chandrasekhar S 1961 *Hydrodynamic and Hydromagnetic Stability* (Oxford: Clarendon Press)
[18] DiTaranto R A 1965 *ASME J. Appl. Mech.* **32** 881–886
[19] Mead D J and Markus S 1969 *J. Sound Vib.* **10** 163–175
[20] Christov C I 1982 *Ann. Univ. Sof., Fac. Math. Mech.* **76** 87–113
[21] Papanicolaou N C, Christov C I and Homsy G M 2009 *Int. J. Num. Meth. Fluids* **59** 945–965
[22] Boyd J P 2000 *Chebyshev and Fourier Spectral Methods* 2nd ed (Mineola, NY: Dover Publications)
[23] Shen J, Tang T and Wang L L 2011 *Spectral Methods: Algorithms, Analysis and Applications* (*Springer Series in Computational Mathematics* vol 41) (Berlin/Heidelberg: Springer-Verlag)
[24] Oron A, Davis S H and Bankoff S G 1997 *Rev. Mod. Phys.* **69** 931–980
[25] Craster R V and Matar O K 2009 *Rev. Mod. Phys.* **81** 1131–1198
[26] Leal L G 2007 *Advanced Transport Phenomena: Fluid Mechanics and Convective Transport Processes* (*Cambridge Series in Chemical Engineering* vol 7) (New York, NY: Cambridge University Press)
[27] Stone H A 2017 Fundamentals of fluid dynamics with an introduction to the importance of interfaces *Soft Interfaces*

(*Lecture Notes of the Les Houches Summer School* vol 98) ed Bocquet L, Quéré D, Witten T A and Cugliandolo L F (New York, NY: Oxford University Press) pp 3–79

[28] Howell P, Kozyreff G and Ockendon J 2009 *Applied Solid Mechanics* (Cambridge, UK: Cambridge University Press)

[29] Lister J R, Skinner D J and Large T M J 2019 *J. Fluid Mech.* **868** 119–140

[30] Martínez-Calvo A, Sevilla A, Peng G G and Stone H A 2020 *J. Fluid Mech.* **885** A25 (*Preprint* arXiv:1902.07167)

[31] Duprat C, Aristoff J M and Stone H A 2011 *J. Fluid Mech.* **679** 641–654 (*Preprint* arXiv:1008.3702)

[32] Anand V, Muchandimath S C and Christov I C 2020 *ASME J. Appl. Mech.* **87** 051012 (*Preprint* arXiv:1910.03513)

[33] Coddington E A and Levinson N 1955 *Theory of Ordinary Differential Equations* (New York, NY: McGraw-Hill)

[34] Wolfram Research, Inc 2023 Mathematica, Version 13.3 Champaign, IL URL https://www.wolfram.com/mathematica

[35] Gabay I, Bacheva V, Ilssar D, Bercovici M, Ramos A and Gat A 2023 *J. Fluid Mech.* **969** A17 (*Preprint* arXiv:2209.04531)

Journal of Physics: Conference Series **2675** (2023) 012017 doi:10.1088/1742-6596/2675/1/012017

Modeling the viscoplastic transient dynamic process

Todor Zhelyazov[1, a)], Sergey Pshenichnov[2, b)]

[1]Technical University of Sofia, 8, Kliment Ohridski blvd, Sofia, 1000, Bulgaria
[2]Institute of Mechanics, Lomonosov Moscow State University, Michurinsky prospect 1, Moscow, 119192, Russian Federation

[a)] elovar@yahoo.com
[b)] serp56@yandex.ru

Abstract. A hollow viscoelastic right circular cylinder subjected to a time-dependent load is considered. The cylinder is with finite dimensions. A transient load acts upon the inner surface of the cylinder, whereas no loads are applied to outer surfaces. The end surfaces of the cylinder are restrained to move along the cylinder axis. The viscoelastic behaviour of the material is modelled using Boltzmann – Volterra equations. A yield criterion is implemented in the numerical analyses. By assumption, upon reaching the yield surface, the linear viscoelastic model switches to a plastic constitutive relation. The transient dynamic behaviour of the material is simulated for different intensities of the applied load and model parameters.

1. Introduction

Applications of visco-elasto-plastic models range from the simulation of the response of a variety of materials (metals, soils, etc. [1], [2], [3]) to geodynamics [4].

The experimental evidence (tension impact test [5]) showed some systematic discrepancies with the stress-strain relationship formulated without accounting for the strain rate. Specifically, the maximum residual strain was smaller than predicted by the theory, and the observed force-time variation at the fixed end during impact showed that the stress there was greater than the theory predicted. It has been suggested that these discrepancies were due to the use in the theory of an invariant relation between stress and strain independent of strain rate. A considerable deviation from the static stress-strain relationship may be expected at high strain rates, as, for example, those involved in deformation under impact load.

Malvern [6] has performed numerical simulations using a stress-strain relationship assumed to depend on the instantaneous plastic strain and the strain rate. According to [1], the plastic strain evolution is related to the overstress function, which, in turn, depends on the rate-independent yield function. Alternatively, the rate dependency can be incorporated into the yield function [7]. As pointed out by [8], as long as some conditions are satisfied, the Perzyna model [1] and the consistency model [7] predict the same behavior in the case of plastic loading, whereas their predictions differ upon unloading. Valuable contributions have been proposed by [9]-[11].

The contribution focuses on the implementation of a viscoelastic-viscoplastic model. A plastic flow condition, referred to as a local failure criterion, is defined in the stress space. In line with the classical point of view, the plastic flow onset takes place when the second invariant of the stress tensor reaches the material yield stress.

Two constitutive relations are employed. Thus, as long as the equivalent stress is less than the local failure criterion, the material response is modeled by superimposing linear Boltzmann-Volterra equations with linear elastodynamics. When the local failure criterion is met, the material behavior, by assumption, is modeled by a rate-dependent plasticity constitutive relation.

Numerical simulations are performed to investigate the model predictions for various values of the model parameters (specifically, the rate-dependent model constants are varied) and for various intensities of the applied load. To validate the proposed modeling approach, a three-dimensional finite element model is built to analyze an axisymmetric problem.

2. Problem formulation

Linear viscoelasticity is modeled within classical elastodynamics defined by the equation of motion

$$\nabla \cdot \overline{\overline{\sigma}} + \vec{f} = \rho \frac{\partial^2 \vec{u}}{\partial t^2}, \tag{1}$$

the constitutive relations

$$\overline{\overline{\sigma}} = \hat{\lambda} tr\left(\overline{\overline{\varepsilon}}\right)\overline{\overline{I}} + 2\hat{\mu}\overline{\overline{\varepsilon}}, \tag{2}$$

along with the strain tensor definition

$$\overline{\overline{\varepsilon}} = \frac{1}{2}\left[\nabla\vec{u} + \left(\nabla\vec{u}\right)^T\right], \tag{3}$$

by modifying the Lamé coefficients $\hat{\lambda}$ and $\hat{\mu}$. In equation (1), $\overline{\overline{\sigma}}$ is the stress tensor, $\vec{f}$ is the body force vector, ρ is the material density, and $\vec{u}$ is the displacement vector. In equation (2), $\overline{\overline{\varepsilon}}$ is the strain tensor, $tr\left(\overline{\overline{\varepsilon}}\right)$ is the trace of the strain tensor, and $\overline{\overline{I}}$ is the unit tensor. The modified Lamé coefficients are defined as

$$\begin{gathered} \hat{\lambda} = K_0\left(1-\hat{T}_v\right) - \frac{2}{3}G_0\left(1-\hat{T}_s\right), \\ \hat{\mu} = G_0\left(1-\hat{T}_s\right), \end{gathered} \tag{4}$$

where K_0 and G_0 are the instantaneous (initial) bulk and shear moduli, respectively; the volumetric and shear relaxation kernels are defined as

$$\hat{T}_j\xi(t) = \int_0^t T_j(t-\eta)\xi(\eta)d\eta, \quad j = v, s, \tag{5}$$

and the instantaneous (initial) Lamé coefficients as

$$\lambda_0 = \frac{E_0 \nu_0}{(1+\nu_0)(1-2\nu_0)},$$
$$G_0 \equiv \mu_0 = \frac{E_0}{2(1+\nu_0)}, \tag{6}$$

where E_0 and ν_0 are the instantaneous (initial) Young's modulus and Poisson's ratio, respectively. For the sequel, definitions for the deviators of the stress and the strain tensors will be needed. The stress tensor deviator is defined componentwise,

$$s_{ij} = \sigma_{ij} - \frac{1}{3}(\sigma_{kk})\delta_{ij}, \quad i,j = 1..3 \tag{7}$$

where σ_{ij} are the components of the stress tensor, and δ_{ij} is the Kronecker symbol, with summation on *k* as implied by the repeated indices. The components of the strain tensor deviator e_{ij} are

$$e_{ij} = \varepsilon_{ij} - \frac{1}{3}(\varepsilon_{kk})\delta_{ij}, \quad i,j = 1..3 \tag{8}$$

where ε_{ij} are the components of the strain tensor, suggesting summation on the index k.

Using a classical definition of stress and strain intensities,

$$\sigma_I = \left(\frac{3}{2} s_{ij} s_{ij}\right)^{1/2}, \quad \varepsilon_I = \left(\frac{2}{3} e_{ij} e_{ij}\right)^{1/2} \quad i,j = 1..3, \tag{9}$$

and their relationship

$$\sigma_I = 3\hat{\mu}\varepsilon_I, \tag{10}$$

where $\hat{\mu}$ has been defined in equation (4), by assumption, the stain has only a viscoelastic component as long as the stress intensity is less than the material yield stress, σ_Y,

$$\sigma_I < \sigma_Y. \tag{11}$$

and inelastic components of stress appear when the inequality (11) is no longer satisfied. Thus, the total strain can be decomposed into a viscoelastic and an inelastic part,

$$\varepsilon_{ij} = \varepsilon_{ij}^{(ve)} + \varepsilon_{ij}^{(pl)}, \quad \varepsilon_{kk}^{(pl)} = 0. \tag{12}$$

The plastic strain rate is generally defined as

$$\dot{\varepsilon}_{ij}^{(pl)} = \dot{\lambda}\frac{\partial Q}{\partial \sigma_{ij}}, \tag{13}$$

where Q is a function defining the plastic potential, and $\dot{\lambda}$ is the magnitude of the plastic strain increment. If the plastic potential coincides with the yield surface, denoted conventionally $f\left(\sigma_{ij},\{\zeta_i\}\right)$, the plastic strain increment is normal to the yield surface. This assumption is generally used to model metals, as presumed by the material constants employed in the model material definition further in the text. In the definition of the yield surface, $\{\zeta_i\}$ denotes a collection of history-dependent internal variables generally employed to describe the evolution of the yield surface (e.g., isotropic or kinematic hardening). Thus, for example, the time rate of change of the inelastic strain deviator can be defined (see, for example, [12])

$$\dot{e}_{ij}^{(pl)} = \frac{3}{2}\Phi\left(\frac{\sigma_I}{\sigma_Y} - 1\right)\frac{s_{ij}}{\sigma_I}. \tag{14}$$

A similar approach is employed also in [13].

The relationship provided by equation (14) appears in a more general form in [1]

$$\dot{\varepsilon}_{ij}^{(pl)} = \gamma\left\langle \Phi\left(F\right)\right\rangle \frac{\partial f}{\partial \sigma_{ij}}, \tag{15}$$

with

$$\left\langle \Phi\left(F\right)\right\rangle = \begin{cases} \Phi\left(F\right) & if \quad F > 0 \\ 0 & if \quad F \leq 0 \end{cases}. \tag{16}$$

In equation (15), γ is referred to as the material viscosity parameter. The function $\Phi\left(F\right)$ should be chosen to fit the best experimental data. In a number of studies [9-11, 14], $\Phi\left(F\right)$ is defined as a power-law function. In the presented investigation, a ready-to-use model is implemented to define the inelastic strain evolution

$$\dot{\lambda} = \dot{\varepsilon}_I^{(pl)}, \quad \dot{\varepsilon}_I^{(pl)} = \gamma\left(\frac{\sigma_I}{\sigma_Y} - 1\right)^{1/m}, \quad \dot{\varepsilon}_{kk}^{(pl)} = 0. \tag{17}$$

In equation (17), γ stands for the material viscosity parameter, and m - for the strain rate hardening parameter.

To model the transition from viscoelastic to viscoplastic behaviour, it is assumed that the viscoelastic model applies as long as the stress intensity remains lower than a predefined value (material yield stress). The mechanical parameters of the material are gradually modified until the yield surface is reached

$$if \quad \sigma_I < \sigma_Y \quad \begin{cases} K = K_0\left(1 - \int_0^{t^*} T_v(t-\eta)d\eta\right) \\ G = G_0\left(1 - \int_0^{t^*} T_s(t-\eta)d\eta\right) \\ \overline{\overline{\sigma}} = \widehat{\lambda} tr\left(\overline{\overline{\varepsilon}}\right)\overline{\overline{I}} + 2\widehat{\mu}\overline{\overline{\varepsilon}} \end{cases}. \tag{18}$$

Upon reaching the yield criterion,

$$\sigma_I = \sigma_Y. \tag{19}$$

at a specified time $t = t_\varepsilon$, the modified model parameters calculated for this specific time

$$\begin{aligned} K_\Omega &= K_0\left(1 - \int_0^{t_\varepsilon} T_v(t-\eta)d\eta\right) \\ G_\Omega &= G_0\left(1 - \int_0^{t_\varepsilon} T_s(t-\eta)d\eta\right) \end{aligned}. \tag{20}$$

are employed in the subsequent analysis without further modification. Within the inelastic domain, the inelastic strain rate is defined by equation (17).

3. Geometry, boundary, and initial conditions.

The hollow cylinder with outer and inner radii R_o and R_i, respectively, and with a length $2L$ ($L = 6R_o$) is represented in nondimensional quantities

$$\tilde{r} = \frac{r}{R_0}, \quad \tilde{r}_i = \frac{R_i}{R_0}, \quad \tilde{r}_i \le \tilde{r} \le 1, \quad \tilde{z} = \frac{z}{R_0}, \quad -6 \le \tilde{z} \le 6. \tag{21}$$

The cylinder is analyzed for the following boundary

$$\begin{gathered} \sigma_{rr}\left(\tilde{r}_i, \tilde{z}, \tau\right) = q\left(\tilde{z}, \tau\right) \text{ for } -6 \le \tilde{z} \le 6, \\ \sigma_{rz}\left(\tilde{r}_i, \tilde{z}, \tau\right) = 0 \text{ for } -6 \le \tilde{z} \le 6, \\ \sigma_{rr}\left(1, \tilde{z}, \tau\right) = 0 \text{ and } \sigma_{rz}\left(1, \tilde{z}, \tau\right) = 0 \text{ for } -6 \le \tilde{z} \le 6, \\ u_z\left(\tilde{r}, \pm 6, \tau\right) = 0 \text{ and } \sigma_{zr}\left(\tilde{r}, \pm 6, \tau\right) = 0 \text{ for } \tilde{r}_i \le \tilde{r} \le 1, \end{gathered} \tag{22}$$

and initial conditions

$$\begin{gathered} u_r\left(\tilde{r}, \tilde{z}, 0\right) = 0, \quad u_z\left(\tilde{r}, \tilde{z}, 0\right) = 0, \quad \frac{\partial u_r}{\partial \tau}\left(\tilde{r}, \tilde{z}, 0\right) = 0, \quad \frac{\partial u_z}{\partial t}\left(\tilde{r}, \tilde{z}, 0\right) = 0, \\ \text{for } -6 \le \tilde{z} \le 6 \text{ and } \tilde{r}_i \le \tilde{r} \le 1. \end{gathered} \tag{23}$$

The time-dependent axisymmetric load is defined as the product of a dimensionless constant F_0, the distribution function $p(\tilde{z})$, and the smoothed Heaviside function $\varphi(\tau)$

$$q(\tilde{z},\tau)=F_0 p(\tilde{z})\varphi(\tau), \quad \tau>0,$$
$$p(\tilde{z})=\frac{1}{2}\left(1+\cos\left(\frac{8\pi\tilde{z}}{L}\right)\right) \text{ for } -0.75\le\tilde{z}\le 0.75, \text{ and } p(\tilde{z})=0 \text{ otherwise,} \tag{24}$$
$$\varphi(\tau)=1-e^{-50\tau}.$$

Normalized time is obtained by using the time needed for the compression wave to travel the distance R_0

$$\tau=\frac{t}{t_0}, \quad t_0=\frac{R_0}{c}, \tag{25}$$

where the speed of the propagation of front of the compression wave is

$$c=\sqrt{\frac{\lambda_0+2\mu_0}{\rho}}. \tag{26}$$

In the performed analyses, it is assumed that

$$T_v(t)\equiv T_s(t)\equiv T(t), \quad t_0 T=ae^{-b\tau}, \quad a=0.2, \quad b=0.9. \tag{27}$$

A model material with instantaneous (initial) constants (E_0=210000MPa, $\nu_0=0.3$) is employed.

4. Results and discussion

The non-stationary wave process is simulated by performing transient finite element analysis. Although the test model is axisymmetric, a three-dimensional finite element model is built (Figure 1).

Figure 1. The finite element mesh.

Figure 2 displays the evolution of the normalized equivalent stress (i.e., the stress intensity divided by the material's yield stress, $\sigma_{I,n} = \frac{\sigma_I}{\sigma_Y}$). The transient wave process is monitored at point P $\left(\tilde{r}=1, \quad \theta=0, \quad \tilde{z}=3\right)$ at the cylinder's outer surface, equally distanced from the cylinder's median section and the right-end surface. The dimensionless time at which the predefined yield surface is first reached is denoted as τ_ε. As previously mentioned, by assumption, upon reaching the yield surface, the viscoelastic model defined by equations (1) - (5) is replaced by the constitutive relation outlined in equation (17).

Inelastic strain evolution at the same point for various values of the strain rate hardening parameter (m) and the viscosity parameter (γ), as referenced in Table 1, are shown in Figure 3 and Figure 4, respectively.

Table 1. Various sets compiled by varying the constant F_0 and the model parameters, specifically, the strain rate hardening parameter (m), and the viscosity parameter (γ).

Set	F_0	γ	m
A	5.3E-3	0.9	5
B	5.3E-3	0.9	6
C	5.3E-3	0.9	4
D	5.3E-3	0.7	5
E	5.3E-3	0.5	5
F	9.3E-3	0.9	5
G	1.3E-2	0.9	5

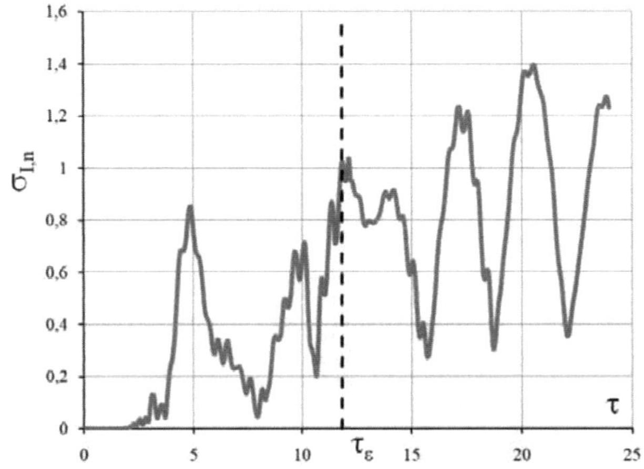

Figure 2. The normalized stress intensity evolution at point P for a given applied load intensity.

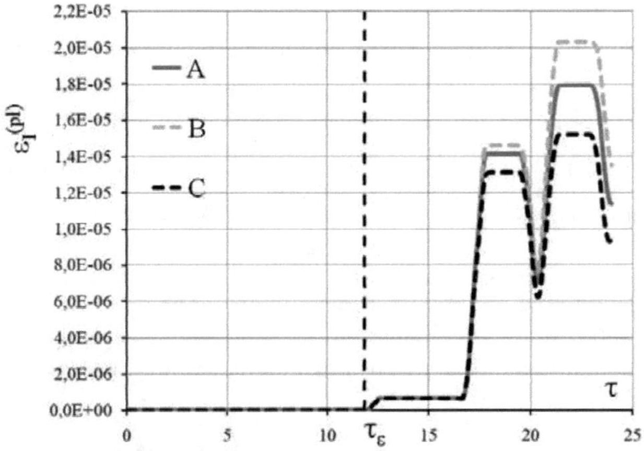

Figure 3. Inelastic strain intensities evolution at the point P, obtained for various strain rate hardening parameter values.

The inelastic strain onset coincides with the reaching of the static yield surface defined by σ_y=350MPa, at the time t_ε when normalized equivalent stress first reaches the value 1.

The obtained results are compared with the inelastic strain evolution obtained for m=5. With the increase of the strain rate hardening parameter, the developed inelastic strain increases and, vice-versa, the diminution of the parameter yields a lower equivalent inelastic strain provided the other model parameters remain unchanged.

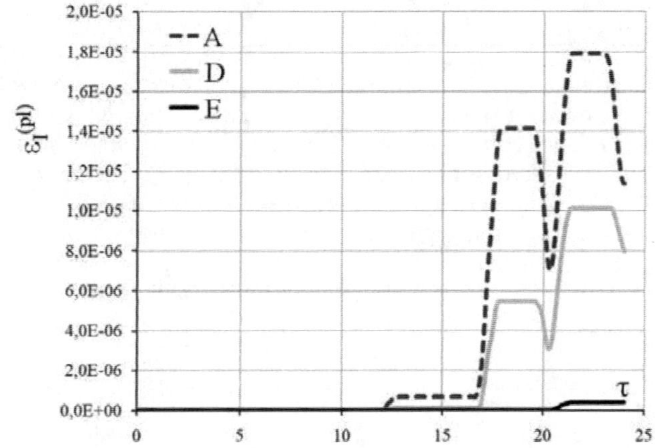

Figure 4. Strain intensity time histories for various values of the viscosity parameter.

With the decrease of the viscosity parameter (γ), the developed inelastic strain, specifically, the equivalent inelastic strain, decreases (see Figure 4); the reference evolution is that for γ=0.9. In Figure 5, the normalized times at which the failure criterion is firstly met are displayed as $t_{\varepsilon,A}$ $t_{\varepsilon,F}$, and , $t_{\varepsilon,G}$. The subscripts *A*, *F*, and *G* relate the equivalent stress evolution depicted in Figure 5 with the sets listed in Table 1.

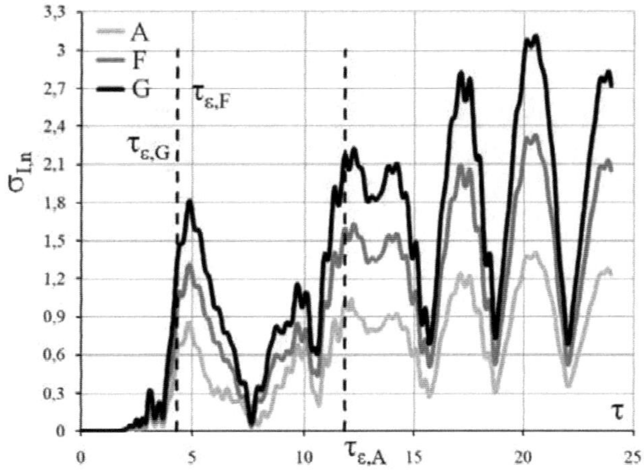

Figure 5. Normalized stress intensity evolution for various applied load intensities (sets A, F, and G in Table 1).

The inelastic strain accumulations starts earlier with the increase of the impact load intensity I related to the constant F_0 , defining the transient load in equation (24) $\left(F_0 = \frac{I}{2G} \right)$. It can be noticed that the normalized tie at which the local failure criterion is met for the responses obtained for sets F and G is almost the same, quite smaller than the time coordinate for which yielding is predicted for set A. On the one hand, the normalized time at which the local failure occurs, decreases while on the other hand, the mechanical waves need some finite time interval to reach the monitored location. The equivalent inelastic strain evolution plotted in Figure 6 confirms the above observation.

Qualitatively, the 'normalized time-equivalent inelastic strain' graphs, in both Figure 5 and Figure 6, are identical. It can be mentioned that the unloading phases, the stress state being in the inelastic domain, are better pronounced for larger applied load intensities (Figure 6).

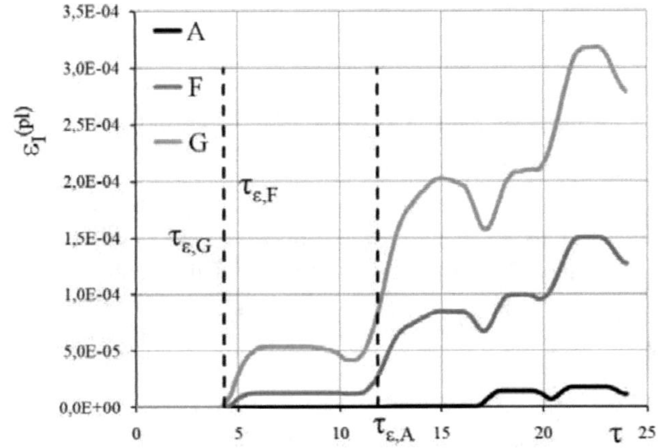

Figure 6. Evolution of the inelastic strain intensities for the various applied load intensities (sets A, F, and G in Table 1).

5. Concluding remarks

The transient wave process in a finite cylinder made of a viscoelastic-viscoplastic material has been investigated.

The implemented two-stage constitutive modeling approach presumes a coupling of viscosity with linear elasticity until the plasticity threshold is reached. Upon reaching the yield surface, the viscoelastic model is replaced by a viscoplastic constitutive relationship. The model constants are continuously modified during the first stage. When the yield surface is reached for the first time, the modified values of the model constants are retained for the subsequent analysis.

The influence of the viscoplastic model parameters has been studied. To this end, numerical simulations for various intensities of the applied transient load and model parameters have been performed.

The numerically obtained transient wave process is visualized through the stress intensity and inelastic strain evolution at the monitored location.

Acknowledgments
The financial support of the Bulgarian National Science Fund, project number KP-06- Russia/5 from 11.12.2020 is gratefully acknowledged.

References

[1] Perzyna P 1966 Fundamental problems in viscoplasticity *Advances in applied mechanics* **9** 243-377
[2] Sokolovsky V V 1969 *Plasticity Theory* (Moscow: Vysshaya Shkola, in Russian)
[3] Hill R 1998 *The mathematical theory of plasticity Vol. 11* (Oxford university press)
[4] Duretz T, de Borst R, Yamato P and Le Pourhiet L 2020 Toward robust and predictive geodynamic modeling: The way forward in frictional plasticity *Geophysical Research Letters* **47**(5) e2019GL086027
[5] Duwez P E and Clark D S 1947 An experimental study of the propagation of plastic deformation under conditions of longitudinal impact *A.S.T.M. Proc.* **47** 502-532.
[6] Malvern LE 1951 Plastic wave propagation in a bar of material exhibiting a strain rate effect *Quarterly of Applied Mathematics* **8**(4) 405-411
[7] Wang W M 1997 Stationary and propagative instabilities in metals – a computational point of view *Dissertation* Delft University of Technology
[8] Heeres OM, Suiker AS and de Borst R 2002 A comparison between the Perzyna viscoplastic model and the consistency viscoplastic model *European Journal of Mechanics-A/Solids* **21**(1) 1-12
[9] Desai C S and Zhang D 1987 Viscoplastic model for geologic materials with generalized flow rule *International Journal for Numerical and Analytical Methods in Geomechanics* **11**(6) 603-620
[10] Simo J C 1989 Strain softening and dissipation: a unification of approaches *Cracking and damage: strain localization and size effect* ed Mazars J and Bazant Z P (London: Elsevier) p 440-461
[11] Simo J C and Hughes T J R 1998 *Computational Inelasticity* (New York: Springer)
[12] Netrebko A V (2002) *Elastic-plastic two-component waves in cylindrical metal shells* DSc. Dissertation, Moscow State University [in Russian]
[13] Bulichov G G (1996) *Mathematical modeling of the dynamics and dynamic destruction of composite materials and structures* DSc. Dissertation, Moscow State University [in Russian]
[14] Wang W M, Sluys L J, De Borst R (1997). Viscoplasticity for instabilities due to strain softening and strain-rate softening *International Journal for Numerical Methods in Engineering* **40**(20) 3839-3864

Design of a system for bioinformatics processing of metagenomic data

A Kirilov[1, a)] and B Kirov[2,3 b)]

[1]Institute of Information and Communication Technologies - BAS, Acad. G. Bonchev St., Block 25 A, 1113 - Sofia, Bulgaria
[2] Faculty of Automatics, Technical University of Sofia, 8 "Kl. Ohridski" Blvd., Sofia, Bulgaria
[3] BioInfoTech lab, Sofia Tech Park, Tsarigradsko Shosse Blvd. 111, Laboratory Building

[a)] Corresponding author: *alex.kirilov@acad.bg*
[b)]*boris.kirov@tu-sofia.bg*

Abstract. Advancements in acquiring, storing and processing information have led to a significant accumulation of large databases in the field of metagenomics—a branch of biology that deals with the study of genetic material directly recovered from environmental or clinical samples. The data collected from metagenomic experiments are massive and noisy, often containing fragmented information representing over 10,000 species. Processing and analyzing this data require software installed on supercomputers. In this paper, the authors present the design, development, and creation of a computational server system aimed at processing and analyzing bioinformatics data in metagenomics and 16S rRNA gene studies. They utilized the specialized software product Galaxy Europe for this purpose. The analysis was performed using the metagenomics-based method Targeted Amplicon Sequencing. The data was divided into three test collections in a special fasta file format. These collections were processed using the web-based Galaxy platform and two configuration variants on a high-performance cluster equipped with NVIDIA Tesla V100 32 GB GPU cards. The analysis process was optimized to reduce information processing time, and the parallel performance of the three different system configurations was investigated for the data used.

1. Introduction to bioformatics processing of metagenomic data

Metagenomics is the study of genetic material obtained directly from environmental samples using modern genetic technologies such as PCR, DNA sequencing, DNA microarrays, FISH, or methylation-specific analysis. Metagenomics focuses on the study of microbes, which are unicellular organisms closely related to both animal and plant life, invisible to the naked eye. These microorganisms can be either beneficial or harmful to us, accompanying us throughout our lives[3].

Metagenomics offers insights into the functional gene composition of microbial communities, providing a broader description than phylogenetic studies, which often rely on the diversity of a single gene, like the 16S rRNA gene. The 16S rRNA gene is a complex of multiple proteins and RNA subunits found in all living cells that play an important role in biological protein synthesis (translation). The ribosome consists of two main components: (1) the small ribosomal subunit (30S ribosomal subunit in prokaryotic cells) and (2) the large subunit (50S ribosomal subunit in prokaryotic cells) [5]. Metagenomics provides genetic information for potentially novel biocatalysts or enzymes, genomic relationships between function and phylogeny for uncultured organisms, and evolutionary profiles of community function and structure. It allows for the study of microbial ecology on a larger scale due to the significant cost reduction in DNA sequencing. Moreover, metagenomics is a powerful tool for generating new hypotheses about microbial function, widely used for analyzing microbial communities without the need for culturing strains. Much of the research utilizes the "Shotgun" method or PCR sequencing, obtaining unbiased samples of all genes from the defined communities in the sample.

There are two commonly used methods for research and data collection in high-performance computing and analytics:

- *Targeted sequencing*

This involves a widely used bacterial 16S rRNA gene technique, which relies on hypervariable regions V1-V9 (see Fig. 1), ranging from about 30 to 100 base pairs (English language "base pair, bp"), enabling the identification of different groups of organisms. The 16S hypervariable region analysis is a powerful tool for bacterial taxonomic studies. The 16S rRNA gene serves as a standard for classifying and identifying microbes, as it is present in most of them and exhibits specific changes.

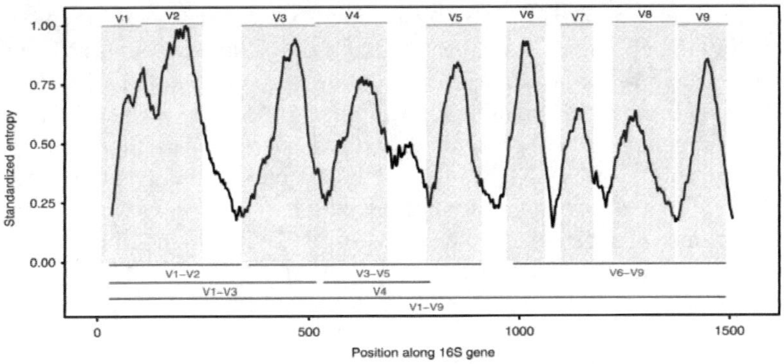

Figure 1. Variable regions (V1-V9) of the 16s rRNA gene [6]

- *"Shotgun" metagenomic sequencing*

This approach involves sequencing many cultured microorganisms and the human genome, randomly cutting the DNA sample, sequencing very short sequences, and reconstructing them into a consensus sequence. In "Shotgun" sequencing (see Figure 2), the entire genome is randomly cut into small fragments (of suitable size for sequencing) and then reassembled [4].

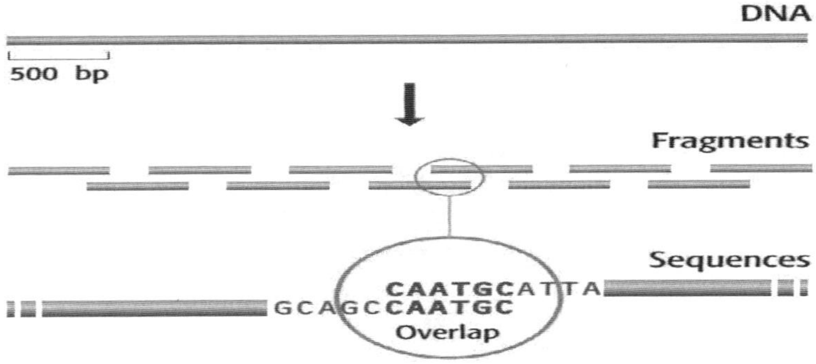

Figure 2. Sequencing execution process using the "Shotgum" method [4*/*

Metagenomics uses tools to facilitate and analyze complex microbial communities and their functional potential. These tools enable taxonomic profiling, functional annotation, and the identification of microbial interactions, contributing to the understanding of the roles of microbiomes in health and disease. A bioinformatics tool refers to software or applications designed to analyze, interpret, and manipulate biological data using computational methods.

The term bioinformatics refers to the use of computers for processing various types of biological information, such as the analysis of biological data, particularly DNA, RNA, and protein sequences. In the field of bioinformatics, computational methods from applied mathematics, statistics, and other exact sciences are utilized, including the implementation of mathematical algorithms related to biological objects and information, along with information and communication disciplines adopted in biological research. It aims to develop and optimize methods and software tools for understanding biological processes and complex datasets. The field of bioinformatics has undergone tremendous development since the mid-1990s, largely influenced by the Human Genome Project and rapid advances in DNA sequencing technology. The main applications of bioinformatics include [1]: (1)Modeling of biological systems and functions; (2) Analyzing laboratory data; (3) Generating models based on accumulated experimental data; (4)Exploring new data using mathematical models; (5) Pattern recognition in experimental data; (6) Predicting gene and protein functions; (7) Conducting in silico experiments.

Bioinformatics research involves the identification of candidate genes and single nucleotide polymorphisms (SNPs) to better understand the genetic basis of diseases, unique adaptations, desirable traits (especially in agricultural species), or differences between populations [1]. The primary goal of bioinformatics is to enhance the understanding of biological processes, with a focus on the development and application of computationally intensive techniques such as pattern recognition, data mining, machine learning algorithms, and visualization. It encompasses the creation and development of databases, algorithms, computational and statistical techniques, and theory to solve formal and practical problems arising from the management and analysis of biological data. Analyzing biological data to extract meaningful information involves writing and running software programs that utilize algorithms from graph theory, artificial intelligence, soft computing, data mining, image processing, and computer simulation.

Databases play a crucial role in bioinformatics research and applications. There are many databases covering different types of information, such as DNA and protein sequences, molecular structures, phenotypes, and biodiversity. These databases may contain empirical data obtained directly from experiments, predicted data derived from analysis, or both. They can be organism-specific or include data collected from multiple other databases. Databases vary in format, access mechanisms, and whether they are public or not. Some commonly used databases include: (1) Genbank and UniProt for biological sequence analysis; (2) Protein Data Bank (PDB) for structural analysis; (3) InterPro and Pfam for finding protein families and motifs; (4) Sequence Read Archive for next-generation sequencing data; (5) Metabolic pathway databases (KEGG, BioCyc) and interaction analysis databases for network analysis; (6) GenoCAD for designing synthetic genetic circuits.

Ribosomal RNA (rRNA) gene sequencing is currently the preferred method for nucleic acid-based phylogenetic reconstruction and the discovery and quantification of microbial diversity. The SILVA ribosomal RNA databases, version 138, dated 12/16/2019, contain more than 9,000,000 small subunit (SSU-16S/18S) and 1,000,000 large subunit (LSU-23S/28S) sequences [2]. The ARB software package with corresponding rRNA databases has become a standard tool for large-scale ribosomal RNA analysis, adopted by researchers worldwide.

The size and number of available datasets are growing rapidly with the collection of data from all over the world. Analyzing big datasets poses challenges in data capture, storage, analysis, sharing, transfer, visualization, search, updating, and protection. Data analysts can discover new correlations that may lead to scientific breakthroughs and disease prevention. Analyzing these vast databases requires software products installed on high-performance clusters.

The purpose of this article is to design, develop and create a Galaxy Europe computing server system on a high-performance cluster with NVIDIA TESLA V100 32GB GPU cards for data processing and analysis in bioinformatics in the field of metagenomics and 16s rRNA gene. The article is divided into several sections. The current section is bioformatics processing of metagenomic data The second section focuses on designing the Galaxy software on the corresponding high-performance cluster. The Experiments section is divided into two subsections, pertaining to the choice of method and data and the results obtained. Some conclusions are reflected in the Conclusion section.

2. Galaxy europe software design on the high performance cluster with NVIDIA TESLA V100 32 GB GPU cards

The Galaxy software is developed and accessible through a browser-based desktop environment for scientific computing. This environment enables a global community of scientists and researchers to achieve accessible, reproducible, and collaborative research, allowing them to share, analyze, and visualize their data with minimal technical difficulties. Together, this community further develops and enhances the Galaxy software framework, integrating analysis and visualization tools. The software runs on public servers, making Galaxy accessible via web-based access and enabling analysis and result publication. Users interact with Galaxy over the network by uploading and analyzing their data. Galaxy interacts with the underlying computing infrastructure (servers for analysis and disks for data storage) without exposing them to the user (see Figure 3) [7].

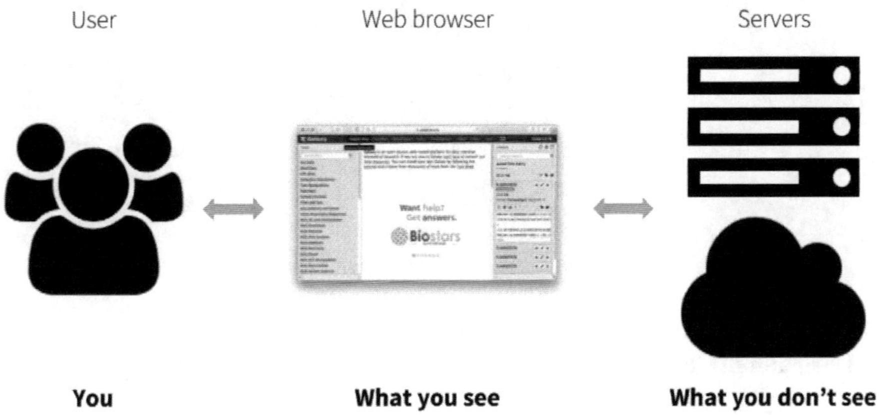

Figure 3. User interaction with the Galaxy Europe platform [7]

The main public Galaxy server (https://usegalaxy.org) offers a rich set of tools for large-scale genomic analyses, providing terabytes of public data for use, along with various shared versions of already analyzed data and visualized workflow steps[8].

2.1. System Architecture

The system used for performance testing possesses the following characteristics (see Fig. 4)

- 16 cores (CPUs), 32 threads computing processor
- 4.8TB disk space for data storage, with the ability to add more memory disks
- 32GB of operating memory, with slots for additional memory.
- 2 x 10 SFP+ ports for fast data transfer in the network
- Ubuntu 22.04 OS

Figure 4. The GPU computing system

2.2. Galaxy software installation

Docker is an open platform for developing, shipping, and running applications. It enables the decoupling of applications from the underlying infrastructure, facilitating rapid software delivery. Using Docker, a Galaxy installation becomes an easy-to-distribute and fully-featured solution suitable for testing and creating production environments. One of the primary objectives is to simplify the access to complete toolkits by providing a publicly available web service, allowing users to set up their own Galaxy server or have administrative access to a local server [19]. Docker proves particularly advantageous in high-density environments and small to medium-sized deployments, where resource optimization is essential (see Fig. 5).

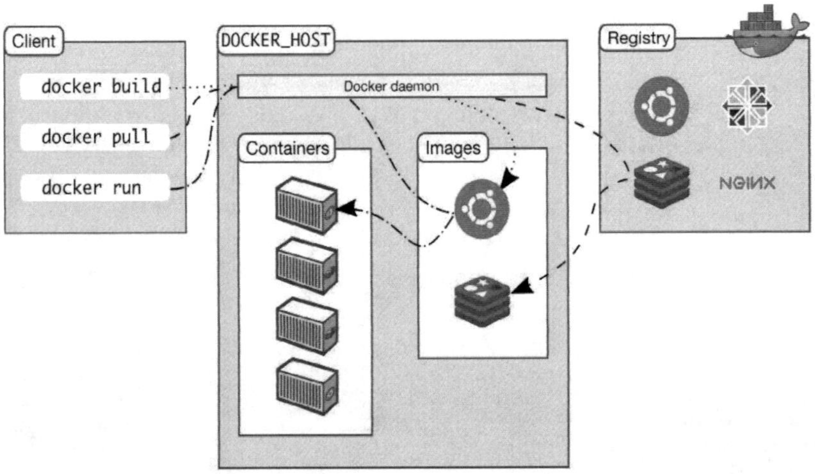

Figure 5. The client-server architecture used by Docker [11]

The Docker Client communicates with the Docker Daemon, which handles tasks such as building, running, and distributing Docker containers. Both the client and daemon can run on the same system, or one can connect a Docker client to a remote Docker daemon. The communication between Docker's Client and Daemon occurs via a REST API over UNIX sockets or a network interface [11]. To install Docker on a Linux operating system, such as Ubuntu version 20.04 [12], one can execute the following commands after successfully accessing the server with the username and password:

$ wget get.docker.com -O install.sh (see Fig. 6)

```
root@agalaxy01:~# wget get.docker.com -O install.sh
--2022-08-19 08:57:46--  http://get.docker.com/
Resolving get.docker.com (get.docker.com)... 52.85.5.66, 52.85.5.127, 52.85.5.10,
 ...
Connecting to get.docker.com (get.docker.com)|52.85.5.66|:80... connected.
HTTP request sent, awaiting response... 301 Moved Permanently
Location: https://get.docker.com/ [following]
--2022-08-19 08:57:46--  https://get.docker.com/
Connecting to get.docker.com (get.docker.com)|52.85.5.66|:443... connected.
HTTP request sent, awaiting response... 200 OK
Length: 20009 (20K) [text/plain]
Saving to: 'install.sh'

install.sh          100%[===================>]  19.54K  --.-KB/s    in 0s

2022-08-19 08:57:46 (165 MB/s) - 'install.sh' saved [20009/20009]
```

Figure 6. The output of the command $ wget get.docker.com -O install.sh

Everything is downloaded and ready to install by typing the following commands

$chmod +x install.sh

$./install.sh

Command *$ curl -fsSL https://get.docker.com -o get-docker.sh*
$ sh get-docker.sh calls an installation script and the job is done (see Fig. 7).

```
root@agalaxy01:~# ./install.sh
+ sh -c DEBIAN_FRONTEND=noninteractive apt-get install -y -qq docker-ce-rootless-
extras >/dev/null
+ sh -c docker version
Client: Docker Engine - Community
 Version:           20.10.17
 API version:       1.41
 Go version:        go1.17.11
 Git commit:        100c701
 Built:             Mon Jun  6 23:02:46 2022
 OS/Arch:           linux/amd64
 Context:           default
 Experimental:      true

Server: Docker Engine - Community
 Engine:
  Version:          20.10.17
  API version:      1.41 (minimum version 1.12)
  Go version:       go1.17.11
  Git commit:       a89b842
  Built:            Mon Jun  6 23:00:51 2022
  OS/Arch:          linux/amd64
  Experimental:     false
 containerd:
  Version:          1.6.7
  GitCommit:        0197261a30bf81f1ee8e6a4dd2dea0ef95d67ccb
 runc:
  Version:          1.1.3
  GitCommit:        v1.1.3-0-g6724737
 docker-init:
  Version:          0.19.0
  GitCommit:        de40ad0

================================================================================
```

Figure 7. Installed Docker for client-server after running the $./install.sh command shows the software versions

The purpose of the installation script is to provide convenience and facilitate a quick installation of the latest versions of Docker [13]. Additionally, we will require another Docker client called Docker Compose, which needs to be installed to finalize our configuration. Docker Compose enables working with applications that consist of a set of containers. The installation of Docker Compose is done manually with a command in the command line [14].

$ curl -SL https://github.com/docker/compose/releases/download/v2.7.0/docker-compose-linux-x86_64 -o /usr/local/bin/docker-compose

To test the installation we write (see Fig. 8):
$ docker compose version

```
root@agalaxy01:~# docker compose version
Docker Compose version v2.6.0
root@agalaxy01:~#
```

Figure 8. The Docker compose version

Docker and Docker Compose have been successfully installed, and the Galaxy installation is now in progress [15]. To obtain Galaxy Docker Compose, one needs to manually download it using the command:

$ git clone https://github.com/bgruening/docker-galaxy-stable.git export

All working data, including the database and virtual environment, will be exported to the *EXPORT_DIR*, which is set to *./export*. Upon the initial launch, the Galaxy container will copy a collection of files to *EXPORT_DIR* and include basic settings set in the *docker-compose.yml file*. These initial credentials can be modified using the following variables:
GALAXY_DEFAULT_ADMIN_USER,GALAXY_DEFAULT_ADMIN_EMAIL, GALAXY_DEFAULT_ADMIN_PASSWORD and GALAXY_DEFAULT_ADMIN_KEY.

Modifications to these settings can be made in the galaxy-configurator, the central configuration location used to configure Galaxy and its additional services. This is achieved through variables set in the docker-compose file, specifically for a Galaxy server, they appear as follows:

```
galaxy-server:
image:${DOCKER_REGISTRY:-quay.io}/${DOCKER_REGISTRY_USERNAME:-bgruening}/galaxy-
server:${IMAGE_TAG:-latest}
build: galaxy-server
environment:
        - GALAXY_DEFAULT_ADMIN_USER=admin
        - GALAXY_DEFAULT_ADMIN_EMAIL=admin@galaxy.org
        - GALAXY_DEFAULT_ADMIN_PASSWORD=password
        - GALAXY_DEFAULT_ADMIN_KEY=fakekey
hostname: galaxy-server
privileged: True
ports:
   - 5555:5555
```

Docker installs the latest version of Gakaxy's Docker *image* with *IMAGE_TAG:-latest*. The *environment* field is for the administrator settings. In *hostname: galaxy-server* write the name of the server, and *ports* - the port on which the communication is performed. The Galaxy Configurator does all the configuration and takes center stage for configuration:

```
galaxy-configurator:
image:${DOCKER_REGISTRY:-quay.io}/${DOCKER_REGISTRY_USERNAME:-bgruening}/galaxy-
configurator:${IMAGE_TAG:-latest}
build: galaxy-configurator
environment:
        - EXPORT_DIR=${EXPORT_DIR:-./export}
        - HOST_PWD=$PWD
        - GALAXY_OVERWRITE_CONFIG=true
        - GALAXY_DEPENDENCY_RESOLUTION=conda
        - GALAXY_JOB_RUNNER=local
        - GALAXY_CONFIG_ADMIN_USERS=admin@galaxy.org
   -
GALAXY_CONFIG_GALAXY_INFRASTRUCTURE_URL=${GALAXY_CONFIG_GALAXY_INFRAS
TRUCTURE_URL:-http://localhost}
    - GALAXY_CONFIG_CONDA_AUTO_INSTALL=true
```

-
GALAXY_CONFIG_AMQP_INTERNAL_CONNECTION=amqp://galaxy:vaiJa3ieghai2ief0jao@rabbitmq/galaxy
- *GALAXY_PROXY_PREFIX=${GALAXY_PROXY_PREFIX:-}*
- *GALAXY_CONFIG_CLEANUP_JOB=onsuccess*
- *NGINX_OVERWRITE_CONFIG=true*

Contains various settings. As we already know *IMAGE_TAG:-latest* installs for us the latest version of Galaxy image. The line *build*: galaxy-configurator performs the functions specified in the configurator. *Environment* settings define the working directories, the admin user, where the tasks run, the database, the web server, etc. The database looks like this::

postgres:
image: postgres:12
hostname: postgres
environment:
- *POSTGRES_PASSWORD=chaopagoosaequuashie*
- *POSTGRES_USER=galaxy*
- *POSTGRES_DB=galaxy*

The database version is set in the field *image: postgres:12* and its name in *hostname: postgres*. The remaining lines are the password (Note: this line must be changed) and the main (admin) user. All network traffic goes through the Nginx server:

nginx:
image:
${DOCKER_REGISTRY:-quay.io}/${DOCKER_REGISTRY_USERNAME:-bgruening}/galaxy-nginx:${IMAGE_TAG:-latest}
build: galaxy-nginx
ports:
- *80:80*

The version being downloaded is also determined by the *IMAGE_TAG:-latest* line, and the server itself communicates on port 80. RabbitMQ is the message brokering software intended to control the message queue for better performance:

rabbitmq:
image: rabbitmq:alpine
container_name: galaxy-rabbitmq
hostname: rabbitmq
environment:
- *RABBITMQ_DEFAULT_USER=galaxy*
- *RABBITMQ_DEFAULT_PASS=vaiJa3ieghai2ief0jao*
- *RABBITMQ_DEFAULT_VHOST=galaxy*

We have RabbitMQ name and version in the *image* and *container_name* fields. The lines in *environment* define the user and the password (Note: this line must be changed). The command that will start the Galaxy server is run:

$ docker compose up -d

Together they run Nginx as a proxy, Postgres as the underlying database, and RabbitMQ as a queue for receiving and processing messages (see Fig. 9).

```
root@alexgalaxy:~/export/docker-galaxy-stable/compose# docker compose up -d


[+] Running 5/5
 :: Container galaxy-rabbitmq                   Started          2.9s
 :: Container compose-postgres-1                Started          3.0s
 :: Container compose-galaxy-configurator-1     Started          2.8s
 :: Container compose-galaxy-server-1           Started          7.1s
 :: Container compose-nginx-1                   Started          9.2s
```

Figure 9. With the docker-compose up -d command, we start Galaxy, with the processes accompanying it running in the background. We can see that they are up and running normally.

After starting all the server processes and configuring the settings, one can access the Galaxy web environment at *http://localhost:5555* (see Fig. 10). The Galaxy graphical user interface (GUI) is user-friendly and offers a wide range of features, including data importing, tool launching, working with histories, creating workflows, and sharing your work. The Galaxy interface is divided into three main parts: (1) The left panel lists all the available tools; (2) The central panel allows you to perform analyses and view the results.; (3)The right panel displays the history of the analysis. With this intuitive interface, users can seamlessly navigate and leverage the full potential of the Galaxy platform for their scientific endeavors.

Figure 10. Galaxy homepage on a local server platform.

3. Experiments conducted

3.1. Model and data selection

For conducting the experiments, the main method for metagenomics, Targeted Sequencing Amplicon Analysis, was used [9]. The structural model of this analysis is shown in Appendix A. The model follows several steps for data processing and analysis. Each step includes the tools used for the specific process: (1) Preparation of data sets - analysis of multiple samples with mothur software [10]. Merginge them into one file; (2) Optimization of compute files - sequencing from the same organisms, expecting that many of our sequences are duplicated; (3) Quality control - every analysis requires the quality of its data to be checked and improved; (4) Sequence alignment - aligning sequences to the V4 variable region of 16S rRNA; (5) Extraction of taxonomic information - determining which microorganisms are present in the samples, in what proportion, and the community structure; (6) Visualization - visualizing the results. Amplicon analysis is a highly targeted approach to analyze genetic variation in specific genomic regions. It refers to the capture and sequencing of rRNA data in a sample, such as 16S for bacteria, archaea, or 18S for eukaryotes. The 16S rRNA gene has several properties that make it ideally suited for our purposes: (1) it is present in all living organisms; (2) it has highly conserved and highly variable regions; and (3) there is a large set of reference databases.

The data needed to test the Galaxy Europe web-based platform and local server system will be sourced from the Human Microbiome Project (HMP) [16]. HMP performed 16S rRNA and metagenomic sequencing of samples from healthy adult men and women between the ages of 18 and 40 years. The datasets come from human fecal samples taken from the Human Microbiome Project Data Portal (HMP) [17], for which the 16S rDNA V3-V5 variable region window (V35) was sequenced for all samples using Roche-454 FLX Titanium. They are divided into two test collections in a special FASTA file format:

- Test Analysis of Collection Database 300 MB - Data Series Collection: SRS072900.fsa,, SRS072858.fsa, SRS072877.fsa, SRS072881.fsa, SRS072906.fsa, SRS066753.fsa
- Test Analysis of Collection Database 700 MB - Data Series Collection:: SRS072010.fsa, SRS072027.fsa, SRS071990.fsa, SRS071992.fsa, SRS071970.fsa, SRS072001.fsa
- Test Analysis of Collection Database 1500 MB - Data Series Collection: SRS071994.fsa, SRS072872.fsa, SRS071984.fsa, SRS071995.fsa, SRS072009.fsa, SRS071975.fsa, SRS072893.fsa, SRS070372.fsa, SRS071979.fsa, SRS072010.fsa, SRS072027.fsa, SRS071990.fsa, SRS071992.fsa, SRS071970.fsa, SRS072001.fsa

Table 1. Test data: Number of sequences (Seqs); Size (Size)

Source	Dataset	Seqs	Size (MB)
HMP Ileum	Test_300	527135	300
HMP Ileum	Test_700	1121140	700
HMP Ileum	Test_1500	2577144	1500

3.2. Results obtained

The tests of the local server system and the Galaxy Europe web-based platform will be divided into two stages: (1) for a load with the same parameters on both systems and (2) performance tests of the server system. The description of the experimental design and data analysis steps will follow the targeted sequencing Amplicon analysis. The collection and comparison of times are taken from the metric data of the processed tasks. In the server system performance test, additional software adjustments were made to improve its performance in computing larger tasks. The settings themselves were performed by changing parameters in the configuration files of the Galaxy Europe server. Data loading takes place on the home page of Galaxy Europe (see Fig. 11).

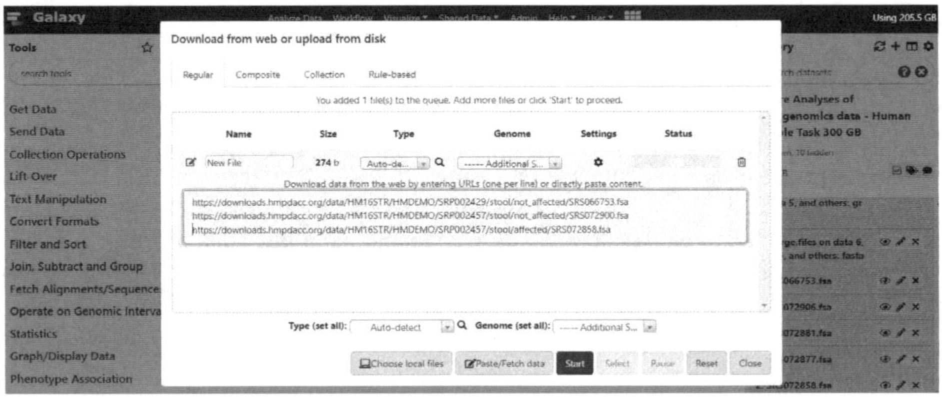

Figure 11. Entering the data links and starting the procedure

With the *Merge.files* and *Make.group* tools, a group file from the Test_300 collection was created and used to analyze the multiple samples. As samples often sequence many of the same organisms, it can be expected that many of our sequences are duplicated. By using the *Unique.seqs* command, we establish the unique sequences to save time and resources in recalculations. This will also reduce the amount of disk space. To further reduce file size and streamline analysis, we use the *Count.seqs* tool. It summarizes the number of times each unique sequence was observed in each of the samples. Verification and improvement of data quality is run with the *Summary.seqs* tool, and the results obtained are stored in a *logfile*. The output data is filtered by base and homopolymer length with the *Screen.seqs* tool, which removes all sequences with ambiguous bases.The following tool will remove all sequences with ambiguous bases (maxambig parameter), homopolymeric stretches of 9 or more bases (maxhomop parameter), and any sequences longer than 530 base pairs (bp) or shorter than 350 bp.

	Start	End	NBases	Ambigs	Polymer	NumSeqs
Minimum:	1	3	3	0	1	1
2.5%-tile:	1	61	61	0	4	13179
25%-tile:	1	411	411	0	4	131784
Median:	1	487	487	0	5	263568
75%-tile:	1	506	506	0	5	395352
97.5%-tile:	1	530	530	1	6	513957
Maximum:	1	947	947	18	31	527135
Mean: 1	430.816	430.816	0.120937		4.77364	
# of unique seqs:		410278				
total # of seqs:		527135				

Figure 12. The logfile results also contain some summary statistics: We have a total of 410,278 unique sequences out of a total of 527,135 that range in length between 3 and 947 bases.

Sequence alignment against the V4 variable region reference of 16S rRNA generated from the Silva database is an important step to improve the clustering of operational taxonomic units (OTUs) [18]. This is performed with the tools *Align.seqs* and *Summary.seqs*. The quality of the alignment can be seen in the source file.

```
                Start   End     NBases  Ambigs  Polymer NumSeqs
Minimum:        1       1       1       0       1       1
2.5%-tile:      1       1231    3       0       1       9715
25%-tile:       1       1231    3       0       1       97150
Median:         1       13425   15      0       2       194299
75%-tile:       1       13425   293     0       4       291448
97.5%-tile:     13391   13425   294     0       5       378883
Maximum:        13425   13425   321     0       8       388597
Mean:     862.515 7447.1  133.255 0       2.71207
# of unique seqs:       290570
total # of seqs:        388597
```

Figure 13. Start and End columns show that the majority of sequences align between positions 1 and 13425

Screen.seqs is run again to ensure everything overlaps the same region. Filters are applied to the columns in the alignment because they contain a lot of extraneous characters like "." and "-" and are not included in our sample. The *Filter.seqs* tool collects the information from them and adds it to the main data. The results of analyzing the data using the Amplicon method provide answers regarding which microorganisms are present in the samples, in what proportion, and what is the structure of their community. This is determined by grouping sequences based on their similarity to defined operational taxonomic units (groups of similar sequences that can be treated as a single 'genus' or 'species' (OTUs)). We further denoise the sequences from potential sequencing errors by pre-clustering them using the *Pre.cluster* command, allowing up to 2 differences between them. We would like to classify the sequences (see Fig.14) using a training set, which we add additionally in the form of files: trainset18_062020.pds.fasta; trainset18_062020.pds.tax.

name	taxonomy
GGSPTC101EFO4Z_cs_nbp_rc	Bacteria(100);Bacteroidetes(100);Bacteroidia(100);Bacteroidales(100);Rikenellaceae(100);Alistipes(100);
GGSPTC101BWHND_cs_nbp_rc	Bacteria(100);Bacteroidetes(100);Bacteroidia(100);Bacteroidales(100);Bacteroidaceae(100);Phocaeicola(99);
GGSPTC101BWLH5_cs_nbp_rc	Bacteria(100);Bacteroidetes(100);Bacteroidia(100);Bacteroidales(100);Bacteroidaceae(100);Phocaeicola(85);
GGSPTC101AS576_cs_nbp_rc	Bacteria(100);Firmicutes(100);Clostridia(100);Clostridiales(100);Ruminococcaceae(95);Ruminococcaceae_unclassified(95);
GGSPTC101CAJVZ_cs_nbp_rc	Bacteria(100);Bacteroidetes(100);Bacteroidia(100);Bacteroidales(100);Bacteroidaceae(100);Bacteroides(100);
GGSPTC101EPNXD_cs_nbp_rc	Bacteria(100);Firmicutes(100);Erysipelotrichia(100);Erysipelotrichales(100);Erysipelotrichaceae(100);Holdemanella(100);
GGSPTC101CZSRS_cs_nbp_rc	Bacteria(100);Firmicutes(100);Clostridia(100);Clostridiales(100);Ruminococcaceae(100);Faecalibacterium(93);
GGSPTC101DEC6Z_cs_nbp_rc	Bacteria(100);Firmicutes(100);Clostridia(100);Clostridiales(100);Lachnospiraceae(100);Lachnospiraceae_unclassified(100);
GGSPTC101CWTS9_cs_nbp_rc	Bacteria(100);Bacteroidetes(100);Bacteroidia(100);Bacteroidales(100);Rikenellaceae(100);Alistipes(100);
GGSPTC101DVJ2W_cs_nbp_rc	Bacteria(100);Bacteroidetes(100);Bacteroidia(100);Bacteroidales(100);Rikenellaceae(100);Alistipes(100);
GGSPTC101BEU6J_cs_nbp_rc	Bacteria(100);Bacteroidetes(100);Bacteroidia(100);Bacteroidales(100);Bacteroidaceae(100);Bacteroides(100);

Figure 14. Taxonomy output where each sequence is classified

The information is used to determine the abundance of the various taxa found. With the tools *Cluster.split* (see Fig. 15), *Make.shared*, *Classify.otu*, all individual sequences are classified and assigned a confidence score (0-100%). Sequences are grouped at a 97% identity threshold, and a consensus classification is determined for each cluster.

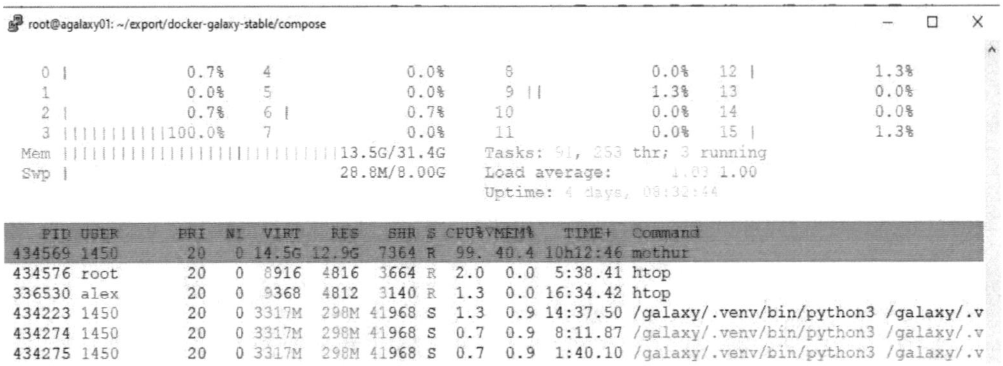

Figure 15. The system load when calculated with the *Cluster.split* tool. The task is calculated over 10 hours and uses 1 core with 13 GB of RAM..

The visualization of the information is done with two tools Taxonomy-to-Krona and Krona pie chart, and the result is an HTML file with interactive visualization (see Fig. 16).

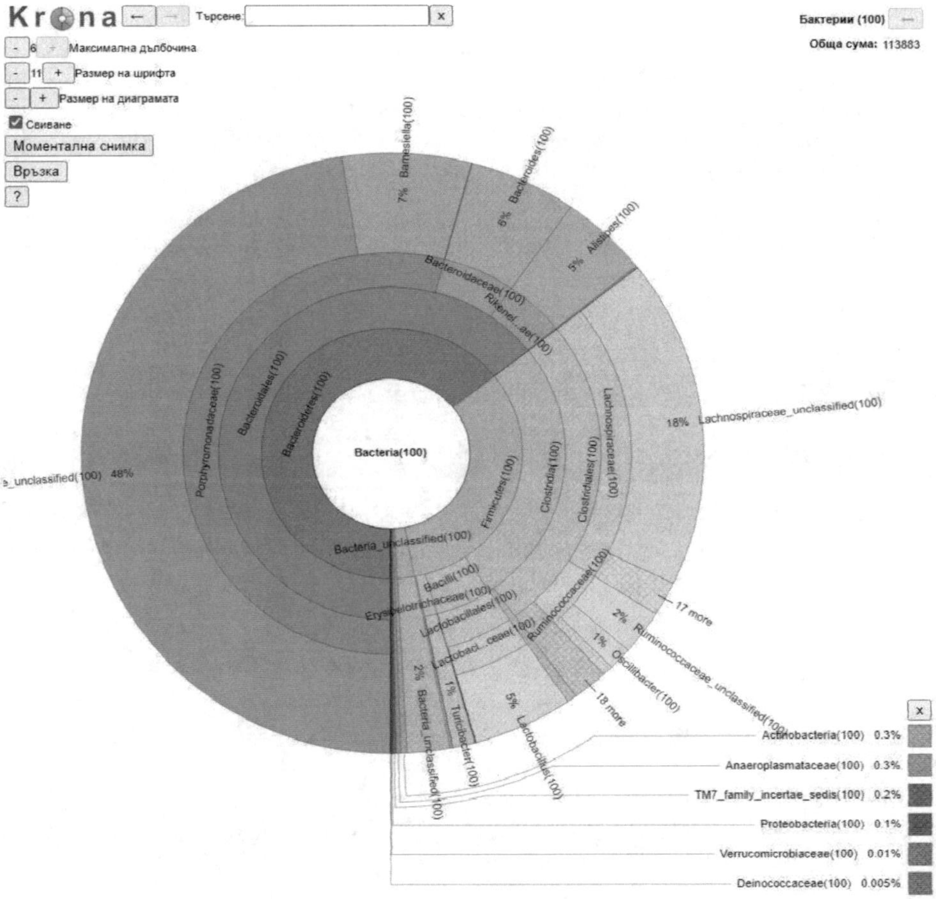

Figure 16. Example visualization of processed sample data

3.3. Experiments

Performance testing was initiated on three sets of test data using the standard Galaxy Europe software workflow tools for processing and analyzing 16S rRNA sequencing data. The data were processed in a web-based Galaxy Europe platform and two server system configurations. The task processing times were taken directly from the platforms' metadata and recorded in Excel spreadsheets for each individual workflow task, as well as the overall analysis time.

The tests were carried out in two stages:

- The first stage - system performance of the web-based Galaxy Europe platform and the local server system. The parameters were chosen identically. The configuration, as an architecture, was predetermined by the Galaxy Europe platform, and the limitation came from the use of resources as ordinary users: 2 cores (CPUs), 4 threads of a computing processor; 250 GB disk space for data storage; 24 GB of operating memory.
- The second stage - local server system performance: 8 cores (CPUs), 16 processor threads; 4.8 TB disk space for data storage; 32 GB of operating memory.

Names of systems in tests:

- Web-based Galaxy Europe Platform - 2 Core Free Galaxy Platform
- Local server system configuration A - 2 Core server system
- Local server system configuration B - 8 Core server system

The three types of systems were tested with a series of data sets that vary in size: 300 MB (527,135 sequences); 700 MB (1,121,140 sequences); 1500 MB (2,577,144 sequences). The aim was to investigate the behavior of the systems under different loads in data processing and analysis, as well as to record the times for each of the flows. The results obtained are recorded, compared, and analyzed.

The model selected for all three configurations of the performance evaluation systems was tested with structurally identical data sets - 16S rRNA sequences in FASTA file format. For evaluation purposes, the systems were compared with data in a single variable region (V4) of the test samples. The obtained test results are recorded in tables and graphs, and diagrams are derived from them.

Table 2. Time performance of the two 2-core systems

	Time (sec)	
Dataset Size (MB)	2 Core server system	2 Core Free Galaxy Platform
300	2618	3223
700	12724	14145
1500	42550	47833

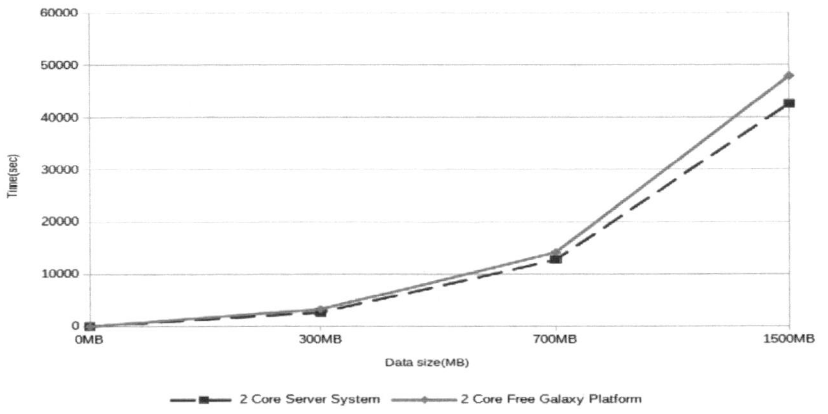

Figure. 17. Graphical distribution of time performance at different data sizes for different system configurations

Using the values from comparative Table 2 on different data sizes, on Figure 17 graph is drawn, which shows the time performance of the two systems that use 2 cores configuration. It shows that the 2-core server system processes the data in less time than the open Galaxy platform. For a smaller amount of data, the time required by both systems is almost identical, but as the data size increases, faster data processing is observed when using the server system. This is also shown by the Charts on Figure 18.

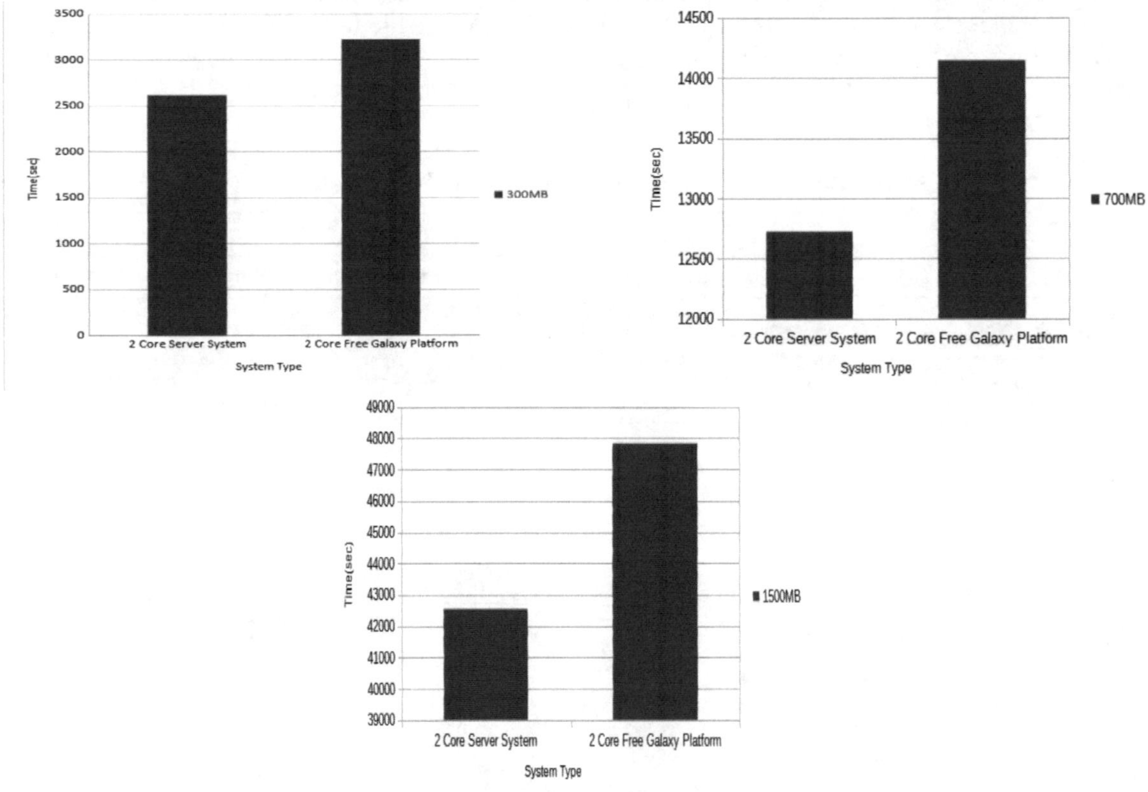

Figure. 18. Comparative graph for the time performance of the three experiments for the two types of systems

Using the values from the comparative Table 4 (see Appendix B), on different data sizes, Graphs drawn (see Fig. 19), which show the time performance of the two systems using 2 cores. They show the performance of each individual task and where the system takes the most time to compute. Both systems on all three data types waste the most time on the *Align.seqs* and *Cluster.split* tool steps in all three data types. We observe the same peak for the server system with 8 processor cores.

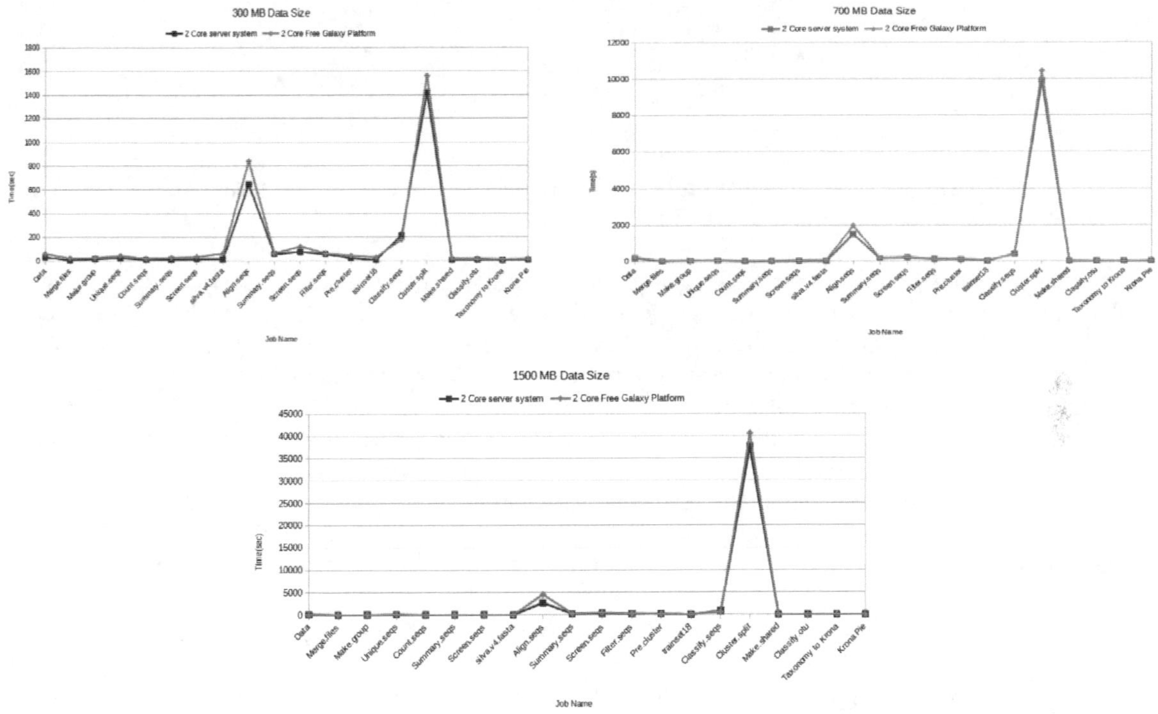

Figure 19 Graph comparing the execution times for running the experiments (300 MB, 700 MB, 1500 MB) on the two types of systems

Table 3 compares the time performance the two systems using 2 cores and 8 cores for the server system in each of the three experiments (300MB, 700 MB, 1500 MB). We should note that the free access to the Galaxy software allows the use of 1 or 2 cores.

Table 3. Times performance of the three systems

Dataset Size (MB)	Time (sec)		
	2 Core server system	2 Core Free Galaxy Platform	8 Core server system
300	2618	3223	1053
700	12724	14145	7433
1500	42550	47833	25112

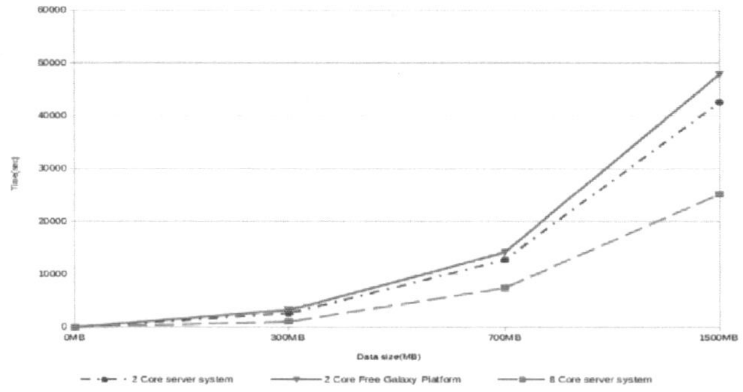

Figure 20. Graphical distribution of time performance at different data sizes for different system configurationsсистемите

Using the numerical values from the comparison table, Graph on Figure 20 is drawn. It clearly shows that the server system is faster to process the data both when using 2 cores and for 8 cores as compared to the open platform. For both 2-core and 8-core systems, the number of threads is the biggest contributor to performance. The speed of data processing increases with increasing number of cores.

Microbial diversity in samples of different data sizes was analyzed and classified using Galaxy Europe software. We visualize the samples with the Krona pie chart tool. The result is an interactive visualization HTML file (see Figure 21), which is a single plot for all our samples in the corresponding data set. Bacterial communities in the analysis of the three types of data were largely identical. Three large classes, Firmicutes, Proteobacteria, and Bacteroidetes, stand out. Some bacterial groups, such as Bacteroides, constitute the most essential part of the gastrointestinal microbiota of mammals, where they play a major role in the processing of complex molecules to simpler ones in the host intestine. Phocaeicola microbe in the human intestine is responsible for various functions such as carbohydrate synthesis or metabolism. Clostridium sensu stricto is one of the important anaerobes in the human intestine. They can metabolize various compounds such as carbohydrates, amino acids, alcohols, and purines.

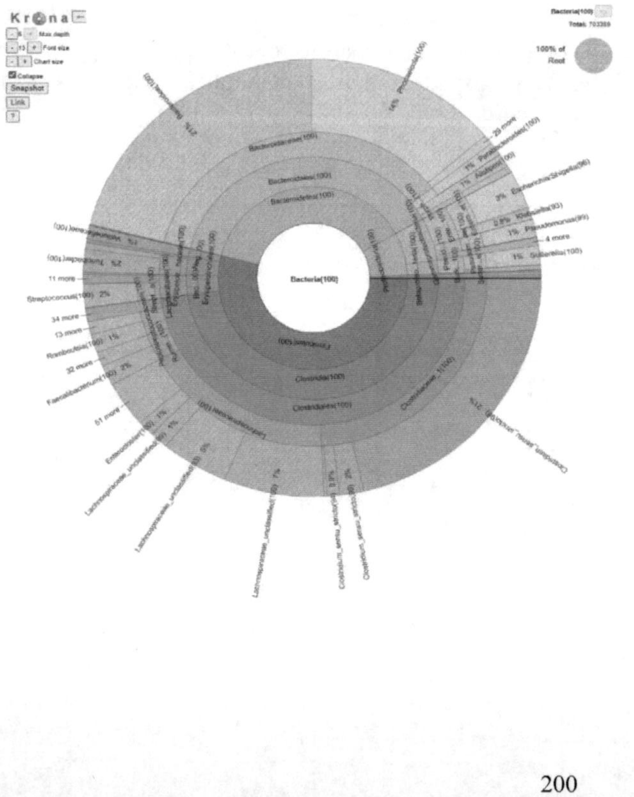

Figure 21. Krona Pie HTML interactive graph. Research result with 1500 MB of data.

3.4. Conclusion

The article describes the configuration and creation of a server system for processing, computing, and analysing metagenomic data using Galaxy software on a high-performance cluster with an NVIDIA TESLA 32GB GPU card.

The reasons for choosing Galaxy software are mainly because it is easy and convenient to use for researchers with no experience in computer programming or system administration. It supports a large set of different types of data and integrates over 8,500 software packages for reformatting, manipulating data, analysis, and results without the need for programmatic manipulation. It also works with the Mothur software package used in the model to process 16S rRNA metagenomic data.

The Galaxy software suite is widely used for analyzing microbial communities using data obtained from a range of NGS technologies. This study analyzed metagenomic samples of 16S rRNA sequences in fasta format (Test_300, Test_700, Test_1500) with the Galaxy Europe software package. The size and number of data sequences were chosen in such a way as to test three different system configurations at different levels. The study aims to track the behavior of the system under different loads when performing individual tasks in the data processing process.

The open Galaxy platform imposes limits to users as far as the system configuration is concerned and only provides access to 1 or 2 processor cores along with 250 GB of disk memory. Therefore, the time to process data over 1000 MB exceeds 10 hours. In our case, processing 1500 MB took 13 hours, and trying to process 3500 MB of data reached values of 70 hours. In some cases, our jobs were put on hold until a resource was available on the processing platform. Although not included in the study, these data further increase the analysis time. Often in bioinformatics, data samples come in raw form and exceed 5000 MB, even 10000 MB, and even processed data reaches such values. This would mean processing times of up to and over 300 hours could be reached. This limits us a lot as scientists, researchers, and students, as time is the most valuable resource we have and we shouldn't waste it waiting.

When comparing the Galaxy open platform and a server system with identical parameters, We obtained similar results when processing 300 MB and 700 MB of data, but when increasing to 1500 MB, we noticed a more noticeable difference in time. Performance tests were conducted with 8 processor cores. The reasons that led not to use the full potential of 16 cores were: not to load the system to full capacity; the system itself has additional processes and libraries that also require resources; some of the tools use algorithms and settings designed to utilize 30% to 50% of the configuration resource; and the possibility to launch additional parallel tasks to reduce the testing time. The results obtained reduced the data processing time by almost one half.

The analysis of samples from the metagenomic data showed several bacterial groups that make up the most essential part of the gastrointestinal microbiota of mammals, where they play a major role in the processing of complex molecules to simpler ones in the host gut.

The experiments performed demonstrate that there is a significant relationship between the characteristics of the input sequence data and the computational time required to process this data.

The processing of large databases requires the use of parallel computations on computer clusters or supercomputers. This will significantly reduce processing and research time.

3.5. Acknowledgments

This work has been accomplished with the partial support by the Grant No. BG05M2OP001-1.001-0003, financed by the Science and Education for Smart Growth Operational Program (2014-2020) and co-financed by the European Union through the European structural and Investment funds and with the financial support of the MES by the Grant №DO1-168/28.07.2022 for providing access to e-infrastructure of the NCHDC – part of the Bulgarian National Roadmap on RIs.

References

[1] A reference Arthur M. Lesk, *Introduction to Bioinformatics*, University of Cambridge, 2002
[2] High quality ribosomal RNA databases SILVA, (https://www.arb-silva.de/documentation/)
[3] National Research Council (US) Committee on Metagenomics: Challenges and Functional Applications. The New Science of Metagenomics: Revealing the Secrets of Our Microbial Planet. Washington (DC): National Academies Press (US); 2007. PMID: 21678629.
[4] Brown, T. A. *GENOMES 3*, 2007 by Garland Science Publishing
[5] Kazumasa Fukuda, Midori Ogawa, Hatsumi Taniguchi and Mitsumasa Saito(2016) *Molecular Approaches to Studying Microbial Communities: Targeting the 16S Ribosomal RNA Gene*, 223J UOEH 38（3）：223－232（2016)
[6] Yang B, Wang Y, Qian PY (March 2016). *"Sensitivity and correlation of hypervariable regions in 16S rRNA genes in phylogenetic analysis"*
[7] Galaxy community hub, Galaxy project tutorial (https://galaxyproject.org/tutorials/g101/)
[8] The Galaxy Community, Nucleic Acids Research (2022) *The Galaxy platform for accessible, reproducible and collaborative biomedical analyses*, Pages W345–W351
[9] Galaxy Training, *Analyses of metagenomics data*, Saskia Hiltemann, BéréniceBatut(https://training.galaxyproject.org/training-material/topics/metagenomics/tutorials/general-tutorial/tutorial.html)
[10] Mothur software, (https://mothur.org/)
[11] Dockerdocs, Docker platform, (https://docs.docker.com/get-started/overview/)
[12] Dockerdocs, Install Docker Desktop on Ubuntu (https://docs.docker.com/desktop/install/ubuntu/)
[13] Installing the latest Docker (https://github.com/docker/docker-install)
[14] Install Docker Compose, (https://docs.docker.com/compose/install/compose-plugin/#install-the-plugin-manually)
[15] Galaxy Docker Image, Björn Grüning, University of Freiburg , (https://github.com/bgruening/docker-galaxy-stable)
[16] The Human Microbiome Project (HMP), (https://hmpdacc.org/hmp/overview/)
[17] Human Microbiome Project Data Portal, (https://portal.hmpdacc.org/)
[18] Schloss, P. D., 2012 Secondary structure improves OTU assignments of 16S rRNA gene sequences. The ISME Journal 7: 457–460
[19] Galaxy community hub, Galaxy Docker, (https://galaxyproject.org/use/galaxy-docker/)

APPENDIX A

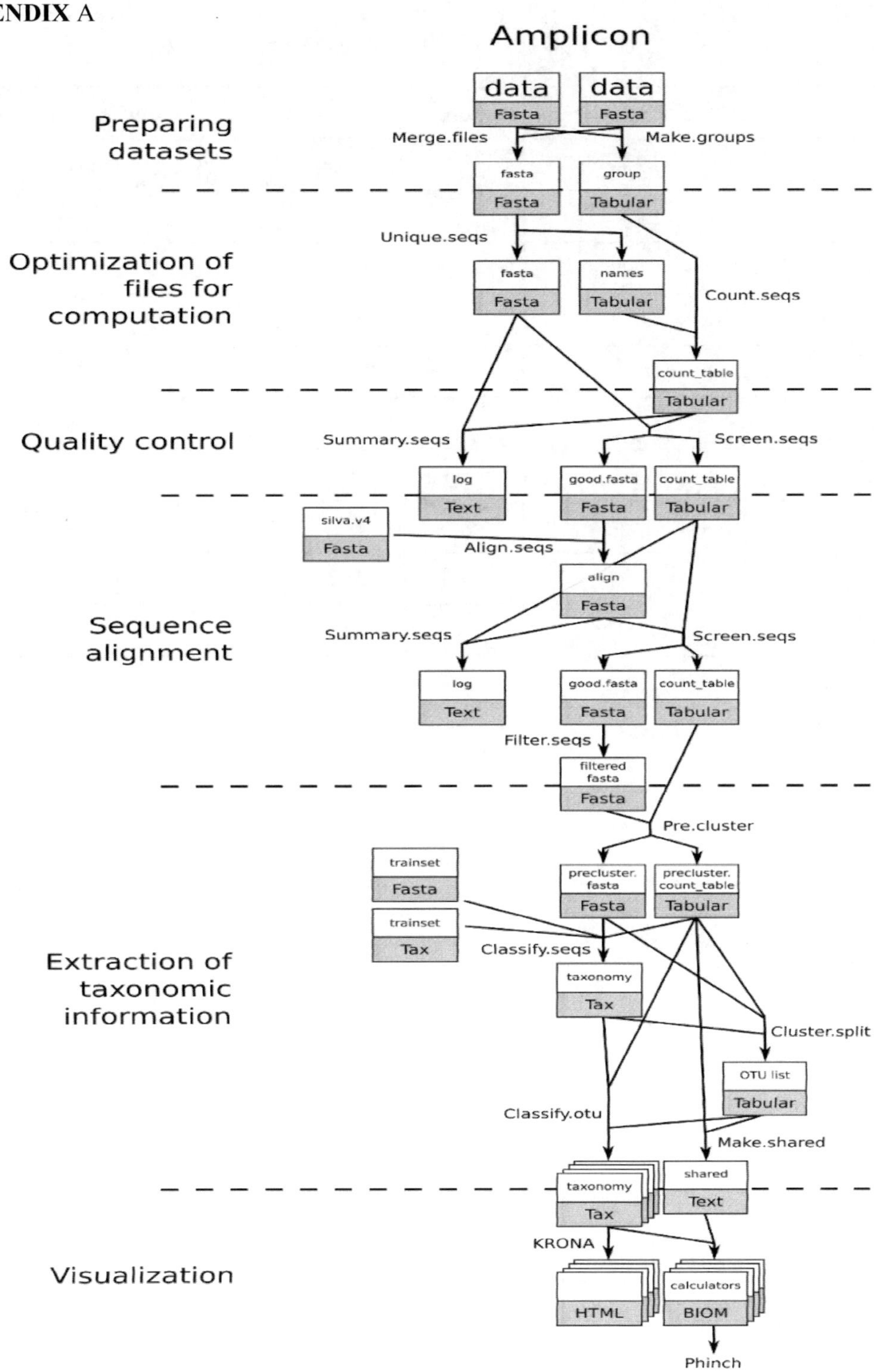

Figure 22. Structural model of Amlicon analysis of metagenomic data [9]

APPENDIX B

Table 4. Test with 1500 MB database

	2 Core server system	2 Core Free Galaxy Platform	8 Core server system
Task	**Time(sec)**	**Time(sec)**	**Time(sec)**
Data	137	180	100
Merge.files	15	60	14
Make.group	59	60	56
Unique.seqs	101	180	86
Count.seqs	28	42	28
Summary.seqs	35	57	16
Screen.seqs	61,2	60	49
silva.v4.fasta	5	38	23
Align.seqs	2676	4560	1055
Summary.seqs	255	300	63
Screen.seqs	350	480	152
Filter.seqs	221	300	54
Pre.cluster	200	180	95
trainset18	5	60	5
Classify.seqs	825	660	201
Cluster.split	37560	40560	23100
Make.shared	5	17	5
Classify.otu	8	16	6
Taxonomy to Krona	3	11	3
Krona Pie	3	12	3

Logistic probability function: an analysis and modification from the perspective of reaction network theory

Meglena Lazarova, Svetoslav Markov and Andrey Vassilev

Faculty of Applied Mathematics and Informatics, Technical University of Sofia, 8 St. Kliment Ohridski Blvd., 1000 Sofia, Bulgaria

E-mail: meglena.laz@tu-sofia.bg

Institute of Mathematics and Informatics, Bulgarian Academy of Sciences, Sofia, Bulgaria

E-mail: smarkov@bio.bas.bg

Faculty of Economics and Business Administration, Sofia University "St. Kliment Ohridski", Sofia, Bulgaria

E-mail: A.Vassilev@feb.uni-sofia.bg

Abstract. In this work we study some characteristics of growth functions of the logistic type. The standard logistic function and the 2-logistic growth function are solutions of ordinary differential equations derived from the perspective of reaction network theory. These solutions are compared in terms of their shape. We are interested in the new 2-logistic probability distribution and its characteristics. Using the tools of reaction network theory and numerical methods we derive some properties of this distribution.

1. Introduction

The logistic distribution is a continuous probability distribution that has become highly relevant in many fields. Often used to describe growth processes, it has roots in the work by Verhulst [1], from which the alternative name "Verhulst model" derives. This distribution serves as the foundation for the logistic function, which is frequently employed to characterize phenomena that exhibit a sigmoidal behavior pattern. Such patterns can be observed in scenarios like population growth or the spread of diseases, making the logistic distribution a versatile and valuable tool in various disciplines. It has also found application in statistical analysis and machine learning, where it has proved suitable for modeling the underlying probability of binary outcome events.

Details on the logistic distribution can be found in the monograph of Balakrishnan [2] and also in Johnson, Kotz and Balakrishnan [3]. The standard logistic distribution is an exponential mixture of extreme value distributions [4]. The shape of the logistic distribution is similar to the shape of the normal distribution. It has heavier tails than the normal distribution and this unique property is usually used for the purposes of the data analysis.

The standard logistic distribution has important uses in describing growth phenomena [5]. This fact has attracted many applications in the modeling of physicochemical processes, weight gain data and etc. Growth models are often used in modeling of various dynamical processes in biology, ecology, epidemiology, social and economic sciences, engineering and natural sciences, demography and also in many others scientific fields. They are frequently formulated in the terms of a system of ODEs (see [6] for a recent example).

Reaction network theory provides a powerful toolkit for formalizing models of growth phenomena. Both the logistic (Verhulst) and the Gompertz models can be derived in a reaction-theoretic framework, as shown in [7]. In this paper we show how the logistic growth model, as derived from a reaction network, can be modified to obtain a version of the logistic distribution. Moreover, we show how a reaction-theoretic formulation naturally suggests a generalization of the logistic model that we call the 2-logistic model. This generalization, as well as the associated probability distribution derived from it, is studied analytically and numerically.

2. The logistic growth model

The logistic growth function can be obtained as a solution of the following differential equation:

$$x' = k_L x \left(1 - \frac{x}{K_L}\right),$$

where k_L and K_L are positive parameters, respectively called (intrinsic) *reaction rate* and *carrying capacity.*

For our purposes it is more convenient to rewrite this differential equation in the form

$$x' = kx(c - x),$$

with $k := k_L/K_L$ and $c := K_L$.

The solution of this equation is a sigmoidal growth function $x = x(t), t \in R$. The related decay function is implicitly involved in the right-hand side of the differential equation as $c - x$, where $c > 0$. Usually $c = 1$.

The logistic growth model can be obtained from the following reaction network:

$$S + X \xrightarrow{k} 2X. \tag{1}$$

Under the assumption of homogeneity and mass action kinetics, the network (1) induces the following system of ordinary differential equations:

$$s' = -ksx, \ x' = +ksx \tag{2}$$

with the initial conditions

$$s(0) = s_0 > 0, \ x(0) = x_0 > 0$$

The system (2) satisfies a conservation relation, which is given by:

$$s' + x' = 0 \quad \Rightarrow \quad s + x = c = const$$

In addition, the dynamical system implies that:

$$s' = -ks(c - s), \ x' = kx(c - x), \ c = s_0 + x_0.$$

The solution of the dynamical system for $x(t)$ is given by the following equation:

$$x(t) = \frac{cx_0}{x_0 + (c - x_0)e^{-ckt}} \tag{3}$$

3. The logistic distribution

Using the expression of solution (3) for $c = 1$ as the cumulative distribution function (CDF), we can define a probability distribution which is based on the logistic model. This has the potential drawback that the CDF has a discontinuity at 0. One approach to circumventing this problem would be to set $x_0 = \epsilon > 0$ and $c = 1 + \epsilon$.

Then the CDF can be defined as

$$F_L(t) := x(t) - \epsilon$$

and the associated probability density function is

$$f_L(t) = F'(t).$$

This is a version of the classical logistic distribution.

Illustrative shapes of the CDF of the obtained logistic distribution are presented in Figures 1 and 2. Depending on the combination of parameter values, one can arrive at CDFs with or without inflection points.

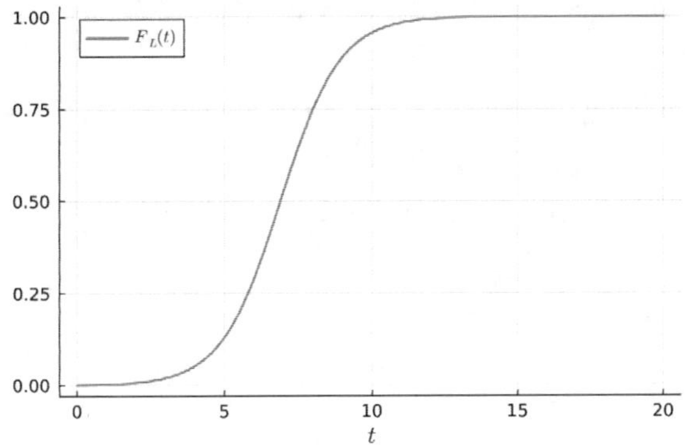

Figure 1. The logistic cumulative distribution function, $\epsilon = 0.001,\ k = 1.0$

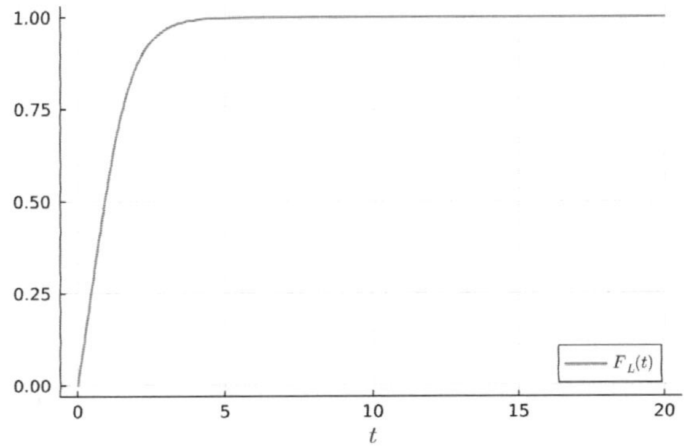

Figure 2. The logistic cumulative distribution function, $\epsilon = 0.5,\ k = 1.0$

4. The 2-logistic growth-decay model

The reaction network derivation of the logistic model suggests natural avenues of generalization of this approach by varying the number of reactants. One possibility is to require a higher number of representatives of species S to enter as reactants in order to trigger a transformation. In the particular case of requiring the presence of two representatives instead of one, we obtain a model that we refer to as the 2-logistic model.

To derive the 2-logistic growth-decay model, we consider the following reaction network:

$$2S + X \xrightarrow{k_1} 3X. \tag{4}$$

The dynamical system induced (assuming homogeneity and mass actions kinetics) by the reaction network (4) is:

$$s' = -2k_1 x s^2, \quad x' = +2k_1 x s^2 \tag{5}$$

and we can define $k = 2k_1$ to simplify notation.

The initial conditions required to define an initial-value problem are given by

$$s(0) = s_0 > 0, \; x(0) = x_0 > 0.$$

The system (5) satisfies the conservation relation: $s(t) + x(t) = c$

Hence for $x(t)$ we have

$$x' = kx\,(c - x)^2$$

5. The 2-logistic distribution

Similarly to the classical logistic distribution, setting $c = 1$ results in a CDF with a discontinuity at 0. We can again set $x_0 = \epsilon > 0$ and $c = 1 + \epsilon$, and define the CDF as

$$F_{2L}(t) := x(t) - \epsilon,$$

with the probability density function given by $f_{2L}(t) = F'(t)$. In this way the 2-logistic distribution is obtained.

The 2-logistic cumulative distribution function for different parameter values is presented in Figures 3 and 4. The figures suggest substantial flexibility in terms of the shapes obtainable in this framework.

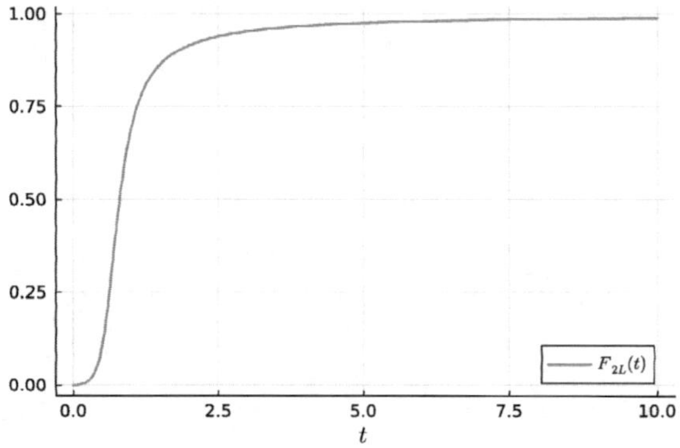

Figure 3. The 2-logistic cumulative distribution function, $\epsilon = 0.001,\; k = 5.0$

Figure 5 shows that the 2-logistic CDF possesses a monotonicity property with respect to the parameter k.

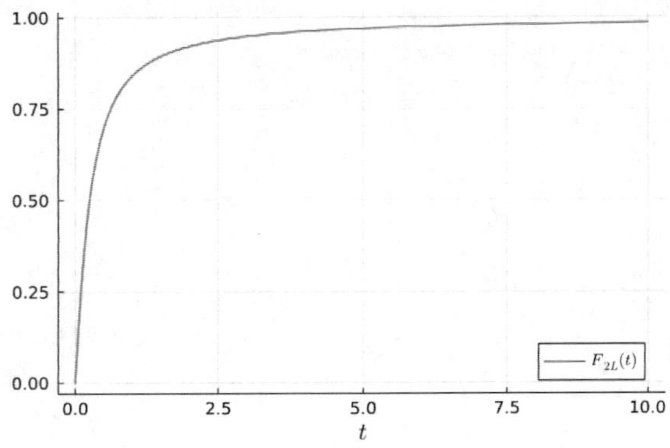

Figure 4. The 2-logistic cumulative distribution function, $\epsilon = 0.7,\ k = 2.0$

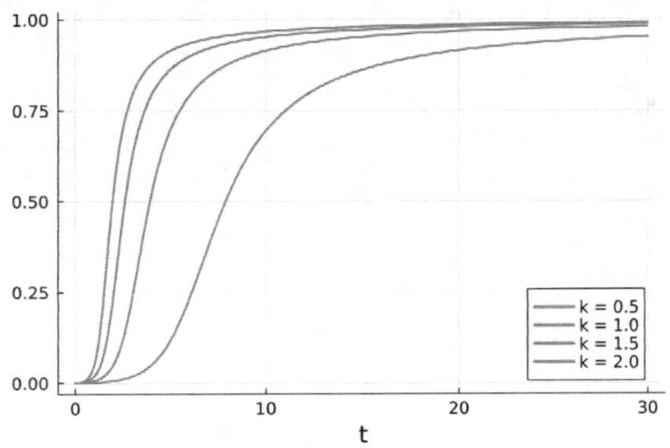

Figure 5. The 2-logistic cumulative distribution function for different values of k, with $\epsilon = 0.001$

To illustrate how the classical logistic and the 2-logistic CDFs differ, Figure 6 compares the shapes of instances of the two CDFs for common values of the parameters ϵ and k. As evidenced by the figure, while the two distributions are derived from common principles and seem to share the "same" parameters, their behavior can differ substantially.

The 2-logistic distribution can be partially characterized by its moments. In order to understand how the latter depend on the parameters of the distribution, we compute and plot the mean and the standard deviation of the 2-logistic distribution for different combinations of values of ϵ and k. Figure 7 demonstrates that the mean is a decreasing function of both ϵ and k, and illustrates the form of the dependence. A similar observation holds with respect to the standard deviation, as shown in Figure 8.

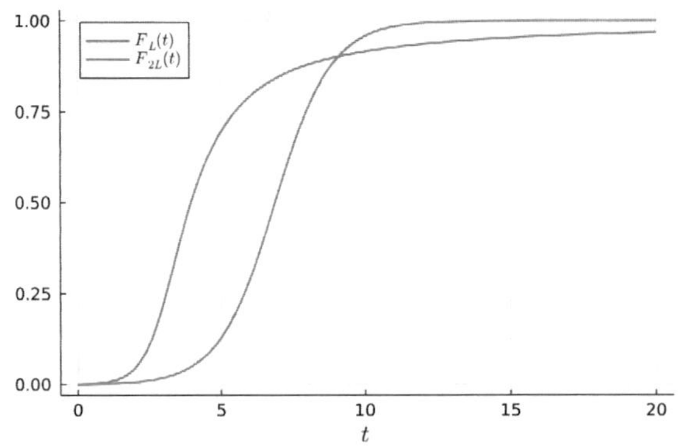

Figure 6. The logistic vs. 2-logistic cumulative distribution functions, common $\epsilon = 0.001$, $k = 1.0$

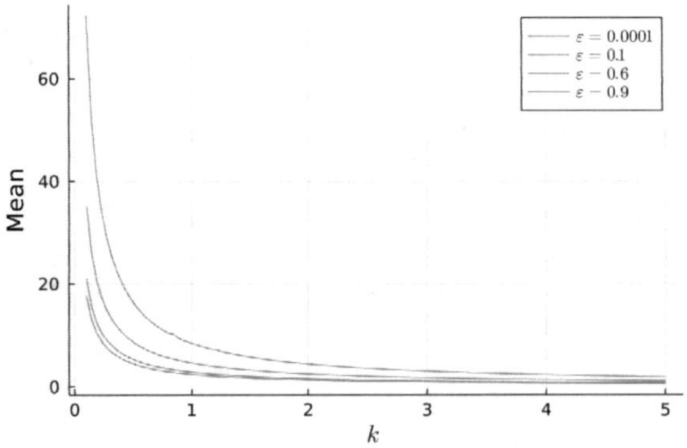

Figure 7. Mean of the 2-logistic distribution for different values of ϵ and k

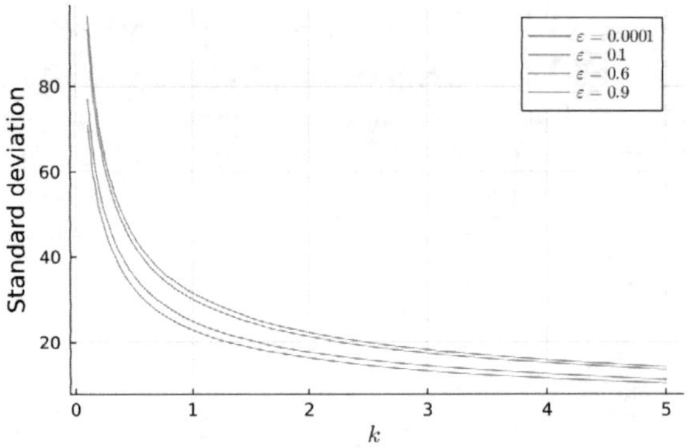

Figure 8. Standard deviation of the 2-logistic distribution for different values of ϵ and k

6. Conclusions

In the present work we formulated a mathematical model which is a modification of the classical logistic (Verhulst) model. We call this model the 2-logistic model. We show how we can derive probability distributions from the classical and the 2-logistic growth-decay models and graphically illustrate some of their properties.

A natural extension of this work would be to delve deeper into the statistical properties of the 2-logistic distribution and, in particular, develop methods for estimating its parameters. Both classical and Bayesian approaches are of interest, especially given the latter's ability to constrain parameter values by means of appropriate priors.

Acknowledgments

The work of Meglena Lazarova is supported by the Bulgarian National Science Fund under Project KP-06-M62/1 "Numerical deterministic, stochastic, machine and deep learning methods with applications in computational, quantitative, algorithmic finance, biomathematics, ecology and algebra" from 2022.

References

[1] Verhulst P F 1838 Notice sur la loi que la population poursuit dans son accroissement *Correspondance mathematique et physique* **10: 113-121**

[2] Balakrishnan N (ed) 1992 *Handbook of the Logistic Distibution* (Dekker, New York)

[3] Johnson N L, Kotz S and Balakrishnan N 1995 *Continuous Univariate Distributions* **vol. 2**, 2-nd edition

[4] Gupta R D and Kundu D 2010 *Journal of Applied Statistical Science* **18.1 : 51**

[5] Nadarajah S and Kotz S 2005 A generalized logistic distribution *International Journal of Mathematics and Mathematical Sciences* **19** pp 3169-3174

[6] Prodanov D 2023 Asymptotic analysis of the SIR model and Gompertz distribution, *Journal of Computational and Applied Mathematics* **422** 114901

[7] Markov S 2019 Reaction networks reveal new links between Gompertz and Verhulst growth functions *Biomath* **8** 1903167

Mathematical and computer modelling of age-related changes of basic anthropometric and mass-inertial characteristics of Bulgarian females

G Nikolova[1,*], D Dantchev[1], M Tsveov[1] and V Kotev[1,2]

[1] Institute of Mechanics, Bulgarian Academy of Sciences, Acad. G. Bonchev str. Bl. 4, 1113 Sofia, Bulgaria
[2] University of National and World Economy, Student Town, UNWE, 1700, Sofia, Bulgaria
[*] Corresponding Author E-mail: gergana1973@gmail.com

Abstract. We study the age changes of the anthropometric as well as of the mass-inertial characteristics (MIC) of Bulgarian women aged 18-25 versus those aged 30-40. We do that via mathematical modelling of the human body (HB). Nowadays mathematical modelling is one of the modern methods for the determination of the MIC of different segments of the body, as well as of the whole body, and for studying the changes of the inertial characteristics during a specific motion of the considered person. The current work presents a 3D mathematical model (MM) of the HB, that is also generated in a computer environment. Both the model and its computer generation allow for the calculation of the MIC of all segments of the body, as well as for the whole HB in specific body positions. We consider such body postures which are presented in the corresponding classification of astronauts' standard positions of interest for NASA. The obtained results are compared with those reported in the literature by other authors. The proposed models shall be also helpful in problems appearing in medicine (orthopaedics and traumatology), computer simulations, rehabilitation robotics, and sports, as well as in areas such as ergonomics, simulation of human behaviour in space, forensics (body fall, car crash) and more.

Introduction

For investigating the kinematics and the dynamics related to human motion naturally the HB's geometric and MIC have to be acquainted. In addition to the classical issues pertinent to this set of problems connected to work, everyday activities, sports, etc., there are also new areas of research where these topics are important like, say, movements in microgravity, or the construction of wearable and rehabilitation robots.

A specific set of problems is related to how the parameters of the body, and so, how the body moves, stem from the changes of the HB with age. Currently, investigations dedicated to the scientific subject of anthropometric measures and how they vary with age have already been an issue of different studies, including, for example, the significance of anthropometric dimensions and how ageing affects them for workplace design [1]. In these studies, one pays attention to the anthropometric differences in the elderly across a variety of age groups and genders [2-5]. With respect to Bulgarians, differences in their stature and weight as taken from the last decades of the 20th century and the beginning of the 21st century have been reported in Ref. [6,7].

The problem for determining the anthropometric and MIC of the HB has a long history. In the 1960s and 1970s, more than a few studies reported anthropometric and MIC of the HB and its segments of male cadavers [8-10]. Afterwards, various methods for investigation of body segment parameters of the living males and females have been used: gamma mass scanning [11,12], computerized tomography [13], magnetic resonance imaging [14], geometrical modelling [15,16], etc.

The current article presents a 3D MM of the HB, that is also generated in a computer environment. Both the model and its computer generation allow for the calculation of the MIC of all segments of the body, as well as for the whole HB in specific body positions. We consider such body postures which are presented in the corresponding classification of astronauts' standard positions of interest for NASA [17]. We study and compare the MIC and basic anthropometric measures (weight and height) of Bulgarian women in the 18–25 age group with those in the 30–40 age group. The obtained results are compared with those reported in the literature.

The data of women aged 18-25 years are taken from our own anthropometric measurements of mass and weight of 36 Bulgarian women, while those aged 30-40 years are taken from an anthropological survey of 2855 women [7]. As of right now, the data collection described in [7] is the most descriptive anthropological survey of the Bulgarian people carried out between 1988 and 1993. The authors of this study measured 5290 people in total (2435 men and 2855 women), 284 of which were from Sofia) aged between 30 and 40 years.

Let us point out that [7] has an extraordinarily large total number of people studied. In fact, the subset of female athletes studied by [18] is limited to 15, whereas in a different foundational study, [19] included 25 Chinese and 25 German women. Therefore, in comparison to the study conducted by Yordanov et al. [7], our sample of 36 women from the city of Sofia may appear small but was greater than those in the other respective studies mentioned above.
We agree, of course, that an effort shall be made to increase the number of studied women in the future.

Finally, let us also mention that the suggested models shall be also useful for issues that arise in ergonomics, forensics (body fall, car crash), computer simulations, rehabilitation robotics, sports, and other fields such as orthopaedics and traumatology.

Model and method

In the current study we will make use of a 16-segmental MM of the HB originally proposed in Ref. [20]. In this article, based on that model, the anthropometric and MIC of the various body segments of a 30-40-year-old woman have been obtained and reported. Thus, to avoid repetition of the information given in [20], we just direct the reader to [20] for details, which we will skip here. Because of that here, and in the following sections, we will discuss only a few differences between the model used in this work in comparison with that of [20]. In order to make it easier for the reader to understand what follows in the current article, we also recall the definition of the geometric parameters measured for women between the ages of 18 and 25 - see Figure 1.

As the figure shows, we decompose the body into 16 segments: head + neck; the torso is divided into three parts: the upper part of the torso, the middle part of the torso, and the lower part of the torso; then we distinguish the upper arm and the forearm, as well as a hand, thigh, shank and foot. Assuming that every segment is a geometrical body, our goal is to be able to ascertain the MIC and, thus, assumed to be reasonably simple– see Figure 1.

To make a further simplification, and as it is customarily assumed in mathematical modelling of the HB, we consider the body to be fully symmetric concerning the sagittal plane, i.e., complete "left–right" symmetry. The decomposition of the body segments is made according to anthropometric points used in [18]. The three sections of the torso's densities are taken from [21], and the densities reported in [22] have been applied to the remaining segments. Details of the required anthropometric points and the densities of the individual segments are given in [20].

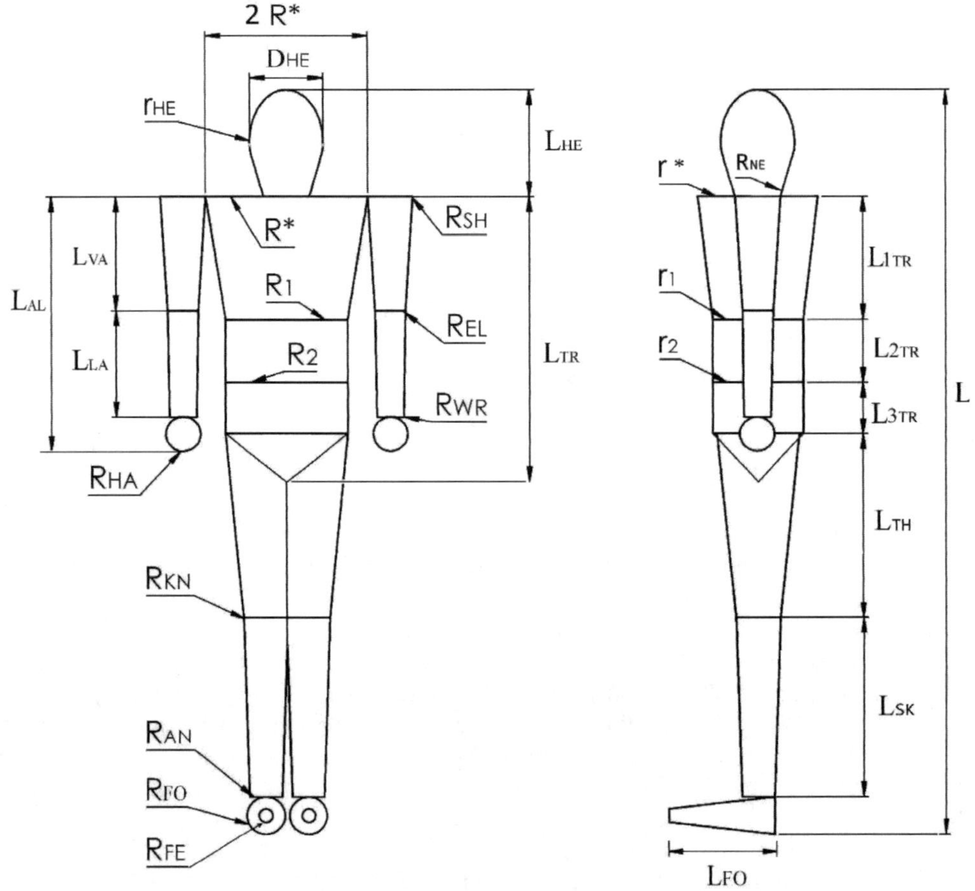

Figure 1. 16-segmental MM of the woman aged 18-25 years.

The next step is to calculate the lengths of the individual segments. For that aim, we use the regression equations following [19]. In other words, a regression equation of the type:

$$L_S = b_0 + b_1 \times BM + b_2 \times BH \tag{1}$$

was used for each segment. Taking into account that the total mass of an average Bulgarian "female aged 18-25" is $BM = 54.3$ [kg] and the body height is $BH = 165$ [cm], we calculated the expected length values of all body segments. We computed the expected length values of all body segments, taking into account that the average Bulgarian "female aged 18-25" has a body height of 165 [cm] and a total mass of BM=54.3 [kg].

We use the following geometrical parameters for the different segments of the body:

A) The head + neck is modelled with two parts:

a) the upper part - as a half of a spheroid with height R_{HE} = 12 [cm], having a circular cross-section with main radii $R^* = r_{HE}$ = 8.14 [cm];

b) the lower part - as a frustum of a cone with a circular cross-section. The upper cross section is with radius R*, while the lower one is characterized by a radius R_0= 5.0 [cm]. The height of the cone is 12.0 [cm]. The length of the head is given by L_{HE}=24 [cm].

This segment is the only one modified compared to the model of a woman aged 30-40 [20], where the head plus neck is assumed to be an ellipsoid. Therefore, we will give a separate scheme only to this segment – see Figure 2.

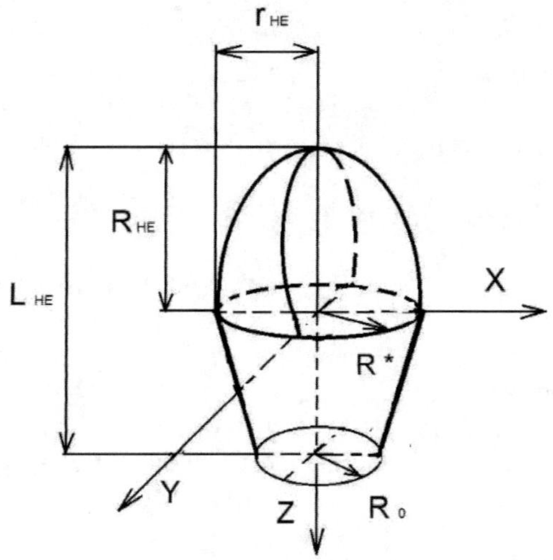

Figure 2. Geometrical model of head+ neck

A) Length L_{TR} of the torso, where $L_{TR} = L_1+L_2+L_3+L_4$. We consider that a reverted right elliptical cone can reasonably represent the upper torso. Here $L_1 = 27.6$ [cm] is the length of the upper part of the torso. The corresponding radii are $R^* = 18.4$ [cm], $r^* = 13.3$ [cm], $R_1 = 13.8$ [cm], $r_1 = 10.0$ [cm].
B) The middle torso is shaped with an elliptical cylinder with $L_2 = 14.1$ [cm], $R_2 = 13.8$ [cm] and $r_2 = 10.0$ [cm].
C) The lower part of the torso consists of two parts and respectively from two geometrical figures: elliptical cylinder + reverted elliptical cone. L_3+L_4 is the length of the lower part of the torso, where $L_4 = R_4 \tan 37^0$. the lengths and radii are as follows: $L_3 = 11.4$ [cm], $L_4 = 10.9$ [cm], $R_3 = 13.8$ [cm], $R_4 = 13.8$ [cm], $r_3 = 10.0$ [cm] and $r_4 = 10.0$ [cm].
D) The upper arm is accepted to be the frustum of a cone with length $L_{UA} = 25.6$ [cm] and radii $R_{SH} = 5.0$ [cm] and $R_{EL} = 3.4$ [cm].
E) The lower arm is also modelled with the frustum of a cone with length $L_{LA} = 23.4$ [cm], $R_{EL} = 3.4$ [cm] and $R_{WR} = 3.0$ [cm].
F) The hand is modelled with a sphere with a radius $R_{HA} = 3.8$ [cm].
G) The thigh is also assumed to be cone frustum with length $L_{TH} = 41.2$ [cm], $R_{TH} = 9.1$ [cm] and $R_{KN} = 4.9$ [cm].
H) The shank is shaped with the frustum of a cone with length $L_{SK} = 40.2$ [cm], $R_{KN} = 4.9$ [cm] and $R_{AN} = 3.6$ [cm].
I) The foot is accepted to be the frustum of the cone with length $L_{FO} = 23.4$ [cm], $R_{FO} = 4.2$ [cm] and $R_{FE} = 1.5$ [cm].

A rough approximation of the body segments' shape, and consequently of their volume and associated properties like mass and moments of inertia, is imposed by modelling the body segments using simple geometrical bodies. Therefore, it is extremely desired to alter the measured geometrical parameters to provide the best estimate of the body segment mass characteristics. The regression mass equations that were developed [19] can be used to accomplish this. However, doing so requires that variations of the adjusted parameters from the corresponding measured values that are greater than 10% be avoided.

Derivation of the MIC

We now go ahead and analytically determine all the relevant parameters for each segment, including mass, volume, and moments of inertia. We can do that using the once-determined geometrical parameters of the segments. Of course, for achieving this goal we rely on the mathematical characteristics of the bodies that are being modeled.

First, let us observe that, for each of the segments, regression equations of the kind

$$M_S[kg] = A_1 + A_2 M[kg] + A_3 L[cm] \tag{2}$$

were employed as is widely known. Considering that the whole mass of an average Bulgarian "female aged 18-25" is M = 54.3 [kg] and the body height is L = 165 [cm], and that the average Bulgarian "female aged 30-40" is characterized by mass M = 65.3 [kg] and body height L = 158 [cm], we calculate the expected mass values of all body segments.

Table 1 compares our data from 2018 to 2019 with the average body weight [kg] and height [cm] (along with their minima, maxima, and standard deviation) of Bulgarian women aged 30 to 40 during the years 1989 to 1993 (see Ref. [7]). Based on the data presented in Table 1, it can be inferred that over a 35-year period, the height of the Bulgarian woman has increased by approximately 7 centimetres, and her weight has increased by 11 kilograms.

Table 1. Comparison of weight and height of "female aged 18-25" (2018-2019) versus "female aged 30-40" (1989-1993).

Age	**n**	**Height**			**Weight**		
		Average	Minimum	Maximum	Average	Minimum	Maximum
18-25	36	165.11 (SD = 6.35)	148	176	54.3 (SD = 7.91)	40	78
30-40 (Ref. 7)	2855	158.12 (SD = 5.8)	150	182	65.3 (SD = 12.0)	45	85

Some results for the mass, centre of mass and moments of inertia

It is possible to derive most of the expressions required for the mass centres in a rather simple way.

- For a frustum of a cone with upper radius R and lower radius r, the corresponding result for the position of the mass centre, measured from the upper cross-section of that segment, is

$$M_{CM} = h\frac{\frac{1}{4}R^2 + \frac{1}{2} + \frac{3}{4}r^2}{R^2 + Rr + r^2} \tag{3}$$

We use this result for the considered set of segments, since the foot, upper and lower arm and shank, are approximated via a frustum of a cone.

- The centre of mass for the upper torso is given by the expression

$$M_{CM}^{UT} = L_2\frac{r_2(R_1 + 3R_2) + r_1(R_1 + R_2)}{2[+r_2(R_1 + 2R_2) + r_1(2R_1 + R_2)]} \tag{4}$$

- For the middle torso, the centre of mass is at $M_{CM}^{MT} = L_2/2$.
- After some relatively elementary mathematics, for the position of the centre of mass of the lower torso, one obtains

$$M_{CM}^{UT} = \frac{\frac{L_3}{2} + (L_3 + L_4/4)\frac{L_4}{3L_3}}{1 + \frac{L_4}{3L_3}} \tag{5}$$

Here, the upper cross-section of the relevant torso region serves as the starting point for measuring the coordinates of each of the three mass centres. We now move on to calculating the inertial moments of the various body segments. Let us start by reminding the well-known fact that any 3D body's inertia tensor with regard to a common coordinate system is [23]

$$\hat{\boldsymbol{I}} = \begin{pmatrix} \iiint_V \rho(\vec{r})(y^2+z^2)d\vec{r} & -\iiint_V \rho(\vec{r})(xy)d\vec{r} & -\iiint_V \rho(\vec{r})(xz)d\vec{r} \\ -\iiint_V \rho(\vec{r})(yx)d\vec{r} & \iiint_V \rho(\vec{r})(x^2+z^2)d\vec{r} & -\iiint_V \rho(\vec{r})(yz)d\vec{r} \\ -\iiint_V \rho(\vec{r})(zx)d\vec{r} & -\iiint_V \rho(\vec{r})(zy)d\vec{r} & \iiint_V \rho(\vec{r})(x^2+y^2)d\vec{r} \end{pmatrix}, \tag{6}$$

where the integrals are over $\vec{r} \in V$, where V here indicates the area that the three-dimensional body is occupied by. The aforementioned computations can be carried out explicitly when the various body segments are modelled using comparatively simple geometrical bodies. The equivalent expressions for the principal moments of inertia for a few of the bodies, such as a sphere, ellipsoid or cylinder, are directly available in the literature (see, for example, [23]). Summarizing, one has:

i) for a sphere with radius R and mass μ

$$I_{XX} = I_{YY} = I_{ZZ} = \frac{2}{5}\,\mu R^2; \tag{7}$$

ii) for a cylinder with radius R, height h and mass μ

$$I_{XX} = I_{YY} = \frac{1}{4}\mu\left(R^2 + \frac{h^2}{3}\right), I_{ZZ} = \frac{1}{2}\,\mu\, R^2; \tag{8}$$

iii) for an ellipsoid with semi-axes a, b, c and total mass μ

$$I_{XX} = \frac{\mu}{5}(b^2+c^2), I_{YY} = \frac{\mu}{5}(a^2+c^2), I_{ZZ} = \frac{\mu}{5}(a^2+b^2). \tag{9}$$

Some effort is needed for the calculations of the principal moments of inertia for the other segments of the body. The following expressions for the principal moments of inertia can be obtained in the case of a frustum of an elliptic cone carrying out the computation described in Eq. (6):

a) For I_{XX}^{CM}

$$I_{XX}^{CM} = \frac{1}{240}\pi h\rho\Bigg(R_2(4h^2+3r_2^2)(3r_1+2R_1)+3r_2(4h^2+r_2^2)(4r_1+R_1)+3r_2R_2^2(2r_1+3R_1)+3R_2^3(r_1+4R_1) - \frac{10\big(h\,r_2(3r_1+R_1)+h\,R_2(r_1+R_1)\big)^2}{r_1(2r_2+R_2)+R_1(r_2+2R_2)}\Bigg), \tag{10}$$

b) For I_{YY}^{CM}

$$I_{YY}^{CM} = \frac{1}{240}\pi h\rho\Bigg(4h^2\big(3r_1(4r_2+R_2)+R_1(3r_2+2R_2)\big)+r_1^2R_1(9r_2+6R_2)+3r_1^3(4r_2+R_2) + 3\,\mathrm{R1}^3(r_2+4R_2) - \frac{10\big(h\,r_2(3r_1+R_1)+h\,R_2(r_1+R_1)\big)^2}{r_1(2r_2+R_2)+R_1(r_2+2R_2)} + 3r_1R_1^2(2\,r_2+3R_2)\Bigg), \tag{11}$$

and

c) For I_{ZZ}^{CM}

$$I_{ZZ}^{CM} = \frac{1}{80}h\pi\rho\Big(r_1^3\,(4\,r_2+R_2)+r_1^2\,R_1(3r_2+2R_2)+r_1\big(4r_2^3+3r_2^2\,R_2+2r_2\,R_2^2+R_2^3\,+R_1^2(2r_2+3R_2)\big) + R_1(R_1^2(r_2+4R_2)+r_2^3+2\,r_2^2\,\,R_2+3r_2\,R_2^2 + 4R_2^3)\Big). \tag{12}$$

In this instance, h denotes the height of the cone's frustum, ρ is the density of the corresponding segment that has been experimentally determined, R_1 and R_2 are the semi-major principal axes of the cone's base and r_1 and r_2 are the semi-axes on the cone's top.

Generation of segment MM of HB using CAD system SolidWorks

Here we are interested in the MIC of the total body. Of course, they will depend on the particular position. We will pick up only such positions which have been presented in the corresponding classification of astronauts' standard positions as selected by NASA. One of the ways to achieve this goal is to generate the considered model in the CAD system – SolidWorks. We have performed such a generation of the model. Furthermore, after calculating, as explained above, the MIC of the various segments we have, by comparing these results with the corresponding ones delivered by the CAD system, we *verified* the computer realization of our model. This is done in the current study for the group of women individuals aged 18-25 and, similarly, it has been earlier done in Ref. [20], for the age 30-40 years old.

We recall that the basic positions of the body, usually of interest in biomechanics, and those are the ones we have in mind here, have been described in the literature long ago – see, e.g., ([10], [15], [17], [24]).

Eight basic body positions are generally considered. Unfortunately, because there are so few female astronauts, data from the aforementioned NASA studies only pertains to men. This indicates that they are insufficient for plausible estimates that can be verified statistically. We will attempt to add information for three of these positions for the female HB to the literature in the sections that follow:

a) first, the "standing position" (Figure 3a illustrates this;
b) the seated position (see Figure 3b with the forearms down);

c) the position of standing with arms raised overhead; refer to Figure 3c.

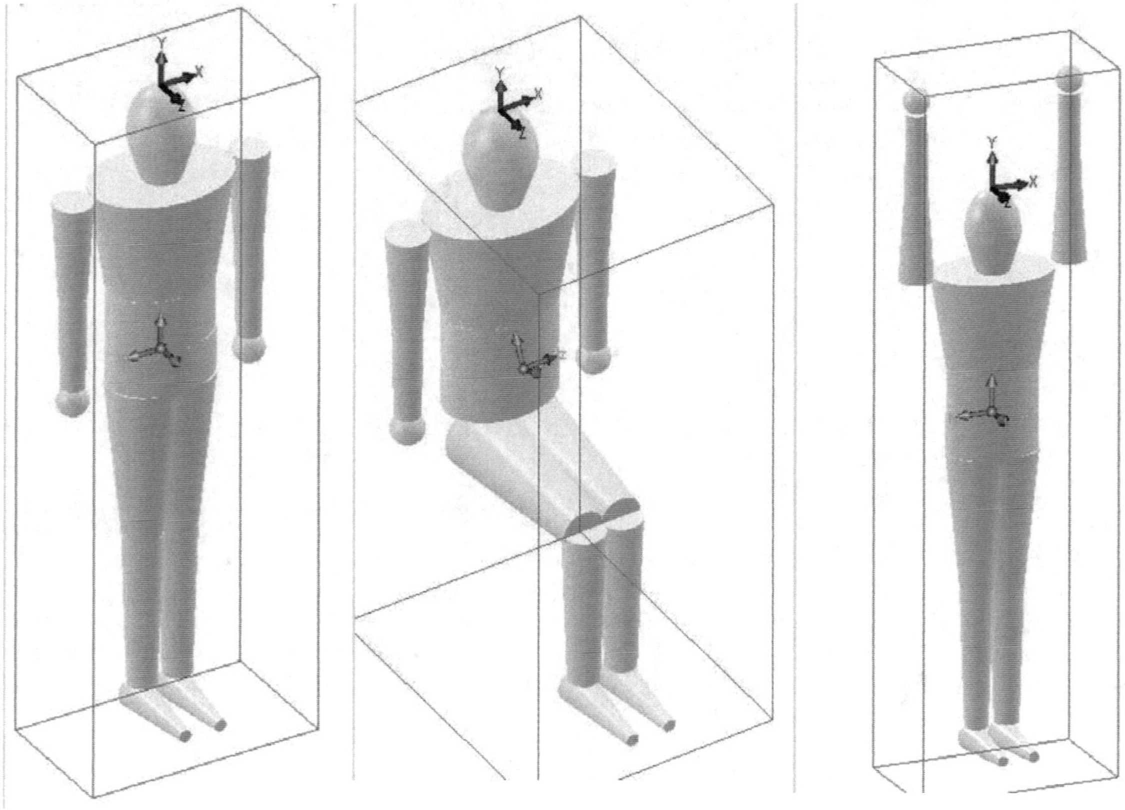

Figure 3. a) Standing, b) Sitting, with forearms down, c) Standing, arms overhead positions.

Standing position

In this standard position, which is usually short-hand termed the “standing position”, the corresponding individual to be studied, stands upright; the arms naturally hang at the sides parallel to the body, with the head positioned in the Frankfurt plane (see Figure 3a). For this position the MIC of the female HB in standing positions we determined for the group 18-25-years-old, together with the ones we are aware of already reported in the literature, are presented in Table 2. The data for the centre of mass in the corresponding posture of the body is measured from the anthropometric point *vertex* of the head (the “x” coordinate) and the sagittal plane through the middle of the body when in this position.

Table 2 also provides a comparison between the average Bulgarian "female aged 30-40" and the data we have collected for the average "female aged 18-25" in this role, with the respective data taken from the literature [25], [26].

Upon closer examination of the data, we find a fairly good agreement between our findings and those previously published in the literature. Observing the two studied groups (see Table 2 – sixth and seventh columns) one can conclude that the biggest difference is with respect to the I_{XX} principal moment of inertia. It is about 13.6%. This is obviously due to the fact that the women in the younger group are taller and thinner.

Table 2. MIC for standing position

Characteristic	Young et al. (1983) [Ref. 25]			Abracham et al. (1979)[Ref.26]	Our data 18-25 years old	Our data 30-40 years old [Ref. 27]
	Min	Max	Mean			
I_{XX} [kg.cm^2×10^3]	5.8	24.0	11.6	-	5.9	6.7
I_{YY} [kg.cm^2×10^3]	49.1	135.0	85.0	-	76.9	78.3
I_{ZZ} [kg.cm^2×10^3]	53.0	135.0	91.9	-	80.7	81.9
Height [cm]	161.20			162.60	165.11	160.51
Mass [kg]	63.90			64.64	54.3	60.65
Centre of mass [cm]	-			-	70.5	67.3

Sitting, with forearms down position

The body segments are the same as in the "standing position" when the person is "sitting, with forearms down," with the exception that the wrist axes and forearms are parallel to the X-axis (see Figure 3b). The centre of mass [cm] and principal moments of inertia [kg.cm^2] of the body model at the respective position are given in Table 3.

Table 3. MIC for sitting, with forearm down position

Characteristic	Our data 18-25 years old	Our data 30-40 years old [Ref. 27]
I_{XX} [kg.cm^2]	10.5	13.0
I_{YY} [kg.cm^2]	55.8	56.9
I_{ZZ} [kg.cm^2]	58.7	59.5
Centre of mass [cm]	64.7	60.7

Observing the two studied groups (see Table 3) one can conclude that the biggest difference in the principal moments of inertia is with respect to the I_{XX} principal moment of inertia. It is about 23.8%. This is quite a substantial difference. It is again obviously due to the fact that the women in the younger group are taller and thinner.

Standing, arms over head position

Positioning the upper extremities over the head, parallel to the X and Y axes, with the hands slightly clenched, is known as the "Standing, arms over head" position. This posture also involves the torso, legs and head, as shown in Figure 3(c). Data for the body model's centre of mass and principal moments of inertia in the appropriate position are provided in Table 4. We observe that in the two studied groups, the biggest difference is with respect to the I_{XX} principal moment of inertia. It is about 12.4%.

Table 4. MIC for standing, arms over head position

Characteristic	Our data 18-25 years old	Our data 30-40 years old [Ref. 27]
I_{XX} [kg.cm^2]	59.7	67.1
I_{YY} [kg.cm^2]	98.0	99.8
I_{ZZ} [kg.cm^2]	103.3	101.8
Centre of mass [cm]	66.0	63.0

Conclusions

The current paper examines the age-related changes in Bulgarian females' basic anthropometric measurements (body weight and height) and MIC using 3D MM that was created in a computer environment. This is done after splitting the measured population into two age groups: I) women aged 18 to 25 and II) women aged 30 to 40. Examining the data reveals a significant increase in each person's height, of roughly 7 [cm], in the category of "female aged 18-25" as opposed to "female aged 30 – 40."

One comes to the conclusion that the body's height rapidly reaches its maximum because of the acceleration. At the same time, however, the weight change is also essential – the average weight of a "female aged 30-40" is about 11 [kg] heavier than that of the average "female aged 18-25". Undoubtedly, both genders—males and females—need to understand the geometric and MIC.

Regrettably, for a variety of reasons, the majority of the data found in the literature relates to men. Furthermore, nearly all of the data that is currently available is specific to the various body segments rather than the entire body, even in cases where such measurements have been done on females. Consequently, it would be ideal to take a strategy that attempts to close that gap. In the current article, a 3D model of a female body that can provide such data is presented along with its computer realization in SolidWorks media. Let us stress that the model allows, in principle, the determination of the mass-inertial parameters of the female body in any posture of the body that is of interest.

We finish this article by noticing that the obtained results are useful for obtaining such mass and anthropometric characteristics which is the case in problems appearing in rehabilitation robotics, computer simulations, medicine, sports and areas such as simulation of human behaviour in space, ergonomics, criminology and other areas.

Acknowledgements

This work has been accomplished with the financial support by the Grant No BG05M2OP001-1.002-0011-C02 financed by the Science and Education for Smart Growth Operational Program (2014-2020) and co-financed by the European Union through the European Structural and Investment funds.

References

[1] Topp S G, Cook J, Holland W W and Elliott A 1970. Influence of environmental factors on height and weight of schoolchildren *Br. J. Prev. Soc. Med* **24** 154–162

[2] Perissinotto E, Pisent C, Sergi G, Grigoletto F and Enzi G 2002 Anthropometric measurements in the elderly: age and gender differences *Br J Nutr* **87** 177-186

[3] Dunsky A, Zach S, Zeev A, Goldbourt U, Shimony T, Goldsmith R and Netz Y 2014 *Eur. Re. Aging Phys. Act.* **11** 149–157

[4] Kuhle S, Maguire B, Ata N and Hamilton D 2015 Percentile curves for anthropometric measures for Canadian children and youth *PLoS One* **10** (7): e0132891

[5] Singh K, Singh S P, Kaur G and Bose K 2020 Comparison of age-related anthropometric changes among elderly Punjabi males residing in old-age homes with those residing with their families. *J Anthropol Surv India* **69** 102–116
[6] Yordanov Y et al 1999 Basic antropological data of the Bulgarian population at the end of the XX-th century *J. Antropology* **2** 19-28
[7] Yordanov Y, Nacheva A, Tornjova S, Kondova N, Dimitrova B and Topalova D 2006 *Anthropology of the Bulgarian population at the end of the 20-th century (30-40 years old persons)* (Prof. Marin Drinov Academic Publishing House, Sofia, Bulgaria)
[8] Dempster W T 1955 Space requirements of the seated operator, WADC Technical Report 55-159 Ohio, Wright-Patterson Air Force Base
[9] Clauser C E, McConville J T, Young J M. 1969 Weight, volume and center of mass of segments of the human body. Technical report AMRL-TR-69-70. Dayton, OH: Wright-Patterson Air Force Base
[10] ChandlerR F, Clauser C E, McConville, J T, Reynolds H M, Young J W 1975. Investigation of inertial properties of the HB. Technical Report AMRL-TR-74-137, Dayton, Ohio: Aerospace Medical Research Laboratory, Aerospace Medical Division, Air Force Systems Command, Wright-Patterson Air Force Base.
[11] Zatsiorsky V M and Seluyanov V N, 1983 The mass and inertia characteristics of the main segments of the HB *Biomechanics VIII-B*, ed H. Matsui and K. Kobayashi (IL, Human Kinetics) pp. 1152-1159
[12] Zatsiorsky V M and Seluyanov V N 1985 Estimation of the mass and inertia characteristics of the human body by means of the best predictive regression equations *Biomechanics IX-B*, ed D. Winter et al. (Human Kinetics, Champaign, IL) pp. 233–239
[13] Wei C and Jensen R K 1995 The application of segment axial density profiles to a HB inertia model. *J. Biomech.* **28** 103-108
[14] Martin P E and Munigiole M 1990 Estimating-segment inertial properties: comparison of magnetic resonance imaging with existing methods, *J. Biomech.* **23** 1039–1046
[15] Hanavan E P 1964 A mathematical model of the human body, AMRLTR-64-102 Ohio, Aerospace Medical Research Laboratories, Wright-Patterson Air Force Base
[16] Yeadon M R 1990 The simulation of aerial movement-II. A mathematical inertia model of the HB. *J. Biomech.* **23** 67-74
[17] NASA-STD-3000, 2000. The Man-System Integration Standards, Volume I, Section 3. Anthropometry and Biomechanics. at: https://msis.jsc.nasa.gov/sections/section03.htm#3.2.1
[18] Zatsiorsky V M 2002 *Kinetics of human motion* (Human Kinetics, Champaign, IL, USA)
[19] Shan G and Bohn C 2003 Anthropometrical data and coefficients of regression related to gender and race. Appl. Ergon. 34 327-337
[20] Nikolova G and Toshev Y 2007 Estimation of male and female body segment parameters of the Bulgarian population using a 16-segmental mathematical model *J. Biomech.* **40** 3700-3707
[21] Erdmann W S 1997 Geometric and inertial data of the trunk in adult males. *J. Biomech.* **30**, 679-688
[22] Bjornstrup J 1996 Estimation of HB Segment Parameters-Statistical Analysis of Results from Prior Investigations, Technical report ISSN 0906- 6233, LIA 96-7, Laboratory of Image analysis, Institute of Electronic Systems, Aalborg University, Denmark.
[23] Landau L D and Lifshitz E M 1976. *Mechanics* (Pergamon Press, New York)
[24] Santschi W et al. 1963 Moments of Inertia and Centers of Gravity of the Living Human Body. AMRL TR 63-36. Wright-Patterson Air Force Base, Ohio: 6570th Aerospace Medical Research Laboratories.
[25] Young R 1983 Anthropometric and mass distribution characteristics of the adult female. FAA-AM-83-16, Oklahoma City.
[26] Abraham S. et al. 1979 Weight and height of adults 18 - 74 Years of Age. US1971-74, (PHS) 79-1659, HANES.

[27] Nikolova G, Kotev V and Dantchev D 2017 Computer and mathematical modelling of the female human body: determination of mass-inertial characteristics in basic body positions. *Proc.7th Int. Conf. on Simul. and Model. Method. Technolog. and Applic.* (SIMULTECH 2017) (SCITEPRESS – Science and Technology Publications, Lda.) 416-421.

An application of some machine learning methods for biological data modeling

Fatima Sapundzhi[1], Meglena Lazarova[2], Tatyana Dzimbova[1,3], Slavi Georgiev[4,5]

[1] Department of Communication and Computer Engineering, Faculty of Engineering, South-West University "Neofit Rilski", 66 Ivan Myhailov Str., 2700 Blagoevgrad, Bulgaria
[2] Faculty of Applied Mathematics and Informatics, Technical University of Sofia, 8 "St. Kliment Ohridski", Blvd., 1000 Sofia, Bulgaria
[3] Institute of Molecular Biology, Bulgarian Academy of Sciences, Acad. G.Bonchev Str, bl. 21, 1113 Sofia, Bulgaria,
[4] Department of Information Modeling, Institute of Mathematics and Informatics, Bulgarian Academy of Sciences, Acad. Georgi Bonchev Str., bl. 8, 1113 Sofia, Bulgaria
[5] Department of Applied Mathematics and Statistics, Faculty of Natural Sciences and Education, University of Ruse, 8 Studentska Str., 7004 Ruse, Bulgaria

sapundzhi@swu.bg

Abstract. The development of fast and reliable methods for predicting the biological activity of the substances in computational biology is of a great importance. This improves the development of some new compounds while keeping costs low. Among many scientists, an attractive target for docking experiments is the Delta-opioid receptor (DOR) and delta-opioid ligands (DOL). Their biological efficacy can be measured by various techniques, which could facilitate the establishment of the relationship between the structure of the compounds and their biological effect. The relationship between the results of the computer experiments and the biological activity of these compounds is modelled by using machine learning regressors. The primary goal of this study is to determine the most appropriate neural network for modelling the relationship between *in vitro* and *in silico* results for DOR and delta-opioid ligands.

1. Introduction

In our society there is a growing need to treat pain, but usually the current treatments are often ineffective and have side effects for the patients. Morphine is a powerful analgesic that can lead to physical dependence and addiction which drives the search for new pain medications with minimal side effects, see [1-3]. The delta-opioid receptor (DOR) is an attractive target for the development of new pain therapies. DOR is highly selective for encephalins and produces both spinal and supraspinal analgesia. The encephalin and its analogs have been extensively studied in order to discover potent and selective DOR ligands because they are endogenous opioid pentapeptides with DOR preference [4-9].

During the early stage of drug's discovery, the computer-aided drug's usage design (CADD) techniques was very attractive in comparison to the manual, time-consuming, and labor-intensive methods of the traditional drug design and testing. The use of CADD is possible to reduce the time and the costs associated with bringing a drug to market and accelerate research. Generally, CADD can be classified into two types: ligand-based methods and structure-based methods. Among these, the ligand-based methods assume that a drug's chemical structure has an effect on its biological activity while the structure-based methods assume that a target's 3D structure has an effect on its biological activity. In these methods, small molecules are matched to select compounds for further study based on their properties. In the early stages of drug discovery, the molecules are evaluated based on some characteristics such as similarity, correlation with specific properties or binding energy [10-16].

The main purpose of this study is to construct a model structure-activity relationship between a series of delta-opioid ligands and molecular docking results with a model of the delta-opioid receptor (DOR) by using machine learning. The next section briefly mentions the software we use to obtain the regression data. Section 3 begins with the description of the data itself. Then the available polynomial regression results are recalled, and the new machine learning methods are explained. Further the calibration algorithm is visually described, and eventually the results are discussed. The paper finishes with a conclusion section.

2. Experimental

2.1. Docking procedure

The docking experiments were carried out with the software GOLD and GoldScore algorithm, see [13,17]. The scale of the score gives an information how good the pose is.

2.2. Receptor

A theoretical model of DOR (PDBe:1ozc) is obtained from Protein Data Bank in Europe (https://www.ebi.ac.uk/pdbe/) [13].

2.3. Ligands

The data for the DOR analogs from *in vitro* test used in this investigation is presented in Table 1, see [13, 18, 19]. The ligand's preparation was done by the program Molegro Molecular Viewer V 2.5 (MMV), (www.clcbio.com). The prepared structures were used for molecular docking by the software GOLD 5.2. The total energies of the formed ligand-receptor complex after docking were calculated by MMV 2.5, using the MolDock scoring function.

3. Results and discussions

The computer modeling and the docking experiments with DOR (PDBe:1ozc) and delta-opioid ligands, presented in Table 1, were carried out with software GOLD 5.2 and all optimization algorithms in the program: GoldScore, ChemScore, ASP and ChemPLP [17]. In order to establish the relationship between the docking results (scoring function) and the values of the biological activity of the delta-opioid ligands, a correlation between these two sets of data is searched for. The correlation analysis was performed by GraphPad Prism software (http://www.graphpad.com/). In some previous studies, a significant correlation between the values of the optimization function GoldScore and the values of the biological activity of the investigated ligands, see [13] where the Pearson correlation coefficient's value is R=-0.72730 was found.

Table 1. Data for the ligands from in vitro test and docking studies (GoldScore function from GOLD and MolDock function from MMV), see [13]

Ligand	*GoldScore*	*Total energy*	*Ligand efficacy*
[Cys(Bzl)^2-Leu^5]-enk	64.68	-107.022	9.3

[Cys(Bzl)^2-Met^5]-enk	81.49	-89.091	3.5
[Cys(O_2NH_2)^2-Leu^5]-enk	67.72	-97.619	29.2
[Cys(O_2NH_2)^2-Met^5]-enk	73.91	-91.246	7.3
[DCys(O_2NH_2)^2-Leu^5]-enk	74.73	-84.852	7.4
[DCys(O_2NH_2)^2-Met^5]-enk	75.13	-86.221	7.1
DPDPE [D-Pen^{2,5}]-enkephalin	57.67	-109.709	30.2
[HCys(O_2NH_2)^2-Leu^5]-enk	68.43	-62.774	3.4
[HCys(O_2NH_2)^2-Met^5]-enk	78.65	-93.301	4.5
[Leu^5]-enkephalin	73.42	-81.869	5.8
[Met^5]-enkephalin	73,26	-118.971	3.6

For determining the relationship between the biological activity and the docking results we use the MATLAB software and the Surface Curve Fitting Toolbox. The Surface Curve Fitting Toolbox provides applications and functions for fitting curves and surfaces to data. A fitting of the experimental data for DOR (PDBe:1ozc) with a polynomial function $z = f(x, y)$ was carried out, see [13]. The values $(z_1, z_2, \dots, z_n)$ of the dependent variable z represent the values of the biological activity of the delta-opioid ligands. The values $(x_1, x_2, \dots, x_n)$ of the independent variable x represent the results from the docking, where the values of the Gold Score function is calculated by GOLD. The values $(y_1, y_2, \dots, y_n)$ of the independent variable y represents the total energies for the ligand-receptor complex formed after the docking procedure, where the values of the MolDock function is calculated by MMV.

In our previous analysing investigation the behaviour of the dependent variable, which depends on more independent variables is applied by using the Surface Fitting Toolbox of MATLAB (http://www.mathworks.com/products/matlab), see [13]. The obtained model could be interpreted as a surface-fitting function and analysed as an experimental data by using the least squares method. To evaluate the goodness of fit, we use the following statistical parameters:

SSE (Sum of squares due to error),
$R - Square$ (R^2), $Adjusted\ R^2$,
$RMSE$ (Root Mean Squared Error).

The results are presented in the following Table 2:

Table 2. Goodness of fit for the polynomial models obtained by the least squares method.

Models Poly (x,y)	Degree of x	Degree of y	SSE	R^2	Adj R^2	RMSE
Poly11	1	1	443.5817	0.5446	0.4308	7.4463
Poly22	2	2	167.1000	0.8285	0.6569	5.7810
Poly33	3	3	0.0092	1.000	0.9999	0.0960

The third-degree polynomial model for fitting the experimental data showed good fitting properties and significant predictive ability. It is suitable for determining the relationship structure and biological activity ($R^2 = 1.0$, $SSE = 0.0092$, $adj\ R^2 = 0.9999$, $RMSE = 0.0960$). Similar studies had been conducted with other compounds, see [13-16].

In this research, our aim is to obtain the relationship $z = f(x, y)$, where z represents the biological activity values of delta-opioid ligands, and the independent variables x and y correspond to the values of the GoldScore function and respectively to the MolDock function. Our intent is to define this nonlinear relationship by utilizing machine learning techniques, see [19].

The regression models that we employed are the k-Nearest Neighbors, Gradient Boosting, Random Forest and the Extra Trees. The latter three are ensemble methods which are anticipated to offer superior performance in comparison to the first traditional method incorporated here as a standard of comparison, see [20-23].

The **k-Nearest Neighbors** (k-NN) technique is an adaptable and an intuitive algorithm that is employed in numerous machine learning applications including classification and regression tasks. More precisely, the k-NN regression, a derivative of this method, is a potent instrument for predicting continuous variables.

The fundamental principle of the k-NN regression stems from the belief that the data points in the dataset, which are close to each other in the feature space, are expected to have similar target values. The method works by pinpointing the 'k' data points in the training set that are closest to a new unseen data point based on a certain distance metric – typically such metric is the Euclidean distance in the multi-dimensional feature space. Once these 'k' nearest points are found, the algorithm generates a prediction for the new data point. In regression scenarios this prediction is commonly the mean of the target variable for the 'k' closest points. It means that the predicted value for an unseen data point is the mean of the values of its nearest neighbors. One of the major strengths of the k-NN regression is its simplicity and intuitive nature. It does not enforce any explicit assumptions about the underlying structure of the data. Hence it is a non-parametric method. This flexibility enables the k-NN regression to adapt effectively to complex non-linear data. However, the k-NN regression has its share of complications. It is a complex task to decline the optimal 'k', because the lower 'k' might result in a model that is overly susceptible to noise whereas the larger 'k' might lead to over-smoothing the model and ignoring the key patterns in the data. Also, the k-NN regression may encounter difficulties with high-dimensional datasets, which is a problem known as the "curse of dimensionality".

The Gradient Boosting represents a potent machine learning method, which is applied across diverse regression and classification applications. Essentially, the gradient boosting regression is an ensemble technique that develops a predictive model by sequentially aligning a sequence of weak learners to the data. Each learner attempts to amend the mistakes of the one before it. Usually, these weak learners are decision trees. Of course, alternative base models can also be applied. The fundamental premise is to amalgamate the outputs of the multiple simplistic models in order to generate a singular, highly precise prediction. This "boosting" principle stems from the idea that the assembly of the weak learners when it is appropriately combined could be transformed into a robust learner. The progressive incorporation of weak learners in gradient boosting effectively acts as a method of the steepest descent. The algorithm approximates the gradient of the loss function (the metric that evaluates how accurately the model's predictions align with the actual values) with respect to the model parameters and how accurately it integrates the new models pointed in the direction that reduces the loss. For the regression tasks, the objective is to forecast a continuous outcome variable. In the context of the Gradient Boosting Regression the ensemble of the trees is trained to predict the residuals or errors of the prior trees. Consequently, each subsequent tree is effectively inching closer to the true, unknown function that we are trying to approximate. Of course, the Gradient Boosting has its merits and demerits like all models. It is an exceedingly effective technique which is capable to fit intricate and non-linear data. Frequently it performs well even on datasets with a blend of categorical and numerical features. However, the gradient boosting models can be susceptible to overfitting, especially if the data is noisy. They can also be computationally demanding and requiring a diligent tuning of several hyperparameters, such as the number of estimators, the depth of the trees, and the learning rate.

Random Forest is another renowned machine learning technique, commonly used in a variety of regression and classification tasks. The Random Forest model, as its name implies, is a conglomerate of individual decision trees each constructed to ensure a rich diversity that consequently guarantees resilience in their combined predictive capability. The Random Forest Regressor functions develop numerous individual decision trees for each regression tasks during the training phase. Each tree is learning from a random subset of the given data and is making its own prediction. When a new prediction is called for, each tree in the forest produces its own prediction. The final output is the average of these individual predictions. This methodology renders the Random Forest Regressor proficient at capturing intricate and non-linear relationships within data. The foundational concept of the Random Forest rests on the proposition that a collective of "weak learners" can combine to form a "strong learner". Every decision tree within the forest is a weak learner trained on a random subset of the data using a random subset of the features at each bifurcation. Such technique is called a bootstrap aggregating or "bagging" along with feature randomness. The potency of the Random Forest is anchored in its simplicity and adaptability. It is relatively resistant to overfitting thanks to the randomness integrated into its creation. It handles both numerical and categorical data effectively, and also manages missing data proficiently. Additionally, it includes an inherent mechanism for assessing feature importance which can aid in deciphering the model. However, like all models, the Random Forest has its constraints. It can be computationally heavy and slower in both training and prediction, particularly as the number of trees escalates. Moreover, it may underperform with very high dimensional sparse data such as text data and can be less decipherable compared to the individual decision trees.

Extra Trees Regressor is an abbreviation for "Extremely Randomized Trees". It is another machine learning technique employed for regression tasks and it is a part of the ensemble learning class. Like the Random Forest, it creates multiple decision trees during the training phase. However, it adds an additional level of randomness and enhance its robustness and reduce its susceptibility to overfitting. In a conventional decision tree, the optimal split from a random feature subset in the node is chosen when expanding the tree. But in Extra Trees the random value is selected for the split for each considered feature rather than the best split point. Consequently, the Extra Trees Regressor injects randomness not only at the sample level but also at the level of each split in every decision tree. The prediction procedure in regression tasks mirrors that of the Random Forest Regressor i. e. each tree in the ensemble makes a prediction. So the final output is the average of these individual predictions. The randomness in the Extra Trees Regressor enables it to capture complex, non-linear relationships and rendering it a highly effective tool for regression. Nonetheless, it is crucial to distinguish the differences between the Extra Trees and the Random Forest methods. While they both utilize bagging and random feature subsets, the primary distinction lies on how they split the nodes. The Extra Trees method is quicker due to its randomness at each split. This results in faster training process. However, this might occasionally lead to a slight performance dip, indicating a trade-off to consider. The Extra Trees is less susceptible to overfitting. It is capable to manage both numerical and categorical data and can effectively handle the missing data. Similar to the Random Forest, it offers a built-in method for feature importance estimation. On the other hand, similar to other tree-based methods, it may not perform well with very high-dimensional sparse data like text data. Its ensemble nature makes it less interpretable compared to the individual decision trees.

In this work, we employ the four machine learning algorithms to create a multivariate regression. During the computing process we performed 10-fold cross-validation for each algorithm's indicators. Furthermore, we implemented a hyperparameter optimization routine to select the optimal values. Table 3 contains the optimal hyperparameters, and Fig. 1 illustrates the algorithm's flow.

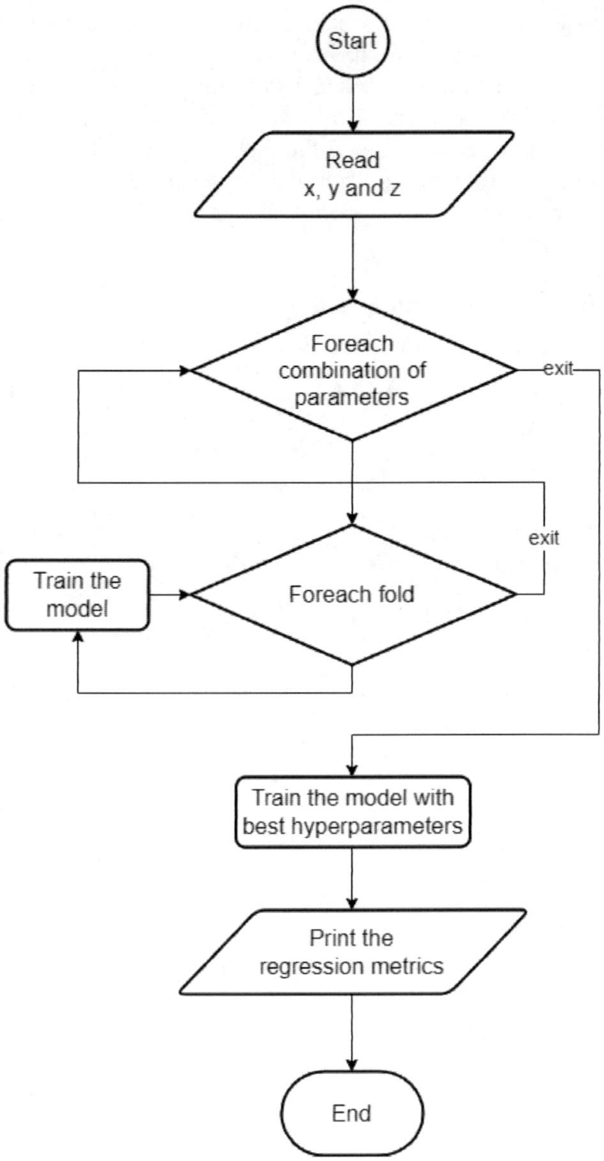

Fig. 1. Flowchart of the regression algorithm.

Cross-validation serves as a technique to gauge how the machine learning model is likely to perform on an unfamiliar dataset (i. e. the data is not previously encountered by the model). This method assists in preventing overfitting – a situation, where a model becomes overly specialized to the training data and performs poorly on some new unseen data. We utilize the most prevalent form of the cross-validation, known as k-fold cross-validation. For this method, the initial sample is randomly divided into k equally sized segments. Out of these k segments, a single one is kept as validation data for the model testing while the remaining k-1 segments serve as training data. This cross-validation procedure is repeated k times. Each time a different segment as validation data is used. The results from the k iterations can then be averaged (or otherwise merged) to yield a single measure of model performance.

On the other hand, hyperparameter optimization involves the locating of the best set of hyperparameters for a machine learning model. Hyperparameters are those parameters that the model does not learn from the data but are preset before the learning procedure commences. For instance, in a Random Forest model the hyperparameters could consist of the count of decision trees in the forest or the count of the features each tree examines when partitioning a node. Unlike model parameters, the hyperparameters are preset before the learning process starts and remain static during the learning. Hyperparameters can have a significant impact on the model performance, hence the importance of the thoughtful selection. For our hyperparameter optimization, we employ a 'grid search' strategy where we define a set of possible values for each hyperparameter and then assess the performance of each possible combination.

Table 3: Modeling by using machine learning regressors.

Regressor	Best hyperparameters	SSE	R-square	RMSE
k-Nearest Neighbors	{'n_neighbors': 2, 'weights': 'uniform'}	351.0150	0.6397	5.6489
Gradient Boosting	{'max_depth': 2, 'n_estimators': 50}	7.9934	0.9918	0.8525
Random Forest	{'max_depth': 2, 'n_estimators': 50}	232.3306	0.7615	4.5958
Extra Trees	{'max_depth': 2, 'n_estimators': 100}	299.7497	0.6923	5.2202

In the second column of Table 3, the best combination of hyperparameters is shown for each method. The same metrics as in [14] are provided in order to compare them. Obviously, **Gradient Boosting** performs best. Its $\mathbf{R}^2$ is almost 1 and the errors are very small. Compared to Table 2, it is almost as good as the third-order polynomial Poly33. The other methods exhibit larger unacceptable errors. This approach could be successfully used to forecast the biological activity of new ligands via the optimization function GoldScore.

4. Conclusions

We conducted docking experiments to explore the interactions between delta-opioid ligands and DOR (PDBe: 1ozc). The established relationship between docking results and in vitro testing allowed us to predict the biological activity of newly synthesized analogs, exhibiting different activity than other compounds in the same series. This analysis provides deeper insights into the correlation between the biological effects of compounds and in silico experiments.

Furthermore, our investigation serves as a valuable tool to verify the alignment between the biological macromolecule (DOR) models and the actual three-dimensional molecular structure. The derived results equip us with the ability to anticipate the effectiveness of compounds with known docking scores and total energies. Such predictive capabilities will play a crucial role in shaping the direction of future studies.

By enhancing our understanding of these relationships and validating our models' accuracy, we are better positioned to make informed predictions about the biological activity of new compounds based on their docking scores and total energies. This valuable knowledge paves the way for future research and discovery in this field, potentially leading to the development of more effective pharmaceutical compounds.

Acknowledgments

This research is supported by the Bulgarian National Science Fund under Project KP-06-M62/1 "Numerical deterministic, stochastic, machine and deep learning methods with applications in computational, quantitative, algorithmic finance, biomathematics, ecology and algebra" from 2022.

References

[1] Rosenblum A., Marsch L., Joseph H., Portenoy R. 2008 *Experimental and clinical psychopharmacology* **16(5)**, 405–416 https://doi.org/10.1037/a0013628
[2] Portenoy R., Payne R., Passik S., Lowinson J., Ruz P., Millman R., Langrod J. 2004 *Acute and Chronic Pain,* 863–904 https://doi.org/10.1037/a0013628
[3] Valentino R, Volkow N 2018 *Neuropsychopharmacology* **43(13)** 2514-2520 https://doi.org/10.1038/s41386-018-0225-3
[4] Darcq E, Kieffer B. 2018 *Nat Rev Neurosci* **19(8)** 499–514 https://doi.org/10.1038/s41583-018-0028-x
[5] Kane B., Svensson B., Ferguson D. 2006 *The AAPS journal* **8(1)** E126-E137. https://doi.org/10.1208/aapsj080115
[6] Lord J, Waterfield A., Hughes J., Kosterlitz H 1977 *Nature* **267** 495-499 https://doi.org/10.1016/0024-3205(80)90312-4
[7] Maldonado R., Severini R., Hans W., Kiefer B., Melchiorri P., Negri L 2001 *British Journal of Pharmacology* **132** 1485–1492.
[8] Pradhan A., Befort K., Nozaki C., Gavériaux-Ruff C., Kieffer B. 2011 *Trends in pharmacological sciences* **32(10)** 581–590. https://doi.org/10.1016/j.tips.2011.06.008
[9] Le Merrer J, Becker J, Befort K, Kieffer L 2009 *Physiol Rev.* **89(4)** 1379–1412 https://doi.org/10.1152/physrev.00005.2009
[10] Morgan M, Christie M 2011 *British journal of pharmacology* **164(4)** 1322–1334 https://doi.org/10.1111/j.1476-5381.2011.01335.x
[11] Dorahy G., Chen J., Balle T. 2023 *Molecules* **28** 1324. https://doi.org/10.3390/molecules28031324
[12] Sapundzhi F, Dzimbova T. 2019 *International Journal of Online and Biomedical Engineering* **15(15)** 39–56 https://doi.org/10.3991/ijoe.v15i15.11566
[13] Sapundzhi F., Dzimbova T., Milanov P., Pencheva N., 2015 *J. Comput. Methods Molec. Design,* **5 (2)**, 98-108.
[14] Sapundzhi F 2019 *Bulgarian Chemical Communications* **51 (4),** 569-579.
[15] Sapundzhi F, Prodanova K, Lazarova M 2019 *AIP Conference Proceedings*, 2172, 100008 1-6
[16] Sapundzhi F 2019 *International Journal of Online and Biomedical Engineering* **15(11)**, 39-145.
[17] Jones G, Willett P, Glen R, Leach A, Taylor R 1997 *J. Mol. Biol.* **267**, 727-748. https://doi.org/10.1006/jmbi.1996.0897
[18] Pencheva N., Bocheva A, Dimitrov E., Ivancheva C 1996 *Eur. J. Pharmacol.* **304** 99-108.
[19] Pencheva N., Milanov P., Vezenkov L., Pajpanova T., Naydenova E. 2004 *Eur. J. Pharmacol.* **498** 249-256.
[20] Traykov M, Trenchev I, 2016 *Genetika* **52(9)**, 1089–1096, https://doi.org/10.7868/s0016675816080130
[21] Russel S, Norvig P 2020 *Artificial Intelligence: A Modern Approach* 4th Ed Pearson USA
[22] Kelleher J, Namee B, D'Arcy A 2020 *Fundamentals of Machine Learning for Predictive Data Analytics: Algorithms, Worked Examples, and Case Studies* 2nd Ed The MIT Press USA
Nedyalkov I, Stefanov A, Georgiev G, 2018 *IX National Conference with International Participation (ELECTRONICA)*, Sofia, Bulgaria, 1-4, https://doi.org/10.1109/ELECTRONICA.2018.8439127

Christov expansion method for nonlocal nonlinear evolution equations

M A Christou[1] and I C Christov[1,2]

[1] Department of Computer Science, University of Nicosia, 46 Makedonitissas Avenue, CY-2417 Nicosia, Cyprus
[2] School of Mechanical Engineering, Purdue University, West Lafayette, Indiana 47907, USA

E-mail: `christou.ma@unic.ac.cy`, `christov@purdue.edu`

Abstract. Christov functions are a complete orthonormal set of functions on $L^2(-\infty,\infty)$ that allow us to expand derivatives, nonlinear products, and nonlocal (integro-differential) terms back into the same basis. These properties are beneficial when solving nonlinear evolution equations using Galerkin spectral methods. In this work, we demonstrate such a "Christov expansion method" for the Benjamin–Ono (BO) equation. In the BO equation, the dispersion term is nonlocal, given by the Hilbert transform of the second spatial derivative of the unknown function. The Hilbert transform of the Christov functions can be computed using complex integration and Cauchy's residue theorem to obtain simple relations. Then, a Galerkin spectral expansion can be used to the solve the BO equation. Time integration is performed using a Crank–Nicolson-type scheme. Importantly, the Christov expansion method yields a banded matrix for the spatial discretization, even though the spatial terms are nonlocal. To demonstrate the approach and its implementation, we perform numerical experiments showing the steady propagation of single and the overtaking interaction of multiple BO solitary waves.

1. Introduction

The goal of this work is to develop and benchmark a Galerkin spectral method [1, 2] for a model nonlinear and *nonlocal* evolution equation.

Recently, there has been significant interest in nonlinear nonlocal partial differential equations (PDEs), from the point of view of integrable systems [3], but also in the context of water waves [4]. Classically, perhaps the most well-known nonlinear nonlocal PDE is the Benjamin–Ono (BO) equation [5, 6, 7, 8]. The BO equation describes the evolution of long, weakly-nonlinear waves on deep water (e.g., internal water waves). The BO equation, made suitably dimensionless [8], reads

$$u_t + uu_x + \mathscr{H}\{u_{xx}\} = 0, \qquad x \in (-\infty, +\infty), \quad t > 0. \tag{1}$$

Here, $u = u(x,t)$, x and t subscripts denote partial derivatives, and some suitable initial condition $u(x,0) = u^0(x)$ is supplied. The operator $\mathscr{H}\{\,\cdot\,\}$ introduced in (1) is the *Hilbert transform*, which is defined as a principal value integral [9]:

$$\mathscr{H}\{f\}(x) = \frac{1}{\pi}\,\mathrm{PV}\int_{-\infty}^{+\infty} \frac{f(y)}{y-x}\,\mathrm{d}y \tag{2}$$

for any function $f(x) \in L^p(-\infty,+\infty)$ for $1 < p < \infty$.

For the present purposes, we would like to observe that the BO equation (1) is much like the Korteweg–de Vries (KdV) equation [10, 8]:

$$u_t + uu_x + u_{xxx} = 0, \qquad x \in (-\infty, +\infty), \quad t > 0. \tag{3}$$

The key difference between KdV and BO is that the local dispersion term u_{xxx} in the KdV equation (3) is replaced by the nonlocal dispersion operator $\mathscr{H}\{u_{xx}\}$ in the BO equation (1). Nevertheless, like KdV, the BO equation (1) is completely integrable, in the sense that there exists an inverse scattering transform that solves it exactly [11, 8]. The BO equation is also well-posed in a suitable Hilbert space [12]. With these mathematical results in mind, we are comfortable discussing the BO equation's numerical solution.

To solve a nonlinear nonlocal PDE, such as (1), on $x \in (-\infty, +\infty)$ using a Galerkin method, a suitable complete set of orthonormal basis functions is needed. Fourier and Chebyshev bases are well-suited to periodic and finite domains [1, 2], but they are not so easily employed on the whole real line. Though certain re-mapping methods from $(-\infty, \infty)$ to $[-1, 1]$ can be used [1], these methods can affect the convergence or might require using a pseudospectral method.

Christov functions, introduced in section 2, are a complete orthonormal set of functions on $L^2(-\infty, \infty)$ that allow us to expand derivatives, nonlinear products, and even nonlocal (integro-differential) terms back into the same basis. In section 3, we exploit these unique properties of Christov functions to develop a Galerkin spectral method for the model nonlinear nonlocal PDE (1). The stationary problem for the KdV equation (3) was previously solved using such a "Christov expansion method" [13]. The novelty of the present work is to solve the BO equation (1) using a Christov expansion method that specifically makes use of the serendipitous properties of the Hilbert transform of Christov functions, which we discuss at the end of section 2. We consider both stationary (sections 3.1 and 4.1) and dynamic (sections 3.2 and 4.2) problems. The numerical results are presented in section 4. Finally, conclusions are stated in section 5.

2. The Christov functions and their properties

J. P. Boyd [14] termed the rational functions introduced in 1982 by C. I. Christov [15] as "Christov functions." These real-valued sets of orthonormal functions on $L^2(-\infty, +\infty)$, denoted by $\{S_n, C_n\}$ [15], are defined as

$$S_n(x) = \frac{\rho_n(x) + \rho_{-n-1}(x)}{\mathrm{i}\sqrt{2}}, \tag{4a}$$

$$C_n(x) = \frac{\rho_n(x) - \rho_{-n-1}(x)}{\sqrt{2}}, \qquad n = 0, 1, 2, \dots, \tag{4b}$$

in terms of Wiener's orthonormal complex-valued functions $\{\rho_n\}$ [16, p. 35],

$$\rho_n(x) = \frac{(\mathrm{i}x - 1)^n}{\sqrt{\pi}(\mathrm{i}x + 1)^{n+1}}. \tag{5}$$

The decay of the functions (4) as $|x| \to \infty$ makes them particularly well suited for problems on the real line with asymptotic boundary conditions. Further, based on the properties of the Wiener functions, various fruitful properties of the Christov functions can be obtained, as shown in particular in the original works [15, 17] and further in [13, 18].

Here, we summarize the key results needed to tackle the construction of a Galerkin spectral method for the BO equation (1). First, we need the double product formulæ for the Christov functions (4):

$$C_n C_k = \frac{1}{2\sqrt{2\pi}}\big[\ C_{n+k+1} - C_{n+k} - C_{n-k} + C_{n-k-1}\big] = \sum_{m=1}^{\infty} \beta_{nk,m} C_m, \tag{6a}$$

$$S_n S_k = \frac{1}{2\sqrt{2\pi}}\big[\ C_{n+k+1} - C_{n+k} + C_{n-k} - C_{n-k-1}\big] = \sum_{m=1}^{\infty} \alpha_{nk,m} C_m, \tag{6b}$$

$$S_n C_k = \frac{1}{2\sqrt{2\pi}}\big[- S_{n+k+1} + S_{n+k} + S_{n-k} - S_{n-k-1}\big] = \sum_{m=1}^{\infty} \gamma_{nk,m} S_m, \tag{6c}$$

where

$$\alpha_{nk,m} = \frac{1}{2\sqrt{2\pi}}\left\{\delta_{m,n+k+1} + \delta_{m,|n-k|} - \delta_{m,n+k} - \operatorname{sgn}\left[|n-k| - \tfrac{1}{2}\right]\delta_{m,[|n-k|-1/2]}\right\}, \tag{7a}$$

$$\beta_{nk,m} = \frac{1}{2\sqrt{2\pi}}\left\{\delta_{m,n+k} + \delta_{m,|n-k|} - \delta_{m,n+k+1} - \operatorname{sgn}\left[|n-k| - \tfrac{1}{2}\right]\delta_{m,[|n-k|-1/2]}\right\}, \tag{7b}$$

$$\gamma_{nk,m} = \frac{1}{2\sqrt{2\pi}}\left\{\delta_{m,n+k} + \operatorname{sgn}(n-k)\delta_{m,|n-k|} - \delta_{m,n+k+1} - \operatorname{sgn}(n-k)\delta_{m,|n-k|-1}\right\}, \tag{7c}$$

and $\delta_{m,n}$ is Kronecker's delta.

Second, we need the first and second derivatives expansion formulæ for the Christov functions (4):

$$C_n' = \sum_{m=0}^{\infty} \theta_{nm} S_m, \qquad S_n' = -\sum_{m=0}^{\infty} \theta_{nm} C_m, \tag{8a}$$

$$C_n'' = \sum_{m=0}^{\infty} \chi_{nm} C_m, \qquad S_n'' = \sum_{m=0}^{\infty} \chi_{nm} S_m, \tag{8b}$$

where the matrices, θ_{mn} and χ_{mn}, are symmetric tridiagonal and pentadiagonal, respectively. Their components can be explicitly given by:

$$\theta_{mn} = -\frac{1}{2} n\delta_{m,n-1} + \left(n + \tfrac{1}{2}\right)\delta_{m,n} - \frac{1}{2}(n+1)\delta_{m,n+1}, \tag{9a}$$

$$\begin{aligned}\chi_{mn} &= -\frac{1}{4}n(n-1)\delta_{m,n-2} + n^2\delta_{m,n-1} - \frac{1}{4}(n+1)(n+2)\delta_{m,n+2} \\ &\quad -\frac{1}{4}n^2 + (2n+1)^2 + (n+1)^2\delta_{m,n} + (n+1)^2\delta_{m,n+1}.\end{aligned} \tag{9b}$$

Finally, we need the Hilbert transform of the Christov functions (4). To evaluate the integrals involved in taking the Hilbert transform of the Christov functions, we can use complex contour integration. Both the upper and lower half-plane can be used [19]. Similar work was done by Weideman [20]. The result is:

$$\mathscr{H}\{C_n\}(x) = \frac{1}{\pi}\,\mathrm{PV}\int_{-\infty}^{+\infty} \frac{C_n(y)}{y-x}\,\mathrm{d}y = \quad S_n(x), \tag{10a}$$

$$\mathscr{H}\{S_n\}(x) = \frac{1}{\pi}\,\mathrm{PV}\int_{-\infty}^{+\infty} \frac{S_n(y)}{y-x}\,\mathrm{d}y = -C_n(x). \tag{10b}$$

This serendipitous result was exploited by Boyd and Xu [21] to construct a pseudospectral method for the BO equation (1).

3. Spectral expansion and implementation details

3.1. Stationary problem

Let us first consider the stationary problem for the BO equation (1) by rewriting it in the moving frame $\xi = x - ct$:

$$(u - c)u' + \mathscr{H}\{u''\} = 0, \tag{11}$$

where $c > 0$ is the phase speed of the traveling wave, and derivatives with respect to ξ are denoted by primes.

Before proceeding further, we note that rescaling the spatial variable ξ does not change the nature of the asymptotic boundary conditions on $(-\infty, +\infty)$. Upon introducing $\xi = \zeta\hat{\xi}$ we recast (11) as

$$(u - c)u' + \zeta\mathscr{H}\{u''\} = 0, \tag{12}$$

The scaling parameter ζ can be used to optimize the method in the sense that it allows one to control the typical length scales of the employed system of functions and to also control the width of the sought localized solution (see (13) below). Naturally, such a coordination between the scales will results in a faster convergence of the Fourier–Galerkin series. Henceforth, for the sake of brevity, we omit writing $\hat{\xi}$ and just write ξ. For further discussion on the use of the scaling parameter ζ, see [18].

In this context, we are interested in solitary waves, which satisfy asymptotic boundary conditions, $u(\xi) \to 0$ as $|\xi| \to \infty$. As is well-known, the stationary problem for the BO equation (12) on the real line, $\xi \in (-\infty, +\infty)$, admits the analytical solution [21] in the form:

$$u(\xi) = \frac{4c}{1 + c^2\zeta^2\xi^2}, \qquad c > 0. \tag{13}$$

The analytical solution (13) is an even function, $u(-\xi) = u(\xi)$, thus in the Galerkin spectral method we may expand the solution of the stationary problem (12) using only the $\{C_n\}$ functions:

$$u(\xi) = \sum_{k=0}^{\infty} a_k C_k(\xi). \tag{14a}$$

Now, for the nonlinear term, using (6), we have

$$u(\xi)u'(\xi) = \sum_{k=0}^{\infty}\sum_{m=0}^{\infty}\sum_{n=0}^{\infty}\sum_{l=0}^{\infty} a_k a_m \theta_{mn} \gamma_{kn,l} S_l. \tag{14b}$$

And, for the first and second derivatives of u with respect to ξ, using (8) we have:

$$u'(\xi) = \sum_{k=0}^{\infty}\sum_{m=0}^{\infty} a_k \theta_{km} S_m(\xi), \tag{14c}$$

$$u''(\xi) = \sum_{k=0}^{\infty}\sum_{m=0}^{\infty} a_k \chi_{km} C_m(\xi). \tag{14d}$$

Finally, for the Hilbert transform of u'', using (10), we have:

$$\mathscr{H}\{u''\}(\xi) = \sum_{k=0}^{\infty}\sum_{m=0}^{\infty} a_k \chi_{km} S_m(\xi). \tag{14e}$$

We insert the aforementioned spectral expansions (14) into the stationary BO equation (11). After integrating on $\xi \in (-\infty, +\infty)$ and introducing iterations (on p) for the nonlinear algebraic system, we have:

$$-c\sum_{m=0}^{\infty} a_k^{p+1}\theta_{km} + \zeta \sum_{m=0}^{\infty} a_k^{p+1}\chi_{km} = -\sum_{m=0}^{\infty}\sum_{n=0}^{\infty}\sum_{l=0}^{\infty} a_k^p a_m^p \theta_{mn}\gamma_{kn,l}. \tag{15}$$

To control the convergence of the iterations (and, specifically, avoid divergence), we introduce under-relaxation as:

$$\tilde{a} = \omega a^{p+1} + (1-\omega)a^p, \tag{16}$$

where $\omega = 0.5$ is the relaxation parameter. Then, $a^p = \tilde{a}$ and the process continues until $\max_k |a_k^{p+1} - a_k^p| / \max_k |a_k^p| < 10^{-12}$.

The summations in (15) are all truncated at finite value to solve the system. Below, we use the first 500 Christov functions. But, one could use even functions in the expansion, for example 200, since the decay of the spectral coefficients is rapid. We will comment further on this point in the coming section, when the numerical results are presented.

3.2. Dynamic problem and time stepping

For the general dynamic case, we discretize the BO equation (1), subject to asymptotic boundary conditions $u(x,t) \to 0$ as $|x| \to \infty$, in time using a three-level Crank–Nicolson-type scheme:

$$\frac{1}{2\tau}u^{p+1} + \frac{1}{2}\mathscr{H}\{u_{xx}^{p+1}\} = \frac{1}{2\tau}u^{p-1} - \frac{1}{2}\mathscr{H}\{u_{xx}^{p-1}\} - u^p u_x^p, \tag{17}$$

where $u^p \approx u(x, t^p)$ with $t^p = p\tau$, and τ is the time step. As in the stationary problem, we rescale x as $x = \zeta\hat{x}$ to get

$$\frac{1}{2\tau}u^{p+1} + \frac{1}{2}\zeta^2\mathscr{H}\{u_{xx}^{p+1}\} = \frac{1}{2\tau}u^{p-1} - \frac{1}{2}\zeta^2\mathscr{H}\{u_{xx}^{p-1}\} - \zeta u^p u_x^p. \tag{18}$$

Note that in the last equation, as before, we omit writing $\hat{x}$ and just write x.

In space, we expand the solution in a spectral series using Christov functions:

$$u^p(x) = \sum_{k=0}^{\infty} \left[a_k^p C_k(x) + b_k^p S_k(x)\right]. \tag{19}$$

As before, we can calculate:

$$u^p u_x^p = \sum_{k=0}^{\infty}\sum_{m=0}^{\infty}\sum_{n=0}^{\infty}\sum_{l=0}^{\infty} \big[a_k^p a_m^p \theta_{mn}\gamma_{kn,l} S_l(x) - b_k^p b_m^p \theta_{mn}\gamma_{kn,l} S_l(x) - a_k^p b_m^p \theta_{mn}\beta_{kn,l} C_l(x) + b_k^p a_m^p \theta_{mn}\alpha_{kn,l} C_l(x)\big]. \tag{20a}$$

$$u_x^p = \sum_{k=0}^{\infty}\sum_{m=0}^{\infty} \left[a_k^p \theta_{km} S_m(x) - b_k^p \theta_{km} C_m(x)\right], \tag{20b}$$

$$u_{xx}^p = \sum_{k=0}^{\infty}\sum_{m=0}^{\infty} \left[a_k^p \chi_{km} C_m(x) + b_k^p \chi_{km} S_m(x)\right], \tag{20c}$$

$$\mathscr{H}\{u_{xx}^p\} = \sum_{k=0}^{\infty}\sum_{m=0}^{\infty} \left[a_k^p \chi_{km} S_m(x) - b_k^p \chi_{km} C_m(x)\right]. \tag{20d}$$

Journal of Physics: Conference Series **2675** (2023) 012022 doi:10.1088/1742-6596/2675/1/012022

Next, we insert the spectral expansions (20) into the time-discrete scaled BO equation (18). After integration, we reduce the problem to the solution of a nonlinear coupled (but *banded*) algebraic system for the unknown expansion coefficients $\{a_k, b_k\}$:

$$\frac{1}{2\tau}a_k^{p+1} - \frac{1}{2}\zeta^2 \sum_{m=0}^{\infty} b_k^{p+1}\chi_{km} = \frac{1}{2\tau}a_k^{p-1} + \frac{1}{2}\zeta^2 \sum_{m=0}^{\infty} b_k^{p-1}\chi_{km} + \frac{1}{2}\zeta \sum_{m=0}^{\infty}\sum_{n=0}^{\infty}\sum_{l=0}^{\infty} a_k^p b_m^p \theta_{mn}\beta_{kn,l} - \frac{1}{2}\zeta \sum_{m=0}^{\infty}\sum_{n=0}^{\infty}\sum_{l=0}^{\infty} b_k^p a_m^p \theta_{mn}\alpha_{kn,l} \quad (21a)$$

and

$$\frac{1}{2\tau}b_k^{p+1} + \frac{1}{2}\zeta^2 \sum_{m=0}^{\infty} a_k^{p+1}\chi_{km} = \frac{1}{2\tau}b_k^{p-1} - \frac{1}{2}\zeta^2 \sum_{m=0}^{\infty} a_k^{p-1}\chi_{km} - \frac{1}{2}\zeta \sum_{m=0}^{\infty}\sum_{n=0}^{\infty}\sum_{l=0}^{\infty} a_k^p a_m^p \theta_{mn}\gamma_{kn,l} - \frac{1}{2}\zeta \sum_{m=0}^{\infty}\sum_{n=0}^{\infty}\sum_{l=0}^{\infty} b_k^p b_m^p \theta_{mn}\gamma_{kn,l}. \quad (21b)$$

The initial conditions (for $p = 0$) are obtained by numerical quadrature on the analytical solution or on the numerical solution when solving the BO equation in the moving frame. As above, the summations in (21) are all truncated at finite value to solve the system.

4. Numerical examples and results

4.1. Verification with exact solution of steadily propagating a solitary wave

In figure 1, we show the shapes of solitons evaluated in the moving frame when (a) $c = 2.5$ and (b) $c = 0.5$. The numerical solutions are presented along with the exact solution (13). As we can see in both cases the numerical and exact results are in a very good agreement. In both cases, the scaling parameter was $\zeta = 4$.

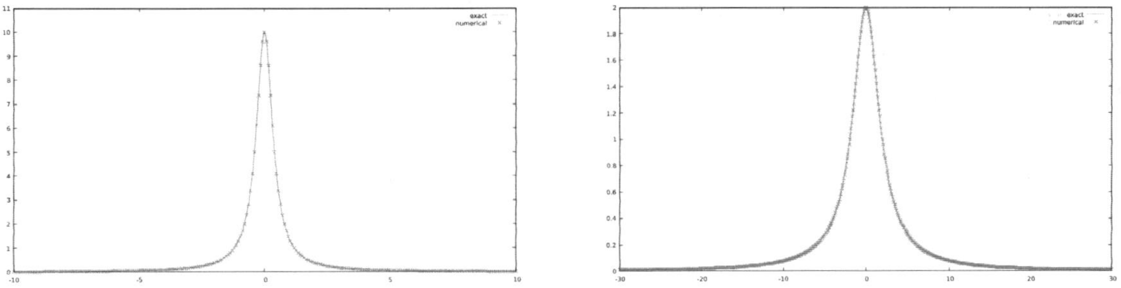

(a) Stationary solitary wave with $c = 2.5$. (b) Stationary solitary wave with $c = 0.5$.

Figure 1: Comparison of the exact and numerical solutions of the solitons in the moving frame for different values of c. Note the different vertical and horizontal scales in the two panels because both the height and the width of the solitons depends on c, see (13).

Next, in figure 2, we present the convergence of the spectral coefficients for the solutions shown in figure 1. We evaluated the stationary shapes using 500 Christov functions. We can see that in both cases the coefficients quickly decay below 10^{-12}. From the graph, we conclude that we could use an even small number of functions, say 100, and still obtain a very good approximation of the exact solution.

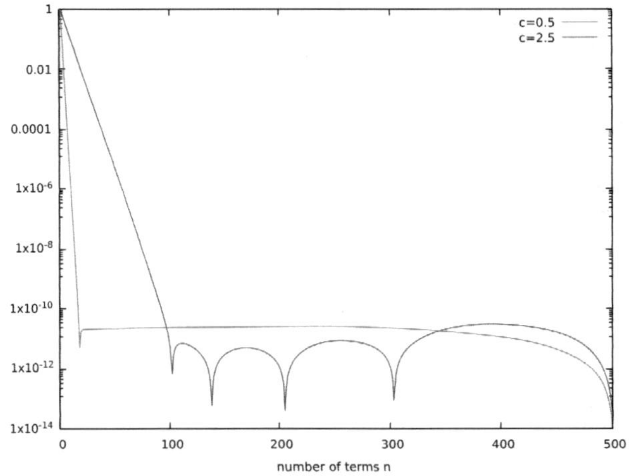

Figure 2: Decay of the spectral coefficients of the Christov function expansion of the stationary BO solution (13).

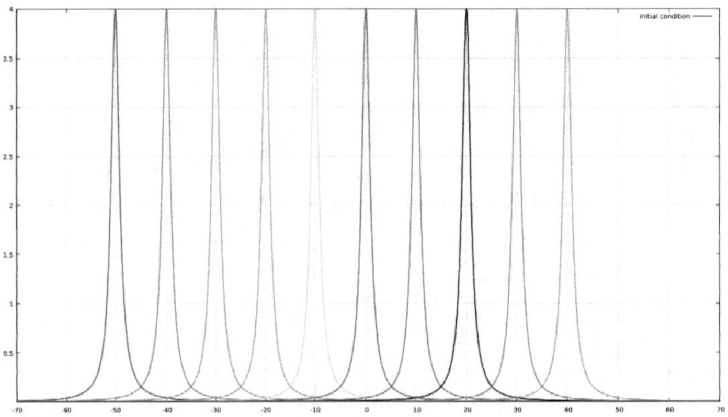

(a) $c = 1$

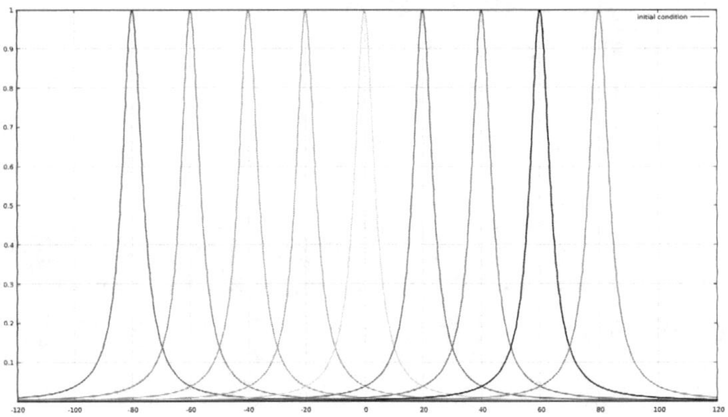

(b) $c = 0.25$

Figure 3: Simulation of single soliton propagation (left to right) for (a) $c = 1$ and (b) $c = 0.25$, both with $\tau = 0.01$.

4.2. Dynamic problem: Soliton propagation and overtaking interaction

In figure 3(a), we show the propagation of a single soliton with $c = 1$ and $\tau = 0.01$. An initial condition of the form (13) was placed at $x = -50$ and let propagate in the right direction. As we can see the shape of the soliton remains the same for a long period of time. The last shape shown in figure 3(a) has propagated to $x = +40$ after 9 000 time steps.

In figure 3(b), we show the propagation of a single soliton with $c = 0.25$ and $\tau = 0.01$. Here, we should mention that in comparison with the previous case, the soliton for $c = 0.25$ is shorter and wider. An initial condition of the form (13) was placed at $x = -80$ and let propagate in the right direction. As we can see, the shape of the soliton remains the same for a long period of time, the last shape shown in figure 3(b) has propagated to $x = +80$ after 64 000 time steps.

Next, we were interested in demonstrating the numerical method on soliton overtaking interaction. That is, we consider two solitons moving in the same direction (here, left to right), where the one on the left side catching up the one on the right and overtaking it in time. As the initial condition, we use the exact solution (13) of the stationary problem, and linearly superimpose two sufficiently-separated solitons. These two solitons are placed far enough apart to avoid interaction in the initial instant of time. The two solitons have different phase speeds, namely $c_1 = 1.5$ (left one) and $c_2 = 0.5$ (right one). Then, using numerical quadrature we compute the spectral coefficients of the initial condition. Here, we remind the reader that a numerical solution of the moving-frame problem could also be used as initial condition, which does not require the assumption of "sufficient separation"; alternatively, multi-soliton solution formulæ for the BO equation also exist [22, 23] and can be used.

The overtaking interaction is depicted in figure 4. The tall (faster) soliton with $c_1 = 1.5$ is placed on the left at $x = -70$ and the shorter (slower) soliton with $c_2 = 0.5$ is placed at $x = -30$ to the right of the taller soliton. From figure 4, we can see that the taller soliton overtakes the shorter one, and they eventually fully separate and completely recover their initial shapes.

5. Conclusion

In this work, we applied the Christov expansion method for the first time to the Benjamin–Ono equation, which is a *nonlocal* nonlinear evolution equation. A key result was that the Hilbert transform of the Christov functions could be obtained using complex integration, which turned the nonlocal PDE into a banded nonlinear algebraic system. We first considered the stationary problem in the moving frame, for which a simple exact solitary wave solution exists (as also exploited in [21]). This solution was important for the validation of the proposed Galerkin method. We then proceeded to investigate the time-dependent (dynamic) problem, and we demonstrated that the proposed Christov expansion spectral method accurately captures both single soliton propagation, as well as soliton overtaking interactions.

In future work, we are interested in extending the proposed approach to more general nonlocal dispersion operators, such the one appearing in the so-called "full-dispersion" Whitham equation [24] (see also [25]):

$$u_t + uu_x + \mathscr{W}\{u_x\} = 0, \tag{22}$$

where $\mathscr{W}\{\cdot\}$ is the nonlocal operator defined in the Fourier ($\widehat{\cdot}$) domain as $\widehat{\mathscr{W}\{u\}}(k) = \sqrt{\tanh(k)/k}\,\widehat{u}(k)$. Using the long-wave (small-$|k|$) approximation $\sqrt{\tanh(k)/k} \sim 1 - k^2/6 = 1 + (\mathrm{i}k)^2/6$, yields $\mathscr{W}\{u_x\} = u_x + \frac{1}{6}u_{xxx}$ as in the classical KdV equation [10, 8]. Another interesting problem would be to consider nonlinearities within the nonlocal terms themselves, such as those that arise in modeling of short and intermediate (as opposed to long) waves. Recently, Ivanov [4] discussed a nonlocal nonlinear *wave* equation due to Matsuno [26]. Suitably rescaled with various dimensionless groups taken as unity for simplicity, the latter takes the form

$$u_{tt} + \mathscr{H}\{u_x\} - \big[uu_x + \mathscr{H}\{u\mathscr{H}\{u_x\}\} - \mathscr{H}\{(\mathscr{H}\{u_t\})^2\}\big]_x = 0. \tag{23}$$

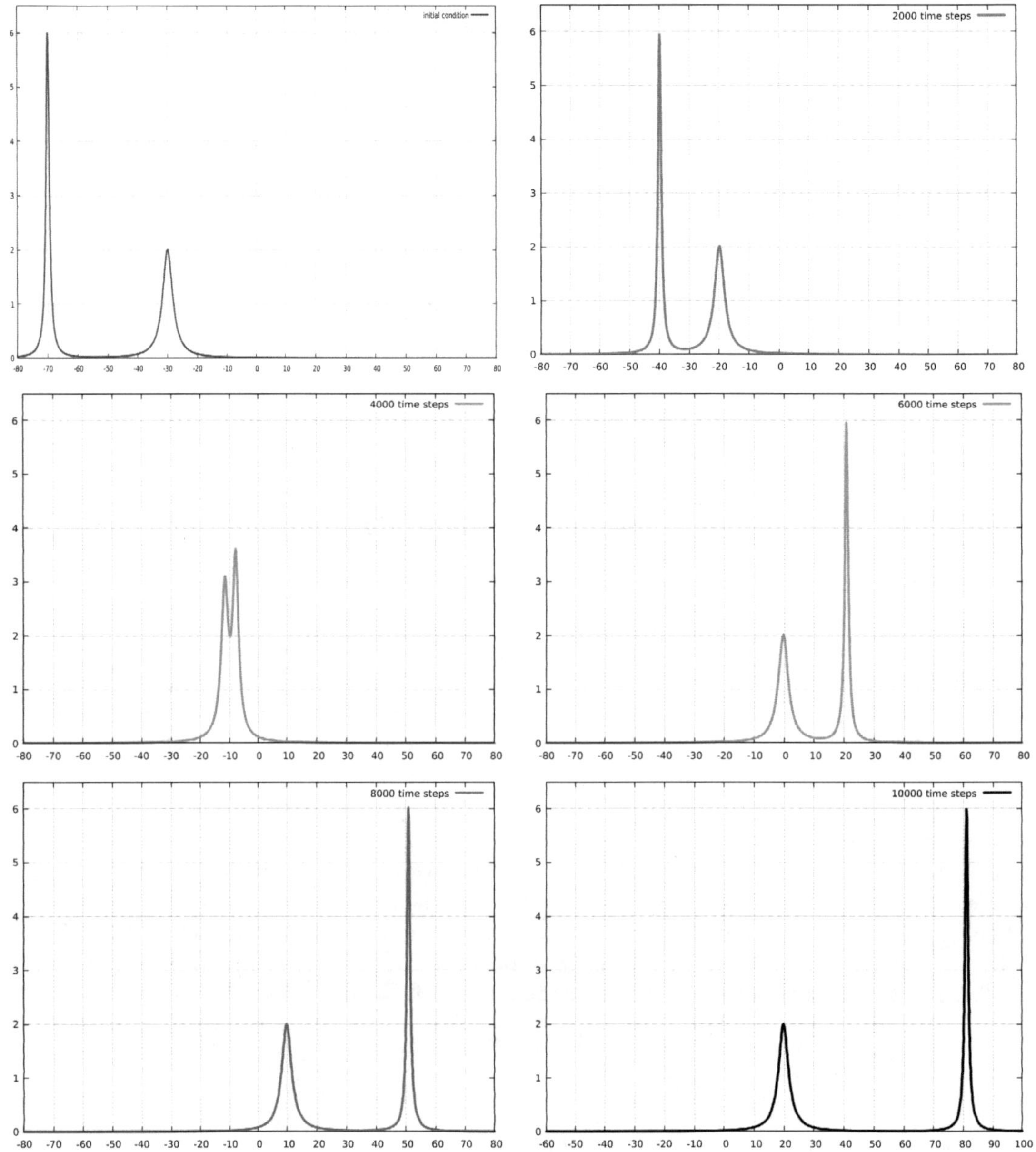

Figure 4: Simulation of a soliton overtaking interaction for $c_1 = 1.5$ and $c_2 = 0.5$.

Ivanov [4] observed that a one-way *evolutionary* counterpart to (23) can also be derived but it "remains highly nonlocal." It would be of interest to attempt to solve these equations numerically via the present approach.

Acknowledgments

ICC would like to acknowledge the hospitality of the University of Nicosia, Cyprus, where this work was completed thanks to a Fulbright US Scholar award from the US Department of State, and the US National Science Foundation, which supports his research on interfacial dynamics under grant CMMI-2029540.

References

[1] Boyd J P 2000 *Chebyshev and Fourier Spectral Methods* 2nd ed (Mineola, NY: Dover Publications)
[2] Shen J, Tang T and Wang L L 2011 *Spectral Methods: Algorithms, Analysis and Applications* (*Springer Series in Computational Mathematics* vol 41) (Berlin/Heidelberg: Springer-Verlag)
[3] Ablowitz M J and Musslimani Z H 2017 *Stud. Appl. Math.* **139** 7–59
[4] Ivanov R I 2023 *Appl. Math. Lett.* **142** 108653
[5] Benjamin T B 1967 *J. Fluid Mech.* **29** 559–592
[6] Davis R E and Acrivos A 1967 *J. Fluid Mech.* **30** 723–736
[7] Ono H 1975 *J. Phys. Soc. Jpn.* **39** 1082–1091
[8] Drazin P G and Johnson R S 1989 *Solitons: An Introduction* 2nd ed (Cambridge, UK: Cambridge University Press)
[9] Calderon A P and Zygmund A 1952 *Acta Math.* **88** 85–139
[10] Korteweg D J and de Vries G 1895 *Phil. Mag. Ser. 5* **39** 422–443
[11] Fokas A S and Ablowitz M J 1983 *Stud. Appl. Math.* **68** 1–10
[12] Tao T 2004 *J. Hyperbol. Differ. Equ.* **1** 27–49
[13] Christov C I and Bekyarov K L 1990 *SIAM J. Sci. Stat. Comput.* **11** 631–647
[14] Boyd J P 1990 *J. Approx. Theory* **61** 98–105
[15] Christov C I 1982 *SIAM J. Appl. Math.* **42** 1337–1344
[16] Wiener N 1964 *Time Series* (Cambridge, MA: MIT Press)
[17] Christov C I 1987 *Ann. Univ. Sofia, Fac. Math. Mech.* **76(2)** 87–113
[18] Christou M A and Christov C I 2002 *J. Comput. Anal. Appl.* **4** 63–77
[19] Christou M A and Papanicolaou N C 2022 *AIP Conf. Proc.* **2522** 030002
[20] Weideman J A C 1995 *Math. Comp.* **64** 745–762
[21] Boyd J P and Xu Z 2011 *Wave Motion* **48** 702–706
[22] Joseph R I 1977 *J. Math. Phys.* **18** 2251–2258
[23] Matsuno Y 1979 *J. Phys. A: Math. Gen.* **12** 619–621
[24] Whitham G B 1967 *Proc. Royal S. Lond.* **299** 6–25
[25] Stefanov A and Wright J D 2020 *J. Dyn. Diff. Eq.* **32** 85–99
[26] Matsuno Y 1992 *Phys. Rev. Lett.* **69** 609–611

Oscillating flow of rarefied viscous gas between cylinders of finite length. Numerical study

P N Gospodinov, D Y Dankov*, M K Mironova

Institute of Mechanics, Bulgarian Academy of Sciences, Acad. G. Bonchev str. Bl. 4, 1113 Sofia, Bulgaria
* Corresponding Author E-mail: dankov@imbm.bas.bg

Abstract. The rarefied viscous gas flow between a stationary inner and an oscillating outer cylinder is studied in this paper. It is a continuation of our previous publications where we study the flow between the infinite length cylinders while here is considered the case of finite length cylinders bounded by two planes with periodic boundary conditions in the cross sections at the ends of the cylinders. So, on the one hand, this changes the flow character creating conditions for periodic self-organization along the axis of symmetry and on the other hand, the current model is reduced to a flow in a limited volume. A continuum model based on Navier-Stokes-Fourier equations for compressible fluid with first order slip boundary conditions on the cylinders wall and statistical model Direct Simulation Monte Carlo (DSMC) method are used to model the flow. The presented numerical results were obtained after reaching sustained oscillations. Considering the model with cylinders of finite length allows to extend its capabilities and obtain a better fit to real conditions.

1. Introduction

The Taylor-Couette problem is a classic fluid dynamics problem in the rarefied gas dynamics that involves the flow between two concentric, independently rotating cylinders. This setup is named after the scientists G.I. Taylor and Maurice Couette, who made significant contributions to the study of fluid dynamics. The study of the Taylor-Couette problem provides valuable insights into the fundamental aspects of fluid dynamics, such as stability, transition to turbulence, and the influence of boundary conditions on flow patterns. The problem has been extensively studied both experimentally and theoretically, and it continues to be a subject of active research in fluid mechanics [3, 5, 6, 7, 15, 16, 17].

In Micro-Electro-Mechanical Systems (MEMS), the Taylor-Couette problem can be relevant in certain applications involving rotating components or fluidic systems. MEMS devices are miniaturized mechanical and electronic systems fabricated through microfabrication techniques. While the Taylor-Couette problem is traditionally associated with macroscopic fluid dynamics, its principles can be applied or adapted to certain MEMS applications where fluidic effects and rotation play a role. Understanding the behaviour of fluids in confined spaces and around rotating microscale components is crucial for the design and optimization of MEMS devices in various fields, ranging from biomedical applications to sensors and actuators.

The acoustic waves in the Taylor-Couette flow, which involves the flow between two rotating concentric cylinders of finite length, can be a complex and interesting phenomenon. Acoustic waves are pressure waves that propagate through a fluid medium, and their interaction with fluid can lead to various effects [7, 8, 10, 11].

The influence of boundary conditions (diffusive on the cylinders walls and periodical on the bounding planes cross-section) and gas rarefaction on fluid flow is particularly significant in scenarios where traditional continuum assumptions break down, such as in rarefied gas flows or microscale systems. Understanding these influences is crucial for accurately modeling and predicting flow behaviour, especially in microscale applications, low-pressure environments, or scenarios where rarefied gas effects are prominent. Specialized numerical methods and experimental techniques are often required to capture these effects accurately [9, 11, 18, 19, 20].
In our previous studies [10, 12, 13, 14], we considered rarefied monatomic gas between two coaxial unconfined cylinders (one dimensional, axisymmetric problem) by using a statistical method – Direct Simulation Monte Carlo (DSMC) and the Navier-Stokes-Fourier equations numerical solving. The present paper is a continuation of a study of Taylor-Couette flow between cylinders of finite length [21, 22] where we consider the Taylor-Couette flow (two dimensional, axisymmetric problem) in rarefied gas between two concentric stationary cylinders with final length. The state of the gas is studied over a long time period close to establishing a steady state in [21], while in [22] is studied the initial moment of occurrence and reflection of the wave from the inner cylinder, when the cylinder temperature suddenly increases. In this paper we study rarefied gas flow between cylinders with different temperature and stationary inner cylinder and oscillating outer cylinder. The two cylinder dimensions have been selected to approximate the actual dimensions of the real Pirani gauges. In accordance with this is the reception the oscillations scheme of the outer cylinder, which in our opinion approaches some of the possible real conditions. The temperature difference and the outer cylinder oscillation causes instability in the gas phase. The flow is modeled numerically by the DSMC method and continuum model – ANSYS-Fluent.

2. Formulation of the problem and methods of solution

We study a rarefied gas between two coaxial cylinders with finite length. The following notations are used to describe the rarefied gas flow: ρ is the density, T is the temperature, P is the thermodynamic pressure and u is the velocity components in r – direction. The inner cylinder has radius R_1 and temperature T_1 and the outer – R_2 and T_2. The inner cylinder is stationary and the outer cylinder velocity is V_2. The figure 1 shows the flow geometry.

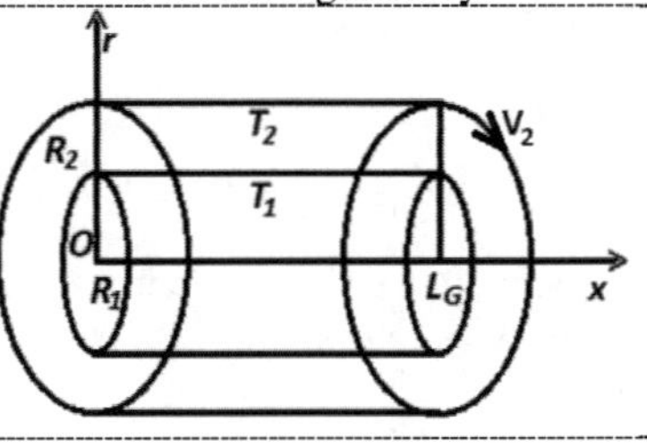

Figure 1. The flow geometry.

2.1 Continuous Model (NSF) and Numerical Simulation

The continuous model is based on the Navier-Stokes-Fourier (NSF) equations for compressible fluid, completed with the equations of continuity and energy transport. For details see [13]. The nonzero stress tensor components [6] can be expressed as:

$$\tau_{rr} = -\mu\left(2\frac{\partial u}{\partial r} - \frac{2}{3}\frac{1}{r}\frac{\partial}{\partial r}(ru)\right), \tag{1}$$

$$\tau_{\varphi\varphi} = \mu\left[\frac{2}{3}\left(\frac{1}{r}\frac{\partial}{\partial r}(ru) - 2\left(\frac{u}{r}\right)\right)\right]. \tag{2}$$

The dissipation function Φ in the equation of energy transport is

$$\Phi = \frac{4}{3}\mu\left[\left(\frac{\partial u}{\partial r}\right)^2 - \frac{\partial u}{\partial r}\frac{u}{r} + \left(\frac{u}{r}\right)^2\right]. \tag{3}$$

For a perfect monatomic gas, the viscosity and the heat transfer coefficients read as [1, 2, 3, 6]:

$$\mu(T) = C_\mu \rho_0 l_0 V_0 \sqrt{T},\ C_\mu = \frac{5}{16}\sqrt{\pi}, \tag{4}$$

$$\lambda(T) = C_\lambda c_P \rho_0 l_0 V_0 \sqrt{T},\ C_\lambda = \frac{15}{32}\sqrt{\pi}. \tag{5}$$

The NSF model is completed with the equation of state for an ideal gas:

$$P = \rho RT. \tag{6}$$

The governing equations are normalized by using the following scales: for density $\rho_0 = mn_0$ (m is the molecular mass, n_0 – the average number density), for velocity $V_0 = \sqrt{2RT_0}$ – R is the gas constant, for length – the distance between the cylinders $L = R_2 - R_1$, for time $t_0 = L/V_0$, for temperature T_0 – the wall temperature of outer cylinders. The Knudsen number is $Kn = l_0/L$ where the mean free path is l_0 and $\gamma = c_P/c_V = 5/3$ (c_P and c_V are the heat capacities at constant pressure and constant volume respectively). In this way, the characteristic number Kn and the constants C_μ and C_λ take part in the dimensionless model. For brevity, the same symbols r, t, P, T, u and R_i, i = 1, 2 are used for the corresponding dimensionless variables. The dimensionless density is denoted with $\bar{\rho}$.
For the problem formulated first-order slip boundary conditions are imposed at both walls, which can be written directly in dimensionless form as follows [14]. Using the slip-velocity, the boundary velocity condition along the ϕ direction can be written:

$$u - u_{wall} = \pm A_\sigma \frac{i_0}{\mu} \tau |_{r=R_{1,2}}. \tag{7}$$

Where $A_\sigma = (2/\sqrt{\pi})\sigma_P$ – slip coefficient, l_0 – mean free path of the molecules, and τ denotes the frictional stress on the wall – in the considered case, this is the stress, which is denoted as $\tau_{r\varphi}$. Because of the axial symmetry, this component can be written as

$$\tau_{r\varphi} = \mu \left[r \frac{\partial}{\partial r}\left(\frac{u}{r}\right)\right] = \mu \left[\frac{\partial u}{\partial r} - \frac{u}{r}\right], \tag{8}$$

$$u - u_{wall} = \pm A_\sigma l_0 \left(\frac{\partial u}{\partial r} - \frac{u}{r}\right) |_{r=R_{1,2}}. \tag{9}$$

Using the slip-temperature, the boundary temperature condition can be written:

$$T \pm \zeta_T Kn_{local} \frac{\partial T}{\partial r} = \overline{T_\iota} \tag{10}$$

at $r = R_i$, i = 1, 2 where for diffuse scattering we have used the temperature jump coefficients ζ_T = 2.1904 calculated, respectively in [12], from the kinetic BGK equation (a form of the collision integral proposed by Bhatnagar, Gross and Krook). The boundary conditions (10) are modeled by using the local Knudsen number Kn_{local}.

$$Kn_{local} = \frac{l}{L} = \left(L\sqrt{2}\pi\sigma^2 \frac{\rho_0}{\rho} n_0\right)^{-1} = \frac{Kn}{\bar{\rho}} \tag{11}$$

In (11) with l is denoted the local mean free path, σ - molecular diameter and ρ – is the dimensional density. For details see [10]. The boundary conditions are periodically at the left (x = 0) and the right (x= L_G) boundaries of both cylinders.
The governing equations along with the specified boundary conditions are solved numerically by the ANSYS-Fluent 21R in the three-dimension form.

2.2 Direct Simulation Monte Carlo (DSMC) Method
The Direct Simulation Monte Carlo (DSMC) method is a computational technique used for simulating the behaviour of rarefied gas flows. It is particularly valuable in scenarios where traditional continuum models, such as the Navier-Stokes equations, become inaccurate due to the breakdown of the continuum assumption. The DSMC method is widely employed in the study of microscale and rarefied

gas flows, as well as in various applications such as spacecraft re-entry, MEMS (Micro-Electro-Mechanical Systems) devices, and vacuum technology.
Particle-Based Simulation: DSMC simulates the gas flow by tracking individual gas particles rather than solving continuum equations for fluid properties. Each particle represents a collection of gas molecules.
Collision Model: The core of the DSMC method lies in the simulation of particle collisions. The method models collisions between gas particles probabilistically, considering factors such as particle velocities, masses, and collision cross-sections. Collision modeling is realized using Bird's scheme "no time counter".
Time Stepping: The simulation advances in discrete time steps. During each time step, particles move freely between collisions, and collision events are simulated based on statistical methods. Flow macro-characteristics (density, velocity, temperature, pressure) are calculated as time-averaged.
Boundary conditions are: diffusive at the outer and inner cylinder walls and periodically at $x = 0$ and $x = L_G$.

3. Numerical results

The ANSYS Fluent pressure based solution system was used for the numerical solution of the continual model in a dimension form. A 3D model was used, for pressure-velocity coupling the SIMPLE scheme, for pressure-PRESTO! schemes are applied. DSMC method is based on the 2D flow geometry (axis symmetrical flow) in a dimensionless form. We study monatomic argon (Ar) gas with pressure $p_0 = 6.43\times10^{-2}(bar)$ and temperature $T_0 = 300(K)$. The flow parameters are: the most portable speed – $V_P = \sqrt{\frac{2RT}{M}} = 353.355(m/s)$, the free molecular path – $l_0 = \frac{RT}{\sqrt{2}\pi\sigma^2 N_A P} = 1.23 \times 10^{-5}(m)$, the molar mass – $M = 40(amu)$, cross section – $\sigma = \pi d^2 = 3.63 \times 10^{-19}(m)$, the inner cylinder radius – $R_1 = 6.25\times10^{-4}(m)$, the outer cylinder radius – $R_2 = 1\times10^{-5}(m)$, the Knudsen number – $Kn = 0.02$, the outer cylinder velocity oscillates by the formula $V_2 = \Delta V\cos(\omega t)$ where $\Delta V = 0.1$ at the dimensionless calculation and $\Delta V = 35.3353(m/s)$ at the dimension calculation. The inner cylinder temperature T_1 suddenly change from 300(K) to 330(K). The length of both cylinders are $L_G = 2.5\times10^{-3}(m)$.
When study oscillating boundary conditions, the Stokes number β is used to describe the frequency and the period of the oscillation. The relationship between the Stokes number and the circular frequency is written in (12).

$$\beta = \sqrt{\frac{\omega L^2}{\nu_0}} = \sqrt{\frac{\omega(R_2-R_1)^2}{\nu_0}},\ \omega = \frac{\beta^2\nu_0}{(R_2-R_1)^2} \tag{12}$$

where $\nu_0 = \frac{\mu_0}{\rho_0} = C_\mu l_0 V_0 \rho_0 \sqrt{\frac{T_{W_1}}{T_{W_1}}}\frac{1}{\rho_0} = C_\mu l_0 V_0 = 2.4074 \times 10^{-3}(m^2/s)$.
The dimensionless circular frequency is $\bar{\omega} = C_\mu Kn\beta^2$.
We use $\zeta = A_\sigma \frac{l}{L} = A_\sigma \frac{l_0}{L}$, $\zeta = 1.1466Kn$ [10], therefore $A_\sigma = \frac{\zeta L}{l_0} = \frac{1.1466KnL}{l_0} = 1.1466$ and replacing in (9) yields

$$u = u_{wall} - A_\sigma l_0\left(\frac{\partial u}{\partial r} - \frac{u}{r}\right) = u_{wall} - 1.1466 l_0 \frac{\tau_{r\varphi}}{\mu} \tag{13}$$

for the outer cylinder were $u_{wall}=V_2$ and

$$u = u_{wall} + A_\sigma l_0\left(\frac{\partial u}{\partial r} - \frac{u}{r}\right) = u_{wall} + 1.1466 l_0 \frac{\tau_{r\varphi}}{\mu} \tag{14}$$

for the inner cylinder were $u_{wall}=0$.
Due to the undefined value of the stress tensor component $\tau_{r\varphi}$ at the initial moments (zero divided by zero) the slip-coefficient on the walls in the velocity boundary conditions is set by accommodation coefficient 0.63 for argon (recommended by ANSYS-Fluent).

We study three cases for Stokes number:

(A) $\beta = 0.5$, $\omega = 1589.648(rad/s)$, $T_{per} = 0.0039526(s)$

(B) $\beta = 2$, $\omega = 25434.3675(rad/s)$, $T_{per} = 0.00024704(s)$

(C) $\beta = 8$, $\omega = 407363.06(rad/s)$, $T_{per} = 0.00000245(s)$

Where T_{per} is the oscillation period. We use the time steps $\Delta t = T_{per}/200$ in the numerical studies.

Some MEMS are cylindrical tubes with rotating outer cylinders, and the need arises to measure the pressure in these devices. The Pirani wire filament is typically operated in a balanced Wheatstone bridge circuit where one leg of the bridge is the Pirani filament and the other three elements of the bridge circuit balance and temperature compensate. For this reason, the temperature of the filament is kept higher compared to the temperature of the outer cylinder.

The figure 2 shows the rotation velocity vector field at the moment when the outer cylinder stops, and figure 3 – when outer cylinder reaches maximal velocity (absolutely value). The figures 4 and 5 show the rotation velocity when the outer cylinder stops for $\beta = 2$ and $\beta = 8$ respectively and the figures 6 and 7 show the rotation velocity when the outer cylinder reaches maximal velocity (absolutely value) for $\beta = 2$ and $\beta = 8$ respectively. The results for $\beta = 0.5$ and $\beta = 2$ are similar, so they for β=0.5 are not shown in the figures. Two layers are observed in the gas at $\beta = 8$. It appears that the inner layer near the stationary cylinder moves slowly, while the second layer near the outer cylinder is entrained by it. No such stratification is observed in the gas at $\beta = 0.5$ and $\beta = 2$. When the velocity of the outer cylinder reaches its maximum (absolute value) the radial velocity distribution is linear at $\beta = 0.5$ and β= 2. while at $\beta = 8$ the linear character is violated. The temperature and density distributions are the similar in all studied cases (see figures 8 and 9). An interesting result is that at $\beta = 0.5$ and $\beta = 2$ at the outer cylinder reaches the maximal velocity the pressure distribution is not constant, while at $\beta = 8$ the pressure in the entire region is approximately constant (see figures 10 and 11).

The obtained numerical results using the ANSYS system were also verified with DSMC. It should be noted that in the cases considered in the present work, almost no development of the current along the axis of symmetry is noticed, and the results obtained for the middle section and the two end sections do not differ significantly. The comparison of the results with our previous work [22] gives grounds for the following conclusion – the pulsation of the temperature of the inner cylinder (filament) leads to a more noticeable formation of periodic structures in the flow than the mechanical oscillations of the inner cylinder.

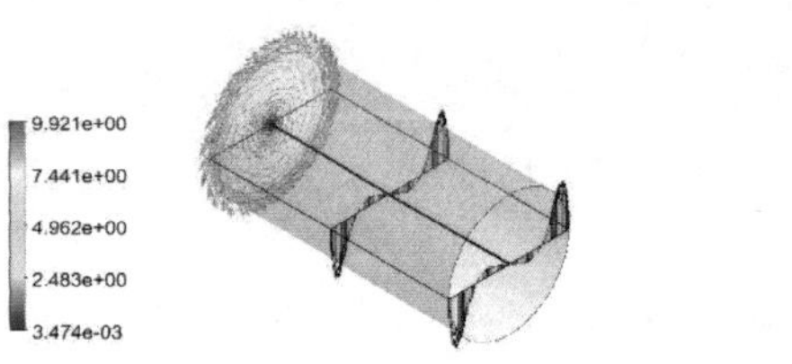

Figure 2. The vector field of the dimension rotation velocity at $\beta = 8$ and $V_2 = 0$.

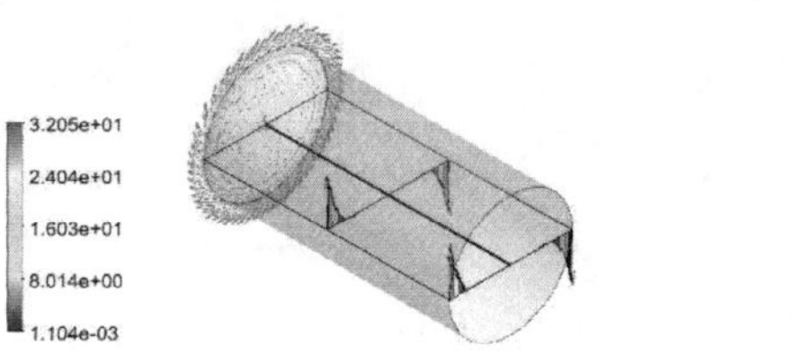

Figure 3. The vector field of the dimension rotation velocity at $\beta = 8$ and $V_2 = -\Delta V$.

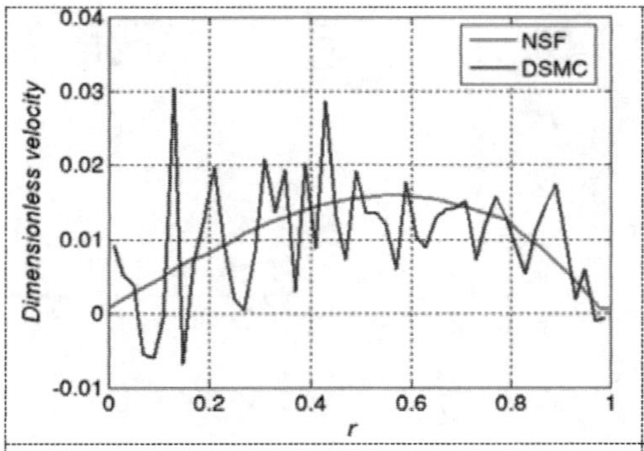

Figure 4. The dimensionless rotation velocity value at $\beta = 2$ and $V_2 = 0$.

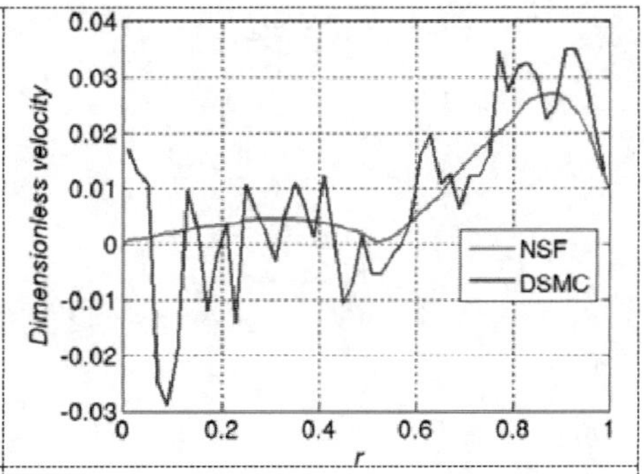

Figure 5. The dimensionless rotation velocity value at $\beta = 8$ and $V_2 = 0$.

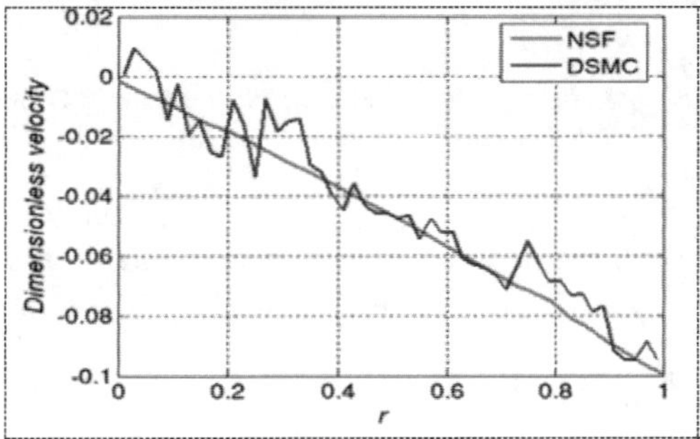

Figure 6. The dimensionless rotation velocity value at $\beta = 2$ and $V_2 = -\Delta V$

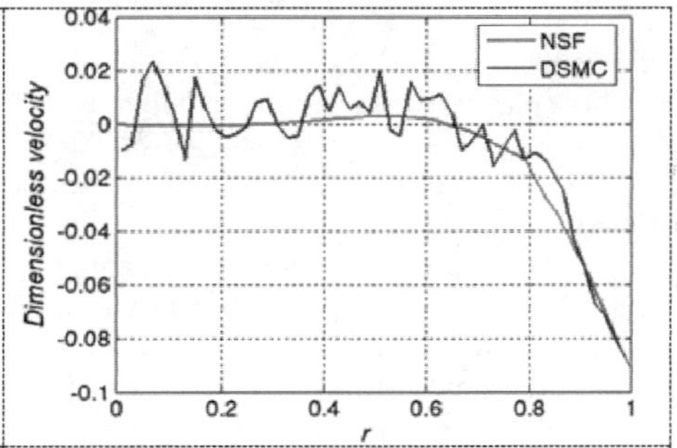

Figure 7. The dimensionless rotation velocity value at $\beta = 8$ and $V_2 = -\Delta V$.

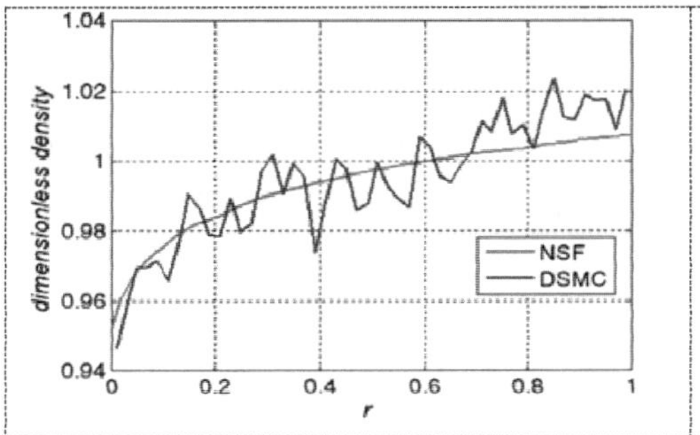

Figure 8. The dimensionless density value at β= 2 and V_2= 0.

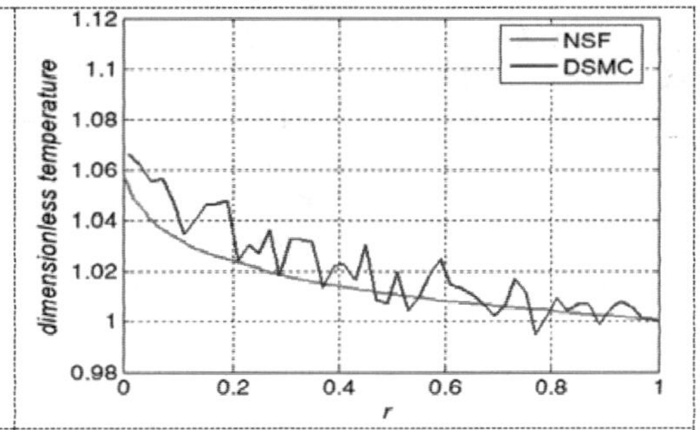

Figure 9. The dimensionless temperature value at $\beta = 2$ and $V_2 = 0$.

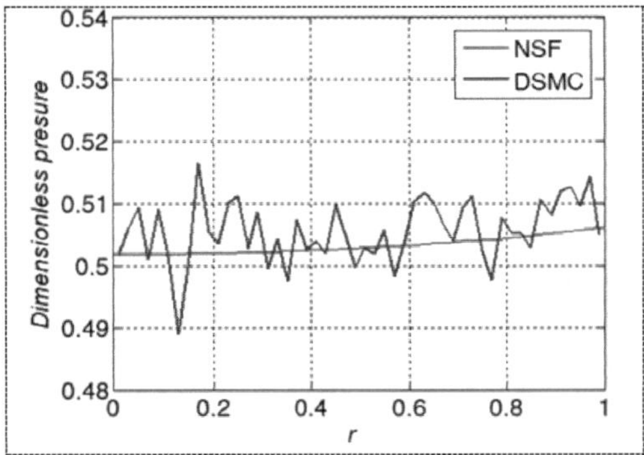

Figure 10. The dimensionless pressure value at $\beta = 2$ and $V_2 = -\Delta V$.

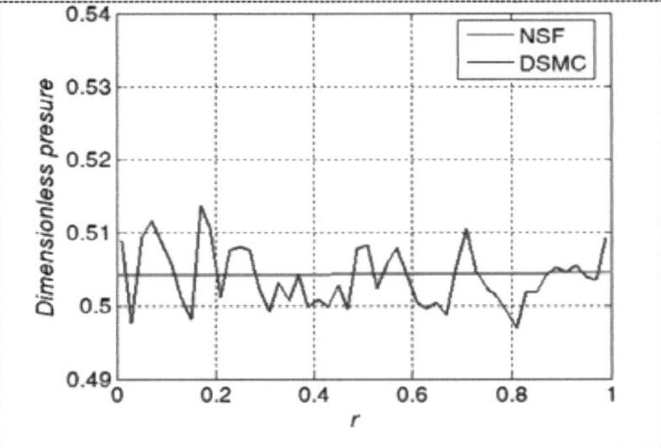

Figure 11. The dimensionless pressure value at $\beta = 8$ and $V_2 = -\Delta V$.

Conclusions

The thermoacoustic waves propagation in a rarefied gas oscillation Taylor-Couette flow between cylinders with finite length is studied on the basis of ANSYS Fluent software and DSMC model, as it should also be noted the very good coincidence of the results obtained by the two methods. The comparison of the results with our previous work [22] gives grounds for the following conclusion – the pulsation of the temperature of the inner cylinder (filament) leads to a more noticeable formation of periodic structures in the flow than the mechanical oscillations of the inner cylinder. The proposed solutions allow to estimate the application limits of various models, and to study the transition processes in some MEMS. The results can be used to resolve the non-stationary wave processes in MEMS some type of Pirani vacuum gauges.

References

[1] Chapman S and Cowling T. 1952 *The Mathematical Theory of Non-Uniform Gases* (Cambridge University Press Cambridge)
[2] Albertoni S, Cercignani C and Gotusso L 1963 Numerical evaluation of the slip coefficient *Phys. Fluids* 6 pp 993–996
[3] Cercignani C and Sernagiotto F 1967 Cylindrical Couette flow of a rarefied gas, *Physics of Fluids* 10 pp 1200–1204
[4] Bird G 1994 *Molecular Gas Dynamics and the Direct Simulation of Gas Flows* (Oxford University Press Oxford)
[5] Cercignani C 2000 *Rarefied Gas Dynamics, From Basic Concepts to Actual Calculations* (Cambridge University Press Cambridge)
[6] Bird R, Stewart W and Lightfoot E 2002 *Transport Phenomena* (John Wiley New York)
[7] Park J, Bahukudumbi P and Beskok A 2004 Rarefaction effects on shear driven oscillatory gas flows: A DSMC study in the entire Knudsen regime, *Phys. Fluids* 16 p 317
[8] Emerson D, Gu Xiao-Jun, Stefanov S, Yuhong S and Barber W 2007 Nonplanar oscillatory shear flow: From the continuum to the free-molecular regime *Phys. Fluids* 19 107105
[9] Sharipov F 2011 Data on the velocity slip and temperature jump on a gas-solid interface *J. Phys. Chem. Ref. Data* 40 023101
[10] Gospodinov P, Roussinov V and Stefanov S 2012 Nonisothermal oscillatory cylindrical Couette gas flow in the slip regime: A computational study *European Journal of Mechanics B/Fluids* 33, pp 68–77
[11] Kalempa D and Sharipov F 2014 Numerical modeling of thermoacoustic waves in a rarefied gas confined between coaxial cylinders, *Vacuum* 109 pp 326–332
[12] Gospodinov P, Rousinov V and Mironova M 2014 Cylindrical nonisothermal oscillatory Couette gas flow in the slip regime: Wall shear stress and energy transfer, numerical

investigation IntellectualArchive.com online publications *Intellectual Archive Bulletin* Feb 19 pp 2–19

[13] Gospodinov P, Dankov D and Roussinov V 2016 Thermo acoustic waves in rarefied gas between two coaxial cylinders at a sudden change of the temperature of the outer cylinder *AMiTaNS'16*, *AIP Conference Proceedings* 1773 (American Institute of Physics Melville NY) 080001 p 7

[14] Gospodinov P, Dankov D, Roussinov V and Stefanov S 2011 Stationary cylindrical Couette flow at different temperature of cylinders: the local Knudsen number effect *AMiTaNS'11 AIP Conference Proceedings* 1404 (American Institute of Physics Melville NY) pp 451–456

[15] Oualli H, Boubdallah A, Hanchi S and Gad-Ei-Hak M 2010 Taylor-Couette flow control using a radially oscillating inner cylinder *Geophysical Research Abstracts EGU General Assembly* 2010 12 EGU2010-1252

[16] Oualli H, Mekadem M, Lebbi M and Bouabdallah A 2016 Taylor-Couette flow control by amplitude variation of the inner cylinder cross-section oscillation *Eur. Phys. J. Appl. Phys.* 71 11102

[17] Lalaoua A and Bouabdallah A 2016 On the Onset of Taylor Vortices in Finite-Length Cavity Subject to a Radial Oscillation Motion *Journal of Applied Fluid Mechanics* vol 9 No. 4 pp 1887-1896

[18] Alkhalidi A *et al.* 2019 Rarefaction and Scale Effects on Heat Transfer Characteristics for Enclosed Rectangular Cavities heated from below *Thermal Science* vol 23 No. 3B pp 1791-1800

[19] Alkhalidi, A., *et al.* 2019 Rarefied flow and heat transfer characteristics of rectangular cavities with heated concave surface *Advances in Mechanical Engineering* vol 11(7) pp 1–14

[20] Eugene I, Kurkin A, Samsonov V and Shakhov V 2017 Simulation of rarefied gas flows in microchannels *Procedia Engineering* 185 pp 160 – 167

[21] Gospodinov P, Dankov D, Roussinov V and Mironova M 2022 Acoustic waves modeling Taylor-Couette flow in rarefied gas at inhomogeneous cylinder wall temperature distribution, *AIP Conference Proceedings* 2522, 080001

[22] Gospodinov P, Dankov D and Mironova M 2023 The continuous and statistical modeling of the Taylor-Couette flow of rarefied viscous gas with suddenly wall temperature change *AMiTaNS'22 AIP Conference Proceedings* 2953 (American Institute of Physics Melville NY) 060004 p 8

Reconstruction coefficient analysis of honeybee collapse due to pesticide contamination

Miglena N. Koleva[1,a], **Lubin G. Vulkov**[2,b]

[1]Department of Mathematics, Faculty of Natural Sciences and Education, University of Ruse, Bulgaria

[2]Department of Applied Mathematics and Statistics, Faculty of Natural Sciences and Education, University of Ruse, Bulgaria

E-mail: [a]`mkoleva@uni-ruse.bg` [b]`lvalkov@uni-ruse.bg`

Abstract. In this paper we consider the inverse problems of identifying space-dependent coefficients of the mortality rate of the bees and the rate of contamination of the forager bees by pesticides. The model is described by a weakly coupled system of two reaction-diffusion equations for the spatial distribution of uncontaminated and contaminated foraging bees. Final time $t = T$ observations of the density of uncontaminant and contaminant forager bees are used. We propose two approaches for studying the problems. The first one uses the overspecified information to transform the problems into non-linear parabolic equations involving the solution values at the final time. This allows us to prove, using fixed-point arguments, existence of solution to the inverse problems. The second study employs the concept of the quasi-solution to establish existence of solution to the inverse problems as minimizers of least-square cost functionals.

Key words: Honeybee population, pesticide contamination, ODE difference equation, inverse problem, quasi-solution, least-square functional, optimal control, adjoint system

1. Introduction

Ecosystem stability and biodiversity greatly rely on the honeybee population, which also plays a vital role in the pollination of bee-dependent crops. Nevertheless, these essential pollinators, including honeybees, are currently facing numerous challenges due to nutritional stress, parasitism, pesticide usage, and the impact of climate change. These factors collectively affect the timing, duration, and variability of seasonal events, see e.g. [1, 2, 4, 8, 12, 18, 19, 25, 26, 28, 30]. Numerous research studies are dedicated to exploring the reasons behind the global decline of the honeybee population, as documented in works like [25, 26, 30]. Among the identified factors are nutritional stress resulting from the scarcity of flowering plants, the presence of roa mites and Nosema, and the impact of diseases like acute bee paralysis/virus mites, which induce paralysis in workers, drones (male bees), larvae, and adults, while sparing the queen [22].

Later, in [23], the authors extended this model by including spatial components to study colony collapse due to pesticides. The presented model operates under the assumption that honeybee colonies exist in an environment with an even food resource gradient in all directions, where the significance of social information conveyed through waggle dancing is diminished. It is posited that a certain proportion of forage bees memorize spatial information from their previous preferred locations and repeatedly return to those areas. On the other hand, the remaining fraction, $1-q$, of bees explore new resources by traveling in random directions. Consequently, the colony's foraging region experiences heterogeneous spatial pesticide contamination, leading to varying fractions of the forager bee population being affected. This model [12, 22, 23] is specifically tailored for application in managed colonies.

The scientific model employed in this study combines both differential and difference equations to characterize the spatial distribution of uncontaminated and contaminated foraging bees. The dynamics of this process are rigorously examined through a comprehensive analysis of ordinary-difference systems of equations, specifically presented as type (8) in Section 2. A honeybee difference equation model that tracks the foragers leaving and entering the hive each day to study the effect of pesticides leading to homing failure is adequately described by this system.

Additionally, the researchers extended their investigation to include considerations regarding the successful return of contaminated foragers to the colony and whether foragers would return to previously visited foraging sites. In this context, the researchers assumed that foragers base their return decisions on their memory, disregarding the waggle dance. Furthermore, they made distinct assumptions regarding the behavior of foragers, distinguishing between two proportions. One proportion is expected to return to the hive and begin anew the following day without relying on memory from the previous day. The other proportion uses their memory from the previous day. The primary focus of the study centers on determining the density of uncontaminated $(u(t,x,y))$ and contaminated forager bees $(c(t,x,y))$ at time t and location $(x,y)\in\mathbb{R}^2$.

In [1], the investigation focuses on studying how local agrometeorological factors impact the fluctuation of beehive weight during the flowering period of sunflowers (Helianthus annuus)

The reasons behind colony losses are attributed to several potential factors, including emerging pathogens and pests, diminished genetic diversity, pesticide usage, scarcity of high-quality food, and alterations in the environment [3, 6, 9, 12]. Notable pathogens and pests known to contribute to colony losses encompass Paenibacillus larvae and Melissococcus plutonius (causing American and European foulbrood, respectively), the parasitic mite Varroa destructor, parasitic microsporidia Nosema species, and various honeybee viruses [4, 12, 19, 20].

The article is organized as follows. We begin with brief formulation of the model of honeybee colony collapse due to pesticide contamination of foraging bees as described in [22], Section 2. Section 3 introduces the inverse problems of identifying space-dependent coefficients of the mortality rate of the bees and the rate of contamination of the forager bees by pesticides. Also, using overspecified information the problems are transformed to non-linear parabolic equations involving the solution values at the final time which provides existence of inverse problems solutions. In Section 4, the quasi-solutions of the inverse problems are studied. Numerical methods play an important role in honeybee population dynamics. The present paper is a theoretical fundament for further developing of honey bee modelling due to pesticide.

2. The model description

In this section, we introduce the spatial model of honeybee colony collapse due to pesticide contamination of foraging bees [23].

2.1. Spatial distribution of forager bees

The models presented in [12, 22, 23] do not incorporate worker bees within the hive, despite them constituting the majority of the bee population and playing a crucial role in caring for juvenile bees. These worker bees stay inside the hive and, therefore, do not become contaminated by coming into contact with pesticides present outside the hive. It is defined $G(\tau,x,y)$, $(x,y)\in\mathbb{R}^2_+$, to be the spacial probability density function of forager bees in the hive, where τ is a given value.

Thus, $G(\tau, x, y)$ satisfies

$$\iint_{R^2} G(\tau, x, y)dxdy = 1, \quad \text{for } \tau > 0.$$

In practice, G is often chosen as the Gaussian function

$$G(\tau, x, y) = \frac{1}{4\pi\varepsilon\tau} \exp\left(-\frac{(x-x_0)^2 + (y-y_0)^2}{4\varepsilon\tau}\right),$$

where (x_0, y_0) is center of the hive and $\sigma = \sqrt{2\varepsilon\tau}$ is the standard derivation of G for a given value of ε and τ. Because the hive is significantly smaller compared to the foraging region, the choice of a Gaussian distribution is suitable. The selection of values for ε and τ is such that G represents a minute region in comparison to the foraging area of forager bees.

Since the foraging region is defined as the entire R^2 space, it is implicit that forager bees stay in proximity to the hive and avoid the boundaries of the foraging region, in accordance with the spatial parameters of the model.

2.2. Model without pesticide contamination

We start with the assuming that in the environment there is no pesticide. As part of their routine, forager bees depart from the hive at sunrise every day and come back to the hive at nightfall.

Denote the density of forager bees by $u(t, x, y)$ at time t and location $(x, y) \in \mathbb{R}^2$. The forager bees' initial distribution, set at the beginning of the first day is

$$u(0, x, y) = u_0(x, y) \geq 0, \quad u_0 \in L^1_+(R^2) \tag{1}$$

and

$$U_0 = \iint_{R^2} u_0(x, y)dxdy,$$

denotes the total population of forager bees at the initial time $t = 0$. Throughout the initial day, forager bees disperse randomly in the R^2 space, leading to the fulfillment of the equation for u at each time $t \in [0, 1)$:

$$\frac{\partial u(t, x, y)}{\partial t} = d\triangle u(t, x, y) - \mu(x, y)u(t, x, y), \quad (x, y) \in \mathbb{R}^2, \tag{2}$$

where

$$U(t) = \iint_{R^2} u(t, x, y)dxdy, \tag{3}$$

stands for the count of forager bees at time $t \in [0, 1)$, $\mu(x, y) \geq 0$ is the mortality rate of the bees, and d is the diffusion rate. The rate of mortality $\mu(x, y)$ includes homing failure and all others factors of forager bees mortality.

At the termination of the first day, the assumption is made that all forager bees not affected by mortality return to their hive. As a result, according to equation (3), The total population of forager bees by the end of the initial day is expressed by:

$$U(1^-) = \iint_{R^2} u(1^-, x, y)dxdy.$$

The forager bees, which have returned to the hive on the first day, play an essential role in caring for the juvenile bees inside the hive

The Alice functional is utilized to characterize their involvement in the generation of forager bees, which ultimately depart from the hive to engage in foraging activities:

$$\frac{\beta U(1^-)^2}{\mathcal{X}^2 + U(1^-)^2}, \tag{4}$$

where β is the maximal production parameter and $\mathcal{X}$ is sigmoidal Hill function production [28].

Forager bees exhibit two distinct types of behavior as they return to the hive at the end of the day and embark on foraging activities the following day.

Some forager bees will return to the hive and begin anew the next day, without any recollection of their previous day's location. Their foraging behavior on the second day is simulated using diffusion.

By integrating these two behavioral patterns and assuming that forager bees disperse within the hive according to a Gaussian probability density function $G(\tau, x, y)$, one derives $u_1(x, y)$, which refers to the distribution of forager bees *at the start of the second day, using* (4), *one obtains*:

$$u_1(x,y) = G(\tau, x, y)\left(\frac{\beta U(1^-)^2}{\mathcal{X}^2 + U(1^-)^2} + (1-q)U(1^-)\right) + qu(1^-, x, y),$$

where $q \in [0, 1]$ represents the proportion of forager bees that exhibit the second behavior type. The fraction of forager bees that conform to this second type of behavior includes individuals who have the capacity to remember a beneficial foraging spot from the previous day. At the onset of the following day, these bees will proceed directly to these locations.

2.3. Model with pesticide contamination

Considering time t and location $(x, y) \in R^2$, let $u(t, x, y)$ represent the density of uncontaminated forager bees, and $c(t, x, y)$ denote the density of contaminated forager bees. The initial distribution at the beginning of day one is as follows:

$$u(0, x, y) = u_0(x, y), \quad C(0, x, y) = C_0(x, y),$$

where $u_0, c_0 \in L^1_+(R^2)$. Throughout the first day, the spatial behavior and movements of the forager bee population is descibed for $t \in [0, 1)$ by

$$\frac{\partial u}{\partial t} = d\triangle u - \mu(x, y)u - \alpha(x, y)u, \tag{5}$$

$$\frac{\partial c}{\partial t} = d\triangle c - \mu(x, y)c + \alpha(x, y)u, \tag{6}$$

where $d > 0$ represents the diffusion rate, $\mu(x, y)$ stands for the mortality rate of the bees, and $\alpha(x, y)$ denotes the contamination level of forager bees by pesticides. Further, we will solve this system with zero Dirichlet boundary conditions.

We assume that α and μ are bounded continuous nonnegative functions on R^2.

At the end of the day the number of uncontaminated and contaminated bees are respectively

$$U(1^-) = \iint\limits_{R^2} u(1^-, x, y)dxdy \text{ and } C(1^-) = \iint\limits_{R^2} c(1^-, x, y)dxdy.$$

In this model is considered that apart apart from the normal homing failure of all foraging bees accounted for in the mortality rate μ, an additional fraction of contaminated forager bees,

$1-p$, also fails to return home at the end of each day. By incorporating this mechanism with the previous one, we arrive at the following model

$$u_1(x,y) = G(\tau,x,y)[\beta + (1-q)U(1^-)] + qu(1^-,x,y)$$

$$C_1(x,y) = G(\tau,x,y)[(1-q)PC(1^-)] + qpc(1^-,x,y),$$

$$B = \frac{\beta(U(1^-)+pC(1^-))^2}{\mathcal{X}^2 + (U(1^-)+pC(1^-))^2}. \tag{7}$$

On the second day, we substitute the initial values $u(0,x,y) = u_0(x,y)$ and $c(0,x,y) = c_0(x,y)$ with $u(1,x,y) = u_1(x,y)$ and $c(1,x,y) = c_1(x,y)$, and then proceed to solve (5), (6) using this revised initial condition on the time interval $t \in [1,2]$. The identical procedure, employing equations such as (7) for the second day, extends to time intervals $[2,3]$, $[3,4]$.

Let $Y = L^1_+(\Omega) \times L^1_+(\Omega)$ under the metric generated by the $L^1 \times L^1$ norm. Define continuous mapping $S : Y \longrightarrow Y$ by

$$S(u_0, c_0) = (u_1, c_1), (u_0, c_0) \in Y_0$$

For studying the model's dynamics, it is essential to explore the solutions of the subsequent *spatial difference equation*:

$$(u_{n+1}, c_{n+1}) = S(u_n, c_n), \quad n = 0, 1, \dots, \quad (u_0, c_0) \in Y_0. \tag{8}$$

3. Inverse problems

In this section, we formulate two inverse problems of identifying reaction space-dependent unknown coefficients in a two-dimensional parabolic equation and in a system of two parabolic equations with overspecified conditions at the final time $t = T$.

In the recent decades, many real phenomena and processes were studied by inverse problems due to their strong mathematical formulation and theoretical study as well as efficient numerical treatment [11, 13, 14, 15, 16, 21, 24]. In the present paper we continue our research on inverse problems modelling the honeybee population dynamics [5, 6, 7].

Let us consider the continuous versions (2) and (5), (6) of the difference equation (10) at final time observation

$$u(T,x,y) = \varphi(x,y), \quad \varphi(x,y) > 0, \quad (x,y) \in \Omega, \tag{9}$$
$$c(T,x,y) = \psi(x,y), \quad \psi(x,y) > 0, \quad (x,y) \in \Omega, \tag{10}$$

where $T = 1$ (first day), $T = 2$ (second day), etc.

Very often, although the fast development of the electronic technology, the beekeepers are not able to provide sufficient observation data for modeling the forager bees behavior concerning food fields with or not pesticide contamination. Now, using the observations (9)-(10) we turn to find $u(t,x,y)$ and $c(t,x,y)$ approximately, solving two inverse problems of the models without and with contamination.

P1. Identification of $\mu(\mathbf{x},\mathbf{y})$ in the model without contamination

The system (5), (6) with known coefficient d, μ, α and with appropriate initial and boundary conditions is called direct problem.

Here we are concentrated on the case of finding $\mu(x,y)$ and $u(t,x,y)$ that satisfy the equation (2) and observation (9)

Let $v(t,x,y) = \frac{\partial u}{\partial t}(t,x,y)$ for $(t,x,y) \in \overline{Q}_T$. We differentiate equation (2) with respect to t to obtain

$$\frac{\partial v}{\partial t} = d\triangle v - \mu(x,y)v = 0. \tag{11}$$

By the condition (9) we see that

$$\mu(x,y) = \frac{-\frac{\partial u}{\partial t}(T,x,y) + d\triangle\varphi(x,y)}{\varphi(x,y)} = \frac{-v(T,x,y) + d\triangle\varphi(x,y)}{\varphi(x,y)}. \tag{12}$$

Thus, the problem (1), (2) and (9) is now equivalent to the following:

$$\frac{\partial v}{\partial t} = d\triangle v + \frac{-v(T,x,y) - d\triangle\varphi(x,y)}{\varphi(x,y)} v, \tag{13}$$

$$v(0,x,y) = d\triangle u_0(x,y) + \frac{v(T,x,y) - d\triangle\varphi(x,y)}{\varphi(x,y)} u_0(x,y), \tag{14}$$

with zero Dirichlet boundary conditions.

Theorem 1 *Let assume that $\varphi(x,y) \in C^{4+\gamma}(\Omega)$, $0 < \gamma < 1$ and $\varphi(x,y) > 0$. Then, there exists at least one solution of the problem* (11)-(14).

The proof is based on the fixed point theory and maximum principle, following the results in [10, 17, 27, 29].

Corollary 1 *Under the assumptions of Theorem 1, the inverse problem* (2), (9) *has at least one solution.*

P2. Identification of $\mu(\mathbf{x},\mathbf{y})$ and $\alpha(\mathbf{x},\mathbf{y})$ in the model with contamination

As before, the direct problem is the system (5), (6) with known coefficient d, μ, α and siutable initial and boundary conditions. Then, using the theory of PDEs, developed in [24, 31], we can obtain existence and uniqueness of classical suficiently smooth solution of the initial boundary value problem for the system (5), (6).

Now we consider the case of finding $\mu(x,y)$, $\alpha(x,y)$ and $u(t,x,y)$, $c(t,x,y)$ that satisfy the system (5), (6) at observation (9) and (10). Let

$$v(t,x,y) = \frac{\partial u}{\partial t}(t,x,y), \quad w(t,x,y) = \frac{\partial c}{\partial t}(t,x,y) \ \text{ for } \ (t,x,y) \in Q_T.$$

We differentiate (5) and (6) with respect to t to obtain :

$$\frac{\partial v}{\partial t} = d\triangle v - \mu(x,y)v - \alpha(x,y)v, \tag{15}$$

$$\frac{\partial w}{\partial t} = d\triangle w - \mu(x,y)w + \alpha(x,y)v. \tag{16}$$

By the conditions (9), (10) and assumption $\varphi(x,y) > 0$ of the Theorem 1, we observe that

$$\mu(x,y) = -\frac{w(T,x,y) + v(T,x,y)}{\varphi + \psi} + d\frac{\triangle\varphi + \triangle\psi}{\varphi + \psi}, \tag{17}$$

$$\alpha(x,y) = \frac{d\triangle\varphi - v(T,x,y)}{\varphi} - \mu(x,y). \tag{18}$$

Thus, the problem (5), (6), (9), (10) is now equivalent to the equations (15), (16) with coefficients (17), (18), zero Dirichlet boundary conditions and initial conditions:

$$v(0,x,y) = \alpha\triangle u_0(x,y) - \mu(x,y)u_0(x,y) - \alpha(x,y)u_0(x,y),$$
$$w(0,x,y) = \alpha\triangle c_0(x,y) - \mu(x,y)c_0(x,y) - \alpha(x,y)c_0(x,y).$$

Similar to Theorem 1 results hold for this initial-boundary value problem.

4. The quasi-solution

In this section, we implement to the inverse final time measured output problem of the system (5), (6) the theory of quasi-solution [11, 13, 14, 15, 16, 21, 24]. This notion of quasi-solution was first proposed in [15].

Let us assume that the coefficients μ, α and the initial data u_0, c_0 satisfy:

$$\mu(x,y), \quad \alpha(x,y) \in C^{\gamma}(\overline{\Omega}) \text{ and } u_0(x,y), \quad c_0(x,y) \in C^{2,\gamma}(\Omega),$$

We define admissible set

$$\mathcal{A} = \left\{\mu(x,y), \alpha(x,y) : 0 < \mu_0 \le \mu \le \mu_1, \ 0 < \alpha_0 \le \alpha \le \alpha_1, \ \nabla\mu, \nabla\alpha \in L^2(\Omega)\right\}.$$

The inverse problem is reduced to optimization of the cost functional: find the pair $(\widehat{\mu}(x,y), \widehat{\alpha}(x,y)) \in \mathcal{A}$, satisfying

$$J(\widehat{\mu}, \widehat{\alpha}) = \min_{\mu,\alpha \in \mathcal{A}} \mathcal{F}(\mu, \alpha),$$

where

$$\begin{aligned} J(\mu,\alpha) \ &= \ \frac{1}{2}\int\limits_{\Omega} \left(\left(u(x,y,T;\mu,\alpha) - \varphi(x,y)\right)^2 + \left(c(x,y,T;\mu,\alpha) - \alpha(x,y)\right)^2 \right) dxdy \\ &+ \frac{R}{2}\int\limits_{\Omega} (|\nabla\mu|^2 + |\nabla\alpha|) dxdy, \end{aligned}$$

and (u, c) is the solution of the system (5), (6) for the given coefficients $\mu(x, y)$, $\alpha(x, y) \in \Omega$. The constants μ_0, μ_1 and α_0, α_1 are given and R is the regularity parameter.

The triple $\{(u, c), \mu, \alpha\}$ is called *quasi-solution of the inverse problem.*

Using the Shauder's theory for the parabolic equations and maximum principle, see e.g. [24, 31], one can prove the following existence result.

Theorem 2 *Let* $0 < \gamma < 1$ *and the coefficients* $\mu(x,y), \alpha(x,y) \in C^{\gamma}(\overline{\Omega})$. *Then, the system* (5), (6) *has a unique solution* $u(x,t), c(x,t) \in C^{2+\gamma,1+\gamma/2}(\overline{Q}_T)$, $Q_T = \Omega \times (0,T)$.

The following theorem provides the existence of the minimizer $(\widehat{\mu}(x,y), \widehat{\ }(x,y)) \in \Omega$ of the cost functional $J(\mu, \alpha)$.

Theorem 3 *Suppose* (u, c) *is the solution of the system* (5), (6). *Then, there exists a minimizer* $(\widehat{\mu}(x,y), \widehat{\alpha}(x,y)) \in \mathcal{A}$, *such that*

$$J(\widehat{\mu}, \widehat{\alpha}) = \min_{\mu,\alpha \in \mathcal{A}} J(\mu, \alpha).$$

The proof uses techniques of the books [13, 14, 21].

In the next theorem, applying the adjoint equation method, see e.g. [13, 14, 21], we derive necessary optimal conditions that satisfy the optimal pair (μ, α). Let the pair (U, C) is the solution of the adjoint system, associated with (5), (6)

$$\begin{aligned} &\frac{\partial U}{\partial t} + \triangle U - \mu(x,y)U - \alpha(x,y)V = 0, \\ &\frac{\partial C}{\partial t} + \triangle C - \mu(x,y)C - \alpha(x,y)U = 0, \\ &U(x,y,T) = u(x,y,T) - \varphi(x,y), \\ &C(x,y,T) = c(x,y,T) - \psi(x,y), \\ &U\Big|_{\partial\Omega} = 0, \ \ C\Big|_{\partial\Omega} = 0, \ \ t \in (0,T). \end{aligned}$$

Theorem 4 *Let (μ, α) be the quasi-solution of the inverse problem* (5), (6), (9), (10)*. Then, there exists quadruplet $(u, c, U, C; \mu, \alpha)$, satisfying*

$$\int_{Q_T} (Uu(\mu - \varphi) + Cc(\alpha - \psi))\, dxdt + R \int_{\partial\Omega} (\nabla\mu.\nabla(\varphi - \alpha) + \nabla\alpha.\nabla(\psi - \alpha))\, dxdy \geq 0,$$

for every $\varphi, \psi \in \mathcal{A}$.

The proof uses techniques from [14].

5. Conclusions

In this paper we considered two reaction coefficient inverse problems, associated with one two-dimensional reaction diffusion equation and a system of two weakly coupled parabolic equations. Following the honeybee modelling, a final-time measurements are posed.

The first approach is based on the reducing the inverse problem to solving nonlocal parabolic ones. The existence of the solution to the nonlocal problems is established. From here follows directly the solvability of the corresponding inverse problems. The second approach is based on the notion of quasi-solutions that is minimizer of a least-square functional. Using optimal control framework, the existence of the minimizer is established.

The results obtained are mainly theoretical and they can serve as a fundament for construction of effective numerical method. For example, obtaining results for the numerical treatment of the inverse problems IP1 and IP2 would be very interesting.

Acknowledgments: This work is supported by the Bulgarian National Science Fund under the Project KP-06-N 62/3 "Numerical methods for inverse problems in evolutionary differential equations with applications to mathematical finance, heat-mass transfer, honeybee population and environmental pollution", 2022.

References

[1] Atanasov A. Hristakov I. Dochev V. 2023 Monitoring agrometeorological factors and beehive weight during sunflower (Helianthus annuus) flowering, 22-th International Scientific Conference Engineering for Rural Development DOI: 10.22616/ERDev.2023.22.TF011
[2] Atanasov A. Georgiev I. Hristakov I. Hristov P. 2022 Application of mathematical model for apiaries location evaluation, *Engineering for rural development, association* 187–193
[3] Atanasov A. Georgiev I. 2021 A multicriteria model for optimal locationof honey bee colonies in regions without overpopulation *AIP Conference Proceedings* **2333** 090008
[4] Atanasov A.Z. Georgiev S.G. Vulkov L.G. 2021 Reconstruction analysis of honeybee colony collapse disorder modeling *Optim. Eng.* **22** 2481–2503
[5] Atanasov A.Z. Koleva M.N. Vulkov L.G. 2023 Numerical optimization identification of a Keller-Segel model for thermoregulation in honey bee colonies in winter *Communications in Computer and Information Science* **1761** Springer Cham 279–293
[6] Atanasov A.Z. Koleva M.N. Vulkov L.G. 2023 Parameter estimation inspired by temperature measurements for a chemotactic model of honeybee thermoregulation *Lecture Notes in Computer Science* **13858** Springer Cham 36–47.
[7] Atanasov A.Z. Koleva M.N. Vulkov L.G. 2023 Numerical analysis of thermoregulation in honey bee colonies in winter based on sign-changing chemotactic coefficient model *Springer Proceedings in Mathematics & Statistics* **412** 269–279
[8] Bagheri S. Mirzaie M. 2019 A mathematical model of honey bee colony dynamics to predict the effect of pollen on colony failure. *PLOS ONE* **14**(11) e0225632
[9] Cao K. Lesnic D 2018 Reconstruction of the perfusion coefficient from temperature measurements using the conjugate gradient method. *Int. J. Comput. Math.* **95**(4) 797–814
[10] Chadam J.M. H.-M. Yin 1990 Determination of an unknown function in a parabolic equation with an overspecified condition *Math Meth Appl Sci* **13** 420–430
[11] Chavent G. 2009 *Nonlinear Least Squares Problems: Theoretical Foundation and Step-by Guide for Applications* Springer Cham

[12] Chen J. DeGrandi-Hoffman G. Ratti V. Kang Y. 2021 Review on mathematical modeling of honeybee population dynamics *Math Biosci Eng* **18**(6) 9606–9650

[13] Hasanov A. Romanov V. 2017 *Introduction to the Inverse Problems for Differential Equations* Springer New York https://doi.org/10.1007/978-3-319-62797-7

[14] Isakov V. 2017 *Inverse Problems for Partial Differential Equations*, Springer Cham, 3-rd ed., p. 406

[15] Ivanov V.K. Vasin V.V. Tanana V.P. 1978 *Theory of Linear Ill-Posed Problems and its Approximations*, Nauka, Moscow (in Russian)

[16] Kabanikhin S.I. 2011 *Inverse and Ill-Posed Problems*, DeGruyer, Berlin

[17] Kamynin V.L. 2003 On the unique solvability of an inverse problem for parabolic equations under a final overdetermination condition *Math Notes* **73**(2) 202–211.

[18] Khoury D.S. Myerscough M.R. Barron, A.B. 2011 A quantative of honey bee colony population dynamics, *PLOS ONE* **6**(4) e 1849

[19] Kribs-Zaleta C.M. Mitchell C. 2014 Modeling colony collapse disorder in honeybees as a contagion, *Math. Biosci. Engn.* **11**6) 1275–1294

[20] Laomettachit T. Liangruksa M. Termsaithong T. Tangthanawatsakul A. Duangphakdee O. 2021 A model of infection in honeybee colonies with social immunity *PLOS ONE* https://doi.org/10.1371/journal.pone.0247294

[21] Lesnic D. 2020 *Inverse Problems with Applications in Science and Engineering* CRC Press London

[22] Magal P. Webb G.F. Wu Y. 2019 Environmental model of honey bee colony collapse due to pesticide contamination. B*ull Math Biol* **81** 4908–4931

[23] Magal P. Webb G.F. 2020 A spatial model of honeybee colony collapse due to pesticide contamination of foraging bees, *J. Math. Biol.* **80** 2363–2393

[24] Marchuk G.I. Shaidurov V.V. 1983 *Difference Methods and their Extrapolations* Springer New York 334 p

[25] Oldroyd B.P. 2007 What's killing American honey bees? *PLOS Biology* **5**(6) 2168

[26] Perry C. J. Sovik E. Myerscough M.R. Barron A.B. 2015 Rapid behavioral maturation accelerates failure of stressed honey bee colonies *Proc. Nat. Academy of Sci.* **112**(11) 3427–3432

[27] Prilepko A.I. Kostin A.B. Solov'ev V.V. 2019 Inverse source and coefficient problems for elliptic and parabolic equations in Hölder and Sobolev Spaces *J Math Sci* **237** 576–594

[28] Ratti V. Kevan P.G. Eberl H.J. 2017 A mathematical model of forager loss in honeybee colonies infested with Varroa destructor and accute bee paralysis virus *Bull. Math. Biol.* **79**(6) 1218–1253

[29] Romanov V.G., Hasanov A.H.2022 Uniquiness and stability analysis of final data inverse source problems for evolution equations *Journal of Inverse and Ill-posed Problems* **30**(3) 425–446

[30] Smith K.M. Loh E.H. Rostal M.K. Zambrana-Torrelio C.M. Mendiola L. Daszak P. 2013 Pathogens, pests, and economics: drivers of honey bee colony declines and losses *Ecohealth* **10**(4) 434–45

[31] Wu Z. Yin J. Wang C. 2006 *Elliptic and Parabolic Equations*, World Scientific London.

Reconstruction of time-dependent right-hand side in parabolic equations on disjoint domains

Miglena N. Koleva[1,a], **Lubin G. Vulkov**[2,b]

[1]Department of Mathematics, Faculty of Natural Sciences and Education, University of Ruse, Bulgaria

[2]Department of Applied Mathematics and Statistics, Faculty of Natural Sciences and Education, University of Ruse, Bulgaria

E-mail: [a]`mkoleva@uni-ruse.bg` [b]`lvalkov@uni-ruse.bg`

Abstract. We study inverse problems of reconstructing the time-dependent right-hand side from point observation in a one dimensional parabolic equation on disjoint intervals. These problems are ill-possed, i.e. very slight errors in the additional input may cause relatively significant errors in the output of the left and right internal right-hand side. In this work, we construct computational algorithms, using the loaded equation method. First, we perform a decomposition with respect to the unknown source of the inverse problem solutions. Then the inverse problems are reduced to a loaded parabolic equation problems. The well-posedness of the inverse problems is studied on the base of loaded equation ones. The numerical performance of the approach is realized by finite difference schemes, solved with decomposition algorithms. Computational experiments show the efficiency of the method.

1. Introduction

A large number of processes in mechanics, physics, heat-mass transfer, biology and mathematical finance can be formulated as inverse differential equations problems, see e.g. [17, 21, 22]. The inverse problems also have a fundamental mathematical significance. An important question is the problem property and to find the conditions that provide the well-posedness.

For problems with unknown coefficients and right-hand side of equations, a number of results are obtained, see e.g. [10, 11, 12, 13, 17, 20, 21, 22]. These are challenging and interesting inverse problems, which has many applications in various fields of physics and mechanics.

In numerous applications, there are layers characterized by material properties that notably contrast with the surrounding medium. These layers serve different purposes such as providing structural support, thermal insulation, or fulfilling electromagnetic and optical functions, etc. In numerous applications, transmission problems of a similar nature emerge, involving solutions defined on two or more disconnected domains. This scenario occurs when the solution within the intermediate region is either already known or can be derived through a simpler equation. The impact of this intermediate region can be simulated using nonlocal jump conditions across the boundary (layer), see e.g. [1, 4, 6, 18]. Numerical methods for such type interface problems are studied in [7, 8, 9, 16]

Differential equations, involving values of the unknown function at domain points very often are called *loaded differential equations*, see e.g. [2, 3, 19, 21]. In this study, we consider source inverse problem for parabolic equations on disjoint domains with point overdetermination. We construct numerical approach, based on loaded equation method.

The subsequent part of the paper is is structured as follows. In the following section, depending on the location of the measurements, we present three inverse problems. Additionally, we develop the loaded equation method for each of these inverse problems. In Section 4, we construct numerical discretizations and discuss the implementation of the methods. Computational results are presented in Section 5. Finally, we provide some concluding remarks.

2. Formulation of the problem

Let $\Omega_i = (a_i, b_i) \times (0, T]$, where $-\infty < a_1 < b_1 < a_2 < b_2 < +\infty$. We consider the following problem [8, 9]

$$\frac{\partial u_i}{\partial t} = \frac{\partial}{\partial x}\left(p_i(x)\frac{\partial u_i}{\partial x}\right) + f_i(x,t), \quad (x,t) \in \Omega_i, \quad i = 1, 2, \tag{1}$$

with interface conditions

$$p_1(b_1)\frac{\partial u_1}{\partial x}(b_1,t) + \alpha_1 u_1(b_1,t) = \beta_1 u_2(a_2,t) + \gamma_1(t), \quad t \in (0,T), \tag{2}$$

$$-p_2(a_2)\frac{\partial u_2}{\partial x}(a_2,t) + \alpha_2 u_2(a_2,t) = \beta_2 u_1(b_1,t) + \gamma_2(t), \quad t \in (0,T), \tag{3}$$

and external boundary conditions and initial conditions

$$u_1(a_1,t) = \varphi_1(t), \quad t \in (0,T), \tag{4}$$

$$u_2(b_2,t) = \varphi_2(t), \quad t \in (0,T), \tag{5}$$

$$u_i(x,0) = u_i^o(x), \quad i = 1, 2. \tag{6}$$

We assume that [8, 9]

$$\begin{aligned} &p_j \in L^\infty(a_j,b_j), \quad \gamma_j(t) \in L^\infty(0,T], \quad f_j(x,t) \in L^\infty(a_j,b_j), \\ &p_j(x) \geq p_{j_0} > 0, \quad \beta_j > 0, \quad \alpha_j > 0, \quad j = 1,2, \quad \beta_1\beta_2 < \alpha_1\alpha_2. \end{aligned} \tag{7}$$

The problem (1)-(7) with known initial and boundary conditions, coefficients, right-hand sides is the *direct problem.*

We consider the inverse problem for the system (1)-(7), in which, in addition to $u_1(x,t)$ and $u_2(x,t)$, the right-hand side $f_1(x,t)$ or $f_2(x,t)$, or both are also unknown. We assume that the function $f_i(x,t)$, $i = 1, 2$ can be represented as

$$f_i(x,t) = \eta_i(t)\psi_i(x), \quad j = 1, 2, \tag{8}$$

where function $\psi_i(x)$, $i = 1, 2$ is given, and the unknown is the time-dependent source $\eta_i(t)$, $i = 1, 2$. The relations can be reconstructed from additional observations of $u_i(x,t)$, $i = 1, 2$ made at some internal points $a_i < x_i^* < b_i$, $i = 1, 2$:

$$u_i(x_i^*, t) = \theta_i(t), \quad 0 \leq t \leq T. \tag{9}$$

Therefore, depending on the observation data, the inverse problem (IP) for identifying the time-dependent right-hand side can be formulated as follows:

IP[1] Given observation $u_1(x_1^*,t) = \theta_1(t)$, $x_1^* \in (a_1,b_1)$. Find (u_1, u_2, η_1);

IP[2] Given observation $u_1(x_2^*,t) = \theta_2(t)$, $x_2^* \in (a_2,b_2)$. Find (u_1, u_2, η_2);

IP[3] Given observation $u_j(x_i^*,t) = \theta_i(t)$, $x_i^* \in (a_i,b_i)$. Find $(u_1, u_2, \eta_1, \eta_2)$.

There exists a large number of investigations considering identification of time dependent source, see e. g. [5, 10, 11, 12, 17, 20, 21], but results for interface problems of type (1)-(6) are scare. In our previous work [14], we reconstruct the right-hand side in the time-fractional generalization of (1)-(6) on disjoint domains, using decomposition technique. In this work the IP is reduced to the loaded parabolic equation [19, 21].

3. IBVPs for the loaded equation

In this section, we implement the loaded equation method (LEM), proposed in [21] for approximate solution of the time-dependent right-hand side in a linear one-dimensional heat equation.

3.1. IP[1]

We describe more in details the LEM for the problem IP[1]. We assume that the function $f_1(x,t)$ is represented in the form (8) and point observations (9) at $i = 1$ are given. We solve the IP[1] under the following assumptions for $\psi_1(x)$:

A1.1 $\psi_1(x_1^*) \neq 0$ and $\psi_1 \in C^2[a_1, b_1]$;

A1.2 $\psi_1(a_1) = \psi_1(b_1) = 0$, $\frac{\partial \psi_1}{\partial x}(b_1) = 0$.

We seek the solution of IP[1] in the form

$$u_1(x,t) = z_1(t)\psi_1(x) + w_1(x,t), \tag{10}$$

where

$$z_1(t) = \int_0^t \eta_1(s)ds. \tag{11}$$

Substituting (10), (11) into (1) for $i = 1$ yields the following equation for $w_1(x,t)$

$$\frac{\partial w_1}{\partial t} = \frac{\partial}{\partial x}\left(p_1(x)\frac{\partial w_1}{\partial x}\right) + z_1(t)\frac{\partial}{\partial x}\left(p_1(x)\frac{\partial \psi_1}{\partial x}\right), \quad (x,t) \in \Omega_1. \tag{12}$$

In view of (10), the condition (9) leads to the following representation for the unknown function $z_1(t)$

$$z_1(t) = \frac{1}{\psi_1(x_1^*)}\left(\theta_1(t) - w_1(x_1^*, t)\right). \tag{13}$$

Substituting (13) into (12) yields the so called loaded parabolic equation

$$\frac{\partial w_1}{\partial t} = \frac{\partial}{\partial x}\left(p_1(x)\frac{\partial w_1}{\partial x}\right) + \frac{\theta_1(t) - w_1(x_1^*, t)}{\psi_1(x_1^*)}\frac{\partial}{\partial x}\left(p_1(x)\frac{\partial \psi_1}{\partial x},\right), \quad (x,t) \in \Omega_1. \tag{14}$$

Using the assumptions A1.1, A1.2 and (10), the boundary and initial conditions for $w_1(x,t)$ take the form

$$\begin{aligned} &w_1(a_1,t) = \varphi_1(t), \quad t \in (0,T), \\ &p_1(b_1)\frac{\partial w_1}{\partial x}(b_1,t) + \alpha_1 w_1(b_1,t) = \beta_1 u_2(a_2,t) + \gamma_1(t), \quad t \in (0,T), \\ &w_1(x,0) = u_1^o(x), \quad x \in [a_1,b_1]. \end{aligned} \tag{15}$$

Thus, the IP[1] is reduced to the IBVP (12)-(15), combined with (1), $i = 2$, (3) and (5), (6). So, this inverse problem belongs to the class of well-posed partial differential equations problems, see e.g. [10, 11, 12, 17, 21]. Despite of the special structure of the equation (14), it can be investigated for existence, uniqueness and continuous dependence from input data of the BVP (1)-(6) and (7), following the methodology in [9].

Theorem 1 (IP[1]) *Assume that the conditions* (7) *and A1.1 with* $L^1[0,1]$ *hold. Then, the problem IP[1] has a unique weak solution* $(w_1, u_2) \in H^{1,0} \equiv L_2(0,T;H^1)$ *and it satisfies a maximum principle in the form:*

$$\textit{If} \;\; w_1(x,0) \geq 0, \;\; u_1(x,0) \geq 0, \;\; \textit{then} \;\; u_i(x,t) \geq 0 \;\; \textit{a.e. in} \;\; \Omega_1 \cup \Omega_2, \;\; i = 1,2.$$

3.2. The IP[2] and IP[3]

We develop LEM for the problem IP[2], in the same manner as IP[1]. For given $f_2(x,t)$ as in (8) and point observations (9) at $i=2$, we solve the IP[2] under the following assumptions:

A2.1 $\psi_2(x_2^*) \neq 0$ and $\psi_2 \in C^2[a_2,b_2]$;

A2.2 $\psi_2(a_2) = \psi_1(b_2) = 0$, $\frac{\partial \psi_2}{\partial x}(a_2) = 0$.

In analogical way as above, we obtain the following loaded equation

$$\frac{\partial w_2}{\partial t} = \frac{\partial}{\partial x}\left(p_2(x)\frac{\partial w_2}{\partial x}\right) + \frac{\theta_2(t) - w_2(x_2^*,t)}{\psi_2(x_2^*)}\frac{\partial}{\partial x}\left(p_2(x)\frac{\partial \psi_2}{\partial x},\right), \quad (x,t) \in \Omega_2, \tag{16}$$

where

$$z_2(t) = \frac{1}{\psi_2(x_2^*)}\left(\theta_2(t) - w_2(x_2^*,t)\right), \quad z_2(t) = \int_0^t \eta_2(s)ds, \tag{17}$$

with boundary and initial conditions for $w_2(x,t)$

$$\begin{aligned} &w_2(a_2,t) = \varphi_2(t), \quad t \in (0,T), \\ &p_2(a_2)\frac{\partial w_2}{\partial x}(a_2,t) + \alpha_2 w_2(a_2,t) = \beta_2 u_1(b_1,t) + \gamma_2(t), \quad t \in (0,T), \\ &w_2(x,0) = u_2^o(x), \quad x \in [a_2,b_2]. \end{aligned} \tag{18}$$

Therefore, the IP[2] is reduced to the IBVP (16)-(18), combined with (1), $i=1$, (2) and (4), (6), $i=1$.

Similar results as in Theorem 1 also hold for IP[2].

Regarding to the IP[3], we apply the same considerations as for IP[1] and IP[2]. As a result, we obtain a new parabolic problem on disjoint intervals involving the equation (14) and initial-boundary conditions (15), where instead of $u_2(t)$ stands $w_2(t)$, as well as equation (18), in which instead of $u_1(t)$ stands $w_1(t)$. Moreover, results similar to Theorem 1 hold.

4. Difference scheme problems

In this section, we develop a finite difference approach for solving inverse problem IP[1]-IP[3]. The more efficient way to realize this approach is to decouple the corresponding discrete systems at each time level.

We introduce nonuniform spatial and temporal meshes

$$\begin{aligned} &\overline{\omega}_{h_i} = \{x_{i,j_i} : x_{i,j_i} = a_i + j_i h_i,\ j_i = 0,1,\ldots,N_i,\quad h_i = (b_i - a_i)/N_i\}, \quad i = 1,2, \\ &\overline{\omega}_\tau = \{t_n : t_n = n\tau,\ n = 0,1,\ldots,M,\quad \tau = T/M\}, \end{aligned}$$

where $x_i := x \in [a_i,b_i]$, $i=1,2$ and denote the function $y_i(x,t)$ at grid node (x_{i,j_i},t_n) by Y_{i,j_i}^n. For simplicity we assume that the observation point x_i^* coincides with internal grid node $j_i = k_i$, $i=1,2$.

4.1. Discretization of IP[1]

Using implicit-explicit time stepping, the discretization of (14) become

$$\frac{W_{1,j_1}^{n+1}}{\tau} - \left(a_{1,j_1}W_{1,\overline{x}j_1}^{n+1}\right)_x + \frac{W_{1,k_1}^{n+1}}{\Psi_{1,k_1}}\left(a_{1,j_1}\Psi_{1,\overline{x}j_1}\right)_x = g_{1,j_1}^n, \quad j_1 = 1,2,\ldots,N_1-1, \tag{19}$$

where, for every grid function y, we introduce the notations

$$y_{i,x_{j_i}} = \frac{y^{n+1}_{i,j_i+1} - y^{n+1}_{i,j_i}}{h_1}, \quad y_{i,\overline{x}_{j_i}} = \frac{y^{n+1}_{i,j_i} - y^{n+1}_{i,j_i-1}}{h_1}, \quad a_{i,j_i} = p_{i,j_i-1/2} = p_j(x_{j,j_i} - h_i/2),$$

and

$$g^n_{i,j_i} = \frac{\Theta^{n+1}_i}{\Psi_{i,k_i}} \left(a_{i,j_i}\Psi_{i,\overline{x}_{j_i}}\right)_x + \frac{W^n_{i,j_i}}{\tau}, \quad j_i = 1,2,\ldots,N_i - 1. \tag{20}$$

$$W^n_{i,k_i} = W_i(x^*_i, t_n), \quad \Psi^n_{i,k_i} = \Psi(x^*_i), \quad i = 1,2.$$

For boundary and initial conditions (15), taking into account assumptions A1.2, we have

$$\begin{aligned} &W^{n+1}_{1,0} = \varphi^{n+1}_1, \\ &\left(\frac{1}{\tau} + \frac{2\alpha_1}{h_1}\right) W^{n+1}_{1,N_1} + \frac{2}{h_1} a_{1,N_1} W^{n+1}_{1,\overline{x}N_1} + \frac{2W^{n+1}_{1,k_1}}{h_1^2 \Psi_{1,k_1}} a_{1,N_1}\Psi_{1,N_1-1} = g^n_{1,N_1}, \\ &W^0_{1,j_1} = u^o_1(x_{1,j_1}), \quad j_1 = 0,1,\ldots,N_1. \end{aligned} \tag{21}$$

where

$$g^n_{1,N_1} = \frac{2\beta_1}{h_1} U^n_{2,0} + \frac{2}{h_1}\gamma^{n+1}_{1,N_1} + \frac{1}{\tau} W^n_{1,N_1} + \frac{2\Theta^{n+1}_1}{h_1^2 \Psi_{1,k_1}} a_{1,N_1}\Psi_{1,N_1-1}.$$

For the discrete domain $\overline{\omega}_{h_2} \times \overline{\omega}_\tau$, from (1), $i = 2$, (3), (5) and (6), $i = 2$, we derive

$$\begin{aligned} &\left(\frac{1}{\tau} + \frac{2\alpha_2}{h_2}\right) U^{n+1}_{2,0} - \frac{2}{h_2} a_{2,1} U_{2,x_0} = \frac{2\beta_2}{h_2} U^n_{1,N_1} + \frac{2}{h_2}\gamma^{n+1}_{2,0} + \frac{1}{\tau} U^n_{2,0}, \\ &\frac{U^{n+1}_{2,j_2}}{\tau} - \left(a_{2,j_2} U^{n+1}_{2,\overline{x}_{j_2}}\right)_x = \frac{U^n_{2,j_2}}{\tau} + f^{n+1}_{2,j_2}, \quad j_2 = 1,2,\ldots,N_2 - 1, \\ &U^{n+1}_{2,N_2} = \varphi^{n+1}_2, \\ &U^0_{2,j_2} = u^o_2(x_{2,j_2}), \quad j_2 = 0,1,\ldots,N_2. \end{aligned} \tag{22}$$

From (13) and (19)-(22), we find

$$Z^{n+1}_1 = \frac{1}{\Psi_{1,k_1}} (\Theta^{n+1}_1 - W^{n+1}_{1,k_1}), \quad n = 0,1,\ldots,M-1, \quad Z^0_1 = 0. \tag{23}$$

Then, for (11) we get

$$\eta^{n+1}_1 = \frac{Z^{n+1}_1 - Z^n_1}{\tau}, \quad n = 0,1,\ldots,M-1. \tag{24}$$

Theorem 2 *Let the assumptions A1.2 and those of Theorem 1 hold. Then, the formula* (23) *is correct and the difference scheme* (19)-(23) *converges in* H^1 *discrete norm to the solution of IP[1] problem with rate of convergence* $O(\tau + h^2)$, $h = \max_i |h_i|$.

4.2. Discretization of IP[2] and IP[3]

For the discretization of IP[2] we apply implicit-explicit discretization as before. Using the notations (20), the finite difference scheme for (16) is

$$\frac{W^{n+1}_{2,j_2}}{\tau} - \left(a_{2,j_2} W^{n+1}_{2,\overline{x}_{j_2}}\right)_x + \frac{W^{n+1}_{2,k_2}}{\Psi_{2,k_2}} \left(a_{2,j_2}\Psi_{2,\overline{x}_{j_2}}\right)_x = g^n_{2,j_2}, \quad j_2 = 1,2,\ldots,N_2 - 1. \tag{25}$$

In view of the assumption A2.2, boundary and initial conditions (18) are approximated as follows

$$\begin{aligned} & W_{2,N_2}^{n+1} = \varphi_2^{n+1}, \\ & \left(\frac{1}{\tau} + \frac{2\alpha_2}{h_2}\right) W_{2,0}^{n+1} - \frac{2}{h_2} a_{2,1} W_{2,x_0}^{n+1} + \frac{2W_{2,k_2}^{n+1}}{h_2^2 \Psi_{2,k_2}} a_{2,1}\Psi_{2,1} = g_{2,0}^n, \\ & W_{2,j_2}^0 = u_2^o(x_{2,j_2}), \quad j_2 = 0, 1, \ldots, N_2, \end{aligned} \tag{26}$$

where

$$g_{2,0}^n = \frac{2\beta_2}{h_2} U_{1,N_1}^n + \frac{2}{h_2}\gamma_{2,0}^{n+1} + \frac{1}{\tau} W_{2,0}^n + \frac{2\Theta_2^{n+1}}{h_2^2 \Psi_{2,k_2}} a_{2,1}\Psi_{2,1}.$$

In the discrete domain $\overline{\omega}_{h_1} \times \overline{\omega}_\tau$, from (1), $i = 1$, (2), (4) and (6), $i = 1$, the corresponding equations are discretized by the same explicit implicit manner

$$\begin{aligned} & U_{1,0}^{n+1} = \varphi_1^{n+1}, \\ & \frac{U_{1,j_1}^{n+1}}{\tau} - \left(a_{1,j_1} U_{1,\overline{x}_{j_1}}^{n+1}\right)_x = \frac{U_{1,j_1}^n}{\tau} + f_{1,j_1}^{n+1}, \quad j_1 = 1, 2, \ldots, N_1 - 1, \\ & \left(\frac{1}{\tau} + \frac{2\alpha_1}{h_1}\right) U_{1,N_1}^{n+1} + \frac{2}{h_1} a_{1,N_1} U_{1,\overline{x}N_1} = \frac{2\beta_1}{h_1} U_{2,0}^n + \frac{2}{h_1}\gamma_{1,N_1}^{n+1} + \frac{1}{\tau} U_{1,N_1}^n, \\ & U_{1,j_1}^0 = u_1^o(x_{1,j_1}), \quad j_1 = 0.1, \ldots, N_1 - 1. \end{aligned} \tag{27}$$

Further, from (17) and (25)-(27), we get

$$Z_2^{n+1} = \frac{1}{\Psi_{2,k_2}}(\Theta_2^{n+1} - W_{2,k_2}^{n+1}), \quad n = 0, 1, \ldots, M-1, \quad Z_2^0 = 0. \tag{28}$$

and for (17) we have

$$\eta_2^{n+1} = \frac{Z_2^{n+1} - Z_2^n}{\tau}, \quad n = 0, 1, \ldots, M-1. \tag{29}$$

Consequently, the numerical discterization of the IP[3] is (19)-(21), (23)-(26), (28), (29).

4.3. Realization
In order to solve the non-local problems (19)-(21) and (25)-(26), we apply the decomposition of the with respect to the solution at point of measurement [21]

$$W_i^{n+1} = V_i^{n+1} + W_{i,k_i}^{n+1} X_i, \quad i = 1, 2. \tag{30}$$

Substituting (30) in (19)-(21) and (25)-(26), we get the following problems for unknown discrete functions V_i and X_i, $i = 1, 2$

$$\begin{aligned} & \frac{V_{1,j_1}^{n+1}}{\tau} - \left(a_{1,j_1} V_{1,\overline{x}_{j_1}}^{n+1}\right)_x = g_{1,j_1}^n, \quad j_1 = 1, 2, \ldots, N_1 - 1, \\ & V_{1,0}^{n+1} = \varphi_1^{n+1}, \\ & \left(\frac{1}{\tau} + \frac{2\alpha_1}{h_1}\right) V_{1,N_1}^{n+1} + \frac{2}{h_1} a_{1,N_1} V_{1,\overline{x}_{N_1}}^{n+1} = g_{1,N_1}^n, \end{aligned} \tag{31}$$

$$\begin{aligned}
&\frac{X_{1,j_1}}{\tau} - \left(a_{1,j_1}X_{1,\overline{x}j_1}\right)_x + \frac{1}{\Psi_{1,k_1}}\left(a_{1,j_1}\Psi_{1,\overline{x}j_1}\right)_x = 0, \quad j_1 = 1,2,\dots,N_1-1,\\
&X_{1,0} = 0,\\
&\left(\frac{1}{\tau} + \frac{2\alpha_1}{h_1}\right)X_{1,N_1} + \frac{2}{h_1}a_{1,N_1}X_{1,\overline{x}N_1} + \frac{2a_{1,N_1}\Psi_{1,N_1-1}}{h_1^2\Psi_{1,k_1}} = 0.
\end{aligned} \tag{32}$$

$$\begin{aligned}
&\frac{V_{2,j_2}^{n+1}}{\tau} - \left(a_{2,j_2}V_{2,\overline{x}j_2}^{n+1}\right)_x = g_{2,j_2}^n, \quad j_2 = 1,2,\dots,N_2-1,\\
&V_{2,N_2}^{n+1} = \varphi_2^{n+1},\\
&\left(\frac{1}{\tau} + \frac{2\alpha_2}{h_2}\right)V_{2,0}^{n+1} - \frac{2}{h_2}a_{2,1}V_{2,x_0}^{n+1} = g_{2,0}^n,
\end{aligned} \tag{33}$$

$$\begin{aligned}
&\frac{X_{2,j_2}}{\tau} - \left(a_{2,j_2}X_{2,\overline{x}j_2}\right)_x + \frac{1}{\Psi_{2,k_2}}\left(a_{2,j_2}\Psi_{2,\overline{x}j_2}\right)_x = 0, \quad j_2 = 1,2,\dots,N_2-1,\\
&X_{2,N_2} = 0,\\
&\left(\frac{1}{\tau} + \frac{2\alpha_2}{h_2}\right)X_{2,0} - \frac{2}{h_2}a_{2,1}X_{2,x_0} + \frac{2a_{2,1}\Psi_{2,1}}{h_2^2\Psi_{2,k_2}} = 0.
\end{aligned} \tag{34}$$

Further, taking (30) at points of measurements, we determine

$$W_{i,k_i}^{n+1} = \frac{V_{i,k_i}^{n+1}}{1 - X_{i,k_i}}, \quad i = 1,2. \tag{35}$$

Similar to Theorem 2 results also holds for IP[2] and IP[3].

5. Computational experiments

We verify the performance of the numerical approach for recovering the right hand sides f_i and solutions u_i, $i = 1,2$. We consider IP[1] and IP[3].

Regarding to the direct problem (1)-(6), it is illustrated in [15], even for more complicated version of the system, that for sufficiently smooth solution, the order of convergence of the numerical scheme (22), (27) is second in space and first.

Within the framework of quasi-real experiments, we are addressing the direct problem with the following parameters, right-hand side, initial and boundary conditions

$$\begin{gathered}
p_1(x) = 2x+2, \quad p_2(x) = 3x^2+1, \quad \alpha_1 = 3, \quad \alpha_2 = 1, \quad \beta_1 = 2, \quad \beta_2 = 0.5,\\
\gamma_1 = 3t, \quad \gamma_2 = 5+t, \quad u_1(a_1,t) = 0, \quad u_2(b_2,t) = -3\cos(\pi b_2) - t,\\
f_1 = 3\sin^2(\pi x)\begin{cases} t, & t \le 0.5,\\ 0, & t > 0.5,\end{cases} \qquad f_2 = \cos^2(\pi x/2)(t+1)^2,\\
a_1 = 0, \quad b_1 = 1, \quad a_2 = 3, \quad b_2 = 5, \quad T = 1, \quad x_1^* = 0.5, \quad x_2^* = 4.
\end{gathered}$$

The computed data are used to:

- generate the perturbed function $\theta_{i\rho_i}$ [21]

$$\theta_{i\rho_i}^n = \theta_i^n + 2\rho_i\left(\sigma_i^n - \frac{1}{2}\right), \quad i = 1,2, \tag{36}$$

where ρ_i denotes inaccuracy level and σ_i^n is a random function normally distributed over the interval $[0,1]$,

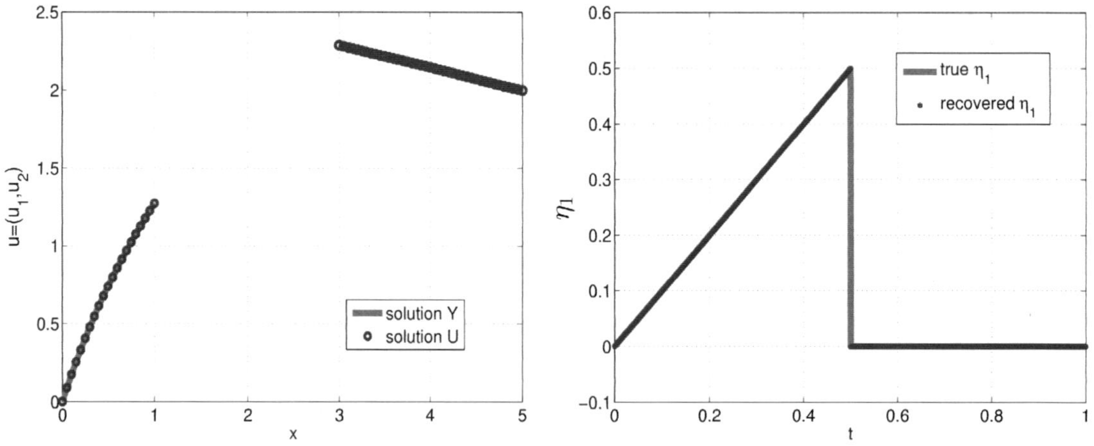

Figure 1. Numerical solutions $Y = (Y_1, Y_2)$ - obtained by direct problem and $U = (U_1, U_2)$ computed by inverse problem IP[1] with noisy free data, $t = T$ (*left*); true and recovered function η_1 (*right*), $\tau = h^2$

- compare the recovered functions η_i, $i = 1, 2$ with the true functions

$$\eta_1(t) = \begin{cases} t, & t \le 0.5, \\ 0, & t > 0.5, \end{cases} \quad \eta_2(t) = (t+1)^2,$$

as well as the computed solution by the inverse problem (U) with the solution (Y) obtained from numerically solving the corresponding direct problem.

The errors of the determined U_i, η_i, $i = 1, 2$ are given by

$$\mathcal{E}_i = \max_{0 \le j_i \le N_i} \max_{0 \le n \le M} |Y^n_{i,j_i} - U^n_{i,j_i}|, \quad \varepsilon_i = \max_{0 \le n \le M} |\eta_i(t_n) - \eta^n_i|.$$

Example 1.(IP[1]) We consider discretizations (19)-(24), where (19)-(21) is realized by (31), (32), (35). In (36), we take $\rho_1 = 0$. The computations are performed for $N_1 = 20$ and $N_2 = 40$, i.e. $h = h_1 = h_2$. The true and recovered solutions at final time are illustrated on Figure 1 for $\tau = h^2$ and on Figure 2 for $\tau = h$. For the test on Figure 1, the corresponding errors are $\mathcal{E}_1 = 4.0385 \times 10^{-13}$, $\mathcal{E}_2 = 1.688 \times 10^{-14}$ and $\varepsilon_1 = 3.161 \times 10^{-13}$, while for the Figure 1, the error values are $\mathcal{E}_1 = 3.913 \times 10^{-13}$, $\mathcal{E}_2 = 1.191 \times 10^{-13}$ and $\varepsilon_1 = 9.770 \times 10^{-15}$. Therefore, for noisy free data, the numerical solution of the inverse problem recovers exactly the numerical solution of the direct problem and the time dependent source η_1 is restored exactly. Only round-off error accumulates during the computations. Consequently, the numerical solution of the IP[1] has the same order of convergence as the discrete direct problem. This confirms the statement of Theorem 2.

Example 2.(IP[3]) Now, we consider the numerical scheme (19)-(21), (25)-(26) realized by (31)-(35). First, we consider $\rho_i = 0$, $i = 1, 2$. On Figure 3 we depict numerical solutions of the direct and inverse problems, as well as recovered and true functions η_1, η_2 for $N_1 = 20$, $N_2 = 40$, $\tau = h$. The errors of the solution and sources for this test are $\mathcal{E}_1 = 2.686 \times 10^{-13}$, $\mathcal{E}_2 = 5.262 \times 10^{-13}$ and $\varepsilon_1 = 1.029 \times 10^{-13}$, $\varepsilon_2 = 4.885 \times 10^{-15}$, i.e. the results are similar as for IP[1].

Now, we consider noisy data. Let $\rho_1 = 0.001$, $\rho_2 = 0.003$ [21]. The computational results for $N_1 = 40$, $N_2 = 80$, $\tau = h$ are illustrated on Figure 4. The corresponding errors are

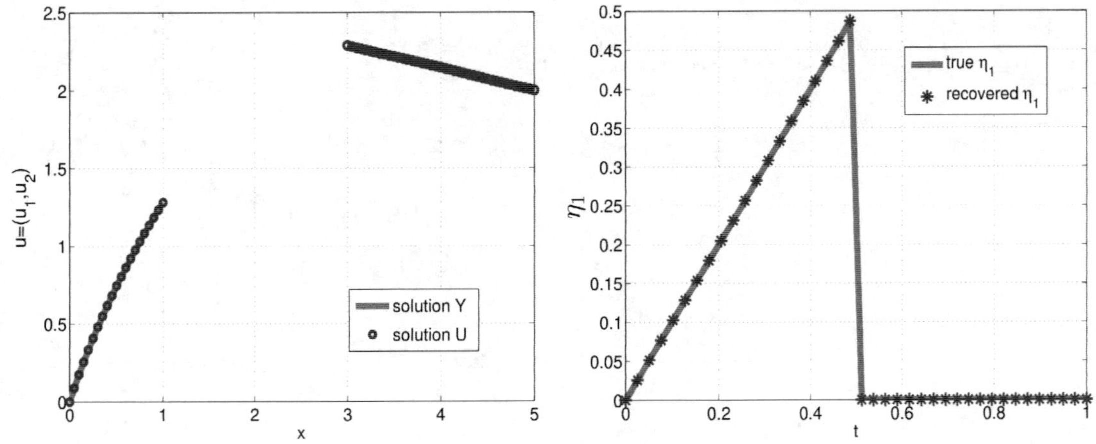

Figure 2. Numerical solutions $Y = (Y_1, Y_2)$ - obtained by direct problem and $U = (U_1, U_2)$ computed by inverse problem IP[1] with noisy free data, $t = T$ (*left*); true and recovered function η_1 (*right*), $\tau = h$

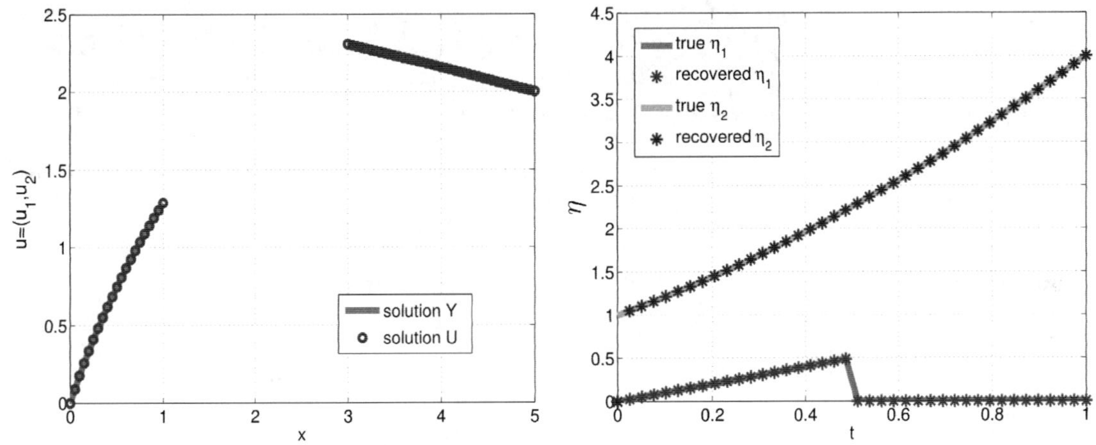

Figure 3. Numerical solutions $Y = (Y_1, Y_2)$ - obtained by direct problem and $U = (U_1, U_2)$ computed by inverse problem IP[3] with noisy free data, $t = T$ (*left*); true and recovered function η_1 (*right*), $\tau = h$

$\mathcal{E}_1 = 9.781 \times 10^{-4}$, $\mathcal{E}_2 = 2.705 \times 10^{-3}$ and $\varepsilon_1 = 5.533 \times 10^{-2}$, $\varepsilon_2 = 6.464 \times 10^{-1}$. As expected, the larger noise leads to a greater deviation. Nevertheless, the recovered solution U demonstrates good accuracy.

6. Conclusions

In this paper, the inverse problem for reconstruction the time dependent right-hand side from point observation in a one-dimensional parabolic equation in disjoint domains is investigated. By using a decomposition with respect to the unknown source, we reduce the inverse problem to a forward loaded well-posed parabolic problem on disjoint intervals. Computational algorithms are realized on test examples.

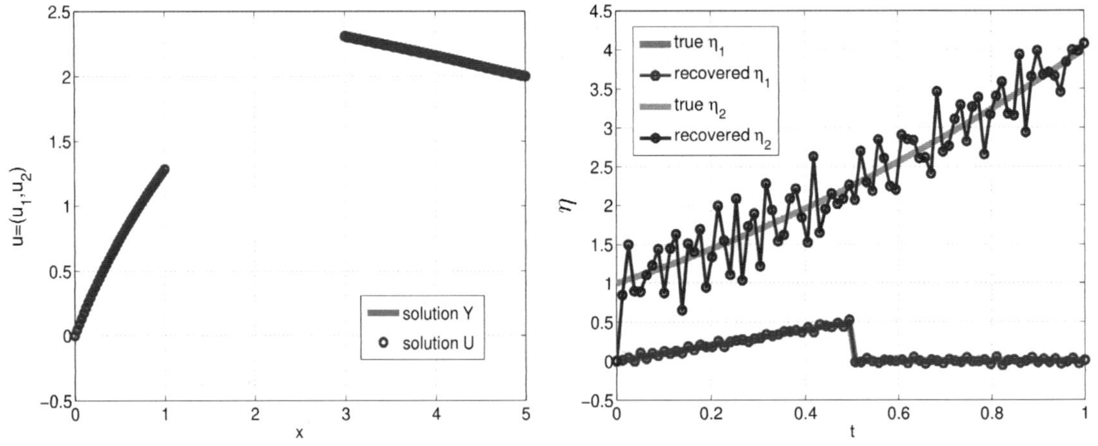

Figure 4. Numerical solutions $Y = (Y_1, Y_2)$ - obtained by direct problem and $U = (U_1, U_2)$ computed by inverse problem IP[3] with smoothed perturbed data, $t = T$ (*left*); true and recovered function η_1 (*right*), $\tau = h$

Numerical results showed that for noisy free measurements, the numerical solution of the inverse problem recovers the solution of the discrete problem and true source term exactly. For perturbed data, the algorithm is still efficient and produces relevant results with satisfactory error.

A natural extension of the present study includes inverse problems for 2D and multidimensional parabolic problems on disjoint domains.

Acknowledgments: This work is supported by the Bulgarian National Science Fund under the Project KP-06-N 62/3 "Numerical methods for inverse problems in evolutionary differential equations with applications to mathematical finance, heat-mass transfer, honeybee population and environmental pollution", 2022.

References

[1] Amosov A.A. 2005 Global solvability of a nonlinear nonstationary problem with a nonlocal boundary condition of radiation heat transfer type *Differential Equations* **41**(1) 96–109

[2] Abdullayev V.M. Aida-zade K.R. 2016 Finite-difference methods for solving loaded parabolic equations *Comput. Math. and Math. Phys* **56** 93–105

[3] Alikhanov A.A. Berezgov A.M. Shkhanukov-Lafishev M.X. 2008 Boundary value problems for certain classes of loaded differential equations and solving them by finite difference method, *Comput. Math. and Math. Phys.* **48** 1581–1590

[4] Datta A.K. 2002 *Biological and Bioenvironmental Heat and Mass Transfer* Marcel Dekker N.Y.

[5] Dimov I. Kandilarov J. Vulkov L. 2018 Numerical solution of direct and inverse problems for degenerate parabolic equations with concentrated sources *AIP Conference Proceedings* **2048** 030014

[6] Givoli D. 2004 Finite element modeling of thin layers *Comput Model Eng Sci* **5** 497–514

[7] Jovanović B.S. Milovanović Z.D. 2014 Numerical approximation of a 2D parabolic transmission problem in disjoint domains *Applied Mathematics and Computation* **228** 508–519

[8] Jovanović, B.S. Vulkov L.G. 2011 Numerical solution of a parabolic transmission problem *IMA Journal of Numerical Analysis* **31**(1) 233–253

[9] Jovanović B.S. Vulkov L. G. 2012 Formulation and analysis of a parabolic transmission problem on disjoint intervals *Publications de l'Institut Mathematique* **9**(105) 111–123

[10] Hasanoglu A. Romanov V.G. 2017 *Introduction to Inverse Problems for Differential Equations* 1st ed Springer Cham 261p.

[11] Hussein M. Lesnic D. Kamynin V. Kostin A. 2020 Direct and inverse problems for degenerate parabolic equations *J Inverse Ill-Posed Probl* **28**(3) 425–448
[12] Kabanikhin S.I. 2011 *Inverse and Ill-Posed Problems* DeGruyer Berlin
[13] Koleva M.N. Milovanović Jeknić Z. D. Vulkov L.G. Determination of external boundary conditions of a stationary nonlinear problem on disjoint intervals at point observation *Study in Computational Inteligence* Springer (accepted)
[14] Koleva M.N. Vulkov L.G. Numerical determination of source from point observation in a time-fractional boundary-value problem on disjoint intervals *Lecture Notes in Computer Science* Springer (accepted)
[15] Koleva M.N. Vulkov L.G. 2023 Weak and classical solutions to multispecies advectiondispersion equations in multilayer porous media *Mathematics* **11**(14) 3103
[16] Milovanovic Z. 2013 Finite difference scheme for a parabolic transmission problem in disjoint domains *Lecture Notes in Computer Science* **8236** 403–410
[17] Lesnic D. 2021 *Inverse Problems with Applications in Science and Engineering* CRC Pres Abingdon UK p. 349
[18] Qatanani N. Barham A. Heeh Q. 2007 Existence and uniqueness of the solution of the coupled conduction radiation energy transfer on diffuse gray surfaces *Surveys in Mathematics and its Applications* **2** 43–58.
[19] Nakhusher A.M. 2012 *Loaded Equations and Applications* Nauka Moskow (in Russian).
[20] Vabishchevich P.N. Vasilev V.I. Vasileva M.V. 2015 Computational identification of the right-hand side of a parabolic equation. *Comput. Math. and Math. Phys.* **55** 1015–1021
[21] Samarskii A.A. Vabishchevich P.N. 2007 *Numerical Methods for Solving Inverse Problems of Mathematical Physics* de Gruyter 452 p.
[22] Zhuo L. Lesnic D. Meng, S. 2020 Reconstruction of the heat transfer coefficient at the interface of a bi-material *Inverse Probl Sci Eng* **28** 374–401.

Journal of Physics: Conference Series **2675** (2023) 012026 doi:10.1088/1742-6596/2675/1/012026

Algebraic Geometry and Elliptic Integrals Approach for Calculation of the Propagation Time of a Signal Between GPS Satellites with Account of General Relativity Theory

Bogdan G. Dimitrov

Abstract. The approach for calculating the propagation time of a signal between GPS satellites will be summarized, based on the proposed new theoretical approach in several previous publications, as well as the perspectives for future development of the theory. Topics include: 1. Basic notions of inter-satellite communications. 2. Shapiro delay formulae in General Relativity Theory - basic formalism and the necessity to extend the formalism by taking into account the satellite motion on a plane elliptic or space-distributed elliptic orbit. 3. Basic facts about the disturbed motion in celestial mechanics and the necessity to incorporate it in the theory of inter-satellite communications, accounting for General Relativity Effects. 4. Propagation time of a signal, emitted by a satellite on a plane and also space-distributed elliptical orbit in terms of zero-order elliptic integrals and respectively of higher order integrals. Proof of the real-valuedness of the propagation time for all cases as one of the criteria for the correctness of the theoretical approach. 5. New analytical algorithms for calculation of zero-order elliptic integrals in the Legendre form. Relation to two representations in the Weierstrass form. 6. The new formalism of intersecting four-dimensional null cones and the resulting physical notions of the (intersecting)) space-time interval (with the property of being positive, negative or equal to zero) and the (intersecting) geodesic distance (being only positive, because is related to the distance, travelled by light or radio signals). Proof of these properties in the general case and in some partial cases. New numerical estimate $E_{\rm lim} > 45.002510943228$ [deg], above which the space-time interval is positive and thus inter-satellite communications between satellites on one plane elliptical orbit are possible. The angular distance of 45[deg] is typical for the disposition of 8 satellites on one orbit in the Russian satellite constellation $GLONASS$, so it might be claimed that such a configuration is favourable from the point of view of inter-satellite communications (with account of GRT effects).

[1] Institute of Nuclear Research and Nuclear Energetics, Bulgarian Academy of Sciences, 72 Tzarigradsko shaussee, 1784 Sofia, Bulgaria

E-mail: [1] `dimitrov.bogdan.bogdan@gmail.com`
[2] Institute for Advanced Physical Studies , Sofia TECH Park, 111 Tzarigradsko shaussee, 1784 Sofia, Bulgaria

E-mail: [2] `bogdan.dimitrov@iaps.institute`

1. Introduction

Autonomous navigation is one of the most important ingredients of satellite communications. The essence of such a concept is that a satellite system such as the Global Navigation Satellite System ($GNSS$), consisting of 30 satellites and orbiting the Earth at a height of 23616 km should be able to function properly without exchanging any signals with ground stations in the course of 6 months (180 days). This means that $GNSS$ should be inter-operable with other navigational satellite systems such as GPS (Global Positioning System) and the Russian system $GLONASS$ (Globalnaya Navigazionnaia Sputnikovaia Sistema) (see the monographs [1] and [2] for a brief introduction into the characteristics of these systems). In other words, satellites on different orbital planes should be able to exchange signals between each other and in such a way, "inter-satellite communications" are realized. Since the exchange of signals takes place in the near-Earth gravitational field (and generally in the gravitational field of the Solar system), the creation of a theory for such processes should take into account the General Relativity Theory effects. In brief, this theory is based on the following two important facts:

1. An electromagnetic, radio or light signal (including a laser signal) is propagating on the null light cone, which can be obtained after setting up the infinitesimal metric element equal to zero $ds^2 = 0$. The metric element ds^2 will be given later in the paper and is chosen to be typical for the near-Earth space. A general theory of the signal propagation in the gravitational field is given in the contemporary monograph [3].

2. From a qualitative and also physical point of view, the significance of the null cone light equation is that it gives the opportunity to calculate the propagation time of the signal - the well-known Shapiro delay formulae, known yet from 1964 [4] and widely applied in various investigations. The Shapiro delay formulae has an important physical meaning - due to the action of the gravitational field, which "curves" the trajectory of the signal, the propagation time increases with a small logarithmic correction. Although very small (at the order of picoseconds), it plays an important role in satellite ranging, presently achieved with great precision.

It should be stressed that the application of General Relativity Theory in the theory of signal propagation between GPS satellites began in the pioneering works of Neil Ashby [5] and [6], where the s.c. increment of the coordinate time Δt was calculated as

$$\int_{path} \Delta t = \int_{path} \left[1 - \frac{(V - \Phi)}{c^2} + \frac{v^2}{2c^2}\right] d\tau \quad , \tag{1}$$

representing a (path) integral over the path of the atomic clock (on board of the satellite), V and Φ are potentials of the Earth and v is the orbital velocity of the satellite. In fact, as noted in [5], "the rate of coordinate time is determined by atomic clocks at rest at infinity, but the rate of the GPS coordinate time, however, is closely related to the International Atomic Time (TAI), which is a time scale computed on the basis of inputs from hundreds of primary time standards, hydrogen masers, and other clocks from all over the world." Because of these path-dependent effects (especially for the case of the inter-satellite communications (ISC) to be considered in this paper), comprising also time dilation (apparent slowing of moving clocks) and frequency shifts due to gravitation, the proper time cannot be used when a signal is being transmitted between two satellites. However, this statement needs to be clarified. If the definition of the atomic time τ is used and also formulae (1), it can be written also [5]

$$d\tau = \frac{ds}{c} = \left[1 + \frac{(V - \Phi)}{c^2} - \frac{v^2}{2c^2}\right] dt \quad , \tag{2}$$

from where it can be concluded that the proper atomic time of fictitious atomic clocks at rest and in a local inertial frame agrees with the coordinate time up to the accuracy of $O(\frac{1}{c^2})$ terms. By definition, an accuracy of $O(\frac{1}{c^2})$ or $O(\frac{1}{c})$ means that terms of the order $\frac{1}{c^2}$ or $\frac{1}{c}$ are accounted in the corresponding expressions. For example, up to $O(\frac{1}{c})$ terms and taking into account the metric element ds^2 in its most general form

$$ds^2 = g_{00}c^2dt^2 + 2g_{0j}cdtdx^j + g_{ij}dx^idx^j = 0 \quad , \tag{3}$$

the increment of the coordinate time Δt can be found first as a solution of a quadratic algebraic equation and subsequently - a differential equation [7]

$$\Delta t = \pm\frac{1}{c}\int_A^B \frac{1}{\sqrt{-g}}\cdot\sqrt{\left(g_{ij} + \frac{g_{0i}g_{0j}}{g_{00}}\right) dx^i dx^j} + \frac{1}{c}\int_A^B \frac{g_{0j}}{-g_{00}}\, dx^j \quad . \tag{4}$$

The coordinate time now is related to the propagation of a light or radio signal from a space-point A to the space-point B and therefore represents the propagation time $\Delta t = T_B - T_A$ with T_B and T_A being the times of reception and emission. Two important conclusions can be made: 1. Since time in satellites is measured by means of atomic clocks and atomic time agrees with the propagation (coordinate) time up to terms of a certain order in $\frac{1}{c}$, it is a measurable quantity. 2. The presence of the differentials dx^i enables to find such a parametrization of the space coordinates, which might not necessarily be related to the signal propagation path and the endpoints A and B. In the present investigation, the coordinates x, y and z will be related to the elliptic motion of the satellite and the endpoints A and B will mark the initial position of the satellite on the orbit (when it emits the signal) and the point B-the final position of the satellite, when the signal is percepted. The reason why the reference system is chosen to be attached to the satellite trajectory is that the s.c. orbital parameters of all the satellite configurations *GPS*, *GLONASS*, *Galileo* and *BeiDou* are known with great accuracy. Further this will be demonstrated. In spite of the fact that the satellite motion will be very important in calculating the propagation time of the signal, it will be proved by simple numerical calculations that the obtained propagation time is typical for the light or radio signals and is not directly related to the celestial time of motion of the satellite. This means that the propagation time is determined by the type of equation used for its calculation (in the case this is the null cone equation) and not by the type of the spatial reference system.

Two cases will be considered further.

1st case This is the most simple case of plane elliptical motion of the satellite on an ellipse, parametrized as

$$x = a(\cos E - e) \quad , \quad y = a\sqrt{1-e^2}\sin E \quad , \tag{5}$$

where E is the eccentric anomaly angle (characterizing the position of the satellite on the orbit) and a and e are respectively the semi-major axis and the eccentricity of the ellipse. This is the problem, investigated in a series of publications. Although in a sense the notion about the "plane elliptical motion" is an idealized one, any further development of the theory (see the proposal in the next "2-nd case") will be based on the applied approach. An interesting fact, which will be proved is that the obtained from the null cone equation propagation time has the dimension of seconds. This is a nontrivial fact, since the propagation time is expressed by a combination of elliptic integrals of the first, second and the third kind, so it is not evident at the beginning that the elliptic integrals are real-valued expressions. This should be so, because time (in the case-the propagation time) as a physical quantity cannot be imaginary. Moreover, further it will be shown that for the case of space-distributed orbits, the corresponding higher-order (second and fourth) order elliptic integrals are imaginary, but since they enter imaginary expressions for the propagation time, it will be again a real-valued quantity.

2. Perturbed motion in celestial mechanics and its relation to the problem about signal propagation

2-nd case. This is the more complicated case of space-distributed (elliptical) orbits, characterized by the full set of 6 Keplerian parameters $(M, a, e, \Omega, I, \omega)$ (called in some monographs such as [8] "the contact elements" or "orbital elements"), where $M = \sqrt{\frac{GM_{\oplus}}{a^3}}$ is the mean motion and

Ω, I, ω are analogous of the Eulerian angles (known from theoretical and celestial mechanics) which will be defined further. Following the same mathematical formalism and assuming that the position of the satellite is determined only by means of the true anomaly angle f (i.e. the other orbital parameters do not change), it will be proved in this paper that the propagation time will depend on more complicated elliptic integrals of the second and fourth order (the order of the elliptic integral is determined by the degree of the integration variable in the nominator of the under-integrand expression). This result is important both from physical and mathematical point of view, because a complete theory how the propagation time changes under the changes $(\delta M, \delta a, \delta e, \delta\Omega, \delta I, \delta\omega)$ of the orbital elements is still lacking. These changes are caused by gravitational perturbations by celestial bodies, tidal forces and etc. A well-known theoretical method in celestial mechanics is the method of "osculating conics", the essence of which is that the orbital elements are assumed to depend on time $\{C_i(t) = (M(t), a(t), e(t), \Omega(t), I(t), \omega(t)), i = 1, 2, 3......, 5, 6)\}$ [8]. It is assumed that the body under consideration moves along a conic, which is osculating (i.e. tangent) to the actual physical trajectory and so it is slowly evolving. In other words, if the orbital vector $\overrightarrow{r}$ and the velocity vector $\overrightarrow{v}$ depend on time and on the orbital parameters $C_i(t)$

$$\overrightarrow{r} = \overrightarrow{r}(t, C_1(t), C_2(t), C_3(t), C_4(t), C_5(t), C_6(t)) \quad ,$$

$$\frac{\partial \overrightarrow{r}}{\partial t} = \overrightarrow{v}(t, C_1(t), C_2(t), C_3(t), C_4(t), C_5(t), C_6(t)) \tag{6}$$

and the parameter t plays the role of time in the perturbed equation of motion

$$\frac{\partial^2 \overrightarrow{r}}{\partial t^2} + \frac{GM}{r^3}\overrightarrow{r} + \sum_{i=1}^{6} \frac{\partial \overrightarrow{v}}{\partial C_i} . \frac{dC_i}{dt} = F_i \quad , \tag{7}$$

then a constraint called " Lagrange constraint" is imposed [8]

$$\sum_{i=1}^{6} \frac{\partial \overrightarrow{r}}{\partial C_i} . \frac{dC_i}{dt} = 0 \quad . \tag{8}$$

The last equality means that locally the physical trajectory, defined by $\{C_i(t), \quad i = 1, 2, 3......, 5, 6)\}$, will coincide with the unperturbed orbit, followed by the body if the perturbations were to cease instantaneously. An important consequence of the theory is the nonlinear system of ordinary differential equations in the Euler-Gauss form [8] with respect to $\frac{dM(t)}{dt}, \frac{da(t)}{dt}, \frac{de(t)}{dt}, \frac{d\Omega(t)}{dt}, \frac{dI(t)}{dt}$ and $\frac{d\omega(t)}{dt}$. If the equations include the s.c. "disturbing gravitational potential" (accounting for the perturbations of outer celestial bodies), then the equations are called "Lagrange planetary equations" [8], [9], [10], [11].

The above formalism constitutes the fundamentals of the celestial mechanics theory of disturbing motion, but the estimates in [5] for the perturbed radius-vector, perturbed eccentric anomaly (based on the celestial mechanics monograph of Richard Fitzpatrick, ch.11.6, [12]), perturbed semi-major axis and perturbed energy are used to estimate the fractional frequency shift

$$\frac{\delta f}{f} = -\frac{v^2}{2c^2} - \frac{GM_E}{c^2 r} + \frac{V'}{c^2} \quad , \tag{9}$$

where V' is the perturbed energy. The estimate is based, however, only on the evaluation of the separate contributions in the perturbed potential and not on any path integration, which in principle can show how the parameters change. In other words, the change of each one of the parameters is not shown to what change in $\frac{\delta \overline{f}}{\overline{f}}$ will correspond. In order to avoid further confusion, we have denoted here the frequency by $\overline{f}$, to distinguish it from the true anomaly angle f, used to denote the position of a satellite on a space-distributed orbit. In fact, the calculation of the propagation time in this paper and the elliptic integrals of second and fourth order will be related only to the true anomaly angle f as a variable of integration. Thus any disturbances in the motion of the satellite will not be taken into account.

Since evidently because of the perturbations of all the 6 Keplerian parameters, it will be natural to assume that the variation of the propagation time δT along all the perturbed orbital elements $(\delta M, \delta a, \delta e, \delta\Omega, \delta I, \delta\omega)$ will give independent contributions

$$\delta T = \frac{\partial T}{\partial M}\delta M + \frac{\partial T}{\partial a}\delta a + \frac{\partial T}{\partial e}\delta e + \frac{\partial T}{\partial \Omega}\delta\Omega + \frac{\partial T}{\partial I}\delta I + \frac{\partial T}{\partial \omega}\delta\omega \quad . \tag{10}$$

3. Keplerian parameters for space-distributed orbits of the satellite configurations *GPS*, *GLONASS*, *BeiDou* and *Galileo*

If a theoretical model is created, the development of a numerical model will be greatly facilitated by the fact that the Keplerian parameters are well-known for the satellite configurations *GPS*, *GLONASS* [13], Galileo (European satellite system) and *BeiDou* (Chinese satellite system). For example, the satellites of the Galileo constellation are situated on three orbital planes with nine-equally spaced operational satellites in each plane. The Galileo satellites are in nearly circular orbits with semi-major axis of 29600 km and a period of about 14 hours [14] and an inclination of the orbital planes 56 degrees. For comparison, the Russian Global Navigation Satellite System *GLONASS*, launched in 1982, consists of 21 satellites in three orbital planes. Each satellite operates in nearly circular orbits with semi-major axis of 25510 km, and the satellites within the same orbital plane are equally spaced by 45 degrees. In the following sections, this fact will be important when discussing the theoretical approach of the space-time interval on two intersecting four-dimensional null cones. Each orbital plane has an inclination angle of $64.8^0 \pm 0.3^0$ (see also [1]), which is more than the inclination angle 56^0 of the orbital planes of the Galileo satellites. The longitude of ascending nodes $\Delta\Omega$ between the orbital planes and the argument of latitude difference Δu are well-known and are correspondingly equal to $\Delta\Omega = 120^0$ and $\Delta u = 45^0$. The satellites of the *GLONASS* constellation within one orbit are situated at an angular distance of 45^0 (8 satellites on one orbit). The eccentricity of the orbit in [1] is estimated to be $e < 0.01$, but according to some other date it can be $e = 0.02$. A *GLONASS* satellite completes an orbit in approximately 11 hours 15 minutes ± 5 sec - less than the orbital period of 14 hours for the Galileo satellite. Consequently, the three characteristic angles of rotation - the eccentric anomaly E, the mean anomaly $M = n(\tau - t)$ and the true anomaly f should be different for the different satellite systems and can be found from the corresponding tables.

The data about the *GPS* satellites can be found in many monographs, but they are briefly summarized for example in [15] - 6 orbital planes with $4-5$ satellites per orbit, semi-major axis $a = 26400\ km$, eccentricity at most $e = 0.02$, but further we shall use the more precise data from the PhD dissertation [16]-$a = 26560.25169632944\ km$ and $e = 0.01323881349526$, the inclination of the orbit I is equal to $I = 55.5^0$ [15] or $I = 0.9614884100802\ rad$ [16].The mean anomaly $M = -0.3134513508155\ rad$ [16] will be used in order to determine the eccentric anomaly angle E, related to the position of the satellite on the orbit and also to the true anomaly f, expressed through E by means of the formulae

$$\tan\frac{f}{2} = \sqrt{\frac{1-\cos f}{1+\cos f}} = \sqrt{\frac{1+e}{1-e}}\tan\frac{E}{2} \quad . \tag{11}$$

The data about the longitude of the ascending node Ω and the argument of perigee ω can also be found in [16], but they will not be used for the concrete calculations in this paper.

4. Description of the most general theoretical model and the general system of algebraic equations

Since the orbital parameters of the satellites from all the satellite constellations are well-known, the following interesting research problem can be formulated: suppose that a light or radio signal is being send from a satellite belonging to one satellite constellation with orbital parameters $(M_1, a_1, e_1, \Omega_1, I_1, \omega_1)$ to another satellite (from the same or from another constellation) on a space-distributed orbit with another orbital parameters $(M_2, a_2, e_2, \Omega_2, I_2, \omega_2)$. Then how can the propagation time of the signal be calculated? This problem is more complicated than it might seem because during the time of flight of the photon or radio signal, the orbital parameters of the second satellite change.Consequently, the curved signal trajectory and the signal, emitted from the first satellite should be "correlated" with the movement of the second satellite. So if one imagines that after its emission the signal is "propagating" on the first null-cone (with origin at the point of emission), then at the moment of reception the signal is propagating not on the first null-cone, but on a second null-cone, defined with origin at the point of reception. So these two points will be constantly changing, but the phenomena of emission and subsequent reception of the signal mathematically will mean that the two four-dimensional null-cones (in terms of the two sets of space-time coordinates T_1, x_1, y_1, z_1 and T_2, x_2, y_2, z_2) will intersect. If the two sets of coordinates are expressed through the changing orbital parameters, then evidently one obtains two seven-dimensional null cones (the coordinate times of emission T_1 and T_2 are the additional seventh coordinates) in terms of the variables $(dT_1, dM_1, da_1, de_1, d\Omega_1, dI_1, d\omega_1)$ and $(dT_2, dM_2, da_2, de_2, d\Omega_2, dI_2, d\omega_2)$, which should intersect. Correspondingly, the algebraic equations of the two four-dimensional null cones in terms of the two sets of four-dimensional coordinates (T_1,x_1, y_1, z_1) and (T_2, x_2, y_2, z_2) will be

$$ds_1^2 = -c^2\left(1+\frac{2V_1}{c^2}\right)(dT_1)^2 + \left(1-\frac{2V_{1.}}{c^2}\right)\left((dx_1)^2+(dy_1)^2+(dz_1)^2\right) = 0 \ , \tag{12}$$

$$ds_2^2 = -c^2\left(1+\frac{2V_2}{c^2}\right)(dT_2)^2 + \left(1-\frac{2V_{2.}}{c^2}\right)\left((dx_2)^2+(dy_2)^2+(dz_2)^2\right) = 0 \ . \tag{13}$$

These two null cones will not intersect, because their coordinates are independent. However, one may intersect these null cones with the 14-dimensional hyperplane, defined as the differential of the Euclidean distance R_{AB}

$$dR_{AB}^2 = d(x_1-x_2)^2 + d(y_1-y_2)^2 + d(z_1-z_2)^2 \ . \tag{14}$$

Now the problem about finding the propagation times T_1 and T_2 is formulated as an algebraic geometry problem of the intersection of two seven-dimensional null cones (12) and (13) (if written in terms of the variables $(dT_1, dM_1, da_1, de_1, d\Omega_1, dI_1, d\omega_1)$ and $(dT_2, dM_2, da_2, de_2, d\Omega_2, dI_2, d\omega_2)$) with the 14−dimensional hyperplane (14), written in terms of all the two sets of variables. It may be noted that since the Euclidean distance R_{AB}^2 between the satellites is changing, the distance R_{AB} depends on the following variables

$$R_{AB} = R_{AB}(T_1, \Gamma_i^{(1)}, T_2, \Gamma_i^{(2)}) \ , \tag{15}$$

where the following new notations are introduced

$$\Gamma_i^{(1)} = (M_1, a_1, e_1, \Omega_1, I_1, \omega_1) \ , \quad \Gamma_i^{(2)} = (M_2, a_2, e_2, \Omega_2, I_2, \omega_2) \ , i = 1, 2, ...6 \ , \tag{16}$$

$$X_j^{(1)} = (x_1, y_1, z_1) \quad , \quad X_j^{(2)} = (x_2, y_2, z_2) \quad , \quad j = 1, 2, 3 \quad . \tag{17}$$

The notations $\Gamma_i^{(1)}$ and $\Gamma_i^{(2)}$ are the previous notations $C_i(t)$ in (8), but here we do not necessarily suppose that they depend on the "perturbed time". The notations (16) and (17) enable us to write equation (12) in the following concise form

$$-c^2\left(1+\frac{2V_1}{c^2}\right)\left[\sum_{i=1}^{6}\frac{\partial T_1}{\partial\Gamma_i^{(1)}}d\Gamma_i^{(1)}\right]^2+\left(1-\frac{2V_{1.}}{c^2}\right)\sum_{j=1}^{3}\left[\sum_{i=1}^{6}\frac{\partial X_j^{(1)}}{\partial\Gamma_i^{(1)}}d\Gamma_i^{(1)}\right]^2=0 \quad . \tag{18}$$

It may easily be proved that

$$(dT_1)^2=\left[\sum_{i=1}^{6}\frac{\partial T_1}{\partial\Gamma_i^{(1)}}d\Gamma_i^{(1)}\right]^2$$

$$=\sum_{i=1}^{6}\left(\frac{\partial T_1}{\partial\Gamma_i^{(1)}}\right)^2\left(d\Gamma_i^{(1)}\right)^2+2\sum_{\substack{i,k=1\\ i\neq k}}^{6}\left(\frac{\partial T_1}{\partial\Gamma_i^{(1)}}\right)\left(\frac{\partial T_1}{\partial\Gamma_k^{(1)}}\right)d\Gamma_i^{(1)}d\Gamma_k^{(1)} \quad . \tag{19}$$

An analogous formulae is valid also for the expression $\left[\sum_{i=1}^{6}\frac{\partial X_j^{(1)}}{\partial\Gamma_i^{(1)}}d\Gamma_i^{(1)}\right]^2$. Thus expression (18) can be written as a six-dimensional quadratic surface with respect to the variables $d\Gamma_i^{(1)}$ $i=1,2,...6$

$$-c^2\left(1+\frac{2V_1}{c^2}\right)\left[\sum_{i=1}^{6}\left(\frac{\partial T_1}{\partial\Gamma_i^{(1)}}\right)^2\left(d\Gamma_i^{(1)}\right)^2+2\sum_{\substack{i,k=1\\ i\neq k}}^{6}\left(\frac{\partial T_1}{\partial\Gamma_i^{(1)}}\right)\left(\frac{\partial T_1}{\partial\Gamma_k^{(1)}}\right)d\Gamma_i^{(1)}d\Gamma_k^{(1)}\right]$$

$$+\left(1-\frac{2V_{1.}}{c^2}\right)\left[\sum_{i=1}^{6}\sum_{j=1}^{3}\left(\frac{\partial X_j^{(1)}}{\partial\Gamma_i^{(1)}}\right)^2\left(d\Gamma_i^{(1)}\right)^2+2\sum_{\substack{i,k=1\\ i\neq k}}^{6}\sum_{j=1}^{3}\left(\frac{\partial X_j^{(1)}}{\partial\Gamma_i^{(1)}}\right)\left(\frac{\partial X_j^{(1)}}{\partial\Gamma_k^{(1)}}\right)d\Gamma_i^{(1)}d\Gamma_k^{(1)}\right]=0 \quad . \tag{20}$$

The same type of an equation can be written with respect to equation (13) for the second propagation time T_2. Similarly, the equation (14) with respect to the 14−dimensional hyperplane can be written as

$$dR_{AB}^2=\sum_{j=1}^{3}2\left(X_j^{(1)}-X_j^{(2)}\right)\sum_{i=1}^{6}\left[\frac{\partial X_j^{(1)}}{\partial\Gamma_i^{(1)}}d\Gamma_i^{(1)}-\frac{\partial X_j^{(2)}}{\partial\Gamma_i^{(2)}}d\Gamma_i^{(2)}\right] \quad , \tag{21}$$

where the differential dR_{AB}^2 should be found from

$$dR_{AB}^2=\sum_{l=1}^{2}\frac{\partial R_{AB}^2}{\partial T_l}\sum_{i=1}^{6}\frac{\partial T_{(l)}}{\partial\Gamma_i^{(l)}}d\Gamma_i^{(l)}+\sum_{l=1}^{2}\sum_{i=1}^{6}\frac{\partial R_{AB}^2}{\partial\Gamma_i^{(l)}}d\Gamma_i^{(l)} \quad . \tag{22}$$

If in the system of equations (20)-(22) the differentials $d\Gamma_i^{(l)}$ are considered to be independent (meaning that the constraint (8) can no longer be applied), then a complicated system of nonlinear differential equations is obtained with respect to $\frac{\partial X_j^{(l)}}{\partial\Gamma_i^{(2)}}$, $\frac{\partial T_{(l)}}{\partial\Gamma_i^{(l)}}$ and $\frac{\partial R_{AB}^2}{\partial T_l}$. In fact, equations (20)-(22) are related to the s.c. intersection theory of algebraic surfaces and they are more peculiar and difficult to deal with in comparison with the examples, given in most of the basic monographs on algebraic geometry (see for example [17] and [18]). In the cited monographs, examples of conics are given [18], also intersection of conics [17]. Basic knowledge about algebraic varieties and intersection theory can be found also in [17], the basic notions of algebraic curves and algebraic geometry in a clear and accessible style are presented in [20], also in [21].

5. The purpose of the present paper

The present paper is a part of a series of papers [22], [23], [24] and [25], each of which represents a part of a systematic approach, aiming at creating a new theoretical formalism for the description of satellite communications between two satellites on different space-distributed orbit, accounting for the effects of GRT. In this final form, the problem is rather complicated and so can be treated, if several other more simple problems are solved. The purposes of this review are several:

1. To present a more general description of the problem, both from physical and mathematical aspects, which have not been given in the previous publications. This has already been performed in the previous sections.

2. To present the mathematical derivation of the propagation time of a signal in terms of zero-order elliptic integrals of the first, second and the third kind for the most simple case of a signal, emitted by a moving along a plane elliptical orbit [24]. This case may seem highly unrealistic, but here the main problem is to justify the assumed elliptical parametrization of the space coordinates in the null cone equation. In order to avoid any doubts that the obtained propagation time is really related to the signal propagation and not to the movement of the satellite, an additional (not previously presented) section has been added with some concrete numerical estimates, based on typical parameters for the GPS orbits, given in the PhD thesis [16].

3. To derive the signal propagation time for the more complicated case of a signal emission by a satellite on a spatially-distributed orbit, characterized by the full set of 6 Keplerian parameters [24]. The obtained propagation time is expressed again in terms of elliptic functions, but they are more complicated, since are of higher order (second and fourth). This case is again not realistic from a physical point of view, because a satellite in its orbital motion may deviate tens and even hundreds of meters due to various disturbing forces, even solar pressure and etc. But although not quite realistic, this method is necessary to be developed, because it constitutes one of the basic elements in the developed in the publications [22], [23] method of intersecting four-dimensional null cones (in terms of the coordinates T, x, y, z in the initially given null-cone metric).

4. In the papers [22], [23] the method of intersecting null cones has been studied on the base of the equations, which are in fact the simple version of the equations (18) - (22) from the previous section. The most simple case of plane elliptical motion of the satellite, described by the changing eccentric anomaly angles E_1 and E_2 have been chosen, and the two propagation times T_1 and T_2. Although such a simplified model might be considered physically unrealistic, most amazingly it helped us to establish some interesting physical consequences. The most important one is that the intersection of the two null cones with the four-dimensional hyperplane defines a distance R^2_{AB}, which is called a space-time interval. It can be positive, negative or equal to zero. Although initially R^2_{AB} was defined as an Euclidean distance, this result is not amazing, because the two null cones are notions from General Relativity Theory, where space-like and time-like vectors (respectively, outside or inside isotropic null cones) and curves are quite natural (see ch. 4 and ch. 6 of the known monograph [26]). Since the real propagation of light or radio signals in GRT is related to time geodesics and to large-scale distances, a new interpretation has been given of the notion of the "geodesic distance"- the space-time interval becomes the positive geodesic distance, if the s.c. "condition for inter-satellite communications" (expressed from the

null cone equation) is substituted into the expression for the space-time interval.

There is also one direct consequence from the proposed formalism of two intersecting null cones, which is important for the inter-satellite communications between the satellites from the $GLONASS$ constellation. It has been proved in [22] and [23] that the limiting value of the eccentric anomaly angle $E_{\lim} = 45.002510943228$ [deg], above which (below) the space-time interval dR^2_{AB} becomes positive (negative) corresponds to the angular distance of 45^0 of disposition of the satellites from the $GLONASS$ constellation within one plane orbit and eccentricity of the orbit $e = 0.02$.

There are a number of other interesting problems, which will be commented in the conclusion of this paper.

6. The Shapiro delay formulae in station–satellite and inter-satellite communications - review of some basic physical facts

We shall begin with some trivial and simple examples in [27], which will help us to understand the physical essence of the phenomena of Shapiro time delay of a signal. In the case of the trivial metric

$$-g_{00} = 1 \ , \ g_{11} = g_{22} = g_{33} = 1 \quad , \ g_{\mu\nu} = 0 \text{ for } \mu \neq \nu \quad , \ \mu, \nu = 0, 1, 2...3 \tag{23}$$

one can obtain the simple expression

$$ds^2 = -c^2(1 - \frac{v^2}{c^2})dt^2 = -c^2 d\tau^2 \ , \tag{24}$$

which implies the time dilation of a moving atomic clock (on board of a satellite) relative to a clock at rest. Note also that t is the time, related to the motion of the satellite (v respectively is the velocity).

In order to obtain the Shapiro time delay, one has to introduce a gravitational potential U in the temporal and spatial parts of the metric

$$-g_{00} = 1 - \frac{2U}{c^2} \quad , \ g_{0j} = 0 \quad , \ g_{ij} = \left(1 + \frac{2U}{c^2}\right)\delta_{ij} \ , \quad i, j = 1, 2, 3 \ . \tag{25}$$

In other words, there is no Shapiro delay in the case of the metric (23).

Now taking into account expression (4) for the coordinate time Δt (which is in fact also the propagation time of the signal), it can be expressed as [7]

$$\Delta t \approx \frac{1}{c}\int\limits_{path} \sqrt{\frac{g_{ij}}{-g_{00}}dx^i dx^j} \approx \frac{1}{c}\int\limits_{path} \left(1 + \frac{2U}{c^2}\right)\sqrt{\delta_{ij}dx^i dx^j} \tag{26}$$

$$= \frac{\rho}{c} + \frac{1}{c^3}\int\limits_{path} 2U d\rho \ , \tag{27}$$

where ρ is called "geometric path"(in the thesis [16] it was called "the slant range") and the second term is the Shapiro delay term. The calculation in (26)-(27) in [7] however presumed that the second contribution $\frac{2U}{c^2}$ in g_{00} in the metric (25) was neglected.

This will turn out to be the correct approach, but the mathematical justification is more evident from the calculation in [16] for the case of the metric (25)

$$\Delta t \approx \frac{1}{c}\int\limits_{path} \sqrt{\frac{g_{ij}}{-g_{00}}dx^i dx^j} \tag{28}$$

$$= \frac{1}{c^2} \int\limits_{path} \sqrt{\frac{1 - \frac{2U}{c^2}}{1 + \frac{2U}{c^2}} \delta_{ij} dx^i dx^j} \approx \int\limits_{path} d\rho - \frac{1}{c^3} \int\limits_{path} 2U d\rho \quad . \tag{29}$$

The result is the same as in (27), but the sign before $\frac{1}{c^3} \int\limits_{path} 2U d\rho$ is negative. So if the gravitational potential is negative, the second Shapiro delay term will increase the propagation time, as should be. Expression (29) was derived by decomposing the under-integral expression $\sqrt{\frac{1-\frac{2U}{c^2}}{1+\frac{2U}{c^2}}}$ in the s.c. weak limit, when $\frac{2U}{c^2} \ll 1$. Further the weak limit will be used in order to obtain the new expression for the propagation time in terms of the eccentric anomaly angle E, but the general case without using this approximation will be briefly considered. The weak-field limit has an important physical meaning, clarified in the PhD thesis [30]: if $\overrightarrow{R}$ is a vector, related to the signal trajectory with respect to a Earth-centered system, $\overrightarrow{r}_A$ and $\overrightarrow{r}_B$ are vectors, related to the points of emission and reception of the signal and $\overrightarrow{r} = \overrightarrow{r}_A - \overrightarrow{r}_B$, then the tangent line dR to the signal trajectory $\overrightarrow{R}$ deviates from the straight line cdt by $O(\frac{1}{c^2})$ terms. This can be written as $\| dr \| = \| dR \| + O(\frac{1}{c^2})$ and follows also from the null cone equation [30]

$$cdt = \left(1 + 2\frac{G_{\oplus} M_E}{rc^2}\right) \| dR \| \quad , \tag{30}$$

where $G_{\oplus} M_E$ is the s.c. geocentric gravitational constant. Further its numerical value will be given in order to estimate numerically the Shapiro delay term. After integration, the Shapiro delay formulae can be found

$$\Delta t = t_B - t_A = T_{rec} - T_{emis} = \int\limits_0^{R_{AB}} \left(1 + 2\frac{G_{\oplus} M_E}{rc^2}\right) \frac{dR}{c} \tag{31}$$

$$= \frac{R_{AB}}{c} + 2\frac{G_{\oplus} M_E}{c^3} \ln\left(\frac{r_A + r_B + R_{AB}}{r_A + r_B - R_{AB}}\right) \tag{32}$$

$$= \frac{| \overrightarrow{r}_B(t_B) - \overrightarrow{r}_A(t_A) |}{c} + 2\frac{G_{\oplus} M_E}{c^3} \ln\left(\frac{r_A(t_A) + r_B(t_B) + R_{AB}}{r_A(t_A) + r_B(t_B) - R_{AB}}\right) \quad . \tag{33}$$

This is the final formulae, frequently applied in many papers, can be found also in the review paper [31] by Sovers, Fanselow and Jacobs on $VLBI$ radio interferometry. The calculation is presented also in the mentioned papers [30], [16], [7], [27] and others. Note one apparent inconsistency of the above formulae (33) - it depends on the Euclidean distance $R_{AB} = | \overrightarrow{r}_B(t_B) - \overrightarrow{r}_A(t_A) |$between the points of emission $r_A(t_A)$ and reception $r_B(t_B)$ of the signal, which are taken at two different moments of time t_A and t_B. However, r_B will be known after the propagation time is calculated, so equation (33) represents a complicated transcendental equation with respect to the reception time t_B.

The Shapiro delay formulae (32) first was widely applied in the initial publications [5], [6] and others, which were dedicated to the problem about communications between stations on the Earth and satellites, accounting for GRT effects. Numerical estimates for the Shapiro delay term are also known [7] - for the signal between a geostationary satellite with orbital radius 42164 km with a clock on the equator at the same longitude, the Shapiro path delay is -27 ps (picoseconds, 1 $ps = 10^{-9}$ sec). Note that the minus sign is in agreement with the minus sign in front of the second (Shapiro) term in (29).

Later on, in the series of three publications [32], [33] and [34] a theoretical modelling has been performed for inter-satellite communications for the purposes of the $GRAIL$ (Gravity Recovery and Interior Laboratory) and the $GRACE$ (Gravity Recovery and Climate Experiment) space missions-each of these missions represent a system of two spacecrafts at a distance around 200 *miles* and orbiting around either the Earth, the Moon or Mars. Precise ranging between the two spacecrafts in $GRAIL$ or $GRACE$ is achieved by applying General Relativity Theory.

7. Is it necessary to apply the light-like geodesic equations for finding the propagation time?

In the paper [32] it was asserted that the points of emission and reception A and B are joined by a time-like geodesics, but yet the Shapiro delay equation (33) was used and was called a "light-like" equation and the geodesic equation has not been used. The reason, from a theoretical point of view, is a theorem, proved in paragraph 38 in the known monograph by Fock [28]:

Theorem: Let

$$L = \sqrt{2F} = \sqrt{2\left(\frac{1}{2}g_{\alpha\beta}\frac{dx_\alpha}{dp}\frac{dx_\beta}{dp}\right)} \tag{34}$$

is the Lagrange function for the variational problem, when the extremum of the integral

$$s = \int_{p_1}^{p_2} Ldp \tag{35}$$

is to be found and p is a space parameter along a curved line (It should be noted that in the proof in [28] it was not presumed that p is parameter along a geodesic line). Let also p is chosen so that the Lagrange equation of motion is satisfied

$$\frac{dF}{dp} = 0 \implies F = const \implies \frac{d}{dp}\frac{\partial F}{\partial \dot{x}_\alpha} - \frac{\partial F}{\partial x_\alpha} = 0 \quad , \tag{36}$$

which has the first integral

$$\dot{x}_\alpha \frac{\partial F}{\partial \dot{x}_\alpha} - F = F = const \quad , \quad \dot{x}_\alpha \equiv \frac{dx_\alpha}{dp} \quad . \tag{37}$$

Then the geodesic line equations $(\alpha, \beta, \nu = 0, 1, 2, 3)$

$$\frac{d^2 x_\nu}{dp^2} + \Gamma^\nu_{\alpha\beta}\frac{dx_\alpha}{dp}\frac{dx_\beta}{dp} = 0 \tag{38}$$

are compatible with the null-cone equation

$$F = g_{\alpha\beta}\frac{dx_\alpha}{dp}\frac{dx_\beta}{dp} = 0 \tag{39}$$

and in (38) $\Gamma^\nu_{\alpha\beta}$ are the affine connection components for a given metric $g_{\alpha\beta}$.

In other words, since the first integral (37) for the variational problem $s = \int_{p_1}^{p_2} Ldp$ is valid also for the partial case $F = 0$, from the fulfillment of the null-cone equation (39) follows the fulfillment of the geodesic system of equations (39).

This theorem in the monograph by Fock has a great importance for the present investigation. If the propagation time T is found from the null-cone equation, it is not necessary to find the propagation time from the complicated nonlinear system of equations (38). In particular, this is the justification for choosing the parameter p not along the curved trajectory of the signal, but for example along the elliptic orbit of the satellite-both for the cases of plane motion of the satellite (characterized by the eccentric anomaly angle E), motion along a space-distributed orbit (characterized by the single parameter f - the true anomaly angle or the whole multitude of parameters $(M, a, e, \Omega, I, \omega)$, characterizing any disturbing motion.

Lastly, it should be mentioned that the relation between the null cone equation and the geodesic equations is a topic, not very well investigated in General Relativity Theory. Some issues will be mentioned in the conclusion part. But now one such issue can be mentioned as an example. After the variational problem is solved and the geodesic equations (38) are obtained, p becomes the parameter along the geodesic line. However, because of the condition $F = 0$ in (39), one may choose any other parameter p_1, not related to the geodesic line, i.e. one may write $F = g_{\alpha\beta}\frac{dx_\alpha}{dp_1}\frac{dx_\beta}{dp_1} = 0$ as well. This is the case, which is related to the further investigation. But $F = 1$ will be another integral of the system (38) and then one can write only $F = g_{\alpha\beta}\frac{dx_\alpha}{dp}\frac{dx_\beta}{dp} = 1$ [29] and not $g_{\alpha\beta}\frac{dx_\alpha}{dp_1}\frac{dx_\beta}{dp_1} = 0$.

8. New results for the propagation time in terms of elliptic functions-signal, emitted by a satellite on a plane elliptical orbit

8.1. The general case without any approximations

Since we have proved in the preceding section that an arbitrary parametrization of the space coordinates in the null cone equation

$$ds^2 = -c^2\left(1 + \frac{2V}{c^2}\right)(dT)^2 + \left(1 - \frac{2V}{c^2}\right)\left((dx)^2 + (dy)^2 + (dz)^2\right) = 0 \tag{40}$$

can be chosen, we shall investigate the most simple example of a satellite, moving along a plane elliptical orbit, parametrized by (5) $x = a(\cos E - e)$, $y = a\sqrt{1 - e^2}\sin E$, $z = 0$. From the null cone equation after expressing dT, subsequent integration and taking into account, that the satellite velocity can be expressed as

$$\mathbf{v} = \sqrt{v_x^2 + v_y^2} = \frac{na}{(1 - e\cos E)}\sqrt{1 - e^2\cos^2 E} \tag{41}$$

($n = \sqrt{\frac{G_{\oplus}M_{\oplus}}{a^3}}$ is the mean motion), the propagation time can be written as [22], [24]

$$T = \int \frac{\mathbf{v}}{c}\sqrt{\frac{(c^2 - 2V)}{(c^2 + 2V)}}\,dE = \frac{a}{c}\int\sqrt{\frac{a_1 y^3 + a_2 y^2 + a_3 y}{b_1 y^3 + b_2 y^2 + b_3 y + b_4}}dy \quad , \tag{42}$$

where the numerical constants a_1, a_2, a_3, b_1, b_2, b_3, b_4 can easily be calculated, y is the variable $y = 1 - e\cos E$. Integrals of the type (42) are not abelian ones (see the monograph by Prasolov and Solovyev [35]), because abelian integrals are related to algebraic curves $F(x, y) := x^2 - P(y) = 0$, where $P(y)$ is an algebraic polynomial. In the case for the integral (42), the under-integral expression is a rational function and not a polynomial. In the monograph [36], where various analytical solutions of most complicated integrals are presented, no analytical solutions of integrals of the type (42) can be found.

8.2. The weak field approximation $\frac{2V}{c^2} \ll 1$-theoretical and numerical considerations

In the book [28] two approximations are proposed for the metric (40)

$$U \ll c^2 \quad , \tag{43}$$

$$\left(\frac{dx}{dt}\right)^2+\left(\frac{dy}{dt}\right)^2+\left(\frac{dz}{dt}\right)^2=v^2\ll c^2 \tag{44}$$

so that the metric (40) differs insignificantly from the metric

$$ds^2=-c^2\left(1+\frac{2V}{c^2}\right)(dT)^2+\left((dx)^2+(dy)^2+(dz)^2\right) \quad . \tag{45}$$

However, it is not stated that the imposition of such assumptions is not absolutely necessary, and the metric (40) is considered to be the more precise metric. From this point of view, the problem about the calculation of the propagation time in the form of the complicated integral (42) is correctly stated. Moreover, the inequality $\frac{U}{c^2}\ll 1$ is valid only to a certain approximation, and it is important to find the precise approximation for its validity.

For the parameters of the GPS orbit - the typical radius $r_s=26561$ $[km]$ of the GPS orbit, in the review paper [37] the constant

$$\beta=\frac{2V}{c^2}=\frac{2G_{\oplus}M_{\oplus}}{c^2 r_s}\ll 1 \tag{46}$$

β was calculated to be 0.334×10^{-9} [24] with the velocity of light taken to be $c=299792458$ $[\frac{m}{\sec}]$ and the mass of the Earth is taken approximately to be $M_E\approx 5.97\times 10^{24}$ $[kg]$. The geocentric gravitational constant $G_{\oplus}M_{\oplus}$ (obtained from the analysis of laser distance measurements of artificial Earth satellites) was taken to be equal to $G_{\oplus}M_{\oplus}=(3986004.405\pm 1)\times 10^8$ $[\frac{m^3}{\sec^2}]$. The value of $G_{\oplus}M_{\oplus}$ can vary also in another range from $G_{\oplus}M_{\oplus}=3986056.75236\times 10^8$ $[\frac{m^3}{\sec^2}]$ to the value $G_{\oplus}M_{\oplus}=3987999.07898\times 10^8$ $[\frac{m^3}{\sec^2}]$ due to the uncertainties in measuring the Newton gravitational constant $G_{\oplus}$. One of the latest values for $G_{\oplus}$ from deep space experiments was reported in the paper [38] to be $(6.674+0.0003).10^{-11}$ $[\frac{m^3}{kg.\sec^2}]$.

Now let us point out one another numerical fact, which is interesting from the point of view of the preceding result. In the book [1] it was pointed out that at an altitude of 20184 km, due to the difference in the gravitational potential, the satellite atomic time runs faster by 45 $\mu\sec/d$ (microseconds per day, 1 $\mu\sec=10^{-6}$ sec). Consequently, for one second the atomic time will run faster by $0.5208333.10^{-9}$ [sec]. In the following sections it will be proved that the dominant part in the propagation time dT (i.e. the part without the Shapiro delay term, which is very small) will be $dT=0.0281341332790419$ [sec]. The corresponding to this propagation time interval dT atomic time is $d\tau=0.0146531934.10^{-9}$ [sec] and can easily be found from 45 $\mu\sec/d$. The ratio of the two time intervals is equal to

$$\frac{d\tau}{dT}=\frac{0.0146531934.10^{-9}}{0.0281341332790419}=0.52083329721.10^{-9} \ . \tag{47}$$

The very small atomic time interval compared to the propagation interval means that the atomic time can serve as a standard for measuring the propagation time, because it will be able to detect changes even at the nanosecond level.

It should be noted that this value is of the order of 10^{-9}, but it is a little greater than the value $\beta=\frac{2V}{c^2}=0.334\times 10^{-9}$

$$\frac{2V}{c^2 r}=\frac{2GM}{c^2 a(1-e\cos E)}=\frac{\beta}{(1-e\cos E)}=\frac{0.334.10^{-9}}{(1-e\cos E)}\ll 1 \ . \tag{48}$$

So due to the changing gravitational potential in our model ($r=a(1-e\cos E)$ and $\frac{1}{(1-e\cos E)}>1$)

$$\frac{2V}{c^2 r}>\beta\sim 10^{-9} \quad . \tag{49}$$

8.3. Propagation time in terms of a sum of elliptic integrals of the first, second and the third kind in the weak field approximation

This section presents the new results in the papers [22], [24]. From the null cone equation (40) and taking into account the weak-field approximation $\frac{2V}{c^2} \ll 1$, after decomposing the expression (29) into a sum and keeping only the first two sums, one can obtain

$$T = \int \frac{\mathbf{v}}{c} \sqrt{\frac{(1-\frac{2V}{c^2})}{(1+\frac{2V}{c^2})}}\, dt \approx \int \frac{\mathbf{v}}{c}(1-\frac{2V}{c^2})dt = I_1 + I_2 \tag{50}$$

$$= \frac{a}{c} \int \sqrt{1-e^2\cos^2 E}\, dE - \frac{2G_{\oplus}M_{\oplus}}{c^3} \int \sqrt{\frac{1+e\cos E}{1-e\cos E}} dE \quad . \tag{51}$$

The formulae in fact gives the propagation time T of the signal, emitted by the satellite at some initial position (given by the eccentric anomaly angle E_{init}), and the final point (given by E_{fin}) of reception of the signal by another satellite.

The coefficient $\frac{a}{c}$ as a ratio of the large semi-major axis of the orbit and the velocity of light $c = 299792458$ $[\frac{m}{\sec}]$ has a dimension $[m/\frac{m}{\sec}] = [\sec]$. The second coefficient $\frac{2G_{\oplus}M_{\oplus}}{c^3}$ has a corresponding dimension $[\frac{m^3}{\sec^2} : \frac{m^3}{\sec^3}] = [\sec]$, the two integrals $\int \sqrt{1-e^2\cos^2 E}$ and $\int \sqrt{\frac{1+e\cos E}{1-e\cos E}} dE$ after the integration are numbers and represent elliptic integrals of the second kind (first term in (51)) and a sum of elliptic integrals of the first and the third kinds respectively (second term in (51)). The general theory of elliptic integrals and curves is given in the monographs [35] and [43], more specific mathematical problems on a higher level are treated in [39], [41] and [42]. Elliptic integrals of first, second and third kinds are defined in the monographs [35], [44] and [45], but the corresponding definitions are briefly summarized also in the paper [24]. Various approaches for analytical calculation are given in the old book [36].

Let us now express the two integrals in (51) in terms of the various types of elliptic integrals. For the purpose, the first integral can be rewritten as

$$T_1 = \int\limits_0^E \sqrt{1-e^2\cos^2 E} dE = -\int\limits_{\frac{\pi}{2}}^{\frac{\pi}{2}-\overline{E}} \sqrt{1-e^2\sin^2 \overline{E}} d\overline{E} \quad , \tag{52}$$

where it was accounted that $\cos E = \sin(\frac{\pi}{2} - E)$ and also $k = e < 1$ (since e is the eccentricity of the orbit) represents the modulus of the elliptic integral.. The integral (52) represents an elliptic integral of the second kind [35].

In order to calculate the second integral $T_2 = -\frac{2G_{\oplus}M_{\oplus}}{c^3} \int \sqrt{\frac{1+e\cos E}{1-e\cos E}} dE$ in (51), let us perform the substitution

$$\sqrt{\frac{1+e\cos E}{1-e\cos E}} = \overline{y} \tag{53}$$

and introduce also the notations

$$-k^2 = \widetilde{k}^2 = \frac{1-e}{1+e} = q \quad , \quad \frac{\overline{y}}{\overline{\overline{k}}} = y \quad , \quad \left(\frac{\overline{y}}{k}\right)^2 = \left(\frac{\overline{y}}{\overline{\overline{k}}}\right)^2 = \widetilde{y} \quad , \quad \widetilde{y} = -\widetilde{\widetilde{y}} \quad . \tag{54}$$

Then the integral T_2 can be represented as a sum of two integrals, i.e. $T_2 = I_2^{(A)} + I_2^{(B)}$. The first integral is

$$I_2^{(A)} = \frac{4GM}{c^3} \frac{1}{\widetilde{k}\, i\sqrt{1-e^2}\, i} \int \frac{d\widetilde{\widetilde{y}}}{\sqrt{\widetilde{\widetilde{y}}\left(\widetilde{\widetilde{y}}+1\right)\left(\widetilde{\widetilde{y}}+\frac{1}{k^4}\right)}} \tag{55}$$

is of zero - order and of the first kind and is in the s.c. Weierstrass form (third-order polynomial under the square root in the under-integral expression). The modulus of this elliptic integral can be calculated if by a suitable linear variable transformation of $\widetilde{y}$ the integral is transformed to an integral in the s.c. Legendre form $\int \frac{dx}{\sqrt{(1-x^2)(1-k^2x^2)}}$, associated with the elliptic curve $y^2 = (1 - x^2)(1 - k^2x^2)$ (k is the modulus of the elliptic integral).

It is important to note that the integral is a real-valued expression, because the propagation time should be real. This fact again confirms the of the applied approach of choosing the space coordinates parametrization to coincide with the orbit of the satellite.

The second integral $I_2^{(B)}$ in the expression for $T_2 = I_2^{(A)} + I_2^{(B)}$ can also be written in the form of a real-valued expression

$$I_2^{(B)} = \frac{4GM}{c^3q^2} \frac{1}{\sqrt{1-e^2}} \int \frac{d\widetilde{y}}{(\widetilde{y} - \frac{1}{q})\sqrt{\widetilde{y}(\widetilde{y}+1)(\widetilde{y}+\frac{1}{q^2})}} \quad . \tag{56}$$

This is an elliptic integral of the third kind.

8.4. Numerical calculation of the propagation time of a signal, emitted by a satellite on a plane elliptical orbit

8.4.1. Numerical calculation of the first six iterations of the eccentric anomaly angle E We shall perform some numerical calculations of the propagation time, based on the derived formulae (51). The numerical data about the parameters of the GPS orbit are taken from the PhD dissertation [16] and are known with great precision. From all the parameters, listed below only the first three will be used in the calculation, performed by the online program web2.0 scientific calculator [46]

semi-major axis a	26560.25169632944 $[km]$,
eccentricity e	0.01323881349526 ,
mean anomaly M	-0.3134513508155 $[rad]$,
inclination I	0.9614884100802 $[rad]$,
longitude of the ascending node Ω	-0.4495096737336 $[rad]$,
argument of perigee ω	-3.001488651204 $[rad]$.

Most important is to find the eccentric anomaly angle E from the Kepler equation $M = E - e\sin E$. The calculation will be based on formulaes (2.54) and (2.55) in Ch.2 of the monograph [11].

The iterative sequence of formulaes are

$$E_{(i+1)} = M + e\sin E_{(i)} \quad , \quad i = 0, 1, 2, ..., \tag{57}$$

where for the first approximation it is assumed $E_0 = M = -0.3134513508155$ $[rad]$ The first three iterative solutions are given according to the following formulaes:

$$E_{(1)} = M + e\sin M \quad , \tag{58}$$

$$E_{(2)} = M + e\sin E_{(1)} = M + e\sin(M + e\sin M) \quad , \tag{59}$$

$$E_{(3)} = M + e\sin E_{(2)} = M + e\sin[M + e\sin(M + e\sin M)] \quad . \tag{60}$$

Thus from the formulaes for $E_{(3)}$, $E_{(5)}$ and $E_{(6)}$ the following numerical values can be found

$$E_{(3)} = M + e\sin E_{(2)} = -0.31758547588467897473 \ [rad] \quad , \tag{61}$$

$$E_{(4)} = M + e\sin E_{(3)} = -0.31758548401096719083 \ [rad] \quad , \tag{62}$$

$$E_{(5)} = M + e \sin E_{(4)} = -0.31758548411317083102 \ [rad] \quad , \tag{63}$$

$$E_{(6)} = M + e \sin E_{(5)} = -0.31758548411445499592 \ [rad] \quad . \tag{64}$$

The approximation $E_{(5)}$ up to the ninth digit is identical with $E_{(4)}$ and $E_{(6)}$ up to the eleventh digit is identical with $E_{(5)}$. The approximation $E_{(6)}$ also up to the seventh digit is identical with $E_{(3)}$, given by (61). However, if the approximate value for $E_{(3)}$ is compared with the value for $E_{(6)}$, the coincidence is only up to the fifth digit.

It is therefore necessary to check what is the impact of the approximations in the eccentric anomaly E on the approximations of the time of propagation of the signal.

8.4.2. Numerical calculation of the first $O(\frac{1}{c})$ correction in the propagation time The corresponding elliptic integral (52) of the second kind for the eccentric anomaly approximation $E_{(3)}$ is

$$T_1^{(E_3)} = \int\limits_0^{E_{(3)}} \sqrt{1 - e^2 \cos^2 E} dE = -0.31755726812593936045 \quad . \tag{65}$$

If we compare this value with the value (61) $E_{(3)} = -0.31758547588467897473 \ [rad]$, then it can be noted that the integration changes the value of $E_{(3)}$ after the fourth digit after the decimal dot.

The same integral with $E_{(6)}$ as an upper integration limit is

$$T_1^{(E_6)} = \int\limits_0^{E_{(6)}} \sqrt{1 - e^2 \cos^2 E} dE = -0.31755856896388663853 6 \quad . \tag{66}$$

This value is different from the value (65) after the fifth digit, so it is a better approximation. However, if compared with $E_{(6)} = -0.31758548411445499592 9 \ [rad]$, the integration in (66) changes the value of $E_{(6)}$ after the fourth digit after the decimal dot.

The calculation of the first $O(\frac{1}{c})$ time correction $\frac{a}{c}T_1$ for the values of the GPS orbit and for the eccentric anomaly $E_{(3)}$ gives the following numerical value

$$\frac{a}{c} T_1^{(E_3)} = -0.0281341332790419 \ [\text{sec}] \quad . \tag{67}$$

Correspondingly, the first $O(\frac{1}{c})$ time correction $\frac{a}{c}T_1^{(E_6)}$ for the eccentric anomaly $E_{(6)}$ is

$$\frac{a}{c} T_1^{(E_6)} = -0.0281342485273829 \ [\text{sec}] \quad . \tag{68}$$

So the two $O(\frac{1}{c})$ time corrections (67) and (68) are identical up to the sixth digit after the decimal dot. Consequently, the important conclusion is that the eccentric anomaly at the sixth approximation level can ensure microsecond stability of the $O(\frac{1}{c})$ time correction. This means also that the third approximation $E_{(3)}$ can be reliably used, at least at the microsecond level.

One more argument in support of the microsecond approximation is that for a satellite, moving along circular orbit (when $e = 0$) the above expression acquires the form:

$$\frac{a}{c} t_{1(circular)}^{(E_3)} = \frac{a}{c} E_{(3)} = -0.028136517830159266 \ [\text{sec}] = -28136.517830159266 \ [\mu\,\text{sec}] \quad . \tag{69}$$

Evidently, the coordinate time for the circular orbit differs from the coordinate time for the GPS elliptic orbit only after the fifth digit after the decimal dot.

8.4.3. Numerical calculation of the Shapiro delay term (the $O(\frac{1}{c^3})$ time correction) Now let us calculate numerically the whole under-integral expression in formulae (51) for the propagation time, taking into account the 25−th digit after the decimal dot

$$\frac{a}{c}\sqrt{1-e^2\cos^2 E} - \frac{2GM}{c^3}\sqrt{\frac{1+e\cos E_{(3)}}{1-e\cos E_{(3)}}} = 0.0885884561709019072629880 \quad (70)$$

$$-0.0000000000299618094618168 = 0.0885884561409400978011712 \quad . \quad (71)$$

The second Shapiro delay term contains ten zeroes after the decimal dot, so the overall result of the calculation of both terms in (70) - (71) is changed by the Shapiro term at and after the picosecond level (1 $p\sec = 10^{-9}$ sec). This result, based on the simple application of the celestial mechanics approach in the calculation of the propagation time is interesting, if compared with the calculations, based on the formulae (32) $T = \frac{R_{AB}}{c} + 2\frac{G_{\oplus}M_E}{c^3}\ln\left(\frac{r_A+r_B+R_{AB}}{r_A+r_B-R_{AB}}\right)$. From the literature it can be seen that relativistic effects on light propagation from Satellite Laser Ranging (SLR) data are measured with an accuracy of 1 μm (1 micrometer), which corresponds to 0.01 $p\sec$ (see also paper [47]). The relativistic observables, defined for the $GRAIL$ mission in [32] are also accurate to 1 μm.

Let us present one final proof for the correctness of the applied in the papers [22], [23], [24] and [25], comparing the calculated propagation time of the signal with the celestial time t_{cel}, which can be calculated from $M = n(t - t_P)$ (M is the mean anomaly with the given numerical data), where t_P is the initial time of perigee passage.

$$\frac{\text{propagation time for electromagnetic signals}}{\text{celestial time from Kepler equation}} = \frac{0.0281179694658269766628639[sec]}{37.508256148[sec]} \quad (72)$$

$$= 0.00074964747372237377033575958\,2 \quad . \quad (73)$$

In other words, the propagation time is 10^4 times greater than the celestial time, which will mean that for this celestial time 37.5082561[sec] the satellite moves at a distance 145.125[km] (the velocity of the satellite is taken to be $v_S = 3.874$ [$km/$sec]) and the light signal will move at a distance 8435.3908[km].

9. New results for the propagation time in terms of elliptic functions-signal, emitted by a satellite on a space-distributed elliptical orbit

9.1. Parametrization of the space coordinates

Now we shall investigate the more complicated case of the propagation time of a signal, emitted by a satellite on a space distributed orbit [24]. The parametrization (with some minor differences) can be found in nearly all books on celestial mechanics, but a particularly clear derivation is given in the monograph [11]. Below we shall use the parametrization in the well-known monograph [8]

$$x = \frac{a(1-e^2)}{1+e\cos f}\left[\cos\Omega\cos(\omega+f) - \sin\Omega\sin(\omega+f)\cos I\right] \quad , \quad (74)$$

$$y = \frac{a(1-e^2)}{1+e\cos f}\left[\sin\Omega\cos(\omega+f) + \cos\Omega\sin(\omega+f)\cos I\right] \quad , \quad (75)$$

$$z = \frac{a(1-e^2)}{1+e\cos f}\sin(\omega+f)\sin I \quad , \quad (76)$$

where $r = \frac{a(1-e^2)}{1+e\cos f}$ is the radius-vector in the orbital plane, the angle Ω of the longitude of the right ascension of the ascending node is the angle between the line of nodes and the direction to the vernal equinox, the argument of perigee (periapsis) ω is the angle within the orbital plane from the ascending node to perigee in the direction of the satellite motion ($0 \leq \omega \leq 360^0$). The angle I is the inclination of the orbit with respect to the equatorial plane and the true anomaly f geometrically represents the angle between the line of nodes and the position vector $\overrightarrow{r}$ on the orbital plane. Since the angle f is related to the motion of the satellite and all the other parameters of the orbit do not change during the motion of the satellite, it can easily be found that

$$\sqrt{(dx)^2 + (dy)^2 + (dz)^2} = \sqrt{(v_f^x)^2 + (v_f^y)^2 + (v_f^z)^2} df = v_f df \quad , \tag{77}$$

where the velocity v_f , associated to the true anomaly angle f is given by

$$v_f = v = \frac{na}{\sqrt{1-e^2}} \sqrt{1 + e^2 + 2e\cos f} \quad . \tag{78}$$

Making use of the null cone equation (40) and also of the approximation $\beta = \frac{2V}{c^2} = \frac{2G_{\oplus}M_{\oplus}}{c^2 a} \ll 1$, one can obtain the general formulae for the propagation time in the form

$$T = \int \frac{v}{c}(1 - \frac{2V}{c^2})dt = \widetilde{T}_1 + \widetilde{T}_2 = \frac{1}{c}\int v dt - \frac{2}{c^3}\int vV dt \quad . \tag{79}$$

As it will be shown, the corrections $\widetilde{T}_1$ has two parts, but the whole correction $\widetilde{T}_2$ does not have two parts. The first part $\widetilde{T}_1$ can be written as

$$\widetilde{T}_1 = \frac{1}{c}\int v_f df = \frac{na}{c\sqrt{1-e^2}} \int \sqrt{1+e^2+2e\cos f} df = \frac{na}{c\sqrt{1-e^2}} (\overline{T}_1^{(1)} + \overline{T}_1^{(2)}) \quad . \tag{80}$$

It is interesting that $\widetilde{T}_1$can be analytically calculated without the use of elliptic integrals, but also with the application of elliptic integrals.

9.2. *Calculation of $\overline{T}_1^{(1)} + \overline{T}_1^{(2)}$ without elliptic integrals*

For the first case without the use of elliptic integrals, the integral $\widetilde{T}_1^{(2)}$ is calculated in [24] as

$$\overline{T}_1^{(2)} = -\frac{(1+e)\sqrt{2}}{\sqrt{q}\sqrt{3e^2+2e+3}} \ln\left[\left(\frac{\sqrt{2}m_1(f_b;r_b) + \frac{1}{2}m_2(f_b;r_b)}{\sqrt{2}m_1(f_a;r_a) + \frac{1}{2}m_2(f_a;r_a)}\right)\left(\frac{m_3(f_a;r_a)}{m3(f_b;r_b)}\right)\right] \tag{81}$$

and $m_1(f;r)$, $m_2(f;r)$, $m_3(f;r)$ are complicated expressions, written in terms either of the initial and final true anomaly angles f_a and f_b or, of the initial distance r_a (at which the emission of the signal takes place) and the final point r_b of reception on the same orbit, corresponding to the propagation time $\overline{T}_1^{(2)}$. The first expression for $\overline{T}_1^{(1)}$ can be written in the same manner, since $\overline{T}_1^{(1)}$ and $\overline{T}_1^{(2)}$ are calculated from the derived integral [24]

$$\overline{T}_1^{(1)} + \overline{T}_1^{(2)} = -\frac{i}{\sqrt{q}}\left[\int \frac{d\overline{m}}{\left(\overline{m} - \frac{q^2}{2}\right)\sqrt{\overline{m} - \frac{q^2}{2}}} + \int \frac{d\overline{m}}{\left(\overline{m} + \frac{q^2}{2}\right)\sqrt{\overline{m} - \frac{q^2}{2}}}\right] \quad , \tag{82}$$

where the expression for $\overline{T}_1^{(2)}$ is obtained after integration of the second integral in (82) and the expression for $\overline{T}_1^{(1)}$- from the first integral in (82). The integration of both integrals is performed by using a formulae from the monograph [36] and given also in the paper [24].

9.3. Calculation of $\widetilde{T}_1$ by means of elliptic integrals of the second order

The integral (80) for $\widetilde{T}_1 = \frac{na}{c\sqrt{1-e^2}} \int \sqrt{1+e^2+2e\cos f} df$ can be calculated by making use of the substitution [24]

$$y = \sqrt{\frac{1}{q}\frac{(1+e\cos E)}{(1-e\cos E)}} \quad , \; q = \frac{1-e}{1+e} \tag{83}$$

and also of the well-known relation (11) $\tan\frac{f}{2} = \sqrt{\frac{1-\cos f}{1+\cos f}} = \sqrt{\frac{1+e}{1-e}}\tan\frac{E}{2}$ from celestial mechanics between the eccentric anomaly angle E and the true anomaly angle f [8]. Note that from the integral (80) it is not clear that it will be an elliptic integral, this becomes clear after the application of the convenient transformation (83).

Then we can write the integral $\widetilde{T}_1$ in the form of an elliptic integral of the second order (because of the term y^2 in the nominator), which is also of the first kind in the Legendre form

$$\widetilde{T}_1 = -2i\frac{na}{c}q^{\frac{3}{2}} \int \frac{y^2 dy}{\sqrt{(1-y^2)(1-q^2y^2)}} = -2i\frac{na}{c}q^{\frac{3}{2}}\widetilde{J}_2^{(4)}(y;q) \quad . \tag{84}$$

Note that for this calculation with elliptic integrals, there is no need to represent $\widetilde{T}_1$ in the form of any sum $\widetilde{T}_1 = \widetilde{T}_1^{(1)} + \widetilde{T}_1^{(2)}$. If we compare the last formulae with the formulae (80) $\widetilde{T}_1 = \frac{1}{c}\int v_f df = \frac{na}{c\sqrt{1-e^2}}(\overline{T}_1^{(1)} + \overline{T}_1^{(2)})$ (expressed by functions, which are not elliptic), two important conclusions can be made:

A. The elliptic integral $\widetilde{J}_2^{(4)}(y;q)$ can be expressed analytically. The paper [25] is dedicated to this problem (see the references there), and further some basic facts about a new approach will be mentioned briefly.

B. The elliptic integral $\widetilde{J}_2^{(4)}(y;q)$ is imaginary, so that the expression for the propagation time $\widetilde{T}_1$ is real-valued, as this was proved for expressions (55) and (56). Further this fact for $\widetilde{T}_1$ shall also be proved.

The integral $\widetilde{J}_2^{(4)}(y;q)$ can also be represented as

$$\widetilde{J}_2^{(4)}(y;q) = \int \frac{y^2 dy}{\sqrt{(1-y^2)(1-q^2y^2)}} = -\frac{1}{q^2}\int \sqrt{1-q^2\sin^2\varphi}\, d\varphi + \frac{1}{q^2}\int \frac{dy}{\sqrt{(1-y^2)(1-q^2y^2)}} \quad , \tag{85}$$

which can be proved very easily [24]. The first integral in (85) is an elliptic integral of the second kind (denoted usually by $E(\varphi) = \int \sqrt{1-q^2\sin^2\varphi}\, d\varphi$, where $y = \sin\varphi$). So we obtain a relation between the second-order elliptic integral $\widetilde{J}_2^{(4)}(y;q)$ of the first kind in the Legendre form, the zero-order elliptic integral of the first kind in the Legendre form $\widetilde{J}_0^{(4)}(y;q)$ (the second integral in (85)) and the elliptic integral $E(\varphi)$. Elliptic integrals of higher order are investigated in the monographs [35], [44] and [45], the other cited above monographs [39], [41] and [42] do not deal with elliptic integrals of higher order, but only with mathematical problems of elliptic integrals and curves. A short summary of the different kinds of higher-order elliptic integrals is given in the paper [24]. For the moment, it is important to mention that the upper subscript in the notation for $\widetilde{J}_2^{(4)}(y;q)$ means that the polynomial in the denominator is of the fourth degree, the lower subscript 2 means that the integral is of the second order. Further analogous notations will also appear, when calculating the $O(\frac{1}{c^3})$ part of the integral (79).

9.4. Calculation of the second part of the integral (79) (the $O(\frac{1}{c^3})$ time correction) by means of elliptic integrals of the fourth order

Let us calculate the second part

$$T_2 = -\frac{2}{c^3}\int vV dt = -\frac{2G_{\oplus}M_{\oplus}}{c^3}\int \frac{v_f}{r}df \tag{86}$$

in the integral (79). The integral now acquires the form

$$T_2 = -\frac{2G_{\oplus}M_{\oplus}}{c^3}.\frac{na}{a\sqrt{1-e^2}}\int (1+e\cos f)\sqrt{1+2e\cos f+e^2}df = T_2^{(1)} + T_2^{(2)} \ , \tag{87}$$

which allows to represent it as a sum of two parts $T_2^{(1)} + T_2^{(2)}$. The first integral is

$$T_2^{(1)} = -\frac{2G_{\oplus}M_{\oplus}}{c^3}.\frac{n}{(1-e^2)^{\frac{3}{2}}}\int \sqrt{1+2e\cos f+e^2}df = -\frac{2G_{\oplus}M_{\oplus}}{c^3}.\frac{n}{(1-e^2)^{\frac{3}{2}}}\widetilde{T}_1 \tag{88}$$

and is therefore equal to the previously calculated in (84) integral $\widetilde{T}_1$, multiplied by the coefficient $-\frac{2G_{\oplus}M_{\oplus}}{c^3}.\frac{n}{(1-e^2)^{\frac{3}{2}}}$. The second term $T_2^{(2)}$ in the second correction T_2 is

$$T_2^{(2)} = -\frac{2G_{\oplus}M_{\oplus}}{c^3}.\frac{ne}{(1-e^2)^{\frac{3}{2}}}\int \cos f\sqrt{1+2e\cos f+e^2}df = -\frac{2G_{\oplus}M_{\oplus}}{c^3}.\frac{ne}{(1-e^2)^{\frac{3}{2}}}\widetilde{T}_2^{(2)} \ , \tag{89}$$

where $\widetilde{T}_2^{(2)}$ is the notation for the more complicated integral

$$\widetilde{T}_2^{(2)} = \int \cos f\sqrt{1+2e\cos f+e^2}df \ . \tag{90}$$

The whole correction $T_2^{(2)}$ is calculated in [24] to be

$$T_2^{(2)} = -inq^{\frac{5}{2}}\frac{2G_{\oplus}M_{\oplus}}{c^3}(1+e^2)\widetilde{J}_2^{(4)}(\widetilde{y},q) + inq^{\frac{3}{2}}\frac{2G_{\oplus}M_{\oplus}}{c^3}\frac{(1+e^2)}{(1-e^2)}\widetilde{J}_4^{(4)}(\widetilde{y},q) \ , \tag{91}$$

where $\widetilde{J}_2^{(4)}(\widetilde{y},q)$ is the previous integral (85) and $\widetilde{J}_4^{(4)}(\widetilde{y},q)$ is the new, fourth-order elliptic integral

$$\widetilde{J}_4^{(4)}(\widetilde{y},q) = \frac{q^5}{i}\int \frac{\widehat{y}^4 d\widehat{y}}{\sqrt{(\widehat{y}^2-1)(1-q^2\widehat{y}^2)}} \ , \tag{92}$$

written in terms of the variable $\widehat{y} = \frac{\widetilde{y}}{q} = \frac{\sqrt{1+2e\cos f+e^2}}{q}$.

9.5. A mathematical proof for the real-valuedness of all the $O(\frac{1}{c})$ and $O(\frac{1}{c^3})$ time corrections for the case of space-distributed orbits. The recurrent system of equations and finding the second and fourth order elliptic integrals

Proving the real-valuedness of all the components of the propagation time is an important step in proving the correctness of choosing the parametrization (74) - (76) of the space coordinates x, y, z in terms of the six Keplerian elements $(M, a, e, \Omega, I, \omega)$. For the previous case of plane orbits the proof was performed also in [24], but for this case of space-distributed orbits the proof will be much more complicated.

9.5.1. Real-valuedness of the expression for $\widetilde{T}_1$ Let us begin first with the proof of the real-valuedness of the first $O(\frac{1}{c})$ correction (80) $\widetilde{T}_1 = \frac{na}{c\sqrt{1-e^2}} \int \sqrt{1+e^2+2e\cos f df}$ [24]. From the representation of the integral (84) $\widetilde{T}_1 = -2i\frac{na}{c}q^{\frac{3}{2}}\widetilde{J}_2^{(4)}(y;q)$ it follows that $\widetilde{T}_1$ will be a real-valued expression if the second order elliptic integral $\widetilde{J}_2^{(4)}(y;q)$ will turn out to be imaginary.

For the purpose, let us perform the variable change

$$\widetilde{y} = \frac{\sqrt{1+2e\cos f + e^2}}{1+e} = \frac{y}{1+e} \quad , \tag{93}$$

after which the integral acquires the form

$$\widetilde{T}_1 = i\frac{2na(1+e)}{cq\sqrt{1-e^2}} \int \frac{\widetilde{y}^2 d\widetilde{y}}{\sqrt{(1-\widetilde{y}^2)\left(1-\frac{\widetilde{y}^2}{q^2}\right)}} = \frac{2na(1+e)}{cq\sqrt{1-e^2}} \int \frac{\widetilde{y}^2 d\widetilde{y}}{\sqrt{(1-\widetilde{y}^2)\left(\frac{\widetilde{y}^2}{q^2}-1\right)}} \quad . \tag{94}$$

Now it is evident that if $\widetilde{T}_1$ is real, then the following identification should be made

$$\widetilde{J}_2^{(4)}(\widetilde{y}, \frac{1}{q}) = \frac{1}{i} \int \frac{\widetilde{y}^2 d\widetilde{y}}{\sqrt{(1-\widetilde{y}^2)\left(\frac{\widetilde{y}^2}{q^2}-1\right)}} = \int \frac{\widetilde{y}^2 d\widetilde{y}}{\sqrt{(1-\widetilde{y}^2)\left(1-\frac{\widetilde{y}^2}{q^2}\right)}} \quad . \tag{95}$$

So let us prove that the expression $(1-\widetilde{y}^2)\left(1-\frac{\widetilde{y}^2}{q^2}\right)$ inside the square in the denominator in the second integral can take negative values. Respectively, then $(1-\widetilde{y}^2)\left(\frac{\widetilde{y}^2}{q^2}-1\right)$ in the first integral will be positive and thus in both cases the elliptic integral $\widetilde{J}_2^{(4)}(\widetilde{y}, \frac{1}{q})$ will not be real-valued. Remembering the definition for the variable change (93) and also that $\cos f \leq 1$, the following series of inequalities can easily be proved

$$\widetilde{y} \leq 1 \ , \rhd 1-\widetilde{y}^2 \geq 0 \ , \quad 1-\frac{\widetilde{y}^2}{q^2} \geq -\frac{(1-q^2)}{q^2} = 1-\frac{1}{q^2} = -\frac{4e}{(1-e)^2} \ . \tag{96}$$

Since $1-\frac{\widetilde{y}^2}{q^2}$ can take negative values from $-\frac{(1-q^2)}{q^2}$ to 0 and keeping in mind the definition $q^2 = \left(\frac{1-e}{1+e}\right)^2 < 1$, it becomes clear that $\widetilde{T}_1$ in (94) is a real-valued expression. Let us clarify also the statement "the expression inside the square in the denominator can take negative values".

It does not mean that only for values from $-\frac{(1-q^2)}{q^2}$ to 0 the expression for $\widetilde{J}_2^{(4)}(\widetilde{y}, \frac{1}{q})$ is imaginary!

Let us present a more elegant and simple proof that the inequalities 96 are fulfilled. Inverting the sign of the last inequality in (96), we obtain the inequality

$$\frac{\widetilde{y}^2}{q^2} - 1 \leq \frac{4e}{(1-e)^2} \quad , \tag{97}$$

should be proved. Keeping in mind the definition for the variable $\widetilde{y}^2$ (93) and also for q^2, the inequality (97) is transformed to

$$\frac{(1+2e\cos f + e^2)}{(1+e)^2} - \frac{(1-e)^2}{(1+e)^2} \leq \frac{4e}{(1-e)^2}q^2 = \frac{4e}{(1+e)^2} \quad . \tag{98}$$

From here it follows

$$1+2e\cos f + e^2 \leq 4e + (1-e)^2 \quad \Rightarrow 2e\cos f \leq 2e \quad , \tag{99}$$

which is fulfilled for all eccentricities e, because of the simple trigonometric inequality $\cos f \leq 1$ for the function cos and also because $0 < e < 1$.

Returning to the defining expression (84) $\widetilde{T}_1 = -2i\frac{na}{c}q^{\frac{3}{2}}\widetilde{J}_2^{(4)}(y;q)$ and after having proved that $\widetilde{T}_1$ is real-valued, it follows that $\widetilde{J}_2^{(4)}(y;q)$ (and also the integral (95) for $\widetilde{J}_2^{(4)}(\widetilde{y},\frac{1}{q})$) are also imaginary expressions.

9.5.2. Real-valuedness of the expression for $T_2^{(2)}$- proof by means of the recurrent system of equations for the elliptic functions According to formulae (91), $T_2^{(2)}$ was represented as a sum of two imaginary expressions $-inq^{\frac{5}{2}}\frac{2G_{\oplus}M_{\oplus}}{c^3}(1+e^2)\widetilde{J}_2^{(4)}(\widetilde{y},q)$ and also $+inq^{\frac{3}{2}}\frac{2G_{\oplus}M_{\oplus}}{c^3}\frac{(1+e^2)}{(1-e^2)}\widetilde{J}_4^{(4)}(\widetilde{y},q)$. Since we proved that $\widetilde{J}_2^{(4)}(\widetilde{y},q)$ is an imaginary expression, it remains to prove that $\widetilde{J}_4^{(4)}(\widetilde{y},q)$ is also an imaginary expression.

The proof now will be made in another way, by using the recurrent system of equations for the elliptic functions (see again the monographs [35], [44] and [45]), obtained after calculating the derivatives

$$\frac{d}{d\widetilde{y}}\left(\sqrt{(1-\widetilde{y}^2)\,(1-q^2\widetilde{y}^2)}\right) \text{ and } \frac{d}{d\widetilde{y}}\left(\widetilde{y}\sqrt{(1-\widetilde{y}^2)\,(1-q^2\widetilde{y}^2)}\right) \tag{100}$$

and integrating the resulting equations from $\widetilde{y}_0$ to $\widetilde{y}_1$. Combining all the three equations, the following two equations can be derived for $\widetilde{J}_4^{(4)}(\widetilde{y},q)$ and $\widetilde{J}_3^{(4)}(\widetilde{y},q)$

$$\widetilde{J}_4^{(4)}(\widetilde{y},q) = \frac{1}{3q^2}\left[\widetilde{y}\sqrt{(1-\widetilde{y}^2)\,(1-q^2\widetilde{y}^2)}\right]\Big|_{\widetilde{y}=\widetilde{y}_0}^{\widetilde{y}=\widetilde{y}_1} + \frac{2(1+q^2)}{3q^2}\widetilde{J}_2^{(4)}(\widetilde{y},q) - \frac{1}{3q^2}\widetilde{J}_0^{(4)}(\widetilde{y},q) \quad , \tag{101}$$

$$\int\sqrt{(1-\widetilde{y}^2)\,(1-q^2\widetilde{y}^2)}d\widetilde{y} = \frac{2}{3}\widetilde{J}_0^{(4)}(\widetilde{y},q) - \frac{1}{3}(1+q^2)\widetilde{J}_2^{(4)}(\widetilde{y},q) + \frac{1}{3}\left[\widetilde{y}\sqrt{(1-\widetilde{y}^2)\,(1-q^2\widetilde{y}^2)}\right]\Big|_{\widetilde{y}=\widetilde{y}_0}^{\widetilde{y}=\widetilde{y}_1} . \tag{102}$$

In the preceding section it was proved that the integral $\widetilde{J}_2^{(4)}(\widetilde{y},q)$ is imaginary, consequently since it enters expression (101) for $\widetilde{J}_4^{(4)}(\widetilde{y},q)$, this integral is also an imaginary one. Also, the integral $\int\sqrt{(1-\widetilde{y}^2)\,(1-q^2\widetilde{y}^2)}d\widetilde{y}$ will also be an imaginary one, because the integral $\widetilde{J}_2^{(4)}(\widetilde{y},q)$ enters the right-hand side of expression (102). Of course, in order to claim that the imaginary contribution $\frac{2(1+q^2)}{3q^2}\widetilde{J}_2^{(4)}(\widetilde{y},q)$ in (101) does not cancel with any of the other two contributions, one should investigate the first and the third terms in (101). Such a cancellation is not likely to happen, but the more rigorous mathematical approach would require to investigate whether the elliptic integral of zero-order in the Legendre form $\widetilde{J}_0^{(4)}(\widetilde{y},q)$ will be real-valued or imaginary. In monographs on elliptic functions, which are more related to the theory of analytical functions [48], [49], [44] such an analysis has been performed.

9.6. Real-valuedness and complex valuedness of elliptic integrals of the zero-order in the Legendre form-basic knowledge about the Christoffel-Schwartz integral

Zero-order elliptic integrals in the Legendre form such as $J_0^{(4)}(y,q) = \int\frac{dy}{\sqrt{(1-y^2)(1-q^2y^2)}}$ are analyzed by means of the well-known Christoffel - Schwartz integral (see especially [48], [49], [44], [50], but also many other monographs on analytical functions)

$$w = f(z) = C\int_0^z (s-a_1)^{\alpha_1-1}(s-a_2)^{\alpha_2-1}.......(s-a_n)^{\alpha_n-1}ds + C_1 \quad , \tag{103}$$

which represents a conformal mapping of the upper complex half-plane onto the inner part of an $n-$polygon. In the last formulae a_1,a_2,......a_n are points on the real axis and α_1, α_2....α_n denote the inner angles, represented by real numbers. Each one of these numbers is not greater than 2 and for them the following equality

$$\alpha_1 + \alpha_2 + ... + \alpha_n = n - 2 \tag{104}$$

is fulfilled. For the four-dimensional case of the zero-order integral $J_0^{(4)}(y,q)$, formulae (103) represents a mapping of the upper complex half-plane onto the rectangle (i.e. the $4-$polygon, which for the case turns out to be the rectangle). So the integral $J(y,q)$ is a partial case of the integral (103) for the following values of the parameters α_n, a_n and the constants C_1 and C:

$$\alpha_1 = \alpha_2 = \alpha_3 = \alpha_4 = \frac{1}{2} \quad , \quad a_1 = -\frac{1}{q} \quad , \quad a_2 = -1 \; , \tag{105}$$

$$a_3 = 1 \quad , \quad a_4 = \frac{1}{k} \; , \quad C_1 = 0 \quad , \quad C = \frac{1}{q} \quad . \tag{106}$$

Therefore, the integral $J_0^{(4)}(y,q)$ can be represented in the form of the following Christoffel-Schwartz integral

$$J_0^{(4)}(y,q) = f(z) = \frac{1}{q}\int\limits_0^z \left(y+\frac{1}{q}\right)^{\frac{1}{2}-1} (y+1)^{\frac{1}{2}-1}(y-1)^{\frac{1}{2}-1}\left(y-\frac{1}{q}\right)^{\frac{1}{2}-1} dy \quad . \tag{107}$$

At the points a_1,a_2,a_3,a_4 the under-integral function in the integral $J_0^{(4)}(y,q)$ (similarly - also in the integral (103) is a branching multi-valued function, which maps the points on the real axis onto the points on the segments of the rectangle, situated on the complex plane. Each expression $(s-a_k)^{\alpha_k - 1}$ in (103) should be interpreted as the branch of the multi-valued function which on the real axis (when $z = s > a_k$) takes real positive values. Indeed, if s is in the segment $[0,1]$, then the function $f(z)$ in [48] takes values from zero to the value of the elliptic integral of the first kind:

$$K = \int\limits_0^1 \frac{ds}{\sqrt{(1-s^2)\,(1-q^2 s^2)}} \quad . \tag{108}$$

In fact, this is an equivalent formulation of the period of the elliptic integral $K = F(\frac{\pi}{2}, e) = \int\limits_0^{\frac{\pi}{2}} \frac{dE}{\sqrt{1-e^2\sin^2 E}}$. Since the function $(1-s^2)\,(1-q^2s^2)$ is a two-valued function at the branching points $s = \pm 1$ and $s = \pm\frac{1}{q}$, one has to choose the under-integral function so that to ensure the non-interruptness of this function, when $1 < s < \frac{1}{q}$. This means that for this case it should be written as [48]

$$\frac{1}{\pm i\sqrt{(s^2-1)\,(1-q^2s^2)}} \quad . \tag{109}$$

Consequently, for $s \subset [1, \frac{1}{q}]$ the function $w = f(z)$ maps the interval $[1, \frac{1}{q}]$ into the interval $[K, K + i\widetilde{K}']$ on the complex plane, where

$$\widetilde{K}' = \int\limits_1^{\frac{1}{q}} \frac{ds}{\sqrt{(s^2-1)\,(1-q^2s^2)}} \quad . \tag{110}$$

For $s \subset [1, \frac{1}{q}]$ the function $w = f(z)$ (107) maps the interval $[1, \frac{1}{q}]$ into the interval $[K, K + i\widetilde{K}\text{'}]$ on the complex plane, where

$$\widetilde{K}\text{'} = \int_{1}^{\frac{1}{q}} \frac{ds}{\sqrt{(s^2-1)(1-q^2s^2)}} \quad . \tag{111}$$

Then the integral in (107) can be written as:

$$\int_{0}^{y} \frac{ds}{\sqrt{(1-s^2)(1-q^2s^2)}} = K + i\widetilde{K}\text{'} = \tag{112}$$

$$= \int_{0}^{1} \frac{ds}{\sqrt{(1-s^2)(1-q^2s^2)}} + i\int_{1}^{y} \frac{ds}{\sqrt{(s^2-1)(1-q^2s^2)}} \quad . \tag{113}$$

In the same way,similarly the analysis can be performed for the other intervals on the real axis. Thus the following theorem about the Christoffel-Schwartz integral can be proved:

Theorem. The intervals $[0, 1]$, $[1, \frac{1}{q}]$, $[\frac{1}{q}, +\infty)$ on the real axis are mapped by the Christoffel - Schwartz integral (107) onto the rectangle with endpoints correspondingly $(0, 0)$, $(K, 0)$, $(K, K + i\widetilde{K}\text{'})$, $(0, i\widetilde{K}\text{'})$ on the complex plane, where $\widetilde{K}\text{'}$ is represented by the integral (111).

The first significance of this theorem is that since the mentioned points are the points of a rectangle, it gives the opportunity to predict when the integral (107) is real-valued or imaginary. Such an analysis will be performed in another publication.

Secondly, it may be noted that the Christoffel-Schwartz theorem is not formulated for elliptic integrals of higher than zero order. So taking into account the results of the theorem for the zero-order integral $\widetilde{J}_0^{(4)}(\widetilde{y}, q)$ and combining the higher-order integrals (101) and (102) for respectively the integrals $\widetilde{J}_4^{(4)}(\widetilde{y}, q)$ and $\widetilde{J}_2^{(4)}(\widetilde{y}, q)$ from the recurrent system of equations, it may be established in another way whether these integrals are real-valued or complex-valued and how the interval $[0, 1]$, $[1, \frac{1}{q}]$, $[\frac{1}{q}, +\infty)$ is mapped by higher-order integrals. This is a problem, not solved for the moment in the theory of elliptic functions.

10. Brief summary of a new analytical algorithm for integrating elliptic integrals of the zeroth-order in the Legendre form

10.1. Transforming an integral in the Legendre form into an integral in the Weierstrass form

A new analytical algorithm for treatment of elliptic integrals in the Legendre form $\widetilde{J}_0^{(4)}(y, q) = \int \frac{dy}{\sqrt{(1-y^2)(1-q^2y^2)}}$ has been proposed in the paper [25] and the main motivation comes from the necessity to treat elliptic integrals of higher order

$$J_n^{(4)}(y) = \int \frac{y^n dy}{\sqrt{a_0y^4 + 4a_1y^3 + 6a_2y^2 + 4a_3y + a_4}} \quad , \quad J_n^{(3)}(x) = \int \frac{x^n dx}{\sqrt{ax^3 + bx^2 + cx + d}} \quad , \tag{114}$$

widely applied in theoretical problems in various areas of physics such as relativistic treatment of perihelion advance (see the monograph [28], but also [51]), light deflection in Schwarzschild geometry and light trajectories around Black Holes, apparent supernova distances in cosmology, soliton equations and nonlinear motion of the pendulum and etc. Both integrals in (114) are of $n-$th order, because of the terms y^n and x^n in the nominators of these integrals. Below we shall only briefly outline the new algorithm, proposed in [25].

The algorithm is based on transforming the integral $\widetilde{J}_0^{(4)}(y,q)$ in the Legendre form into the integral $\widetilde{J}_0^{(3)}(x,g_2,g_3)$ in the s.c.Weierstrass form, where the integrals are defined as

$$\widetilde{J}_0^{(4)}(y,q) = \int \frac{dy}{\sqrt{(1-y^2)(1-q^2y^2)}} \quad , \quad \widetilde{J}_0^{(3)}(x,g_2,g_3) = -\sqrt{a}\int \frac{dx}{\sqrt{4x^3-g_2x-g_3}} \tag{115}$$

and the following transformation is applied

$$x = \frac{a}{y^2} + b \quad . \tag{116}$$

After applying the above transformation, it is proved that if the parameters a and b are determined as

$$a = \frac{9g_3}{g_2}K(q) \quad , \tag{117}$$

$$b = -\frac{a}{3}(1+q^2) = -\frac{3g_3}{g_2}K(q) \quad , \tag{118}$$

then the integrals $\widetilde{J}_0^{(4)}(y,q)$ and $\widetilde{J}_0^{(3)}(x,g_2,g_3)$ are equal. In (117) and (118) $K(q)$ is a rational function, depending on the modulus parameter q of the elliptic integral

$$K(q) \equiv \frac{q^4-q^2+1}{2q^4-5q^2+2} \quad . \tag{119}$$

Note that in the transformation (116) and the subsequently determined parameters (117) and (118), no any restrictions have been imposed on the modulus of the elliptic integral q to be a small quantity, so the analytical method is applicable to highly elliptic orbits, which are used in modern satellite technologies. This will be mentioned also further.

10.2. Another representation of the integral in the Weierstrass form without the conformal coefficient

The second representation is based on the s.c. theorem for "four-dimensional uniformization" [52], according to which an elliptic integral of the zero-order in its general four-dimensional form and in the Legendre representation

$$J_0^{(4)}(y) = \int \frac{dy}{\sqrt{a_0y^4+4a_1y^3+6a_2y^2+4a_3y+a_4}} \tag{120}$$

after a series of transformations (explicitly given in the paper [25], following [52]) can be brought to the elliptic integral in the Weierstrass representation

$$I_0^{(3)}(x) = \int_s^{\infty} \frac{d\sigma}{\sqrt{4\sigma^3-\overline{g}_2\sigma-\overline{g}_3}} \quad , \tag{121}$$

where for the polynomial of the fourth degree in the Legendre form

$$f(y) \equiv (1-y^2)(1-q^2y^2) = 1-(1+q^2)y^2+y^4 \tag{122}$$

the new Weierstrass invariants $\overline{g}_2$ and $\overline{g}_3$ can be exactly calculated [25]. From the two representations of the integral (120)

$$\widetilde{J}_0^{(4)}(y,q) = \int \frac{dy}{\sqrt{(1-y^2)(1-q^2y^2)}} = -\sqrt{a}\int \frac{dx}{\sqrt{4x^3-g_2x-g_3}} = \int \frac{d\sigma}{\sqrt{4\sigma^3-\overline{g}_2\sigma-\overline{g}_3}} \tag{123}$$

the Weierstrass invariants g_2 and g_3 can be calculated to be

$$g_2 = \overline{g}_2 a^{\frac{4}{3}} = (3K(q))^4 \frac{\overline{g}_3^4}{\overline{g}_2^3} \quad , \quad g_3 = -a^2 \overline{g}_3 = -27^2 K(q)^6 \frac{\overline{g}_3^7}{\overline{g}_2^6} \quad , \tag{124}$$

where $\overline{g}_2$ and $\overline{g}_3$ are found to be complicated polynomial functions of the modulus q of the elliptic integral in the Legendre form [25] and $K = K(q)$ is the rational function in terms of q, given by the defining expression (119). We shall give the calculated expressions for $\overline{g}_2$ and $\overline{g}_3$, calculated for the polynomial (122), making use of the formulaes in the monograph [52]

$$\overline{g}_2 = q^2 + \frac{1}{12}(1+q^2)^2 > 0 \ , \ \overline{g}_3 = -\frac{1}{6}q^2(1+q^2) + \frac{(1+q^2)^3}{6^3} \quad . \tag{125}$$

Thus the Weierstrass invariants g_2 and g_3 (also the coefficient function $a(q)$) in the second integral in (123) are known functions. In the Conclusion part of this paper and on the baseof the theorem for "four-dimensional uniformization", it will be shown that an arbitrary elliptic integral (120) of the fourth-degree polynomial $a_0y^4 + 4a_1y^3 + 6a_2y^2 + 4a_3y + a_4$ can thus be parametrized with a complicated function, depending on the Weierstrass elliptic function $\rho(z)$ and its derivative.

10.3. Application of the Weierstrass integral and of the Weierstrass elliptic curve in the parametrizable form

Finding a solution $z - z_0$ of the second integral in (123) in the Weierstrass form $\int \frac{dx}{\sqrt{4x^3 - g_2x - g_3}}$ is equivalent to finding the solution of the cubic equation $y^2 = 4x^3 - g_2x - g_3$. This "equivalency" can be written as [50]

$$z - z_0 = \int_{x_0}^{x} \frac{dx}{\sqrt{4x^3 - g_2x - g_3}} \Leftrightarrow y^2 = 4x^3 - g_2x - g_3 \quad . \tag{126}$$

The cubic curve $y^2 = 4x^3 - g_2x - g_3$ is called also the "uniformization curve", because the functions of the complex variable z

$$x = \rho(z) \quad , \quad y = \rho^{'}(z) \equiv \frac{d\rho}{dz} \tag{127}$$

satisfy the cubic curve (126). In fact, the equality $x = \rho(z)$ is a solution also of the elliptic integral in (126) and this is the known problem of "inversion" for elliptic integrals. In (127) $\rho(z)$ is the Weierstrass function [50]

$$\rho(z) \equiv \frac{1}{z^2} + \sum \left(\frac{1}{(z-\omega)^2} - \frac{1}{\omega^2} \right) = \frac{1}{z^2} + \frac{g_2}{20}z^2 + \frac{g_3}{28}z^4 + ... \tag{128}$$

From the second representation for the Weierstrass function $\rho(z)$ it becomes clear why in the literature it is denoted by $\rho(z; g_2, g_3)$.Taking into account the standard definition for the Weierstrass invariants g_2 and g_3 and also the defining equalities (124), the invariants can be represented as

$$g_2 = 60G_4 = 60 \sum{}^{'} \frac{1}{\omega^4} = (3K(q))^4 \frac{\overline{g}_3^4}{\overline{g}_2^3} \quad , \tag{129}$$

$$g_3 = 140G_6 = 140 \sum{}^{'} \frac{1}{\omega^6} = -27^2 \left(K(q)\right)^6 \frac{\overline{g}_3^7}{\overline{g}_2^6} \quad . \tag{130}$$

The invariants are defined as infinite convergent sums over the period ω of the two-periodic lattice and the prime "$'$" above the two sums means that the period $\omega = 0$ is excluded from the summation, so that the expressions would not tend to infinity.

There is also a unique mapping between the pair of Weierstrass invariants (g_2, g_3) and the points $z = m\omega_1 + n\omega_2$ of the two-dimensional lattice Λ on the complex plane

$$\Lambda \equiv \left\{ m\omega_1 + n\omega_2 \; ; \quad 0 \leq m \leq 1 \; , 0 \leq n \leq 1 \; , Imag\left(\frac{\omega_1}{\omega_2}\right) > 0 \right\} \quad . \tag{131}$$

The following theorem, proved in Ch. 6 of the monograph by Knapp [53] allows us to understand how the result of integration for the elliptic integral (123) $\widetilde{J}_0^{(q)}(y) = \int \frac{dy}{\sqrt{(1-y^2)(1-q^2y^2)}}$ can be represented in an analytical form by means of the Weierstrass representation. This formulae will be explicitly written in another publication.

10.4. A theorem about the unique correspondence between the points on the two-dimensional complex lattice and the Weierstrass invariants (g_2, g_3) and also between elliptic curves and elliptic integrals in the Weierstrass form

Theorem [53] There exists an unique correspondence $(g_2(\Lambda), g_3(\Lambda)) \Leftrightarrow \Lambda$ between the lattices Λ on the complex plane $\mathbb{C}$ and the pair (g_2, g_3) of the complex numbers (the Weierstrass invariants) such that the discriminant of the cubic polynomial $4x^3 - g_2x - g_3$ is different from zero. If a, b, c are the roots of the above polynomial, i. e.

$$4x^3 - g_2x - g_3 = 4(z-a)(z-b)(z-c) \quad , \tag{132}$$

then the periods (ω_1, ω_2) on the complex lattice Λ (131) can be found from the following integrals

$$\omega_1 = \int\limits_{\Gamma_1} \frac{dz}{2\sqrt{(z-a)(z-b)(z-c)}} \quad , \quad \omega_2 = \int\limits_{\Gamma 2} \frac{dz}{2\sqrt{(z-a)(z-b)(z-c)}} \quad . \tag{133}$$

In the first integral the unique branch Γ_1 of the square root is chosen with cuts from a to b and from c to ∞ and in the second integral the branch Γ_2 is with cuts from b to c and from a to ∞.

Moreover, for every lattice Λ on the complex plane, defined according to (131), a biholomorphic mapping $\varphi : \mathbb{C}/\Lambda \mapsto E(\mathbb{C})$, where $E(\mathbb{C})$ denotes the elliptic curve $y^2 = 4x^3 - g_2(\Lambda)x - g_3(\Lambda)$. The mapping φ is defined by means of the Weierstrass elliptic function $\rho(z)$ and the inverse mapping - by the corresponding elliptic integral.

A detailed study whether the periods (ω_1, ω_2) are real and imaginary depending on the roots of the polynomial $4x^3 - g_2x - g_3$ is given in Ch. 6.5 the known monograph [48].

11. New physical and mathematical theory of the space-time interval on intersecting null cones for the case of plane elliptical orbits

11.1. Preliminary statement of the problem and the basic system of algebraic equations

This new approach, developed first in the paper [22] and later on summarized in [23] is a logical continuation of the approach of one null cone. In the preceding sections serious arguments were presented in favour of the fact that one can choose a reference system of space coordinates, parametrizing the null cone equation. In particular, numerical estimates were given which prove the consistency of the theoretical approach with some typical experimental data, related to satellite laser ranging.

The theoretical approach up to now was related to the problem about calculation of the propagation time of a signal, emitted by a moving along an elliptical orbit (plane orbit or also a space-oriented orbit) satellite. However, such an approach does not give an answer to the question: what is the propagation time if the signal is intercepted by a second satellite, which is moving during the time of propagation of the signal.

Evidently, there should be some "consistency" between the events of emission and perception of the signal. Since these two events are treated in the framework of General Relativity Theory, it will be quite natural to construct a new theory, based on two null cones (12) $ds_1^2 = -c^2\left(1+\frac{2V_1}{c^2}\right)(dT_1)^2 + \left(1-\frac{2V_1}{c^2}\right)\left((dx_1)^2+(dy_1)^2+(dz_1)^2\right) = 0$ and (13) $ds_2^2 = -c^2\left(1+\frac{2V_2}{c^2}\right)(dT_2)^2 + \left(1-\frac{2V_2}{c^2}\right)\left((dx_2)^2+(dy_2)^2+(dz_2)^2\right) = 0$ with origins at the signal-emitting and signal-receiving satellite.

For the moment, we shall not be able to give an answer to the question how to calculate the propagation time for a realistic model, when the signal-emitting- and signal-receiving satellites move on different space-distributed orbits, experiencing also various disturbing force and influences on the part not only of the Earth, but also the Moon, the Sun and other planets. However, in equations (20) - (22) a preliminary generalized model was presented, based on the dependence of the propagation times T_1 and T_2 on several celestial- mechanical parameters. Evidently, such a complicated approach will be related to serious mathematical and technical difficulties, without being sure that there will be any new physics or at least any new numerical estimate, being related to the process of emission and reception of the signal.

That is why, the more "idealized" approach which was used in the papers [22] and [23] is based only on the dependence of the two propagation times only on the two eccentric anomaly angles E_1 and E_2, characterizing the (independent) plane elliptical motion of the two satellites. Naturally, the two four-dimensional null cone equations (12) and (13) in terms of the variables $dT_1, dx_1, dy_1, dz_1, dT_2, dx_2, dy_2, dz_2$ are independent. They will become dependent if they are "intersected" by the hyper-plane equation, obtained after taking the differential of the Euclidean distance $R_{AB}^2 = (x_1-x_2)^2 + (y_1-y_2)^2 + (z_1-z_2)^2$. In terms of the two-dimensional parametrization (5) $x = a(\cos E - e)$, $y = a\sqrt{1-e^2}\sin E$ (now there will be two parametrizations for (x_1, y_1) and (x_2, y_2)) this equation for dR_{AB}^2 is written as

$$dR_{AB}^2 = 2(x_1-x_2)d(x_1-x_2) + 2(y_1-y_2)d(y_1-y_2) \quad . \tag{134}$$

In fact, the above equation of the four-dimensional hyperplane in terms of the variables dx_1, dx_2, dy_1, dy_2 is a simplified version of the more general hyper-plane equation (22), written in terms of 14 variables.

Our further purpose will be to solve the system of equations (12), (13) and (134) in terms of the parametrization (5), to find a resulting algebraic equation with respect to dR_{AB}^2 and to solve it as a differential equation in full derivatives with respect to the two set of parameters (e_1, E_1) and (e_2, E_2). The solution for R_{AB}^2 will allow us to make important physical conclusions.

11.2. Finding the solution of the system of algebraic equations

In the paper [22] it has been proved that combining the equations (12), (13) and (134), one can obtain the following differential equation with respect to dR_{AB}^2

$$dR_{AB}^2 = F_1(E_1, E_2)dE_1 + F_2(E_1, E_2)dE_2 \quad , \tag{135}$$

where $F_1(E_1, E_2)$ and $F_2(E_1, E_2)$ are the expressions

$$F_1(E_1, E_2) := 2e_1a_1^2(1 - e_1\cos E_1)\sin E_1 - S_1(E_1, E_2) \quad , \tag{136}$$

$$F_2(E_1, E_2) = 2e_2a_2^2(1 - e_2\cos E_2)\sin E_1 - S_2(E_1, E_2) \tag{137}$$

and $S_1(E_1, E_2)$ is the expression

$$S_1(E_1, E_2) := -2[a_1a_2\sqrt{(1-e_1^2)(1-e_2^2)}\sin E_2\cos E_1 +$$

$$+a_1 a_2 \sin E_1 \cos E_2 - e_2 a_1 a_2 \sin E_1] \quad . \tag{138}$$

The expression for $S_2(E_1, E_2)$ is the same as $S_1(E_1, E_2)$, but with interchanged $E_1 \Longleftrightarrow E_2$, i.e. $S_2(E_1, E_2) = S_1(E_2, E_1)$. Since equation (135) is a differential equation in full derivatives, the following differential equations follow

$$F_1(E_1, E_2) = \frac{\partial R_{AB}^2}{\partial E_1} \quad , \quad F_2(E_1, E_2) = \frac{\partial R_{AB}^2}{\partial E_2} \quad . \tag{139}$$

If the first equation is integrated then

$$R_{AB}^2 = \int F_1(E_1, E_2) dE_1 + \varphi(E_2) = \tag{140}$$

$$= -2e_1 a_1^2 \cos E_1 + \frac{1}{2} e_1^2 a_1^2 \cos(2E_1) +$$

$$+2a_1 a_2 \sqrt{\left(1 - e_1^2\right)\left(1 - e_2^2\right)} \sin E_1 \sin E_2 -$$

$$-2a_1 a_2 \cos E_1 \cos E_2 + 2e_2 a_1 a_2 \cos E_1 + \varphi(E_2) \quad , \tag{141}$$

where $\varphi(E_2)$ is a function, which has to be determined from the second equation $F_2(E_1, E_2) = \frac{\partial R_{AB}^2}{\partial E_2}$ in (139). If from (141) the derivative $\frac{\partial R_{AB}^2}{\partial E_2}$ is calculated and then is set up equal to $F_2(E_1, E_2)$ given by expression (137), the following simple differential equation for $\varphi(E_2)$ can be obtained

$$\frac{\partial \varphi(E_2)}{\partial E_2} = \left(2e_2 a_2^2 - 2e_1 a_1 a_2\right) \sin E_2 - e_2^2 a_2^2 \sin(2E_2) \quad . \tag{142}$$

If the equation is integrated and the result is substituted into (141), then the final expression for R_{AB}^2 is obtained

$$\widehat{R}_{AB}^2 = \left(-2e_1 a_1^2 \cos E_1 - 2e_2 a_2^2 \cos E_2\right) +$$

$$+ \left(2e_2 a_1 a_2 \cos E_1 + 2e_1 a_1 a_2 \cos E_2\right) +$$

$$+\frac{1}{2}\left(e_1^2 a_1^2 \cos(2E_1) + e_2^2 a_2^2 \cos(2E_2)\right) - 2a_1 a_2 \cos E_1 \cos E_2 +$$

$$+2a_1 a_2 \sqrt{\left(1 - e_1^2\right)\left(1 - e_2^2\right)} \sin E_1 \sin E_2 \quad . \tag{143}$$

Note that this expression is symmetrical and does not change under interchange of the indices 1 and 2, as it should be. In the next section it will be explained why the result of integration is denoted in (143) as $\widehat{R}_{AB}^2$ instead of keeping the original notation R_{AB}^2.

12. New physical consequences from the other expression for the Euclidean distance:space-time interval, condition for inter-satellite communications and geodesic distance

12.1. Euclidean distance and space-time interval-comparison between the corresponding formulaes and physical meaning of the space-time interval (distance)

The first important observation which can be made is that the expression (143) for R_{AB}^2 is that it is different from the expression for the Euclidean distance $R_{AB}^2 = (x_1 - x_2)^2 + (y_1 - y_2)^2$

$$R_{AB}^2 = \left[(a_1 \cos E_1 - a_2 \cos E_2) + (a_2 e_2 - a_1 e_1)\right]^2 +$$

$$+ \left[a_1 \sqrt{1 - e_1^2} \sin E_1 - a_2 \sqrt{1 - e_2^2} \sin E_2\right]^2 \quad . \tag{144}$$

written in terms of the two-dimensional elliptic coordinates (5). It is easily seen that when $E_1 = E_2$, $a_1 = a_2$ and $e_1 = e_2$ (the case of coinciding points $x_1 = x_2$ and $y_1 = y_2$), the Euclidean distance R_{AB} (144) is equal to zero, but the distance $\widehat{R}^2_{AB}$ (143) is not equal to zero. In other words, for $e_1 = e_2 = e$, $a_1 = a_2 = a$, $E_1 = E_2 = E$ the expression (143) for $\widehat{R}^2_{AB}$ becomes equal to

$$\widehat{R}^2_{AB} = 4a^2 \sin^2 E.(1-e^2) + a^2(e^2-2) \quad . \tag{145}$$

Now it is interesting to note that $\widehat{R}^2_{AB}$ is positive for

$$\sin^2 E \geq \frac{2-e^2}{4(1-e^2)} \quad , \tag{146}$$

but for

$$\sin^2 E \leq \frac{2-e^2}{4(1-e^2)} \tag{147}$$

the expression (145) for $\widehat{R}^2_{AB}$ is negative. It can be also equal to zero for

$$\sin^2 E = \frac{2-e^2}{4(1-e^2)} \quad . \tag{148}$$

Consequently, (145) for $\widehat{R}^2_{AB}$ possesses the characteristics of the space-time interval, which is a typical notion for Special and General Relativity. That is the reason why in (143) the notation $\widehat{R}^2_{AB}$ was introduced, in order to distinguish the space-time distance $\widehat{R}^2_{AB}$ from the Euclidean distance R^2_{AB}.

Thus, there is no any mistake in obtaining the formulae (143) for $\widehat{R}^2_{AB}$, since it was derived from the intersection of the two four-dimensional null cones (12), (13) with the hyperplane equation (134). Thus, from a theoretical point of view a new result has been obtained - the intersection of the two null-cones (the null-cone equations are a partial case of the space-time interval) with the hyper-plane equation gives a space-time interval, which can be positive, negative or equal to zero.

12.2. The compatibility condition for inters-satellite communications - definition from a physical and mathematical point of view

From a physical point of view, it is clear that the space-time distance cannot be related to the physical process of signal propagation, because any signal (light, radio) propagates along a large-scale distance, which is of course only positive.

On the other hand, relation (143) for $\widehat{R}^2_{AB}$ was obtained from a set of equations, which included also the defining equation for the Euclidean distance $R^2_{AB} = (x_1 - x_2)^2 + (y_1 - y_2)^2$. Therefore, it seems natural to ask whether the propagation of a signal takes place when the large-scale Euclidean distance is compatible with the space-time interval (143). The equality of the two expressions (143) and (144) is possible when the following equation is fulfilled

$$4a_1a_2\sqrt{\left(1-e_1^2\right)\left(1-e_2^2\right)} \sin E_1 \sin E_2 =$$

$$= a_1^2 + a_2^2 + (a_2e_2 - a_1e_1)^2 - \frac{1}{2}\left(e_1^2a_1^2 + e_2^2a_2^2\right) \quad . \tag{149}$$

The above equality shall be conditionally called "the compatibility condition for inter-satellite communications". The earlier established equality (148) $\sin^2 E = \frac{2-e^2}{4(1-e^2)}$ for the partial case

$e_1 = e_2 = e$, $a_1 = a_2 = a$, $E_1 = E_2 = E$ is a consequence of the above equality. The limiting value E_{lim}, for which the equality (148) is fulfilled is

$$E_{\text{lim}} = \arcsin\left[\frac{1}{2}\sqrt{\frac{2-e^2}{1-e^2}}\right] =$$

$$= 45.002510943228 \ [\text{deg}] \ = 0.785441987624 \ [rad] \quad , \tag{150}$$

which is calculated for the eccentricity of the GPS orbit $e = 0.01323881349526$. Note also that E_{lim} does not depend on the large semi-major axis a of the orbit.

12.3. Consistency of the positivity of the space-time interval with the value for E_{lim} from the compatibility condition

The space-time interval (143) for equal eccentricities and semi-major axis, but for different eccentric anomaly angles $E_1 \neq E_2$ can be written as

$$\widehat{R}^2_{AB} = e^2a^2 - e^2a^2(\sin E_1 + \sin E_2)^2 -$$

$$-2a^2\cos(E_1 + E_2) \quad . \tag{151}$$

The space- time interval will be positive (i.e. $\widehat{R}^2_{AB} > 0$), if the following inequality is satisfied

$$e^2 - e^2(\sin^2 E_1 + \sin^2 E_2) +$$

$$+2(1-e^2)\sin E_1 \sin E_2 > 2\cos E_1 \cos E_2 \quad . \tag{152}$$

If we take into account the standard inequalities for the cos-function

$$\cos E_1 \leq 1 \quad , \ \cos E_2 \leq 1 \tag{153}$$

and also the simple trigonometric relations $\sin^2 E_1 = 1 - \cos^2 E_1$, $\sin^2 E_2 = 1 - \cos^2 E_2$, then the first two terms on the first line of the above inequality can be written as

$$e^2 - e^2(\sin^2 E_1 + \sin^2 E_2) =$$

$$= e^2\cos^2 E_1 - e^2 + e^2\cos^2 E_2 \leq e^2 - e^2 + e^2 = e^2 \quad . \tag{154}$$

Substituting into inequality (152), it can be derived

$$2\cos E_1 \cos E_2 < e^2 + 2\sin E_1 \sin E_2 \quad , \tag{155}$$

which can be represented also as

$$\cos(E_1 + E_2) < \frac{e^2}{2} \quad \Longrightarrow \quad E_1 + E_2 > \arccos(\frac{e^2}{2}) \quad . \tag{156}$$

For the typical value of the eccentricity $e = 0.01323881349526$ of the GPS orbit, it can be obtained

$$E_1 + E_2 > 89.994978993712 \ [\text{deg}] \ . \tag{157}$$

Note that the sign is greater because cos is a decreasing function with the increase of the angle (in the first quadrant). For the third and the fourth quadrant cos is an increasing function and the sign should be the reverse one.

If one sets up $E_1 = E_2 = E$ in (156), then

$$\sin^2 E > \frac{1}{2}(1 - \frac{e^2}{2}) \implies E > \overline{E} = \arcsin \frac{\sqrt{2 - e^2}}{2} \quad , \tag{158}$$

where the numerical result for $\overline{E}$ is twice as smaller than the value 89.994978993712 [deg] in (157)

$$\overline{E} = 44.997489496856 \ \text{[deg]} \quad . \tag{159}$$

It should be clarified that this numerical value is a little lower that the limiting value $E_{\text{lim}} =$ 45.002510943228 [deg] (150). So it may seem that in the interval $\overline{E} < E < E_{\text{lim}}$ the space-time interval (143) is negative, but yet the space-time interval (154) in terms of the two eccentric anomaly angles E_1 and E_2will be positive, consequently $E > \overline{E} = 44.997489496856$ [deg] (158) and (159) will be fulfilled. This would have been a contradiction, but this is not the case. The compatibility limiting value E_{lim} is an exact result, while trigonometric approximations are used for derivation of the inequalities (157) and (158).

12.4. Restriction on the ellipticity of the orbit. Consequences for the RadioAstron space project and for satellites on large elliptical orbits

Since $\sin E \leq 1$, one should have also the inequality

$$\sin E = \frac{1}{2}\sqrt{\frac{(2 - e^2)}{(1 - e^2)}} \leq 1 \quad , \tag{160}$$

which is fulfilled for

$$e^2 \leq \frac{2}{3} \quad \text{or} \quad e \leq 0.816496580927726 \quad . \tag{161}$$

Surprisingly, highly eccentric orbits (i.e. with the ratio $e = \frac{\sqrt{a^2-b^2}}{a}$ tending to one, where a and b are the great and small axis of the ellipse) are not favourable for inter-satellite communications. For *GPS* satellites which have very low eccentricity orbits (of the order 0.01) and for communication satellites on circular orbits ($e = 0$), inter-satellite communications between moving satellites can be practically achieved. For the first time the above calculation shows that such communications depend on the eccentricity of the orbit, which should be experimentally checked.

Another example can be given with the RadioAstron space mission (a system of an Earth based radio-telescope and a moving near-Earth space antenna) with a large semi-major axis of $a \approx 2 \times 10^8$ m with a variable orbital eccentricity ranging from $e = 0.59$ to the large value $e = 0.966$, which is higher than the value 0.816496580927726. So from the theory developed in this paper and also in the papers [22] and [23] it will follow that for eccentricity in the interval $0.59 < e < 0.816$, inter-satellite communications of RadioAstron with another satellites on the same orbit will be possible, but this will not be possible for eccentricities in the interval $0.816 < e < 0.966$. We should stress that this result is not applicable for the radio communications between the Earth-based radio telescope and the satellite on orbit, but only for the satellites on highly elliptic plane orbits.

In view of the restriction on the ellipticity of the orbit, it is perhaps interesting to mention that the traditional methods for autonomous navigation inter-satellite orientation (and ranging), which are occasionally suited for nearly circular orbits (and with small eccentricity), are inapplicable and limited when spacecrafts on large elliptical orbits are considered [54]. The reason is that large elliptical orbit satellites (such as SBIRS of the USA and the Russian Molniya series of communication satellites) operate slowly in apogee and operate quickly when passing the perigee, so navigational methods typical for *GPS* cannot be used. Although the analysis in the monograph of Liu [54] is purely classical, it would be interesting to include the General Relativity methods. It should be reminded again that *GPS* satellites (also the *GLONASS* satellites) are very-low eccentricity orbits.

12.5. New results:the notion of the geodesic distance from the compatibility condition and the space-time interval

If the compatibility condition (149) is substituted into the expression (143) for the space-time interval, the following expression is obtained

$$\widetilde{R}^2_{AB} = \frac{1}{2}(a_1^2 + a_2^2) + \frac{1}{2}(a_2 e_2 - a_1 e_1)^2 + \frac{1}{4}\left(a_1^2 e_1^2 + a_2^2 e_2^2\right) -$$

$$-\left(2e_1 a_1^2 \cos E_1 + 2e_2 a_2^2 \cos E_2\right) -$$

$$-\left(e_1^2 a_1^2 \sin^2 E_1 + e_2^2 a_2^2 \sin^2 E_2\right) - 2a_1 a_2 \cos E_1 \cos E_2 +$$

$$+2a_1 a_2 \left(e_2 \cos E_1 + e_1 \cos E_2\right) \quad . \tag{162}$$

The distance $\widetilde{R}^2_{AB}$ is called the "geodesic distance" and it is different from the space-time distance (interval) (143) and from the Euclidean distance (144). In order to understand why $\widetilde{R}^2_{AB}$ is called a geodesic distance, let us find the difference between the square of the geodesic distance and the Euclidean distance

$$R^2_{AB} - \widetilde{R}^2_{AB} = \frac{1}{2}(a_1^2 + a_2^2) - e_1 e_2 a_1 a_2 +$$

$$+\frac{1}{4}(a_1^2 e_1^2 + a_2^2 e_2^2) - 2a_1 a_2 \sqrt{\left(1 - e_1^2\right)\left(1 - e_2^2\right)} \quad . \tag{163}$$

12.6. Equality to zero of the geodesic distance when the compatibility condition is fulfilled-the case of equal eccentricities, eccentric anomaly angles and semi-major axis

For the case $e_1 = e_2 = e$, $a_1 = a_2 = a$, $E_1 = E_2 = E$ the geodesic distance (162) acquires the form

$$\widetilde{R}^2_{AB} = -a^2 + \frac{1}{2}a^2 e^2 + 2a^2(1 - e^2)\sin^2 E \quad . \tag{164}$$

This expression becomes equal to zero for this value of E_{lim}, for which the compatibility condition (149) is fulfilled, i.e. $\sin^2 E = \sin^2 E_{\text{lim}} = \frac{1}{4}\frac{(2-e^2)}{(1-e^2}$. It is easy to check that for this value of E_{lim} the space-time interval (145) $\widehat{R}^2_{AB} = 4a^2 \sin^2 E.(1 - e^2) + a^2(e^2 - 2)$ is also equal to zero. Consequently, for this simplest case the geodesic distance is compatible with the space-time distance.

12.7. Geodesic distance greater than the (non-zero) Euclidean distance - the case of equal eccentricities and semi-major axis, but different eccentric anomaly angles

It is instructive to compare the non-zero Euclidean distance (given by 144 for $e_1 = e_2 = e$ and $a_1 = a_2 = a$, but for $E_1 \neq E_2$) and to compare it with the geodesic distance (162). Now it can be noted however that the difference $R^2_{AB}-\ \widetilde{R}^2_{AB}$ in (163) does not depend on the eccentric anomaly angles. So for $e_1 = e_2 = e$ and for $a_1 = a_2 = a$ one can represent (163) as

$$\widetilde{R}_{AB} = \sqrt{R^2_{AB} + a^2(1 - \frac{3}{2}e^2)} \quad . \tag{165}$$

Taking into account the restriction (161) $e^2 \leq \frac{2}{3}$ on the value of the ellipticity of the orbit, the second term under the square root in (165) is positive. Due to this

$$\widetilde{R}_{AB} \geq R_{AB} \quad , \tag{166}$$

which means that the geodesic distance, travelled by the signal is greater than the Euclidean distance. Importantly, it may be concluded also that we established the same physical meaning of the geodesic distance, but following a different "logical" sequence of definitions, which have nothing to do with any null geodesic distance. The geodesic distance $\widetilde{R}_{AB}$ is in fact a distance, defined on the intersection of two algebraic varieties (two four-dimensional null cones) with a hyper-plane equation. Let us remind also that first the space-time interval $\widehat{R}^2_{AB}$ (143) was defined as a solution of a system of three algebraic equations, then it was required the space-time interval to be compatible with the Euclidean distance R_{AB} (144)- as a result the compatibility condition (149) was obtained and finally-the geodesic distance $\widetilde{R}_{AB}$ (162) was derived.

Since the important property of the geodesic distance $\widetilde{R}_{AB}$ to be greater than the Euclidean distance is proved for some partial cases, it remains to prove it for the general case of different eccentricities, semi-major axis and eccentric anomaly angles. If this happens to be the case, the consistency of the formalism will be proved.

12.8. Geodesic distance greater than the (non-zero) Euclidean distance - the general case of different eccentricities, semi-major axis and eccentric anomaly angles

In order to perform the proof, let us write the condition for inter-satellite communications (149) (also called "compatibility condition") in the form

$$\sin E_1 \sin E_2 = p = \frac{P_1(e_1, a_1; e_2, a_2)}{Q_1(e_1, a_1; e_2, a_2)} \quad , \tag{167}$$

where $P_1(e_1, a_1; e_2, a_2)$ and $Q_1(e_1, a_1; e_2, a_2)$ for given values of the two eccentricities e_1, e_2 and the semi-major axis a_1, a_2 are the numerical parameters

$$P_1(e_1, a_1; e_2, a_2) := a_1^2 + a_2^2 + (a_2 e_2 - a_1 e_1)^2 - \frac{1}{2}(e_1^2 a_1^2 + e_2^2 a_2^2) \ , \tag{168}$$

$$Q_1(e_1, a_1; e_2, a_2) := 4 a_1 a_2 \sqrt{(1 - e_1^2)(1 - e_2^2)} \quad . \tag{169}$$

Since

$$\sin E_1 \sin E_2 \leq 1 \quad , \tag{170}$$

from the preceding relations (167) - (170) it can be obtained

$$-\frac{1}{2}(a_1^2 + a_2^2) + a_1 a_2 e_1 e_2 \geq$$

$$\geq -\frac{1}{4}(e_1^2 a_1^2 + e_2^2 a_2^2) + \frac{1}{2}(e_1^2 a_1^2 + e_2^2 a_2^2)$$

$$-2 a_1 a_2 \sqrt{(1 - e_1^2)(1 - e_2^2)} \quad . \tag{171}$$

Substituting the terms in the left-hand side of the above inequality in the expression (163) for $\widetilde{R}^2_{AB} - R^2_{AB}$ (we change the sign in both sides of the inequality), it can be obtained

$$\widetilde{R}^2_{AB} - R^2_{AB} \geq -\frac{1}{4}(e_1^2 a_1^2 + e_2^2 a_2^2) + \frac{1}{2}(e_1^2 a_1^2 + e_2^2 a_2^2)$$

$$-2a_1a_2\sqrt{(1-e_1^2)(1-e_2^2)}-\frac{1}{4}(e_1^2a_1^2+e_2^2a_2^2)$$

$$+2a_1a_2\sqrt{(1-e_1^2)(1-e_2^2)} \quad . \tag{172}$$

All the terms in the right-hand side of the above inequality cancel, so one obtains for this case again the inequality (166) $\widetilde{R}_{AB} \geq R_{AB}$. So again, in analogy with the previous partial case (165), when $\widetilde{R}_{AB} = \sqrt{R_{AB}^2 + a^2(1-\frac{3}{2}e^2)}$ was fulfilled and in the framework of the two null cones formalism, the geodesic distance retains its property of being greater than the Euclidean distance. Similar was the case with the Shapiro delay formulae (32) $\Delta t = \frac{R_{AB}}{c} + 2\frac{G_{\oplus}M_E}{c^3}\ln\left(\frac{r_A+r_B+R_{AB}}{r_A+r_B-R_{AB}}\right)$.

13. Algebraic treatment of the space-time interval and the geodesic distance

13.1. Zero space-time interval from non-zero Euclidean distance - analysis of fourth-order algebraic equations by means of higher algebra theorems

Now we shall explore the problem when the space-time interval can be zero for the case of non-zero Euclidean distance. For the purpose, expression (151) for the case of equal to zero space-time interval $\widehat{R}_{AB}^2 = 0$ can be written as

$$2\sqrt{(1-\sin^2 E_1)(1-\sin^2 E_2)} =$$

$$= e^2 - e^2(\sin^2 E_1 + \sin^2 E_2) + 2(1-e^2)\sin E_1 \sin E_2 \quad . \tag{173}$$

After some transformations and introducing the notation $\sin^2 E_1 = y$, the above expression can be presented in the form of a quartic (fourth-degree) algebraic equation

$$y^4 + a_1y^3 + a_2y^2 + a_3y + a_4 = 0 \quad . \tag{174}$$

The coefficient functions of this equation are given in Appendix C of the paper [22]. Consequently, the problem about finding those values of the eccentric anomaly angle E_1 for which the space-time interval (151) is zero is equivalent to the algebraic problem of finding all the roots of the above fourth-order algebraic equation, which are within the circle $\mid y \mid = \mid \sin^2 E_1 \mid < 1$ (we exclude the boundary points $y = \sin^2 E_1 = 1$). It is well-known that an algebraic equation of fourth degree will always possess roots. The problem is that these roots should be within the circle $\mid y \mid < 1$. This is a well-known problem in algebra theory and is treated by the well-known Schur theorem. It is interesting to mention that the theorem was published yet in the 1918 year [56] and is mentioned also in most textbooks on higher algebra.

13.2. The Schur theorem from algebra theory for the roots of an $n-th$ degree polynomial within the circle $\mid y \mid < 1$

Below we present the formulation of the Schur theorem, which can be found in the monograph by Nikola Obreshkoff [55] and also in the paper [22]. In Appendix B of this paper the proof of the theorem is given, also taken from the cited monograph. The formulation of the theory is presented also in the paper [23].

Theorem (Schur) The necessary and sufficient conditions for the polynomial of $n-$th degree

$$f(y) = a_0y^n + a_1y^{n-1} + ... + a_{n-2}y^2 + a_{n-1}y + a_n \tag{175}$$

to have roots only in the circle $\mid y \mid < 1$ are the following ones:

1. The fulfillment of the inequality

$$\mid a_0 \mid > \mid a_n \mid . \tag{176}$$

2. The roots of the polynomial of the $(n-1)-$th degree

$$f_1(y) = \frac{1}{y}\left[a_0 f(y) - a_n f^*(y)\right] \tag{177}$$

should be contained in the circle $\mid y \mid< 1$, where $f^*(y)$ is the s.c. "inverse polynomial", defined as

$$f^*(y) = y^n f(\frac{1}{y}) = a_n y^n + a_{n-1}y^{n-1} + ... + a_2 y^2 + a_1 y + a_0 \quad . \tag{178}$$

In case of fulfillment of the inverse inequality

$$\mid a_0 \mid<\mid a_n \mid \tag{179}$$

the $(n-1)$ degree polynomial $f_1(y)$ (again with the requirement the roots to remain within the circle $\mid y \mid< 1$) is given by the expression

$$f_1(y) = a_n f(y) - a_0 f^*(y) \quad . \tag{180}$$

Concerning the necessary conditions, the Schur theorem has one another advantage - if the condition (177) (or (180)) about the roots of the polynomial $f_1(y)$ is not fulfilled, then the polynomial $f(y)$ will not have any roots within the circle. This allows to apply the theorem not only with respect to the space-time interval algebraic equation (which is proved to have roots in Appendix C of the paper [22]), but also with respect to the geodesic equation, which should not have any roots within the circle $\mid y \mid< 1$ (this is proved in Appendix E of the paper [22]). The last fact may be confirmed by independent calculations, since it has been already proved that the geodesic distance is greater than the Euclidean distance, so it cannot become equal to zero. This is fully consistent from a physical point of view, since the light or signal propagation is related to the geodesic distance and not to the space-time interval, which can be equal to zero or even become negative. In this aspect, it is really amazing how the physical interpretation is consistent with the mathematical results about these two algebraic equations.

It is important to mention that these conclusions are valid in view of the fact that the eccentricity e is very small (in celestial mechanics, it is of the order of 0.01), and on the base of this it is possible to compare terms with inverse powers in e in the corresponding inequalities - the higher inverse powers in e will lead to a larger number. For example, a term of the order of $\frac{1}{e^2}$ will give a number of the order of 10000, but as it was shown, there will be terms proportional to $\frac{1}{e^{10}}$, $\frac{1}{e^{12}}$ and even $\frac{1}{e^{14}}$, which are extremely large numbers. It is important that terms which differ by two orders in inverse powers of e will have greatly different numerical values.

13.3. The general strategy for using the Schur theorem - the "chain" of lower-degree polynomials

The basic fact which is a consequence of the Schur theorem and which shall be used further in the proof is that the polynomial of the $(n-1)$ degree (177) or (180) is a sufficient condition for the existence of roots within the unit circle of the initial polynomial of $n-$th degree. But then, if a new polynomial of $(n-2)$ degree is constructed according to formulaes (177) or (180), then this polynomial can become a sufficient condition for the roots of the $(n-1)$ degree polynomial. In such a way, a chain of lower-degree polynomials is constructed - each polynomial represents a necessary and at the same time a sufficient condition for the construction of a lower degree polynomial. The last constructed polynomial will be of first order, and from it the condition for the roots to be contained in the unit circle can easily be found. Note the important role of the necessary and sufficient condition - if from the linear polynomial the condition for the roots is found, then it will be a sufficient condition for the second-order polynomial, further this polynomial will be a necessary and sufficient condition for the third-order polynomial and etc. In such a way, the first-order polynomial will turn out to be a sufficient condition for the roots to remain within the unit circle with respect to the initial n-th degree polynomial, provided also that for each polynomial the corresponding inequalities between the coefficient functions are fulfilled. It can be claimed that this "chain" of lower-degree polynomials, together with the corresponding inequalities between the coefficient functions, represents a modified version of the Schur theorem. So from the point of view of pure mathematics, such a modified version of the Schur theorem without any doubt is interesting, the peculiar moment is that the physical information (availability of roots with respect to the space - time equation and absence of any roots with respect to the geodesic equation) is very important for the proof of such a modified version of the theorem. Of course, the proof is limited for the investigated case of polynomials of fourth degree.

13.4. Some consistent numerical checks, confirming the approach

Now let us discuss some numerical relations, following from the equation for the zero space-time interval (173), which in Appendix C of the paper [22] on the base of the Schur theorem and the "chain" of algebraic equations is proved to have roots within the interval

$$15.64[\text{deg}] < E_1 < 56.88[\text{deg}] \quad , \quad 15.64[\text{deg}] < E_2 < 56.88[\text{deg}] \tag{181}$$

If the two inequalities are summed up, then one can obtain

$$31.28[\text{deg}] < E_1 + E_2 < 113.76[\text{deg}] \quad . \tag{182}$$

Importantly, the earlier found restriction (157) $E_1+E_2 > 89.994978993712$ [deg] falls within the range of the inequality (182). This is a confirmation of the correctness of the formalism, because (157) was obtained on the base of trigonometric estimates, having nothing to do with searching the roots of the cubic equation. The value (150) for $E_{\text{lim}} = \arcsin\left[\frac{1}{2}\sqrt{\frac{2-e^2}{1-e^2}}\right] = 45.002510943228$ [deg] from the condition (149) for inter-satellite communications is also compatible with the inequality (182), which is a second confirmation about the correctness of the approach and the performed calculations.

There is also a third interesting numerical confirmation. From the inequality (170) $\sin E_1 \sin E_2 \leq 1$, represented now in the form $\sin E_2 = \frac{p}{\sin E_1} < 1$ and representing the condition for inter-satellite communications (see equations (171) - (172)) we can write $E_1 > \arcsin p = 30.002899$ [deg], which turns out to be higher than the lower bound in 15.64 [deg] $< E_1 < 56.88$ [deg] (181). Consequently, there is an interval

$$15.64 \ [\text{deg}] < E_1 < 30.002899 \ [\text{deg}] \quad , \tag{183}$$

where the space-time interval can exist (and can have zeroes) but the condition for inter-satellite communications and the resulting from it geodesic distance equation (162) cannot be defined. This confirms the conclusion that the space-time distance is a more broader notion and has a more general meaning in comparison with the Euclidean distance and the geodesic distance. This also means that the notion of space-time distance can be defined independently from the geodesic distance.

In this review paper we shall not treat the other case of the geodesic algebraic equation, which is also a fourth - degree algebraic equation and which is treated in details in the paper [22]. The important property of the geodesic distance $\widetilde{R}^2_{AB}$ to be greater than the Euclidean distance, if "translated" into the algebraic terminology, means that the algebraic equation does not have any roots within the circle $\mid y \mid < 1$. This turns out to be the case.

14. Conclusion

In this paper a new generalized model of the Shapiro delay formulae is presented, which is based on the null cone equations formalism in General Relativity Theory. The paper summarizes the new approaches, proposed in several other previous papers [22], [23], [24] and [25], but at some places new mathematical proofs are added, also some supplementary theorems, which prove the correctness of the applied theoretical approach - this is for example the theorem about the compatibility of the null cone equation with the null geodesic equations, proved in the known General Relativity monograph [28]. Chronologically, the first developed approach was the s.c. "two null cones intersecting formalism" [22], [23] and afterwards the formalism of calculating the propagation time of a signal, emitted by a satellite moving along a space-distributed orbit was proposed in [24]. So the future development of this theory will be 1. to put together all these separate elements in the theory in [24] in the generalizing formalism of the "two null intersecting four-dimensional cones", developed in [22], [23]. 2. Since some consistency arguments have been added (the theorem from [28]) and also some numerical estimates that the approach of parametrizing the space coordinates in the null cone equations in terms of the coordinates of the elliptic orbit of the satellite is consistent, at the beginning of this paper some basic facts have been added, concerning the disturbed motion in celestial mechanics. Also, for the first time the two null cone equations were presented for the most general case of the six Keplerian parameters, parametrizing the orbit of the satellite. This is the formalism, which shall be developed in subsequent papers.

Let us formulate briefly the general purpose of the formalism, which is difficult to develop to a full extent, because knowledge from several areas of physics and mathematics needs to be combined: celestial mechanics, General Relativity theory, theory of elliptic curves and elliptic integrals (including elliptic integrals of higher order), algebraic geometry and especially the theory of intersection of algebraic varieties, theory of analytical functions (Christoffel-Schwartz theorem and any possibilities for practical applications to higher-order elliptic integrals, not just for the zero-order elliptic integrals, for which it is valid). So the general purpose of the formalism is: to describe analytically and numerically the processes of signal exchange (of course, with account of GR effects) between moving satellites on one orbit or on different space-distributed orbits, possibly also taking into account disturbing effects in the motion. In fact, such a research programm requires the incorporation into the formalism of the basic approaches: 1. Finding the propagation times for the signal for different kinds of orbit of the satellites - plane elliptic, space-distributed and orbits, on which act the s.c. "disturbing forces". 2. Calculating the propagation times T_1 and T_2, which are correspondingly the "signal-emitting" propagation time T_1 and the "signal-receiving" propagation time T_2 in the two null-cone equations (12) $ds_1^2 = -c^2\left(1 + \frac{2V_1}{c^2}\right)(dT_1)^2 + \left(1 - \frac{2V_1}{c^2}\right)\left((dx_1)^2 + (dy_1)^2 + (dz_1)^2\right) = 0$ and (13) $ds_2^2 = -c^2\left(1 + \frac{2V_2}{c^2}\right)(dT_2)^2 + \left(1 - \frac{2V_2}{c^2}\right)\left((dx_2)^2 + (dy_2)^2 + (dz_2)^2\right) = 0$. Let us note that orbits of the two satellites may have different characteristics and parametrizations - for example, the second null-cone equation might be replaced by the null cone equation, derived from a metric in a rotating frame around the Earth [5]

$$0 = -ds^2 = -(1 - \frac{\omega_E^2 r^2}{c^2})(cdT)^2 + 2\omega_E^2 r^2 d\phi dT + \left[(dr)^2 + (rd\phi)^2 + (dz)^2\right] \quad , \tag{184}$$

where ω_E is the uniform rotating angular velocity of the Earth. Such a case is not investigated in the cited publications [22], [23], [24] and [25], since this is not related to the problem about inter-satellite communications. But in principle, such a problem can be investigated.

2. Further, applying the formalism of "two intersecting null cones" [22], [23], the equations for the space-time interval $\widehat{R}_{AB}$ (143), for the geodesic distance $\widetilde{R}_{AB}$ (162) and the so called "compatibility condition for inter-satellite communications" (149) can be defined. One of the important conclusions in the papers [22], [23] is related to the clarification of the properties of

these three equations and also the fact that these important physical notions appear in a strictly determined sequential order. Namely, after solving the two null cone equations, intersected by the hyper-plane equation $dR^2_{AB} = (x_1 - x_2)^2 + (y_1 - y_2)^2$ (for the two-dimensional case of a plane orbit, solved in the cited papers), the formulae for the space-time interval $\widehat{R}_{AB}$ (143) was obtained in terms of the variables $a_1, a_2, e_1, e_2, E_1, E_2$ and after comparing it with the Euclidean distance R^2_{AB} (144), it became clear that $\widehat{R}_{AB}$ can be positive, negative or equal to zero. Note that the variables E_1 and E_2 are related to the motion of the satellite- for some other parametrizations of the space coordinates they will be another. Further, it was required that the space-time interval is comparable to the Euclidean distance and thus, the compatibility condition (149) was obtained. Then, if the compatible condition is substituted back into the formulae for the space-time interval $\widehat{R}_{AB}$ (143), the geodesic distance (162) $\widetilde{R}^2_{AB}$ was obtained and it was proved that it is strictly positive for all the partial and also for the general case. This positivity for all the cases is one of the strongest arguments in favour of the consistency of the formalism.

All these facts, established in the papers [22], [23] can serve as a starting point for applying the approach of "two intersecting null cones" in another setting, meaning for more complicated and realistic situations. In other words, it is important to prove all these properties for the space-time interval and the geodesic distance for other cases -for example, for the case of the space-distributed orbits, when the more general parametrization of the space coordinates (74) - (76) [8] has been used. This more general case may be subdivided into two cases: case A. The variable, related to the motion of the satellite is the true anomaly angle f- respectively, in the framework of the two null cone formalism, instead of E_1, E_2 the changing variables will be f_1 and f_2. case B. The whole entity of parameters $(M, a, e, \Omega, I, \omega)$ is changing. This corresponds to the case when a signal is being send by a satellite on a space-distributed orbit with parameters $(M_1, a_1, e_1, \Omega_1, I_1, \omega_1)$ and the orbit is "deformed" to an orbit $(M_2, a_2, e_2, \Omega_2, I_2, \omega_2)$, where the signal-receiving satellite is situated and during the propagation time of travel of the signal, the second satellite is also moving. So for this case, the two null cones formalism displays a "consistency" in the events of emission and reception, but on the level of General Relativity (since two null cone equations are used) and also on the classical "celestial mechanics level" (because of the parametrizations in terms of the space coordinates of the satellites on the two elliptic orbits). We remind that here the mean anomaly $M = n(t - t_{per})$ (where $n = \sqrt{\frac{G_{\oplus} M_{\oplus}}{a^3}}$ is the mean motion) is related to the true anomaly angle f and to the eccentric anomaly angle E (see formulae (11)).

14.1. Proposals for further development of the approach, based on elliptic integrals of higher order, the recurrent system of equations for higher order integrals and some new analytical algorithms for calculation of elliptic integrals of zero-order in the Legendre form

As a first step in this direction, one may use the calculated propagation times in terms of the higher order elliptic integrals for the case of the parametrization (74)-(76) of the space coordinates of the space-distributed orbit. The propagation times for this cased were proved to be expressed by higher-order (second and fourth) elliptic integrals, which by means of the recurrent system for elliptic equations (101) and (102) (see [35]) were expressed through the zero-order elliptic integral $\widetilde{J}_0^{(4)}(\widetilde{y}, q)$ in the Legendre form.

Here an interesting opportunity arizes also to apply the analytical approach, developed in the paper [25] and in the section "Application of the Weierstrass integral and of the Weierstrass elliptic curve in the parametrizable form" of this paper. For the purpose, one may use the representation $\widetilde{J}_0^{(4)}(\widetilde{y}, q)$ (123) in the form of two equivalent, but different integrals in the Weierstrass form. From the comparison of the two equivalent integrals, the Weierstrass invariants g_2 and g_3 were determined as complicated rational functions (124) and moreover, the obtained formulaes did not presume any smallness of the modulus q of the elliptic integral. The knowledge of the functions g_2 and g_3 from the formulae (124) allows to define the Weierstrass function (128) $\rho(z) = \frac{1}{z^2} + \frac{g_2}{20}z^2 + \frac{g_3}{28}z^4 + \ldots.$. Now, since the Weierstrass function is known, but the value of the integral $\widetilde{J}_0^{(4)}(y,q)$ should be expressed in terms of the $\widetilde{y}-$coordinates in the integral $\widetilde{J}_0^{(4)}(y,q)$, two more expressions are needed from the s.c."theorem for four-dimensional uniformization", taken from the monograph [52] and presented also in the paper [25]

$$y = y_0 + \frac{1}{4}\frac{f^{'}(y_0)}{\left[\rho(z;g_2,g_3) - \frac{1}{24}f^{''}(y_0)\right]} \quad , \tag{185}$$

$$Y = -\frac{1}{4}\frac{f^{'}(y_0)\rho^{'}(z;g_2,g_3)}{\left[\rho(z;g_2,g_3) - \frac{1}{24}f^{''}(y_0)\right]^2} \quad , \tag{186}$$

where $Y = f(y)$ is the fourth-order algebraic polynomial

$$Y \equiv f(y) \equiv a_0y^4 + 4a_1y^3 + 6a_2y^2 + 4a_3y + a_4 \Leftrightarrow \tag{187}$$

$$\Leftrightarrow J_0^{(4)}(y) = \int \frac{dy}{\sqrt{a_0y^4 + 4a_1y^3 + 6a_2y^2 + 4a_3y + a_4}} \quad , \quad f(y_0) = 0 \quad . \tag{188}$$

In other words, similar to the uniformization (126) $z - z_0 = \int\limits_{x_0}^{x} \frac{dx}{\sqrt{4x^3 - g_2x - g_3}} \Leftrightarrow y^2 = 4x^3 - g_2x - g_3$ of the cubic curve with the uniformization functions $x \equiv \rho(z)$ and $y \equiv \rho^{'}(z)$ and its equivalence with the elliptic integral, the equalities (187) - (188) mean that the fourth degree polynomial (187) is equivalent to the elliptic integral (188) (in its general form with a fourth-degree polynomial in the denominator) and the corresponding uniformization functions y (185) and Y (186) "uniformize" the fourth-degree algebraic polynomial (187). Note that in the uniformization functions (185) and (186) the Weierstrass invariants g_2 and g_3 (124) are the ones in the second integral $-\sqrt{a}\int \frac{dx}{\sqrt{4x^3 - g_2x - g_3}}$ in (123), which means that additionally the "conformal" function

$$-\sqrt{a} = -\sqrt{\frac{9g_3}{g_2}K(q)} = -\sqrt{\frac{9g_3}{g_2}\frac{(q^4 - q^2 + 1)}{(2q^4 - 5q^2 + 2)}} \tag{189}$$

has to be calculated with g_2 and g_3 defined as $g_2 = (3K(q))^4\frac{\overline{g}_3^4}{\overline{g}_2^3}$, $g_3 = -27^2K(q)^6\frac{\overline{g}_3^7}{\overline{g}_2^6}$ (124). But of course, the uniformization may be performed also with respect to the invariants $\overline{g}_2$ and $\overline{g}_3$ in the third integral $\int \frac{d\sigma}{\sqrt{4\sigma^3 - \overline{g}_2\sigma - \overline{g}_3}}$ in (123), so it is interesting to compare the results from both approaches.

In such a way and if the two null cones formalism is applied, this will enable us to express the propagation times in a complicated way. Further, it remains to solve the corresponding differential equations for dR_{AB}^2, to find the expressions for the space-time interval $\widehat{R}_{AB}^2$ and for the geodesic distance $\widetilde{R}_{AB}^2$ and thus to see whether the important properties of the space-time interval to be positive, negative and zero and for the geodesic distance to be only positive will be valid also for this case.

14.2. Most important experimental consequence from the developed theory of two null intersecting cones and the plane elliptical orbits of the satellites

Now it may be summarized that with respect to the approach of the two null intersecting cones, applied to plane elliptical orbits, there are two important consequences from the theory:

1. The found limiting value for the eccentric anomaly angle (150) $E_{\lim} = \arcsin\left[\frac{1}{2}\sqrt{\frac{2-e^2}{1-e^2}}\right] =$ 45.002510943228 [deg], above which the space-time interval $\widehat{R}^2_{AB}$ is positive and the compatibility condition (149) for inter-satellite communications is fulfilled. Evidently, the corresponding restriction in the above proposed ""two null cones" approach, applied to the two space-distributed orbits, will be with respect to the true anomaly angle $f_{\lim}$. Although $E_{\lim}$ is found from a formalism, which is idealistic from a physical and experimental point of view, it should be taken into account, when inter-satellite communications between satellites on one orbit take place, since it corresponds to satellite configuration of 8 satellites per orbit and at angular distance 45 deg, which is typical for the $GLONASS$ satellite configuration (8 satellites per orbit). Note however that the value $E_{\lim}$ differs slightly from the value 45 deg, which means that there will be a slight difference of 0.002510943228 [deg] from the "exact" angular distance 45 deg. This will result in a slight increase of the distance between the satellites, which can be calculated and perhaps should be taken into account. It is also interesting that for the limiting value $E_{\lim}$ in (150) the true anomaly angle f can be calculated from the known formulae in celestial mechanics $\cos f = \frac{\cos E - e}{1 - e\cos E}$ to be $f = 45.541436900412$ [deg] for the typical GPS orbit eccentricity $e = 0.01323881349526$. We note that this value for f is also near to the value $E_{\lim} = 45.002510943228$ [deg].

2. Quite an interesting consequence from the theory of two intersecting null cones is the restriction on the eccentricity of the orbit (161) $e^2 \leq \frac{2}{3}$ or $e \leq 0.816496580927726$, which follows from the simple inequality (160) $\sin E = \frac{1}{2}\sqrt{\frac{(2-e^2)}{(1-e^2)}} \leq 1$ for the sin −function. It is really amazing how this simple inequality leads to the fact that the geodesic distance $\widetilde{R}_{AB}$ is greater than the Euclidean distance $\widetilde{R}_{AB} \geq R_{AB}$ (166), because of the simple formulae (165) $\widetilde{R}_{AB} = \sqrt{R^2_{AB} + a^2(1 - \frac{3}{2}e^2)}$. That is why in the forthcoming developments of the theory of two intersecting null cones, for example on the base of the space-distributed orbits, it is extremely important to check whether such a restriction on the ellipticity will also hold. Quite probably, it will depend also on the other parameters $(a, e, \Omega, I, \omega)$ of the space-distributed orbit. The restriction on the ellipticity is particularly important in reference to the future developments of satellite constellations on highly elliptical orbits [54]. The communications between such satellites on such orbits, taking into account the General Relativity effects is still an unexplored theoretical topic.

3. An important consequence from the theory about the propagation time of a signal, emitted by a satellite, moving along a plane orbit or a space-distributed orbit is the real-valuedness of the expression for the propagation time. Although for the second case the proof from a mathematical point of view is more complicated, this is a proof also of the correctness of the formalism and also opens the possibility for extending the formalism in the sense, explained in the previous section. It is important to note that the proposed mathematical proofs for the real-valuedness of the integrals $\frac{a}{c}\int\sqrt{1 - e^2\cos^2 E}dE$ and $-\frac{2G_{\oplus}M_{\oplus}}{c^3}\int\sqrt{\frac{1+e\cos E}{1-e\cos E}}dE$ in the expression (51) for the propagation time T and further, in proving the real-valuedness of the expression (84) $\widetilde{T}_1 = -2i\frac{na}{c}q^{\frac{3}{2}}\widetilde{J}^{(4)}_2(y; q)$, expression (88) for $T^{(1)}_2 = -\frac{2G_{\oplus}M_{\oplus}}{c^3}.\frac{n}{(1-e^2)^{\frac{3}{2}}}\widetilde{T}_1$ and the two integrals $-inq^{\frac{5}{2}}\frac{2G_{\oplus}M_{\oplus}}{c^3}(1 + e^2)\widetilde{J}^{(4)}_2(\widetilde{y}, q)$ and $inq^{\frac{3}{2}}\frac{2G_{\oplus}M_{\oplus}}{c^3}\frac{(1+e^2)}{(1-e^2)}\ \widetilde{J}^{(4)}_4(\widetilde{y}, q)$ in the expression (91) for $T^{(2)}_2$ did not imply that any restrictions on the modulus of the elliptic integrals q or on the ellipticity of the orbits are imposed. Concerning the two null cone formalism in this paper, the only restriction on the ellipticity was $e \leq 0.816496580927726$, but there was no requirement for the ellipticity to be small of the order of 0.01. But such an assumption turned out to be useful later on, when the properties of the space-time interval and the geodesic distance were confirmed by the approach of fourth-order algebraic equations.Only the algebraic equation for

the space-time distance was considered in this paper, the full investigation on the two algebraic equations can be found in [22]. The problem however remains, also with reference of the two null cones approach and the use of space-distributed orbits: can the properties of the space-time distance and the geodesic distance from an algebraic point of view be proved without assuming necessarily the smallness of the ellipticity of the orbit?

14.3. Acknowledgements

The author is grateful to the organizers of the Fifteenth Conference of the Euro-American Consortium for Promoting the Application of Mathematics in Technical and Natural Sciences (AMITaNS22) June 21-26, 2023, Albena resourt, Bulgaria and especially to Prof. Michail Todorov for the opportunity to participate in this conference.

15. References

[1] Teunissen P and Montenbruck O Eds 2017 *Springer Handbook of Global Navigation Satellite Systems* (Switzerland: Springer International Publishing AG)
[2] Grewal M, Weil L and Andrews A 2001 *Global Positioning Systems, Inertial Navigation and Integration* (JohnWiley and Sons, Inc)
[3] Straumann N. 2013 *General Relativity, second edition* (Dordrecht: Springer Science+Business Media)
[4] Shapiro I 1964 *Fourth test of General Relativity* Phys. Rev. Lett. **13** 789
[5] Ashby N 2003 *Relativity in the Global Positioning System* Living Reviews in Relativity **6** 1 (*Preprint* https://link.springer.com/content/pdf/10.12942%2Flrr-2003-1.pdf)
[6] Ashby N and Nelson R 2010 *Relativity in Fundamental Astronomy: Dynamics, Reference Frames, and Data Analysis, Proc. of the IAU Symp. 261 2009* ed. by Klioner S Seidelmann P and Soffel M (Cambridge: Cambridge University Press)
[7] *Relativistic time transfer in the vicinity of the Earth and in the Solar System* Recommendation ITU-R (Radiocommunication Sector of the International Telecommunication Union) TF.2018 (08/2012)
[8] Kopeikin S, Efroimsky M and Kaplan G 2011 *Relativistic Celestial Mechanics of the Solar System* (New York:Wiley-VCH)
[9] Danby J 1962 *Fundamentals of Celestial Mechanics* (New York:The Macmillan Company)
[10] Kaula M 1968 *An Introduction to Planetary Physics. The Terrestrial Planets* (New York London:John Wiley and Sons, Inc.)
[11] Murray C and Dermott S 2009 *Solar System Dynamics* (Cambridge:Cambridge University Press)
[12] Fitzpatrick 2012 *An Introduction to Celestial Mechanics* http://mathship.com/900/Mechanics/
[13] Soffel M and Langhans R 2013 *Space-Time Reference Systems* (Berlin Heidelberg: Springer-Verlag)
[14] Guochang X and Xu Y 2003 *GPS Theory, Algorithms and Applications* Third Edition (Berlin: Springer-Verlag)
[15] Calais E *Satellite Orbits. Keplerian Motion. Perturbed Motion. GPS Orbits (*Purdue University - EAS Department*)* https://www.geologie.ens.fr/~ecalais/teaching/gps-geodesy/satellite orbits.pdf
[16] Gulklett M 2003 *Relativistic effects in GPS and LEO* PhD Thesis The Niels Bohr Institute for Physics, Astronomy and Geophysics *Preprint* https://www.yumpu.com/en/document/view/4706552/relativistic-effects-in-gps-and-leo-niels-bohr-institutet
[17] Shafarevich I. 2013 *Basic Algebraic Geometry 1. Varieties in Projective Space* Third Edition (Heidelberg Berlin New York:Springer-Verlag)
[18] Holme A 2012 *A Royal Road to Algebraic Geometry* (Berlin Heidelberg:Springer-Verlag)
[19] Shafarevich I (ed.) 1994 *Algebraic Geometry I. Encyclopedia of Mathematical Sciences vol.23* (Berlin Heidelberg GmbV:Springer-Verlag)
[20] Harris J 1992 *Algebraic Geometry. A First Course. Graduate Texts in Mathematics* (New York: Springer-Verlag Inc.)
[21] Fulton W. 2008 *Algebraic Curves. An Introduction to Algebraic Geometry* (https://dept.math.lsa.umich.edu/~wfulton/CurveBook.pdf)
[22] Dimitrov B 2017 *Two null gravitational cones in the theory of GPS-intersatellite communications between two moving satellites. I. Physical and mathematical theory of the space-time interval and the geodesic distance on intersecting null cones Preprint* https://arxiv.org/abs/1712.01101v3
[23] Dimitrov B 2019 *New Mathematical Models of GPS Intersatellite Communications in the Gravitational Field of the Near-Earth Space* AIP Conf. Proc. **2075** 040007 (*Preprint* gr-qc/1909.12753v1)
[24] Dimitrov B 2022 *Elliptic Integrals for Calculation of the Propagation Time of a Signal, Emitted and Received by Satellites on One Orbit* AIP Conf. Proc. **2522** 978-0-7354-4361-7 (*Preprint* gen-ph/2201.13208v2)

[25] Dimitrov B 2023 *New Methods for Analytical Calculation of Elliptic Integrals Applied in Various Physical Problems* to appear in AIP Conf. Proc. (*Preprint* gr-qc/2301.00643)
[26] Hawking S and Ellis G 1973 *The Large Scale Structure of Space-Time* (Cambridge:Cambridge University Press)
[27] Nelson R 2011 *Relativistic time transfer in the vicinity of the Earth and in the Solar system* Metrologia **48**, S171
[28] Fock V 1959 *Theory of Space, Time and Gravitation* (New York:Pergamon Press)
[29] Zelmanov A and Agakov V 1989 *Elements of the General Theory of Relativity* (Moscow:Nauka Publishing House)
[30] Duchayne L 2008 *Transfert de temps de haute performance : le Lien Micro-Onde de la mission ACES.* Physique mathematique PhD Thesis Observatoire de Paris *Preprint* https://tel.archives-ouvertes.fr/tel-00349882/document
[31] Sovers O Fanselow J and Jacobs C 1998 *Astrometry and geodesy with radio interferometry: Experiments, models, results* Rev. Mod. Phys. **70** 1393
[32] Turyshev S Toth V and Sazhin M 2013 *General relativistic observables of the GRAIL mission* Phys. Rev. **D87** 024020 (*Preprint* gr-qc/1212.0232v4)
[33] Turyshev S Sazhin M and Toth V 2014 *General relativistic laser interferometric observables of the GRACE-Follow-On mission* Phys. Rev. **D89**, 105029 (*Preprint* gr-qc/1402.7111v1)
[34] Turyshev S Yu N and Toth V 2016 *General relativistic observables for the ACES experiment* Phys. Rev. **D93** 045027 (*Preprint* qr-qc/1512.09019v2)
[35] Prasolov V and Solov'yev Y 1997 *Elliptic Functions and Elliptic Integrals. AMS Translations of Mathematical Monographs 170* (Rhode Island: Providence) Russian original: Prasolov V and Solov'yev Y 1997 *Elliptic Functions and Algebraic Equations* (Moscow:Factorial Publishing)
[36] Timofeev A 1948 *Integration of Functions* (Moscow:State Publishing House for Technical-Theoretical Literature)
[37] Sanchez J-F 2007 *Introducing Relativity in Global Navigation Satellite System* Ann. Phys. (Leipzig) **16** 258
[38] Muller J, Dirkx D, Kopeikin S, Lion G, Panet I, Petit G and Visser P 2018 *High Performance Clocks and Gravity Field Determination* Space Sci. Rev. **214** 5
[39] Husemoller D 2004 *Elliptic Curves Graduate Texts in Mathematics 111* Sec. Ed. (New York:Springer-Verlag Inc.)
[40] Silverman J and Tate J 2015 *Rational Points on Elliptic Curves Undergraduate Texts in Mathematics* Sec. Ed. (Cham Heidelberg:Springer)
[41] Silverman J 2009 *The Arithmetic of Elliptic Curves Graduate Texts in Mathematics 106* Sec. Ed. (Dordrecht Heidelberg:Springer Science+Business Media LLC)
[42] Silverman J 1994 *Advanced Topics in the Arithmetic of Elliptic Curves Graduate Texts in Mathematics 151* (New York:Springer Science+Business Media LLC)
[43] Lang S 1978 *Elliptic Curves Diophantine Analysis Grundlehren der Mathematischen Wissenschaften 231* (Berlin Heidelberg:Springer-Verlag GmbH)
[44] Smirnov V 1964 *A Course of Higher Mathematics: vol. 3, part 2: Complex Variables Special Functions. International Series of Monographs in Pure and Applied Mathematics: volume 60* (Pergamon Press)
[45] Fichtenholz G 2003 *Course in Differential and Integral Calculus, vol.2, 8th ed.* (Moscow:Fizmatlit)
[46] online programm *web2.0 scientific calculator* https://web2.0calc.com/
[47] Petit G and Wolf P 1994 *Relativistic Theory for Picosecond Time Transfer in the Vicinity of the Earth* Astron. Astroph. **286** 971
[48] Markushevich A 1950 *Theory of Analytical Functions* (Moscow-Leningrad:State Publishing House) in Russian English Translation: Markushevich A 1977 *Theory of Functions of a Complex Variable* vol.1 and 2 (Chelsea)
[49] Markushevich A 1966 *A Short Course on the Theory of Analytical Functions* (Moscow:Nauka Publishing House)
[50] Lavrentiev M and Shabat B 1978 *Methods of the Theory of Functions of Complex Variable* (Moscow:Science Publishing House) 4th ed. in Russian
[51] Weinberg S 1972 *Gravitation and Cosmology. Principles and Applications of the General Relativity Theory* (Massachusetts:Massachusetts Institute of Technology)
[52] Whittaker E and Watson G 1927 *A Course of Modern Analysis. An Introduction* fourth edition (Cambridge:Cambridge University Press)
[53] Knapp A 1992 *Elliptic Curves Mathematical Notes 40* (Princeton New Jersey:Princeton University Press) Russian Translation Knapp A 2004 *Elliptic Curves* (Moscow: Factorial Press)
[54] Liu F Lu S and Sun Y 2019 *Guidance and Control. Technology of Spacecraft on Elliptical Orbit* (Singapore Beijing:Springer Nature Singapore Pte Ltd. and National Defense Industry Press)

[55] Obreshkoff N 1954, *Higher Algebra* (Sofia:State Publishing House Science and Arts) in Bulgarian ; see also Obreshkoff N 1963 *Verteilung und Berechnung der Nullstellen reeller Polynome* (Berlin:VEB Deutscher Verlag der Wissenschaften) in German
[56] Schur I 1918 Journal f. Math. **148** 122

Convergence of the L1 Two-Term Equation Scheme

Y. Dimitrov(1,2), **S. Georgiev**(2,3), **R. Miryanov**(4), **V. Todorov**(2,5)
[1]Department of Mathematics, Physics and Informatics, University of Forestry, Sofia 1756, Bulgaria
[2]Institute of Mathematics and Informatics, Bulgarian Academy of Sciences, 1113 Sofia, Bulgaria
[3]Department of Applied Mathematics and Statistics, University of Ruse, 7004 Ruse, Bulgaria
[4]Dept of Statistics and Applied Mathematics, University of Economics – Varna, 9002 Varna, Bulgaria
[5]Institute of Information and Communication Technologies, Bulgarian Academy of Sciences, 1113 Sofia, Bulgaria

Corresponding author: yuri.dimitrov@ltu.bg
E-mail: sggeorgiev@math.bas.bg, miryanov@ue-varna.bg, vtodorov@math.bas.bg

Abstract. Fractional derivatives have found application in modeling various processes in different fields of science. Finite difference schemes are a main approach for numerical solution of models using fractional differential equations. In this paper we investigate the convergence and order of the numerical solution of two-term ordinary fractional differential equation which uses the L1 approximation of the fractional derivative. Inequalities for the weights of L1 approximation are derived and used to prove the convergence of the L1 scheme for the two-term equation. Conditions for the parameter of the two-term equation and error estimates of L1 scheme are obtained. Experimental results for the order and error of the L1 scheme are presented in the paper.

1. Introduction

Fractional differential equations are an essential means of modeling complex processes in biology, physics, economics, and engineering [1, 2, 3, 4, 5]. Fractional integrals and derivatives are a generalization of integrals and derivatives of integer order. The Riemann-Liouville fractional integral of order $\alpha > 0$ is defined as

$$_aI_x^\alpha f(x) = \frac{1}{\Gamma(\alpha)}\int_a^x (x-t)^{\alpha-1} f(t)dt.$$

When $0 < \alpha < 1$, the Caputo fractional derivative of order α is defined as

$$_aD_x^\alpha f(x) = f^{(\alpha)}(x) = {_a}I_x^{1-\alpha} f'(x) = \frac{1}{\Gamma(1-\alpha)}\int_a^x \frac{f'(t)}{(x-t)^\alpha}dt.$$

The gamma function satisfies $\Gamma(x+1) = x\Gamma(x)$ and $\Gamma(n+1) = n!$, where n is an integer. The exponential function has a fractional derivative $_0D_x^\alpha e^x = x^{1-\alpha}E_{1,2-\alpha}(x)$, where $E_{1,2-\alpha}(x)$ is the Mittag-Leffler function

$$E_{1,2-\alpha}(x) = \sum_{k=0}^{\infty} \frac{x^k}{\Gamma(2+k-\alpha)}.$$

The value of the fractional derivative at a point x is determined by the values of the function on the interval (a, x), where a is the lower limit of fractional differentiation. Fractional differential equations are suitable to model processes with memory. Finite difference schemes for solving fractional differential equations use approximations of fractional derivatives. Let n be a positive integer and $h = (x - a)/n$ be the step size of a uniform mesh of size n in the interval $[a, x]$. Denote $x_k = a + kh$ and $f_k = f(a + kh)$, for $k = 0, 1, \cdots, n$. L1 approximation is a main approximation of the Caputo fractional derivative

$$\frac{1}{\Gamma(2-\alpha)h^\alpha}\sum_{k=1}^{n-1}\sigma_k^{(\alpha)} f_{n-k} = f_n^{(\alpha)} + O\left(h^{2-\alpha}\right), \tag{1}$$

where $\sigma_0^{(\alpha)} = 1$, $\sigma_n^{(\alpha)} = (n-1)^{1-\alpha} - n^{1-\alpha}$ and

$$\sigma_k^{(\alpha)} = (k-1)^{1-\alpha} - k^{1-\alpha} + (k+1)^{1-\alpha}, \quad (k = 1, \cdots, n-1).$$

The weights of L1 approximation (1) satisfy

$$\sigma_0^{(\alpha)} > 0, \quad \sigma_1^{(\alpha)} < \sigma_1^{(\alpha)} < \cdots < \sigma_{n-1}^{(\alpha)} < 0, \quad \sum_{k=0}^{n}\sigma_k^{(\alpha)} = 0. \tag{2}$$

Finite difference schemes for ordinary and partial fractional differential equations that use L1 approximation are studied in [6, 7, 8, 9]. Approximations of the fractional derivative of order $3 - \alpha$, which are called $L1 - 2$ and $L1 - 2_\sigma$ formulas are constructed by Alikhanov [10] and Gao et al. [11]. In [12, 13] we use the L1 approximation and second derivative approximations for construction of second-order approximations of Caputo fractional derivative whose weights satisfy (2). In [13] we derive the second-order asymptotic formulas of the L1 approximation

$$\frac{1}{\Gamma(2-\alpha)h^\alpha}\sum_{k=1}^{n-1}\sigma_k^{(\alpha)} f_{n-k} = f_n^{(\alpha)} + \frac{\zeta(\alpha-1)}{\Gamma(2-\alpha)} f_n'' h^{2-\alpha} + O\left(h^2\right).$$

The Riemann sum of the fractional integral has an asymptotic formula [14]

$$h^\alpha \sum_{k=1}^{n-1}\frac{f_{n-k}}{k^{1-\alpha}} = \int_0^x \frac{f(t)}{(x-t)^{1-\alpha}}dt + \sum_{k=0}^{\infty}(-1)^k\frac{\zeta(1-\alpha-k)}{k!} f_n^{(k)} h^{k+\alpha} + L_f, \tag{3}$$

where

$$L_f = -\Gamma(\alpha)\sum_{k=0}^{\infty}\frac{B_{k+1}}{(k+1)!}\left(\sum_{m=0}^{k}(-1)^m\binom{k}{m}\frac{x^{\alpha-m-1}}{\Gamma(\alpha-m)} f_0^{(k-m)}\right)h^{k+1}.$$

In this paper we investigate the convergence and order of the L1 scheme for the two-term ordinary fractional differential equation. The paper is organized as follows. In section 2 we use (3) to construct the asymptotic formula for $\sum_{k=1}^{m-1}|\sigma_k^{(\alpha)}|(m-k)^\alpha$ and we derive the inequality

$$\sum_{k=1}^{m-1}|\sigma_k^{(\alpha)}|(m-k)^\alpha < m^\alpha - \frac{3}{4}. \tag{4}$$

In section 3 we construct the numerical solution of the two-term equation which uses the L1 approximation of Caputo fractional derivative. In section 4 we use inequality (4) to prove the convergence and order of the L1 scheme for the two-term equation. We determine conditions for the parameter of the two-term equation and derive estimates for the error of the L1 scheme.

2. Properties of L1 approximation

In this section we prove inequality (4). The proof uses the following properties of the weights of L1 approximation.

Claim 1. *The numbers* $\sigma_k^{(\alpha)}$ *satisfy*

$$(i) \quad \sum_{k=1}^{m-1} |\sigma_k^{(\alpha)}| = 1 + (m-1)^{1-\alpha} - m^{1-\alpha},$$

$$(ii) \quad \sum_{k=1}^{m-1} |\sigma_k^{(\alpha)}| = m(m-1)^{1-\alpha} - (m-1)m^{1-\alpha}.$$

Proof. (i) From (2)

$$\sum_{k=1}^{m-1} |\sigma_k^{(\alpha)}| = \sigma_0^{(\alpha)} + \sigma_m^{(\alpha)} = 1 + (m-1)^{1-\alpha} - m^{1-\alpha}.$$

(ii) Express the numbers $k|\sigma_k^{(\alpha)}|$ as

$$\begin{aligned} k|\sigma_k^{(\alpha)}| &= 2k^{2-\alpha} - k(k-1)^{1-\alpha} - k(k+1)^{1-\alpha} \\ &= 2k^{2-\alpha} - (k-1+1)(k-1)^{1-\alpha} - (k+1-1)(k+1)^{1-\alpha} \\ &= k^{2-\alpha} - (k-1)^{2-\alpha} + k^{2-\alpha} - (k+1)^{2-\alpha} + (k+1)^{1-\alpha} - (k-1)^{1-\alpha}. \end{aligned}$$

Then

$$\begin{aligned} \sum_{k=1}^{m-1} k|\sigma_k^{(\alpha)}| &= \sum_{k=1}^{m-1} \left(k^{2-\alpha} - (k-1)^{2-\alpha}\right) + \sum_{k=1}^{m-1} \left(k^{2-\alpha} - (k+1)^{2-\alpha}\right) \\ &+ \sum_{k=1}^{m-1} \left((k+1)^{1-\alpha} - (k-1)^{1-\alpha}\right) \\ &= (m-1)^{2-\alpha} - (m^{2-\alpha} - 1) + m^{1-\alpha} + (m-1)^{1-\alpha} - 1 \\ &= m(m-1)^{1-\alpha} - (m-1)m^{1-\alpha}. \end{aligned}$$

□

When x and y are positive numbers the beta function is defined as

$$B(x,y) = \frac{\Gamma(x)\Gamma(y)}{\Gamma(x+y)} = \int_0^1 t^{x-1}(1-t)^{y-1}dt.$$

The beta function has a value

$$\begin{aligned} B(1+\alpha, 2-\alpha) &= \int_0^1 x^{\alpha}(1-x)^{1-\alpha}dx = \frac{1}{2}\Gamma(1+\alpha)\Gamma(2-\alpha) \\ &= \frac{1}{2}\alpha(1-\alpha)\Gamma(\alpha)\Gamma(1-\alpha) = \frac{\alpha(1-\alpha)\pi}{2\sin(\pi\alpha)}. \end{aligned} \tag{5}$$

The integral in formula (5) for the value of the beta function $B(1+\alpha, 2-\alpha)$ has singularities of order α and $1-\alpha$ at both endpoints. By applying (3) to both endpoints we find

$$s_m = \sum_{k=1}^{m-1} k^{1-\alpha}(m-k)^\alpha \sim B(1+\alpha, 2-\alpha)m^2 + \zeta(\alpha-1)m^\alpha + \zeta(-\alpha)m^{1-\alpha} - \frac{\alpha\zeta(\alpha-2)}{m^{1-\alpha}} - \frac{(1-\alpha)\zeta(-1-\alpha)}{m^\alpha}. \tag{6}$$

From (6)

$$\sum_{k=1}^{m-1} |\sigma_k^{(\alpha)}|(m-k)^\alpha = m^\alpha + 2s_m - s_{m-1} - s_{m+1},$$

$$\sum_{k=1}^{m-1} |\sigma_k^{(\alpha)}|(m-k)^\alpha \sim m^\alpha - \frac{\alpha(1-\alpha)\pi}{\sin(\pi\alpha)} + \alpha(1-\alpha)\left(\frac{\zeta(\alpha-1)}{m^{2-\alpha}} + \frac{\zeta(-\alpha)}{m^{1+\alpha}}\right).$$

The function

$$\Gamma(1+\alpha)\Gamma(2-\alpha) = \frac{\alpha(1-\alpha)\pi}{\sin(\pi\alpha)}$$

has a minimum value $\pi/4$ which is attained at $\alpha = 1/2$.

Claim 2. *Let $0 < \alpha < 1$ and m is a positive integer. Then*

$$(m-k)^\alpha < m^\alpha - \frac{\alpha k}{m^{1-\alpha}}. \tag{7}$$

Proof. From the binomial formula

$$(m-k)^\alpha = m^\alpha \left(1 - \frac{k}{m}\right)^\alpha = m^\alpha \sum_{i=0}^{\infty} (-1)^i \binom{\alpha}{k} \left(\frac{k}{m}\right)^i.$$

Then

$$(m-k)^\alpha - m^\alpha + \frac{\alpha k}{m^{1-\alpha}} = m^\alpha \sum_{i=2}^{\infty} (-1)^i \binom{\alpha}{k} \left(\frac{k}{m}\right)^i < 0.$$

because the numbers $(-1)^i \binom{\alpha}{k} < 0$. □

Lemma 3. *Let $m \geq 2$. Then*

$$\sum_{k=1}^{m-1} |\sigma_k^{(\alpha)}|(m-k)^\alpha < m^\alpha - \frac{3}{4}. \tag{8}$$

Proof. From Claim 1 and Claim 2

$$\begin{aligned}\sum_{k=1}^{m-1} |\sigma_k^{(\alpha)}|(m-k)^\alpha &< m^\alpha \sum_{k=1}^{m-1} |\sigma_k^{(\alpha)}| + \frac{\alpha}{m^{1-\alpha}} \sum_{k=1}^{m-1} k|\sigma_k^{(\alpha)}| \\ &< m^\alpha \left(1 + (m-1)^{1-\alpha} - m^{1-\alpha} -\right) - \alpha \left(m^\alpha (m-1)^{1-\alpha} - m + 1\right) \\ &= m^\alpha - (1-\alpha)m^\alpha \left(m^{1-\alpha} - (m-1)^{1-\alpha}\right) - \alpha.\end{aligned} \tag{9}$$

From (8)

$$m^{1-\alpha} - (m-1)^{1-\alpha} > \frac{1-\alpha}{m^\alpha}.$$

Inequality (9) holds when

$$(1-\alpha)m^\alpha\left(m^{1-\alpha} - (m-1)^{1-\alpha}\right) + \alpha > \frac{3}{4},$$

$$(1-\alpha)m^\alpha\left(m^{1-\alpha} - (m-1)^{1-\alpha}\right) + \alpha - \frac{3}{4} > (1-\alpha)^2 + \alpha - \frac{3}{4} = \frac{1}{4}(2\alpha-1)^2 \geq 0.$$

□

3. Numerical solution of two-term equation

The two-term equation is an ordinary fractional differential equation

$$y^{(\alpha)} + Dy = F(x), \quad y(a) = y_0. \tag{10}$$

The analytical and numerical solutions of two-term equation are studied in [15, 16, 17, 18]. Let $h = (b-a)/n$ be the step size of a uniform mesh in the interval $[a, b]$. Denote by $E_m h^{2-\alpha}$ the truncation error of the L1 approximation at the point $x_m = a + mh$ for $m = 1, \cdots, n$. By approximating the Caputo fractional derivative at the point x_m we find

$$\frac{1}{\Gamma(2-\alpha)h^\alpha}\sum_{k=0}^{m}\sigma_k^{(\alpha)} y_{m-k} + Dy_m = F_m + E_m h^{2-\alpha}. \tag{11}$$

The numerical solution $\{u_m\}_{m=0}^n$ of two-term equation (1), where u_m is an approximation of the solution y_m, is found from (11) by negating the error term

$$(1 + D\Gamma(2-\alpha)h^\alpha)u_m + \sum_{k=1}^{m}\sigma_k^{(\alpha)} u_{m-k} = \Gamma(2-\alpha)h^\alpha F_m. \tag{12}$$

The numerical solution of equation (10) is computed with $u_0 = y_0$ and

$$u_m = \frac{1}{1 + D\Gamma(2-\alpha)h^\alpha}\left(\Gamma(2-\alpha)h^\alpha F_m - \sum_{k=1}^{m}\sigma_k^{(\alpha)} u_{m-k}\right). \tag{13}$$

Example: Consider the two-term equation

$$y^{(\alpha)} + Dy = \alpha x^{1-\alpha} E_{1,2-\alpha}(\alpha x) + De^{\alpha x}, \quad y_0 = 1. \tag{14}$$

Equation (14) has a solution $y(x) = e^{\alpha x}$. Experimental results for the error and order of L1 scheme (13) for numerical solution of equation (14) are given in Table 1 and Table 2. The results for the error of numerical solution (13) in the first and second columns of Table 2 show that the error of the method may be large for negative values of the parameter D. In section 4 we prove that numerical solution (13) of two-term equation has an order of accuracy $2-\alpha$ for positive values of the parameter D and negative values with a small or large modulus and we derive an estimate for the error. These results are in accordance with the experimental results in Table 1 and the third column of Table 2.

Table 1: Error and order of numerical solution (13) on $[0,1]$.

h	$\alpha=0.25, D=1$		$\alpha=0.75, D=100$		$\alpha=0.4, D=-0.1$	
	Error	*Order*	*Error*	*Order*	*Error*	*Order*
0.005	5.5778×10^{-7}	1.70463	5.5257×10^{-6}	1.2454	1.0222×10^{-5}	1.5741
0.0025	2.3495×10^{-7}	1.71301	2.3275×10^{-6}	1.2473	3.4169×10^{-6}	1.5808
0.00125	9.8889×10^{-8}	1.71965	9.7964×10^{-7}	1.2484	1.1383×10^{-6}	1.5857

Table 2: Error and order of numerical solution (13) on $[0,2]$.

h	$\alpha=0.5, D=-10$		$\alpha=0.6, D=-20$		$\alpha=0.5, D=-100$	
	Error	*Order*	*Error*	*Order*	*Error*	*Order*
0.005	7.0659×10^{94}	150.62	3.7349×10^{170}	−87.96	5.5705×10^{-7}	1.4881
0.0025	2.8119×10^{85}	31.226	4.8464×10^{136}	112.57	1.9807×10^{-7}	1.4917
0.00125	1.3395×10^{82}	11.035	1.1639×10^{127}	31.955	7.0309×10^{-8}	1.4943

4. Convergence of the L1 scheme

In this section we study the convergence of the L1 scheme (13) for the two-term equation. Let $e_m = y_m - u_m$ be the error of the L1 scheme at the point x_m. From (11) and (12) the error e_m satisfies $e_0 = 0$ and

$$e_m = -\frac{1}{1+D\Gamma(2-\alpha)h^\alpha}\sum_{k=1}^{m}\sigma_k^{(\alpha)}e_{m-k} + A_m h^2, \tag{15}$$

where

$$A_m = \frac{\Gamma(2-\alpha)E_m}{1+D\Gamma(2-\alpha)h^\alpha}.$$

The truncation error of the L1 approximation satisfies [15]

$$\begin{aligned} E_m &< \frac{1}{\Gamma(2-\alpha)}\left(\frac{2+\alpha}{2^\alpha(2-\alpha)} - \frac{11+\alpha}{12}\right)\max_{a\le x\le b}\left|y''(x)\right| \\ &< \frac{2}{\Gamma(2-\alpha)}\max_{a\le x\le b}\left|y''(x)\right|. \end{aligned}$$

Denote $C = \max_{a\le x\le b}|y''(x)|$. The numbers A_m satisfy the estimate

$$|A_m| < \frac{2C}{|1+D\Gamma(2-\alpha)|h^\alpha}. \tag{16}$$

The error e_1 of numerical solution (13) at the first step satisfies

$$|e_1| = |A_1|h^2 < \frac{2Ch^2}{|1+D\Gamma(2-\alpha)h^\alpha|}. \tag{17}$$

In Theorem 4 and Theorem 5 we prove the convergence of the L1 scheme for the two-term equation on the interval $[0,1]$ when the solution $y\in C^2[0,1]$. Theorem 4 and Theorem 5 are proved by induction on m, where $1\le m\le n$.

Theorem 4. *Let $D \in (-\infty, -3/h^\alpha) \cup (0, \infty)$. The error of numerical solution (13) satisfies*

$$|e_m| < 3Ch^2 m^\alpha \tag{18}$$

for all $m = 1, \cdots, n$.

Proof. From (15)

$$|e_m| < \frac{1}{|1 + D\Gamma(2-\alpha)h^\alpha|} \sum_{k=1}^{m-1} |\sigma_k^{(\alpha)}||e_{m-k}| + |A_m|h^2. \tag{19}$$

When $0 < \alpha < 1$, the gamma function satisfies [19]

$$0.88 < \Gamma(2-\alpha) < 1.$$

When the parameter $D < -3/h^\alpha$ the denominator of (19) satisfies

$$|1 + D\Gamma(2-\alpha)h^\alpha| > |\frac{3}{h^\alpha} 0.88 h^\alpha - 1| > 1.$$

When $D \in (-\infty, -3/h^\alpha) \cup (0, \infty)$ the denominator of (19) is greater than one. From (17)

$$|e_1| < 2Ch^2.$$

Suppose that inequality (18) for the error e_k holds for all values of $k = 1, \cdots, m-1$. From (16) the numbers A_m satisfy $|A_m| < 2C$. Then

$$\begin{aligned} |e_m| &< \sum_{k=1}^{m-1} |\sigma_k^{(\alpha)}||e_{m-k}| + |A_m|h^2 \\ &< 3Ch^2 \sum_{k=1}^{m-1} |\sigma_k^{(\alpha)}|(m-k)^\alpha + 2Ch^2 \\ &< 3Ch^2 \left(m^\alpha - \frac{3}{4}\right) + 2Ch^2 < 3Ch^2 m^\alpha. \end{aligned}$$

□

Theorem 5. *Let $-\frac{1}{3\Gamma(2-\alpha)} < D < 0$. Then*

$$|e_m| < 8Ch^2 m^\alpha \tag{20}$$

for all $m = 1, \cdots, n$.

Proof. From (17)

$$|e_1| < \frac{2Ch^2}{|1 + D\Gamma(2-\alpha)h^\alpha|} < 3Ch^2. \tag{21}$$

Suppose that inequality (20) for the error e_k holds for all $k = 1, \cdots, m-1$.

$$|e_m| < \frac{1}{1 + D\Gamma(2-\alpha)h^\alpha} \sum_{k=1}^{m-1} |\sigma_k^{(\alpha)}||e_{m-k}| + |A_m|h^2. \tag{22}$$

When $-1/3 < D\Gamma(2-\alpha) < 0$ the following inequality holds

$$\frac{1}{1+D\Gamma(2-\alpha)h^\alpha} < \frac{1}{1-h^\alpha/3} < 1+\frac{h^\alpha}{3}\frac{1}{1-h^\alpha/3} < 1+\frac{h^\alpha}{2}.$$

From (16)

$$|A_m| < \frac{2C}{|1+D\Gamma(2-\alpha)|h^\alpha} < 2C\left(1+\frac{h^\alpha}{2}\right).$$

From (8) and (22)

$$\begin{aligned}|e_m| &< 8Ch^2\left(1+\frac{h^\alpha}{2}\right)\sum_{k=1}^{m-1}|\sigma_k^{(\alpha)}|(m-k)^\alpha + |A_m|h^2\\ &< 8Ch^2\left(1+\frac{h^\alpha}{2}\right)\left(m^\alpha-\frac{3}{4}\right)+2C\left(1+\frac{h^\alpha}{2}\right)h^2\\ &= 8Ch^2m^\alpha + 4Ch^{2+\alpha}m^\alpha - 4Ch^2 - 3Ch^{2+\alpha}\\ &< 8Ch^2m^\alpha - 2Ch^2 - 3Ch^{2+\alpha} < 8Ch^2m^\alpha.\end{aligned}$$

□

From Theorem 4 and Theorem 5 the error of the L1 scheme (13) of the two-term equation in the interval $[0,1]$, satisfies the estimate

$$|e_m| < 8Ch^2m^\alpha \leq 8Ch^{2-\alpha}. \tag{23}$$

for each $m = 1,\cdots,n$. Now we generalize the results of Theorem 6 and Theorem 8 to an arbitrary interval $[a,b]$. Consider the two-term equation

$$y^{(\alpha)}(x) + Dy(x) = F(x), \quad y(a) = y_0, \tag{24}$$

for $x \in [a,b]$. The uniform mesh in the interval $[a,b]$ has a step size $h = (b-a)/n$. Substitute $x = a+(b-a)t$ and

$$z(t) = y(x) = y(a+(b-a)t).$$

When $x \in [a,b]$ the new variable $t \in [0,1]$. The function $z(t)$ satisfies a two-term equation

$$z^{(\alpha)}(t) + (b-a)^\alpha Dz(t) = G(t), \qquad z(0) = y_0, \tag{25}$$

where $G(t) = (b-a)^\alpha F(a+(b-a)t)$. From Theorem 4 and Theorem 5 we get conditions for the parameter D for the convergence and order of L1 scheme for two-term equation (25).

Corollary 6. *The error of numerical solution* (13) *of two-term equation* (24)*, when the solution* $y \in C^2[a,b]$ *satisfies*

$$|E_m| < 8(b-a)^\alpha Ch^{2-\alpha} \tag{26}$$

for each $m = 1,\cdots,n$ *and*

$$D \in \left(-\infty, -\frac{3}{h^\alpha}\right) \cup \left(-\frac{1}{3(b-a)^\alpha\Gamma(2-\alpha)}, \infty\right). \tag{27}$$

5. Conclusions

In this paper we show that L1 scheme for the two-term equation converges and has an order $2-\alpha$, when the parameter D satisfies conditions (27). In [12] and [13] we construct second-order approximations of the Caputo fractional derivative. In future work we will study the convergence of second order and higher order numerical solutions of ordinary and partial fractional differential equations.

Acknowledgements
This study was supported by the Bulgarian National Science Fund under Project KP-06-M62/1 "Numerical deterministic, stochastic, machine and deep learning methods with applications in computational, quantitative, algorithmic finance, biomathematics, ecology and algebra" from 2022.

References

[1] Acay B, Bas E, Abdeljawad T 2020 *Chaos, Solitons & Fractals* **130** 109438
[2] Baleanu D, Balas V E, Agarwal P 2023 *Advanced Studies in Complex Systems*
[3] Singh H, Kumar D, Baleanu D 2019 *Methods of Mathematical Modelling "Fractional Differential Equations"* 1st ed. (New York: CRC Press)
[4] Sun D, Liu J, Su X, Pei G 2022 *Front. Public Health* **10** 1036901
[5] Zhang H, Liu F, Turner I, Yang Q 2016 *Computers & Mathematics with Applications* **71(9)** 1772-1783
[6] Jin B, Lazarov R, Zhou Z 2016 *IMA Journal of Numerical Analysis* **36(1)** 197–221
[7] Lin Y, Xu C 2007 *J. Comput. Phys.* **225(2)** 1533–1552
[8] Ramezani M, Mokhtari R, Haase G 2020 *Applied Numerical Mathematics* **153** 300–318
[9] Yan Y, Khan M, Ford N 2018 *SIAM Journal on Numerical Analysis* **56(1)** 210–227
[10] Alikhanov A A 2015 *Journal of Computational Physics* **280** 424–438
[11] Gao G H, Sun Z Z, Zhang H W 2014 *Journal of Computational Physics* **259** 33–50
[12] Apostolov S, Dimitrov Y, Todorov V 2022 *Lecture Notes in Computer Science* **13127** Springer, Cham
[13] Dimitrov Y 2016 *Journal of Fractional Calculus and Applications* **7(2)** 175–195
[14] Dimitrov Y 2018 *Journal of Fractional Calculus and Applications* **9(1)** 15–44
[15] Cai M, Li C 2020 *Mathematics* **8(1)** 43
[16] Daraghmeh A, Qatanani N, Saadeh A 2020 *Applied Mathematics* **11** 1100–1115
[17] Podlubny I 1999 *Fractional Differential Equations* (New Jersey: Academic Press)
[18] Wang Y 2020 *Advances in Difference Equations* **80** 1687–1847
[19] Deming W, Colcord C 1935 *Nature* **135** 917

Approaches for improvement of reliability of the Prony's method computation

O O Drobakhin[1] and O V Olevskyi[2]

[1]Department of Applied Radiophysics, Electronics, and Nanomaterials, Oles Honchar Dnipro National University, Haharina Ave, 72, Dnipro, Dnipropetrovsk Region, 49045, Ukraine
[2]Department of Computer Technologies, Oles Honchar Dnipro National University, Haharina Ave, 72, Dnipro, Dnipropetrovsk Region, 49045, Ukraine

E-mail: olevskyi.o22@fpm.dnu.edu.ua

Abstract. Parametric spectral analysis methods such as the Prony's method can estimate the frequencies and amplitudes of a signal, conforming to their model, with great precision. At the same time, the addition of noise to the signal can lead to a complete model breakdown which leads to erroneous parameter values. This is especially true for the impulse noise. The article explores several possible algorithms which can be applied to the Prony's method in order to refine the results and make them more noise resistant. Such algorithms include signal segmentation methods where the results of each segment processing influence the final estimate as well as the conceptually related method of point skipping. An approach based on the use of non-Euclidean norms as a measure of the linear algebraic equation system's solution quality is developed and tested. Initially, the methods are applied to model digital signals, comprised of harmonic components with varying complex frequencies and amplitudes. Additive white Gaussian and impulse noise is added to the model signals. The results are then applied to the noisy results of a real-life synthetic aperture synthesis experiment obtained in the intermediate zone of radiation.

1. Introduction

The Prony's method is considered to be a powerful but simple in implementation method for parametric spectral analysis, closely related with the more complex matrix pencil method [1]. This method is based on the representation of a signal in the form of a sum of complex exponential components [2, 3]. It has been successfully used in such fields as electronics [4–6], mechanics [7, 8], medicine [9], and electromagnetics [7, 10]. In the latter field it was specifically successfully used for the problems of radar imaging [11, 12] and non-destructive layer thickness estimation [13].

The classic approach to the Prony's method is based on the fact that a sum of complex exponential functions is a solution of a linear prediction problem. Due to this, one of the crucial steps of the method consists of solving a linear prediction system

$$\begin{bmatrix} s_0 & s_1 & \cdots & s_M \\ s_1 & s_2 & \cdots & s_{M+1} \\ \vdots & \vdots & \ddots & \vdots \\ s_{N-M-1} & s_{N-M} & \cdots & s_{N-1} \end{bmatrix} \begin{bmatrix} c_0 \\ c_1 \\ \vdots \\ c_M \end{bmatrix} = \mathbf{0}. \tag{1.1}$$

There are several standard ways used for solving (1.1) [14]. The most commonly used ones are the Moore-Penrose pseudoinverse method [15, 16] and the eigenvector method [17]. The latter has been demonstrated to be superior in noise resistance when tested on signals with white noise.

2. Signal with the sum of impulse and white noise

In order to generate an impulse noise sequence, an array initiated with zeros of the same length as the test signal would be constructed. After this, for each position in the array a uniformly distributed random value between 0 and 1 would be generated. If the value belonged to the interval $[1-p; 1-0.5p)$, where p is the requested impulse occurrence probability, the corresponding array value would be substituted for -1. If the value belonged to $[1-0.5p; 1)$, the array value would be substituted for 1 (fig. 1(a)). After this, the values of all non-zero array cells would be multiplied by a normally distributed random value with predetermined parameters (fig. 1(b)).

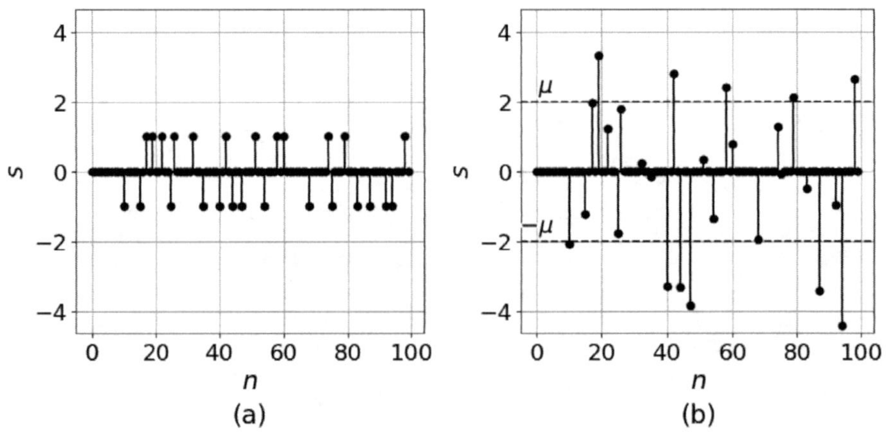

Figure 1. Position an direction (a) generation and white noise multiplication (b) for $p = 0.25$, $\mu = 2$, $\sigma = 1$

Even small amounts of impulse noise can lead to the classic Prony's method implementation being unable to correctly recover the parameters of the signal (fig. 2)

These results lead to the assumption that the parts where the impulse noise occurs should be suppressed or fully disregarded during the calculation. Due to this, two approaches were developed: segmented approach and point skipping approach. The latter method can be considered an extreme case of the former.

3. Segmented approach

Segmented approach to the Prony's method is known to have been used for the tasks of non-stationary signal analysis [18, 19]. In the current paper the segmentation of the signal is used in order to negate the effects of impulse noise on the estimation of a signal with constant parameters. This approach calls for the signal $\boldsymbol{s}$ to be divided into overlapping segments

$$\boldsymbol{s}_k = \begin{bmatrix} s_{\Delta nk} & s_{\Delta nk+1} & s_{\Delta nk+L-1} \end{bmatrix}^T.$$

After the segmentation, the Prony's method system (1.1) is applied to each of the segments. The estimated value of the linear prediction vector can then be calculated as a median of the segment results [20]. The remaining steps of the calculation follow the standard Prony's method

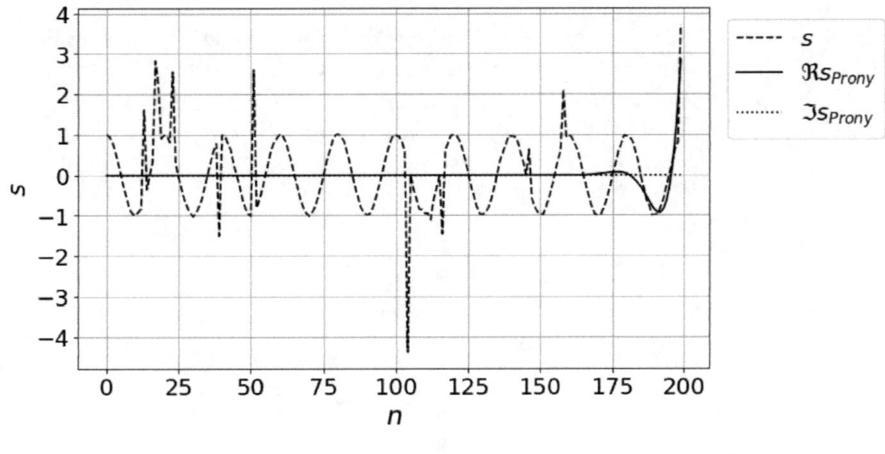

Figure 2. Results of the restoration of a signal with white and impulse noise with the help of the standard Prony's method for $p = 0.25$, $\mu = 2$, $\sigma = 1$, SNR $= 30$ dB.

procedure. The problem remaining with this method is the one of finding the optimal length and overlap fraction of the segments. For a signal with known length and noise parameters this can be done with the help of trial and error experiment. The results of this testing were then used when comparing the method with others.

4. Point skipping approach

Due to the system matrix in (1.1) being overdetermined, the rows containing the impulse noise values can be omitted while preserving an acceptable level of accuracy with regards to white noise (fig. 3).

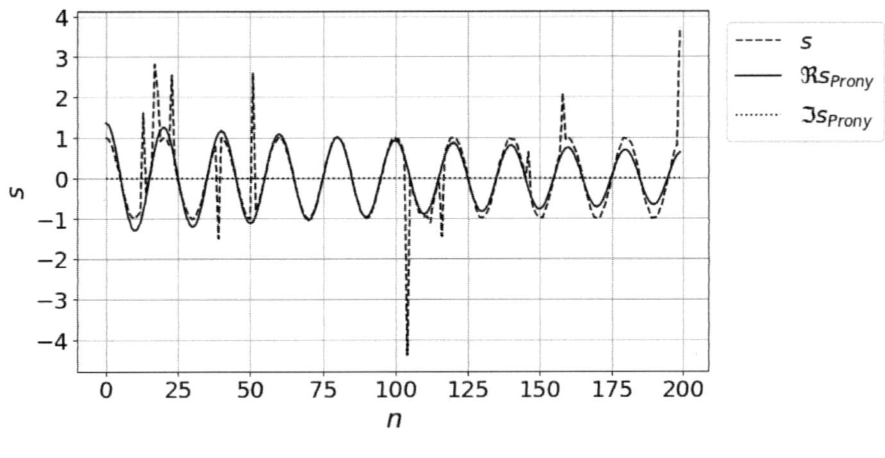

Figure 3. Results of the restoration of a signal with white and impulse noise with the help of the Prony's method with point skipping for $p = 0.25$, $\mu = 2$, $\sigma = 1$, SNR $= 30$ dB.

There are several problems which can arise when using such an approach. The first one is the fact that one can only omit a limited amount of rows from the system matrix (1.1) before it becomes underdetermined. If M is the order of the signal model and N is the amount of points in the signal, the amount of rows in the system can be expressed as $L = N - M$. The maximum amount of rows which may need omitting due to a single point with erroneous data equals to $M+1$. In order for the system to remain solvable with any missing point positions, the amount of remaining rows should remain equal or above $M+1$. Thus, in order to determine the maximum acceptable number of erroneous data points, one can write

$$L - K(M+1) \geq M+1,$$
$$K \leq \frac{N-M}{M+1} - 1, \quad (4.1)$$

where K is the number of the erroneous data points.

According to (4.1), the maximum acceptable fraction of erroneous points in a signal can be expressed as

$$\nu = \frac{K}{N} \leq \frac{1 - MN^{-1}}{M+1} - \frac{1}{N}.$$

For example, for a signal with $N = 200$ and $M = 10$ the actual value is $\nu \approx 0.081$, while the theoretical maximum is $(M+1)^{-1} \approx 0.09$. Such values are generally acceptable as in realistic situations such noise fraction rarely exceeds 0.05.

The second crucial problem which arises if said approach is used is the identification of the impulse noise. This problem can be broken into the search for the acceptable impulse candidates and the estimation of the amount of most likely candidates that should be excluded from the signal.

Investigation of the signal with the help of finite differences can be used as one of the simplest approaches to the peak identification task. In this case the signal points with the highest absolute values of the first finite difference are considered the impulse candidates.

Other method, which can be implemented for the signal investigation is the use of the matched filter-like approach. For this approach, the ideal peak signal is generated. In the simplest form it can be expressed as $\boldsymbol{s}_{imp} = \begin{bmatrix} 0 & 1 & 0 \end{bmatrix}^T$. After this, a segment of the signal with the same length $\boldsymbol{s}_k$ is taken with padding performed if needed. The mean values of each signal are then subtracted from corresponding sequences resulting in the vectors $\boldsymbol{s}'_{imp}$ and $\boldsymbol{s}'_k$The candidacy estimate of the value is the calculated as

$$g_k = \left| \boldsymbol{s}'^T_{imp} \boldsymbol{s}'_k \right| .$$

The problem of the optimal amount of excluded candidates can be solved with the help of the duration pseudonorm. It can be calculated as [21]

$$\Delta s_n = s_n - s_{0n},$$
$$\rho = \sum_{n=0}^{N-1} \left\{ \left(|\Delta s_n|^2 + \sigma^2 \right)^\beta - \sigma^{2\beta} \right\},$$

where σ is the standard deviation of the white noise, and β is a parameter. Generally, $\beta = \frac{1}{16}$. This approach is considered to be a general approach to the problem of impulse noise filtering [22] and is known to work well when applied to the Prony's method [23].

This measure is used because the resultant signal is compared to the input signal which has the impulse noise added. Due to having a very sharp characteristic (fig. 4), the pseudonorm can be used to estimate the accuracy of signal restoration while effectively omitting the peaks.

The appropriate number of skipped candidates can now be estimated by consecutive application of the Prony's method with skipped points to the signal with different amounts of points skipped up to the limit, detailed in the previous section. After this, the amount which gave the minimum values of the pseudonorm value is considered as the true amount (fig. 5).

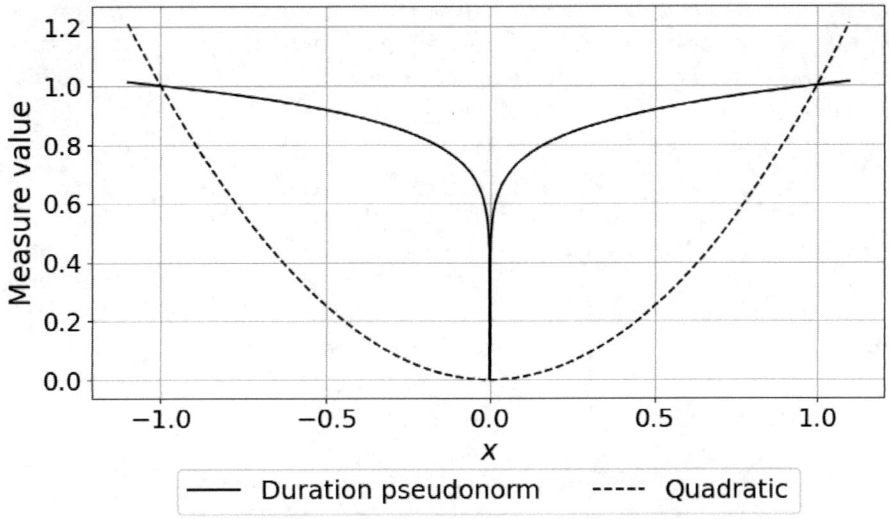

Figure 4. Duration pseudonorm vs. classic quadratic measure.

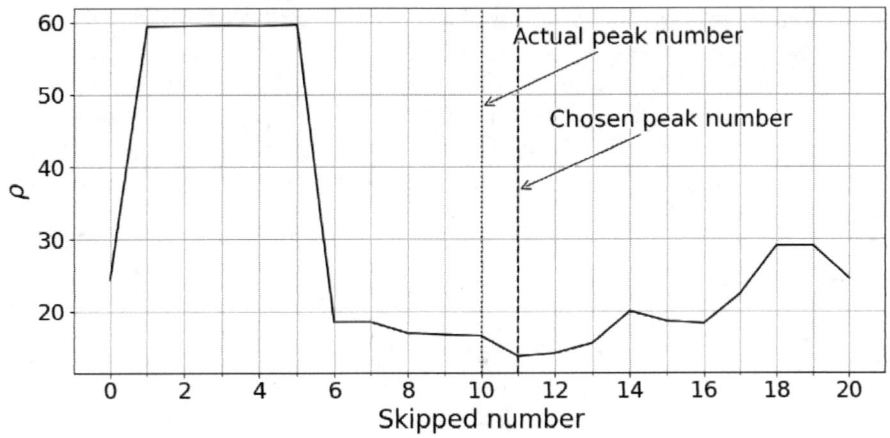

Figure 5. Example of the impulse amount estimation.

It is important to note that the amount of skipped points which leads to the minimum values of the pseudonorm does not necessarily equal to the actual true amount of peaks. This can be caused by some peaks not having a big effect on the results of the calculation, which leads to underestimation, and some white noise points having exceptionally high or just computationally unfortunate values, which leads to overestimation.

5. Testing on signal models

The method in its different interpretations was tested on several signal models. The initial testing was conducted for a signal of the form $s_n = \sin(2\pi f \Delta t n)$.

In this case the normalized frequency was varied in the range between 0.01 and 0.48 with a step of 0.01. For each of the points, the noise was generated 1000 times. During the experiments, the SNR of the white noise was set at 30 dB, which is considered a normal low-noise situation. The parameters of the impulse noise were as follows: $p = 0.05$, $\mu = 2$, $\sigma = 1$. The lengths of the signal was $N = 200$ points, which corresponds with the region in which the standard Prony's method generally works best. As can be seen in fig. 6, the simple finite difference method works the best for a signal consisting of a single sine function. At the same time, the segmented approach works best when the frequency of the signal is high.

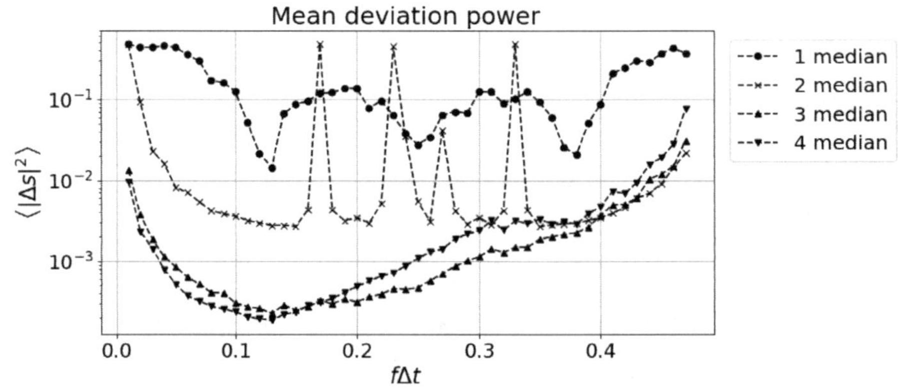

Figure 6. Median values of mean deviation power for different methods. 1 – classic eigenvector method, 2 – segmented approach. 3 – point skipping with finite differences, 4 – point skipping with matched detection.

At the same time, it is important to note that point skipping methods, while being more accurate in general, tend to rarely give divergent solutions with big values of deviations. This can lead to the mean values of the mean deviation power exceeding those of the segmented method (fig. 7).

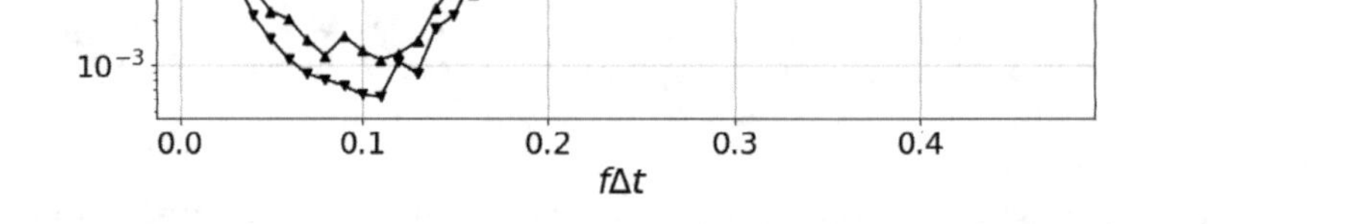

Figure 7. Mean values of mean deviation power for different methods. 1 – classic eigenvector method, 2 – segmented approach. 3 – point skipping with finite differences, 4 – point skipping with matched detection.

This means that the point skipping method requires the researcher to also implement an algorithm for the identification of such divergent solutions. This can be easily done as a signal power threshold due to its the values being extremely high.

The methods were then tested for signals with varying amount of components with random frequencies and amplitudes of the form $s_n = \sum_{m=1}^{M} A_m \exp(-\alpha_m \Delta tn) \cos(2\pi f_m \Delta tn + \varphi_m)$. For each of the component total numbers, the noise was generated with the same parameters as in previous tests. The amplitudes of the components belong to the range from 0.5 to 1.5 and the normalized frequency can wary from 0.01 to 0.48. For the following tests the normalized decay constant was varied in the range between 0 and $-199^{-1}\ln 0.1 \approx 0.0116$ which, in the most extreme case, corresponds to the component reducing its amplitude to 0.1 of the initial value. Due to the difference in the total energy of the signals with different amounts of components, the deviation shown in the following graphs in this section is normalized in relation to the signal mean power. As can be seen on fig. 8, the point skipping methods produce more accurate results for all numbers of components in the signal.

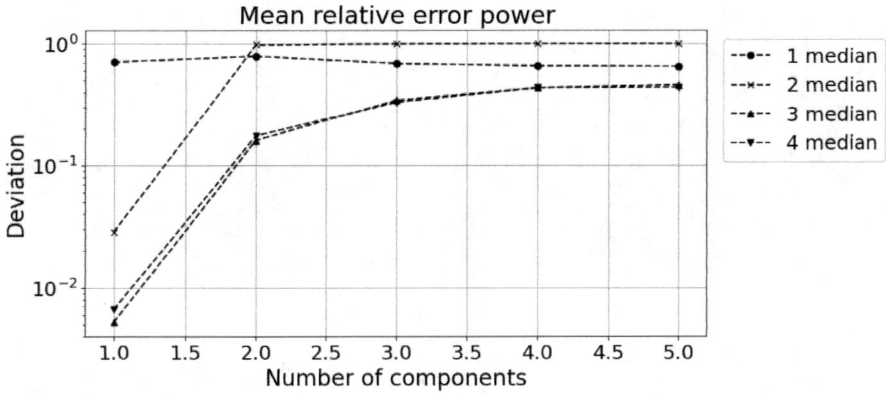

Figure 8. Median values of relative mean deviation power for different methods for signal with decay. 1 – classic eigenvector method, 2 – segmented approach. 3 – point skipping with finite differences, 4 – point skipping with matched detection.

As can be seen from fig. 9, the point skipping method with matched peak identification has shown the best accuracy characteristics in case of the mean normalized deviation values.

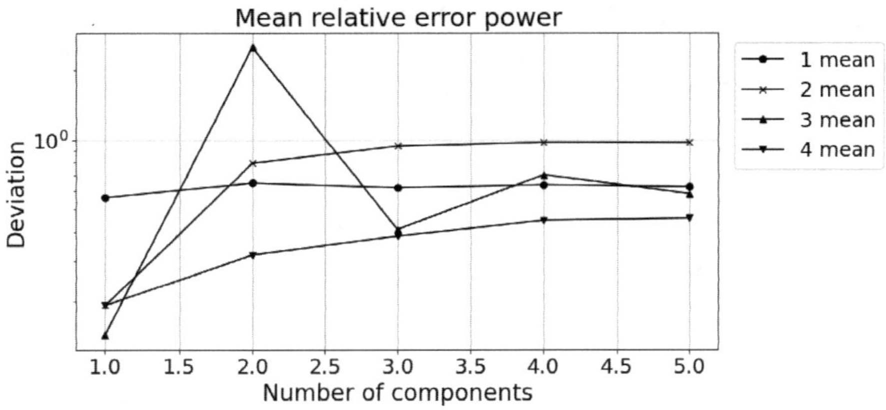

Figure 9. Mean values of relative mean deviation power for different methods for signal with decay. 1 – classic eigenvector method, 2 – segmented approach. 3 – point skipping with finite differences, 4 – point skipping with matched detection.

Based on this data it can be said that the matched detection point skipping method can be considered the best for general application.

6. Testing on real data

The method was tested a longitudinal cross section of a radioimage obtained with the help of aperture synthesis in intermediate radiation zone in order to estimate the distance to the reflector. Such approach can be used in order to estimate the shape of conductive objects instead of the standard laser interferometry [24]. The radar images are usually processed with the help of Fourier transform or Fourier series, which may lead to difficulties during the calculation process [25]. The use of the Prony's method in the frequency domain is a known technique which can be used when the data peak does not correspond with the discrete Fourier transform grid [26]. The tested cross section had a dropout on one of the frequencies, which created the impulse of the exact type the method was developed for (fig. 10).

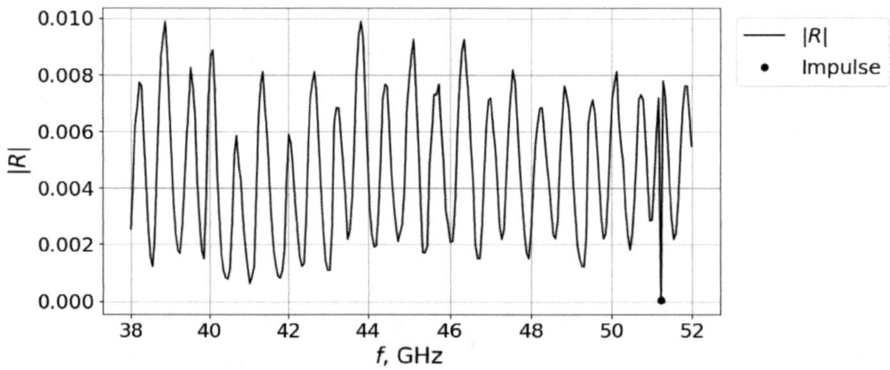

Figure 10. Radar image with impulse-type imperfection.

Fig. 11 shows that the Prony's method with point skipping is able to accurately estimate longitudinal distances on a radar image.

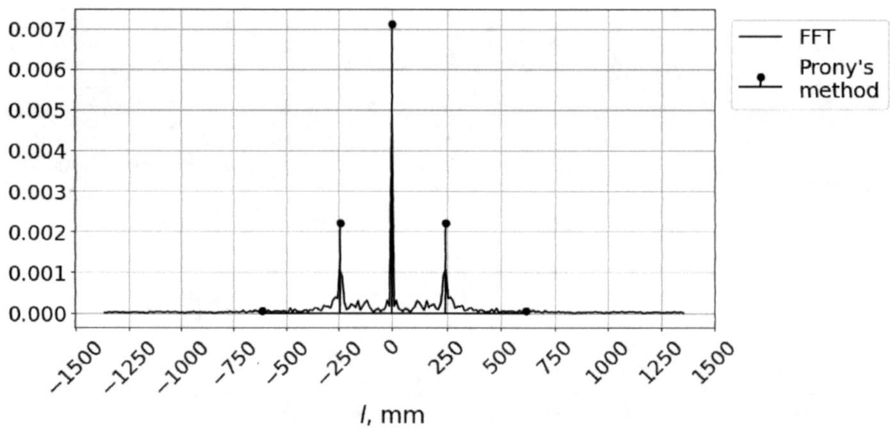

Figure 11. Prony's method and Fourier method estimation of the longitudinal distances on a radar image.

The method was able to estimate the position of the reflector as 242.1 mm with the true distance being 240 mm. The Prony's method without the point skipping approach was able to estimate this distance as 247.3 mm. This result means that the Prony's method with point skipping can be used in order to process real data in cases when the reliability of the sensor output is not guaranteed.

7. Conclusions

In the current paper it was shown that the Prony's method can be modified in order to process the signals with missing data. This can be used in order to estimate the characteristics of a signal in the presence of impulse noise. In order to make this applicable to real world situations, an additional method for impulse noise identification needs to be developed. It was shown that even the simplest forms of such method can lead to drastic improvement of the parameter estimation precision. It was also shown that the method is suitable for the processing of real data with impulse-type noise induced by the receiving equipment dropouts due to unstable power source.

References

[1] Hua Y and Sarkar T K 1990 *IEEE Transactions on Acoustics, Speech, and Signal Processing* **38** 814–24
[2] Hildebrand F B 1974 *Introduction to Numerical Analysis* (New York: McGraw-Hill)
[3] Marple S L 2019 *Digital Spectral Analysis: Second Edition* (New York: Dover Publications) chap 11, p 271 Dover Books on Electrical Engineering
[4] Xu M M, Xiao L Y and Wang H F 2013 *2nd IET Renewable Power Generation Conf. (RPG 2013)* pp 1–4
[5] Tawfik M M and Morcos M M 2005 *IEEE Trans. Power Deliv.* **20** 532–4
[6] Costa F F, Fernandes D A, de Almeida L A L and Naidu S R 2005 *2005 European Conf. on Power Electronics and Applications* pp 9 pp.–P.9
[7] Younan N H and Taylor C D 1991 *Electromagnetics* **11** 223–33
[8] Drobakhin O O, Olevskyi O V and Olevskyi V I 2017 *AIP Conference Proceedings* **1895** 060001 ISSN 0094-243X URL `https://doi.org/10.1063/1.5007386`
[9] Fernández Rodríguez A, De Santiago Rodrigo L, López Guillén E, Rodríguez Ascariz J M, Miguel Jiménez J M and Boquete L 2018 *BMC Bioinform.* **19** 451 ISSN 1471-2105 URL `https://dx.doi.org/10.1186/s12859-018-2473-y`
[10] Van Blaricum M and Mittra R 1975 *IEEE Trans. Antennas Propag.* **23** 777–81
[11] Yoshizawa A and Uchida S 2021 *2020 17th European Radar Conf. (EuRAD)* pp 298–301
[12] Drobakhin O O and Olevskyi O V 2018 *AIP Conf. Proc.* **2025** 050002 ISSN 0094-243X URL `https://doi.org/10.1063/1.5064904`
[13] Drobakhin O O and Korotkaya V G 1987 *The Soviet journal of nondestructive testing* **23**(5) 315–23
[14] Drobakhin O O 1989 *Optoelectronics, instrumentation, and data processing* (4) 37–42
[15] Kumaresan R and Tufts D 1982 *IEEE Transactions on Acoustics, Speech, and Signal Processing* **30** 833–40
[16] Porat B and Friedlander B 1987 *IEEE Transactions on Acoustics, Speech, and Signal Processing* **35** 231–5
[17] Van Blaricum M and Mittra R 1978 *IEEE Trans. Antennas Propag.* **26** 174–82
[18] Barone P, Massaro E and Polichetti A 1988 *Astron. Astrophys.* **209** 435–44
[19] Gu Y, Chen D, Jin T and Ren Z 2011 *Communication Systems and Information Technology* ed Ma M (Berlin, Heidelberg: Springer Berlin Heidelberg) pp 211–7 ISBN 978-3-642-21762-3
[20] Drobakhin O and Olevskyi O 2021 *2021 IEEE 26th Int. Seminar/Workshop on Direct and Inverse Problems of Electromagnetic and Acoustic Wave Theory (DIPED)* pp 209–12
[21] Borulko V F and Vovk S M 2015 *Radio Electronics, Computer Science, Control* **18**(1) 7–14 URL `http://ric.zntu.edu.ua/article/view/66392`
[22] Vovk S M, Antropov O S and Borulko V F 2008 *Radioelectron. Commun. Syst.* **51** 488–94 ISSN 1934-8061 URL `https://doi.org/10.3103/S0735272708090045`
[23] Andreev M V, Drobakhin O O and Lebedev S G 2012 *2012 Int. Conf. on Mathematical Methods in Electromagnetic Theory* pp 161–3
[24] Olevskyi V and Olevska Y 2018 *J. Geom. Symmetry Phys.* **50** 57–69 URL `https://doi.org/10.7546/jgsp-50-2018-57-69`
[25] Olevskyi V and Olevska Y 2018 *Geom. Integr. Quantization* **19** 159–167
[26] Andreev M V and Drobakhin O O 2016 *2016 8th Int. Conf. on Ultrawideband and Ultrashort Impulse Signals (UWBUSIS)* pp 18–20

Flowpattern in hydrocyclones, numerical simulations with experimental verification

Alex C. Hoffmann[1], **Daniel Fonnes**[1] **and Yu-Fen Chang**[1,2,3]

[1] Dept. of Physics and Technology, University of Bergen, Allégaten 55, 5007 Bergen, Norway
[2] Western Norway University of Applied Sciences, Inndalsveien 28, 5063 Bergen, Norway
[3] UiT The Arctic University of Norway, Hansine Hansens veg 18, 9019 Tromsø Norway

E-mail: alex.hoffmann@uib.no, yu-fen.chang@hvl.no, yu-fen.chang@uit.no

Abstract. This paper reports a detailed study of the flow in cyclone separators, with the use of most up to date computational fluid dynamics simulations, which are validated with positron emission particle tracking (PEPT) experiments tracing the movement of particles through the cyclone. The parameters varied were the viscosity of the carrier liquid, the flowrate and, in the numerical simulations, the inlet configurations of the cyclone, namely one and two inlets and, with the two inlets, a) both at right angles to the cyclone axis and b) angled downwards. The study reveals features of the flow, which have not been seen till now, but are necessary for the understanding and modelling of the separation and purification efficiency of cyclones. The results of the simulations and the close agreement with experiment are a testament to the reliability and accuracy of large eddy simulation (LES), even for flow features as difficult to simulate as the confined strongly swirling flows in cyclone separators. The results show that a contiguous, smooth surface of zero axial velocity does exist and has approximately the shape that has been assumed by modellers. The significant effects of fluid viscosity, underflow and modifications to the inlet are also shown.

1. Introduction

Cyclones are used in many important industries as separators to separate solids from fluids, to separate immiscible fluids of different densities and as classifiers to increase or decrease dispersed phase concentrations in multiphase mixtures. Cyclones can also be used for classification of solid particles according to size or density.

They are very widely used as the first and most important step in the cleaning of gaseous or liquid effluents before discharge to the environment, whether the atmosphere or the sea, and knowledge-based design and optimisation of cyclones is therefore of interest to the technological and scientific community, as also evidenced by the large number of recent articles cited in this article.

The focus of this work is on solid-liquid separation, the relevance of which is increasing, especially in the petroleum industry. As the reserves of oil and gas are declining, reserves that previously have been seen as non-commercial due to technological limitations or the cost of extracting and/or processing have now become feasible. Mature oilfields are also producing for a longer time, which presents new challenges due to increased water and sand production. To meet these challenges it is important to know the limitation of the equipment and to ensure that equipment can meet the needed specifications. The move to subsea processing, which allows processing and purification under pressure and makes it possible to limit the material that needs to be transported to the surface also makes it more critical to know the performance of processing equipment precisely, so that intervention in subsea installations, which is costly and time consuming, can be minimised.

The simplicity, robustness and compactness of the cyclone are key properties explaining its extensive use. Though simple in construction, the complex internal flow of the cyclone has been a focus of intensive research using state-of-the-art computational and experimental techniques. A challenge has been to produce good experimental results for the details of the flow pattern of the fluid and the particles inside the cyclone. The approach in this work is to use positron emission particle tracking (PEPT) to extract the trajectory of an actual particle as it moves through the cyclone. This is made possible by the use a positron emission tomography (PET) scanner and a mobile flow loop in which the cyclone is installed, and a positron emitting ion-exchange resin tracer particle.

The primary aim of this project is to improve the understanding of solid-liquid separation in the cyclones, focusing on the flow features controlling the particles' actual paths through the cyclone and their separation/capture or emission/loss from the cyclone, also to throw light on the most successful models prediction cyclone separation performance. Recent technological advances in PEPT have made possible tracking of an individual particle at high velocity with the required accuracy to explicitly reveal the particle trajectory through a cyclone. The tracking of individual particles experimentally reveals how the cyclone is performing and where any trouble areas are located. This work combines the results from PEPT experiments with numerical simulation of an identical setup. Being able to regenerate simulated trajectories qualitatively similar to the PEPT trajectories, this is strong evidence that the numerical simulations are valid and can be used in the further investigation of the flow pattern inside the cyclone.

1.1. Modelling

We review here briefly some of the important features and basic assumptions of the most successful cyclone separation performance models, which will be relevant to the work described in this paper. It is fair to say that the most used cyclone separation models are based on the 'equilibrium orbit' concepts pioneered by Walter Barth [1] and later developed further by Muschelknautz [2], Mothes and Löffler [3], Lorenz [4] and Muschelknautz [5], among others.

The equilibrium-orbit models are based on the concept that the axial flow direction in the outer part of the vortex in reverse-flow cyclones is downwards in the direction of the dust outlet and the axial flow in the inner part of the vortex is upwardly directed, towards the fluid outlet. A particle in the outer part of the vortex will therefore be separated, while one in the inner part will be lost and emitted from the cyclone. A surface of zero axial velocity separates the two regions, and the particle rotating in this surface (in this paper called CS for 'control surface', sometimes called LZVV for 'locus of zero vertical velocity' [6]) will be lost or separated with equal probability, and is therefore the cyclone 'cut size'. The forces acting on the particle rotating in CS are:

- A radially inwardly directed drag force due to the flow across CS (this flow is assumed to be uniform) of the carrier fluid as it moves from the radius of the inlet to that of the carrier fluid outlet
- The outwardly directed 'centrifugal force' due to the rotation of the particle

For the cut size these two are equal.

The question arises where this surface of zero vertical velocity is located. Most assume this surface to be the cylindrical surface obtained by prolonging the vortex finder tube to the bottom of the cyclone, while some, most notably Svarovsky [6], assumed it to be a conical surface ('mantle') with the base at the lower cross-section of the vortex finder and the apex at the bottom of the cyclone. The latter has often been used in the modelling of hydrocyclones, and the former for gas cyclones, although there is no fundamental reason there should be a difference in this respect between the two, at least in the absence of a gas core in the hydrocyclone. This question will be answered by the results presented in this paper.

A large number of papers have been published wherein these models are discussed and used for cyclone performance modelling, some recent ones are refs [5, 7–17]

Although it is still such that the classical analytical cyclone cut-size models are the most successful and the most used for predicting cyclone performance, the trend both in the research literature and in industry has been to model cyclones using computational fluid dynamics (CFD). This was for a long time fraught with difficulties, since it turned out that the classical turbulence models, such as the $k-\epsilon$ model, could not be used for modelling the strongly swirling confined flows in cyclones. However, more advanced turbulence models, such as the Reynolds Stress model (RSM) and the large eddy simulation (LES) turbulence model proved to be suitable for the simulation of the flow in cyclones, and this has made CFD simulation of cyclones very popular. Gronald and Derksen [18] reviewed this issue and some papers discussing the simulation of the flow in cyclones by CFD using the latest turbulence models are refs [7, 10, 11, 19–29]. The major CFD packages even offer a cyclone as a standard test case for new users. Mousavian and Najafi [30] found that LES simulations were superior to other turbulence models in simulating the flow in cyclones. In CFD simulations, the number of grid points required to directly resolve all turbulent scales scales with Re^3, limiting the practicality for very high-Re flows. In LES, the large eddies are explicitly resolved on the grid, while the smaller eddies are modelled using subgrid-scale (SGS) models that account for the turbulent energy transfer from the large to the small scales. Using LES and SGS models makes it possible to capture the essential features of turbulence at a lower computational cost compared to direct numerical simulation (DNS). In this study, the Reynolds numbers are above 3500, except for the tests with the highest viscosity (20 cP) [31]. Using LES and SGS models strikes a balance between capturing the important large-scale features of turbulence and modelling the effects of the small, unresolved scales, making it possible to study turbulent flows in a computationally efficient manner.

The dynamics of particles in cyclones is normally simulated by Lagrangian particle tracking. Tracking enough particles of varying sizes till they leave the cyclone, either through the dust exit (collected) or the carrier fluid exit (lost), will yield a grade-efficiency curve and thus a cut-size as the size of particle that will be collected with 50% efficiency. Eulerian modelling of the particle flow has not been published much, even though this would elucidate important phenomena taking place in cyclones and make easier the simulation of flow features caused by the effects of two-way coupling, which is especially important in the region just after the inlet.

1.2. PEPT

The experimental method for validating the CFD simulations is positron emission particle tracking (PEPT). This method, which was pioneered in Birmingham UK by Seville and Parker [32–35], makes it possible to track a particle moving through a process, located in the field of view (FOV) of a three-dimensional (3D) detector array, by cross-triangulation. The method is based on the radioactive decay of a suitable isotope with which a tracer particle is labelled, in the work shown here ^{18}F, which gives rise to the emission of a positron that travels some small distance through the medium till it annihilates with an electron giving rise to the back-to-back emission of two photons of 511 keV each. These photons are detected by detectors and a straight line between two detectors reacting to the same annihilation event (i.e. both within a very narrow time window) defines a 'line of response' (LOR), somewhere along which the annihilation event must have taken place. Cross triangulation using many LORs within a short time interval allows determination of the centroid of the tracer particle, the accuracy of which depends on the cross-triangulation algorithm and the activity on the tracer particle.

Some scatter is inherent in the process, the reasons for this are discussed in refs [36] and [37] and are briefly listed below:

- The length through which the emitted positron travels before annihilation
- The angle between the two photons may be less than 180 degrees, due to the positron and electron annihilating having significant amounts of kinetic energy
- The photons emitted may be deflected due to interaction with the material (Compton scattering) on their way to the detectors
- Three photons may be generated in one annihilation, or one photon from each of two annihilation events occurring at the same time may be detected, giving rise to false LORs
- 'Parallax blurring' occurs if a photon impacts obliquely on and penetrates a detector, and is detected by the neighbouring detector

The authors of this paper have developed accurate algorithms for determining the particle position, and this, combined with efficient particle labelling, has made it possible to locate the particle centroid with a standard deviation of less than 100 μm in all three coordinate directions. The most reliable way of estimating the standard deviation in the ms positions (meaning that one position is determined every millisecond) of the tracer particle centroid is to determine the position of a stationary particle. In the context of another project, such a determination was carried out, and the standard deviations in six series, each consisting of 1000 ms positions were determined. The mean standard deviations in the positions of the centroid of the tracer particle were 93.51 ± 0.648, 86.51 ± 0.299 and 80.97 ± 0.810 μm in the x, y and z-directions, respectively, where the z axis is the axis aligned with the axis of the cylindrical field of view of the PET scanner. This is much better accuracy than any reported in the published literature [38]. This very high spatial resolution is also evident in the results published in this paper. The standard deviation scales with $1/\sqrt{n}$, where n is the number of LORs used in the cross-triangulation [21, 39].

The algorithm developed by the authors is based on using the analytical expression for the spatial point that minimises the sum of squares of the distances to all the LORs measured within one millisecond as a first approximation, and then considering LOR cutpoints in the plane normal to the z-axis (the x-y plane) in a series of iterative steps, in each step eliminating LORs with only outlying cutpoints in successively narrower spatial windows, the final window size chosen as the one leading to the smallest scatter in the millisecond positions. For the positioning in the y-z plane, only the LORs remaining after the elimination process in the x-y plane are included. This algorithm used in this work can be applied to track multiple particles at a time if the differences between the radioactivity of each particle are adequate. Algorithms on basis of fitting the cutpoint distribution with peak shape functions as well as computing the 3D relations between LORs have been developed in our lab for tracking single and multiple particles. Other algorithms for tracking of multiple particles simultaneously, on basis of density maps for the LORs [40], on basis of machine learning [41] and on basis of identifying voxels with many LOR crossings [39], have been developed. The algorithm used in this paper is illustrated in Figure 1.

Work in various groups is also ongoing to develop new tracer particles, some recent papers discussing or reviewing this are refs [42–47].

2. Scope and Objectives of this work

This work is part of a larger project aimed at determining the suitability of cyclone separators for separating a variety of solids from liquid streams consisting of liquids other than water, focusing on liquids of higher viscosity than that of water.

The specific aims of the work described in this paper are to verify that numerical simulations of the fluid and particle flows can correctly reflect even the unsteady and unstable phenomena

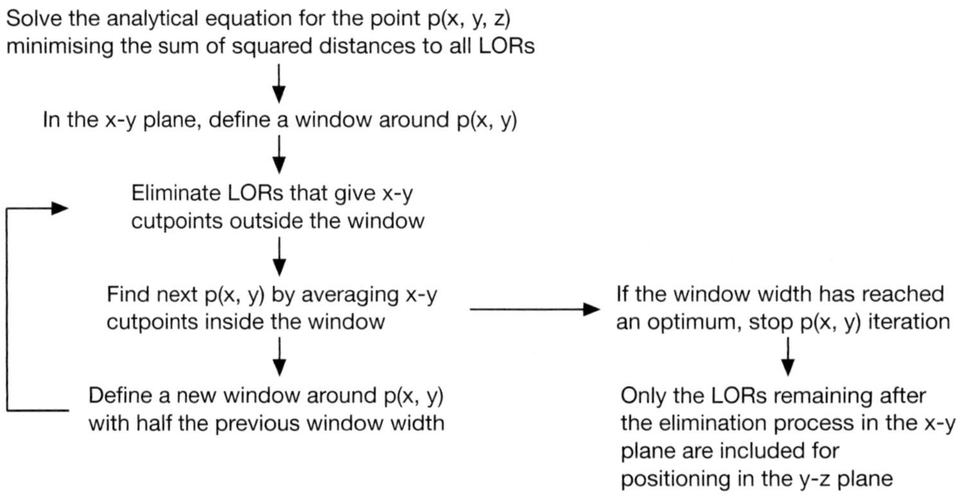

Figure 1. Figure outlining the structure of the PEPT particle positioning algorithm used in this work developed by the authors.

taking place. Furthermore to use numerical simulations to elucidate some of the important assumptions on which the most utilised and successful analytical cyclone performance models are based.

3. Simulations and experiments

A brief account of the numerical simulations and experiments carried out during this work will be given in in this section.

The CFD simulations in this work are performed with the LES turbulence model and the Smagorinsky subgrid turbulence model, which in previous work has proven to be the model most accurately reflecting experimental results [48]. We note that also the RSM model gave results qualitatively agreeing with experiment.

The equations to be solved numerically are the Navier-Stokes equations:

$$\begin{aligned} &\boldsymbol{\nabla} \cdot \mathbf{v} = 0 \\ &\rho \frac{D\mathbf{v}}{Dt} = \rho \mathbf{g} - \boldsymbol{\nabla} p + \mu \boldsymbol{\nabla}^2 \mathbf{v} \end{aligned} \tag{1}$$

A numerical mesh is applied to cover the domain of interest, and discrete representations of the equations are solved on the mesh. The mesh has to be fine enough to obtain an accurate solution, and as coarse as possible to limit the computational effort.

For LES of the flow of incompressible fluids a spatial filter is applied to the flow variables. For an arbitrary flow variable, $\phi(\mathbf{x})$, which could be a component of the fluid velocity or the pressure, the spatial filter is:

$$\bar{\phi} = \int G(\mathbf{x} - \mathbf{x}', \Delta)\phi(\mathbf{x}', t)d\mathbf{x}' \tag{2}$$

where $G(\mathbf{x}, \Delta)$ is some suitable function of the spatial coordinates. Δ is the width of the filter, similar in magnitude to the size of the computational cells. By applying this filter, turbulent eddies smaller than the grid size are filtered out, and their effect on the flow is modelled with a subgrid turbulence model. This model, however, can be much simpler than most turbulence models in use, since this fine-scale turbulence is almost isotropic. In this work the well-known and very simple Smagorinsky subgrid turbulence model was used. The larger eddies are in LES simulated directly.

The geometry of the cyclone, which is the same as that used in the experiments, is shown in Figure 2A, and details of the numerical grid are shown in Figure 2B. A 'trimmer' model is used for the grid together with a Prism layer mesher and Surface remesher. The number of prism layers is two and the base size of the grid-cells is 1.2 mm. To check for grid independence several simulation cases were run with the base size varying from 0.8 to 2.0 mm with increments of 0.2 mm. From this, a base size of 1.2 mm was found to be acceptable. For the Lagrangian particle tracking the drag model of Schiller-Naumann was used. More details about the numerical simulation scheme are given in ref [49]. The simulations were performed using the software package Star-CCM+. This work builds on the work described in [49], but since that work the cyclone used for the experiments has been significantly improved in order to more closely match the geometry used in the CFD simulations. The inlet was modified to make it better match that of a standard slot-inlet cyclone and the CFD geometry. In addition to the modification to the inlet, also the wall of the vortex finder tube was made thinner. Some modifications to the flow loop were also made to avoid vibrations and improve the flow from the holding tank and the flows in the pump and through the cyclone underflow.

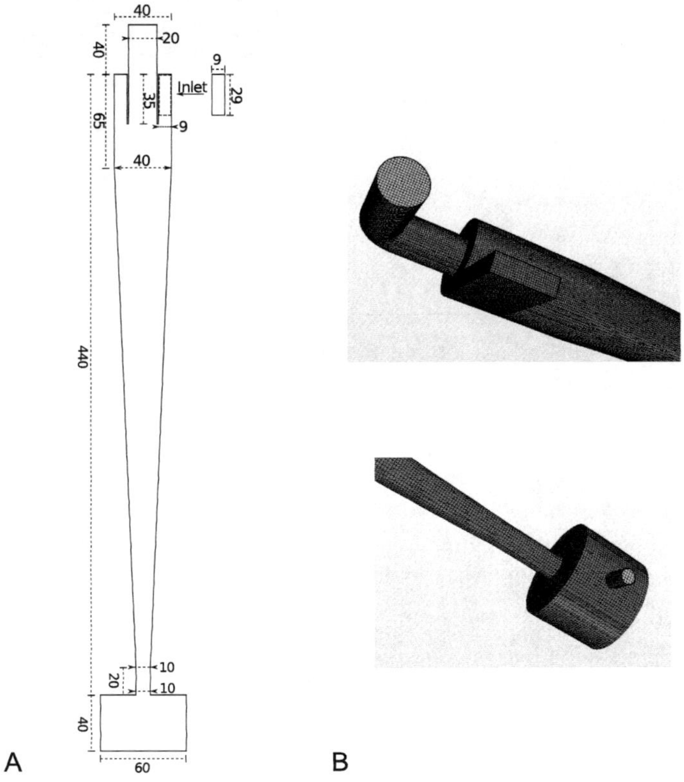

Figure 2. The cyclone used in the experiments and the simulations. A) The geometry of the cyclone, drawn to scale. The dimensions are given in mm. B) Details of the numerical grid used in the CFD simulations.

The experimental rig is shown in Figure 3, in diagrammatic form in Figure 3A and installed in the PET scanner in Figure 3B.

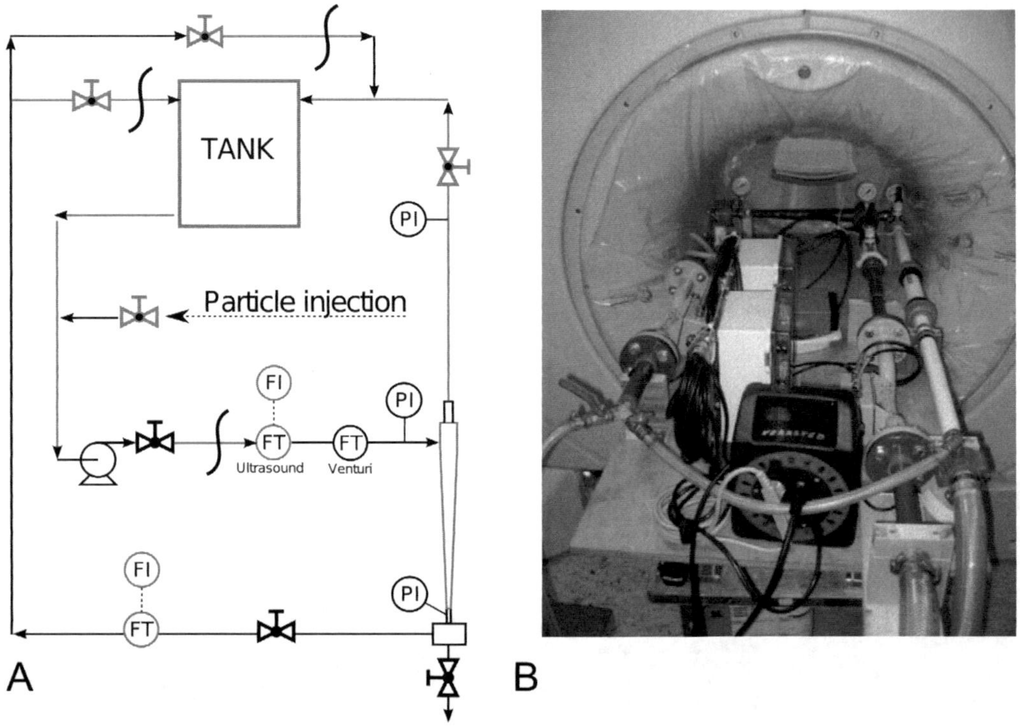

Figure 3. The experimental rig. A) Diagram showing the valving, the holding tank and the cyclone with the underflow provision. PI: pressure indicator. FI: flow indicator. FT: flow transmitter. B) Photo showing the rig installed in the PET scanner.

The experiments were performed at Haukeland University Hospital, the activation of the resin particles was also done there. A number of particles were activated before each experiment was started. The flow loop was test run to detect any leakage and to verify that it was working as designed. After testing, the rig was moved to the location of the PET scanner and placed in the scanner FOV. The flow loop was started and flow rate and underflow ratio were tuned to the desired values for each experiment. Then a particle was injected by the use of a syringe upstream of the pump at the same time as the scanner is set to start logging. The logging was continued for 90 seconds. Then the flow loop was taken out of the scanner and liquid was extracted from the collection chamber until the particles were recovered. This procedure was followed for the experiments in which the particle flowed to the collection chamber. In the experiments where the particle was not collected in the collection chamber and instead transported back to the holding tank, the logging time was increased to 120 seconds to have more particles running through the cyclone in each experiment and have a better chance of logging an individual particle moving through the cyclone without interference from any other tracer particle.

The tracer particles used were strong anion-exchange beads, Amberlyst A26 Hydroxide Form (Acros Organics). The particle size was measured with microscopy to be 430 μm, with a standard deviation of 56 μm The envelope density of the particles immersed in water was determined by settling experiments to be between 1080 and 1120 kg/m^3. This means that the particles will have the the same Stokes number, $\mathrm{Stk} \equiv \frac{(\rho_p-\rho)d_p^2}{18\mu}\frac{v_{\mathrm{in}}}{D}$ as sand particles (density $\rho_p = 2700$ kg/m^3) of diameter $d_p = 104$ μm in water, and therefore behave in the same way in the cyclone as such sand particles. When 'tracer particle' is mentioned in this article, it is a particle with these properties. The dynamic response time of the tracer particle in water, $\frac{(\rho_p-\rho_w)*D^2}{\mu_w}$ is 0.018 s, which probably means that the turbulent dispersion of the particle is somewhat reduced, but it should not be significant compared to the mean motion or the particles' behaviour in the region of the end of the vortex. Transient effects should be revealed in both the numerical simulations and in the PEPT experiments. In the higher viscosity fluids the response time will be shorter.

A sand particle in water of 104 μm is a rather small particle, still considerably larger than the cut-size of the cyclone, but rather small in the distribution of sand particles that will be charged to the cyclone. The low relative density of the tracer particle ensures that it behaves in the same way in the cyclone as a typical sand particle that the cyclones would normally need to separate.

The experiments carried out are listed in Table1. The Reynolds number, Re $\equiv \frac{\rho v D}{\mu}$, depending on viscosity, μ, density, ρ, velocity of the flow, v, and on the diameter of the cyclone body, D, can be calculated as listed in ref [31]. Alginate was chosen as a viscosity modifier, since this turned out to be the best way of modifying the fluid viscosity keeping all else constant. Alginate did give rise to slightly non-Newtonian behaviour, but this was not enough to cause any concern.

Table 1. List of the experiments carried out

Fluid	Flow rate [m^3/hr]	% underflow
Salt soln	3.4	5
Water	3.4	0
Water	3.4	5
Water	3.4	10
Water	3.4	20
Water	2	0
Water	1	0
Alginate 20 cP	3.4	5
Alginate 15 cP	3.4	5
Alginate 10 cP	3.4	5
Alginate 5 cP	3.4	5

4. Results and Discussion

4.1. The fluid flow pattern

Some PEPT experiments were performed to study the fluid flow pattern using a tracer particle with neutral density relative to the fluid. To match the density of the particle, salt was added to the water, since it was not practical to tune the density of the radioactive particle to match that of water. The drawback of this is that the ^{18}F on the tracer particle will leach into the fluid. It was nevertheless possible to obtain a number of useful tracks indicating the flow pattern of the fluid in the cyclone. In the absence of a free surface or stratification in the flow, Reynolds number similarity should ensure dynamic similarity in the cyclone, the fluid flow pattern should therefore not be affected by the density of the fluid except through the Reynolds number, and this is only weakly affected by the slight change in density that is involved here (see also Section 4.3). Some of these tracks are shown in Figure 4A, and may be compared with the streamlines for the fluid flow resulting from CFD simulations shown in Figure 4B. In accordance with the operating conditions a neutral density particle should leave the cyclone through the fluid outlet 95% of the times and the particle outlet at the bottom of the cyclone 5% of the times. It is clear that the fluid streamlines in the CFD simulations agree with the tracks in the PEPT experiments and that the streamlines and the tracks are consistent with what would be expected on basis of the operating conditions. The streamlines shown in this paper are snapshots of the turbulent flow, which is filtered with the spatial LES filter and therefore does not reflect the fine-scale turbulence.

The experiments show that most of the tracer particles moving from the outer part of the vortex with downwards flow to the inner part with upwards flow on their way to the fluid outlet do so in the upper part of the cyclone, and the flow there is therefore the most important in determining the cyclone separation efficiency.

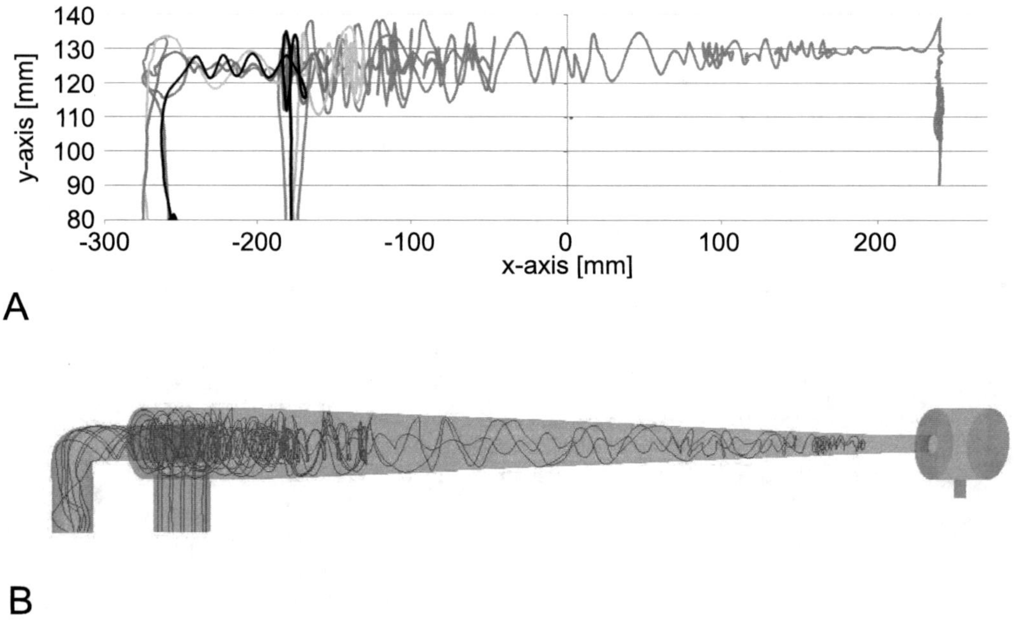

Figure 4. A) PEPT tracks of neutral density particles in the cyclone, the tracks from the different tracer particle passes have been given different colours. B) Streamlines for the fluid flow in the cyclone under the same operating conditions. Flowrate: 3.4 m^3/hr, underflow: 5%.

In the rest of the work presented in this paper the comparison between experiment and CFD will be on basis of the movement of particles that are not neutral density but have the properties described above for the tracer particles, which will be tracked by PEPT experimentally and Lagrangian tracking in the CFD simulations.

4.2. Effect of underflow

Withdrawing a stream from the bottom of the cyclone is known to often have a positive effect on the separation efficiency, at the cost of the separated particles still being obtained in an, albeit much more concentrated, fluid stream, which then has to be treated further, perhaps in a filter installation, before the particles are obtained fully separated. Underflow is often used in cyclones operating on liquids and more rarely on cyclones operating on gases. The simulations and experiments described in this section will show the important effect underflow may have on the fluid and particle flow patterns in cyclones. Figure 5 shows tracks of a tracer particle with PEPT and with CFD. Excellent agreement between PEPT and CFD is seen, even with regards to the instability in the bottom of the cyclone, which is caused by the so-called 'end of the vortex', where the vortex core bends to the cyclone wall and rotates (precesses) at a more or less fixed axial position, which in turn causes the particle to flow in an irregular pattern, having a residence time much longer in this region than in the upper part of the cyclone, something that can clearly lead to clogging of the cyclone if the particle concentration is high. The simulations reflect correctly not only the presence of the instability under these same conditions conditions as the experiments, but even the approximate position of it. The vortex end can also lead to wear of the cyclone wall in consequence of the particles' prolonged residence time. In gas cyclones it has even been known to lead to failure with the part of the cyclone under this position falling off.

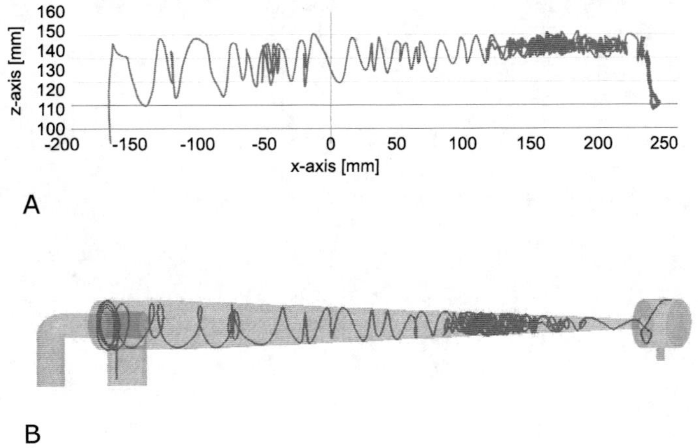

Figure 5. Experimental (A) and CFD (B) tracks of a tracer particle in the cyclone with the underflow closed. Flowrate: 3.4 m^3/hr, underflow 0%.

The pressure distribution in a plane containing the cyclone axis in the CFD simulation shown in Figure 6A confirms that the vortex motion ends at a given position in the separation space, and under this position is much weaker consistent with the idea of the vortex 'ending' at this position. A similar picture can be seen in the pressure profile in Figure 10B later in this paper. There, at the higher flow, a 'secondary vortex' under the main vortex end, described and discussed in ref [50], but not shown explicitly till now, can be distinguished under the position of the vortex end. Under the vortex end the isosurface of zero axial velocity is also less visible, as can be seen in Figure 6B.

Drawing even only a little underflow has a strong effect on this behaviour of the fluid and the particle in the bottom of the cyclone. Figure 7 shows PEPT and CFD tracks for a tracer particle with 5% underflow. The passage of the particle through the cyclone is seen to be much smoother and the severe disturbance in the bottom of the cyclone as shown in Figure 5 has completely disappeared. The agreement between PEPT and CFD is seen to be very good also in this case, confirming that the LES simulations reflect correctly also the effect of underflow on the fluid and particle flow patterns.

The pressure profile and the zero axial velocity isosurface shown in Figure 8 A and B confirm the effect seen in Figure 7 that the underflow has the effect of eliminating the end-of-vortex effects in the cyclone.

4.3. Effect of fluid flow rate
It is interesting to know if the fluid flow rate through the cyclone, and therefore the cyclone Reynolds number, has an effect on the flow pattern in the cyclone. It is normally assumed that there is no such effect, or that the effect is only weak.

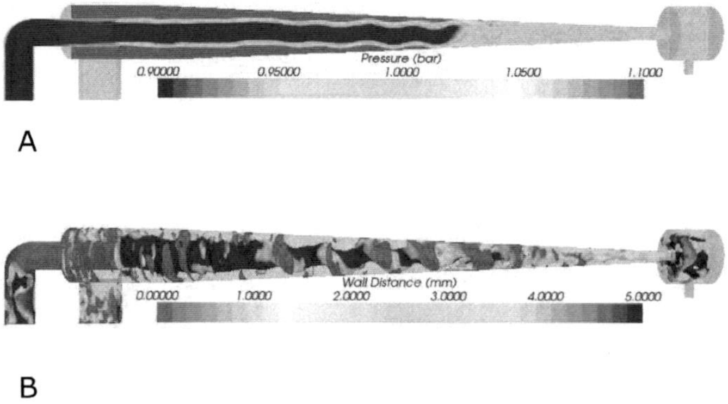

Figure 6. CFD snapshots showing A) the static pressure in a section of the cyclone containing the axis and B) isosurface of zero axial velocity coloured according to the distance from the cyclone wall. Flowrate: 3.4 m^3/hr, underflow: 0%.

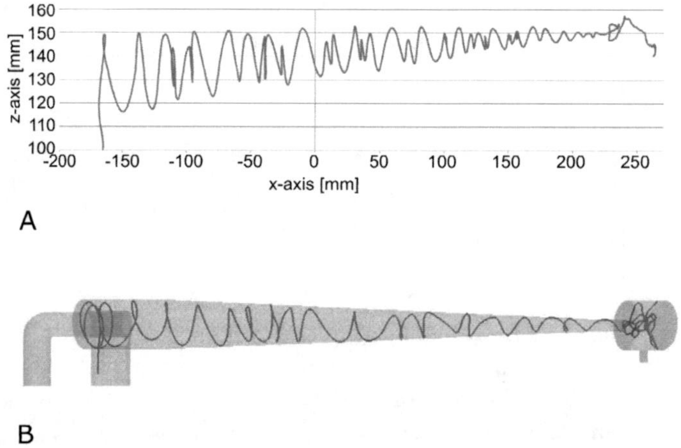

Figure 7. Experimental (A) and CFD (B) tracks of a tracer particle in the cyclone with 5% underflow. Flowrate: 3.4 m^3/hr, underflow 5%.

Figure 9 shows PEPT and CFD results from a case with low fluid flowrate through the cyclone. Some discrepancy between the experiment and CFD can be seen in this case, although both simulations and PEPT tracks indicate some instabilities near the transition from cylindrical to conical and also low in the cyclone separation space close to the particle outlet. The particle path through the cyclone is more disturbed in the experiment than in the CFD.

It is interesting the compare the CFD results with an equivalent case where the fluid flowrate is high. Figure 10 shows such a case. It appears from the static pressure distribution that the vortex is longer in the case with the low flow, this agrees well with the particle flow pattern which shows that the particle passes through the cyclone to the dust outlet relatively unhindered, but should be taken with a grain of salt since, in this respect, the CFD simulation does not agree well with the experiment.

In Figure 9 the isosurface of zero velocity is seen to be scattered to some extent due to the image representing a snapshot of an LES simulation, but a continuous surface with blue

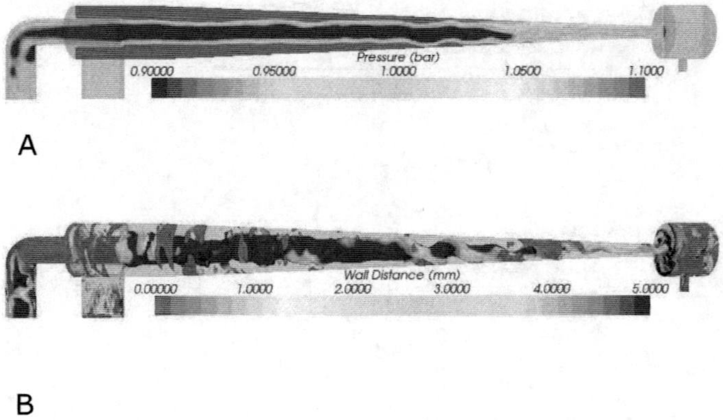

Figure 8. CFD snapshots showing A) the static pressure in a section of the cyclone containing the axis and B) isosurface of zero axial velocity coloured according to the distance from the cyclone wall. Flowrate: 3.4 m^3/hr, underflow: 5%.

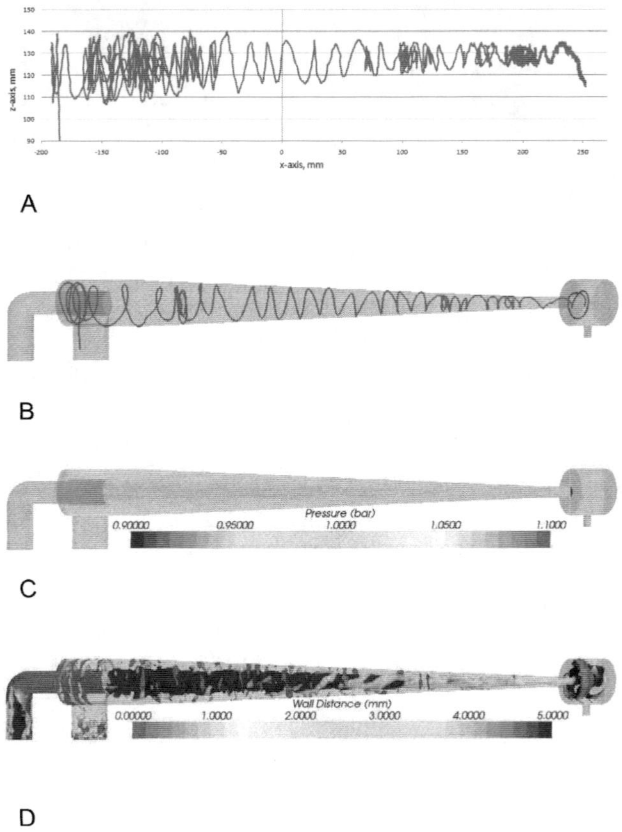

Figure 9. CFD simulations and PEPT tracks of a tracer particle in a cyclone with reduced fluid flowrate. A) PEPT tracks of tracer particle. B) Lagrangian tracks of a tracer particle in a cyclone with the same geometry and operating conditions. C) Pressure distribution in the cyclone by CFD. D) Isosurface for an axial fluid velocity of zero, coloured according to the distance from the cyclone wall. Flowrate: 1.0 m^3/hr, underflow: 0%.

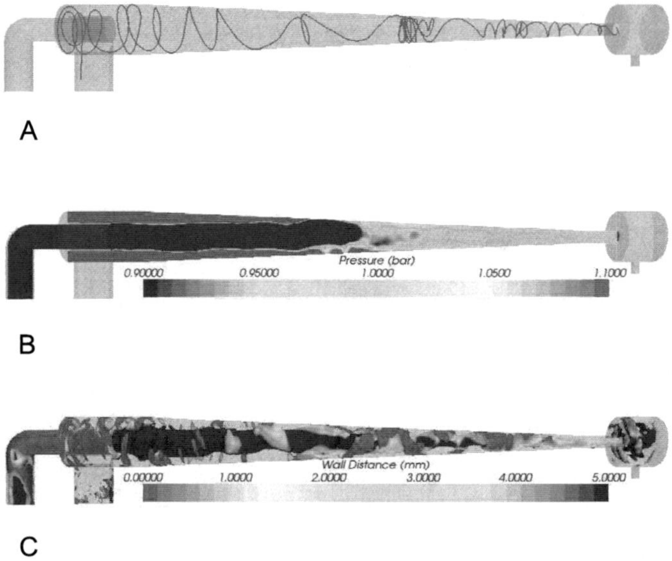

Figure 10. CFD simulations of the flow in a cyclone with high fluid flowrate. A) Lagrangian tracks of a tracer particle. B) Pressure distribution in the cyclone by CFD. C) Isosurface for the axial fluid velocity of zero, coloured according to the distance from the cyclone wall. Flowrate: 7.4 m^3/hr, underflow: 0%.

colour is seen under the fluid outlet. This surface roughly forms a cylinder under the vortex finder wall, but it seems to contract in the radial direction when moving down in the cyclone, maintaining the blue colour, indicating that the surface of zero axial velocity does, in the conical part this cyclone with a long conical section, form a 'mantle' as assumed by some as mentioned in Section 1.1. In Figure 10 the surface is seen to be less stable, indicating that the vortex becomes unstable in the lower part of the cyclone at this high flowrate, a flowrate giving rise to an inlet velocity so high that it is rarely used in practise in cyclones operating on liquids.

4.4. Effect of fluid viscosity

The effect of fluid viscosity on the separation efficiency of the cyclone has been investigated in details in a separate subproject using a separate experimental rig. This is reported elsewhere [51] and experimental results for other aspects of the effect of viscosity on the fluid flow pattern have been discussed in ref [49]. Here the focus is on CFD results, verified by PEPT results, showing the effect of fluid viscosity on the flow patterns of the fluid and the tracer particle, particularly in the bottom of the cyclone.

Figure 11A shows the pressure drop over the cyclone as a function of the viscosity of the fluid flowing in the cyclone. This confirms the, intuitively somewhat surprising, result that the pressure drop decreases with increasing viscosity in a cyclone. This phenomenon, the reason for it, and the related effect that the pressure drop in cyclones decreases with increasing wall roughness, are discussed in ref [50]. It is a consequence of a weaker swirl in the cyclone caused by increasing wall friction in the cyclone body. Figure 11B shows the pressure drop over the cyclone as a function of the liquid (water) flowrate. The pressure drop should be proportional to the square of the liquid flowrate, which is indicated by the dashed curve, and the results bear this out. The numerical results generally agreed with the experimental ones to within around 10%, which is acceptable in view of the difficulties involved in measuring the pressure drop experimentally due to the residual swirl present in the vortex finder [50].

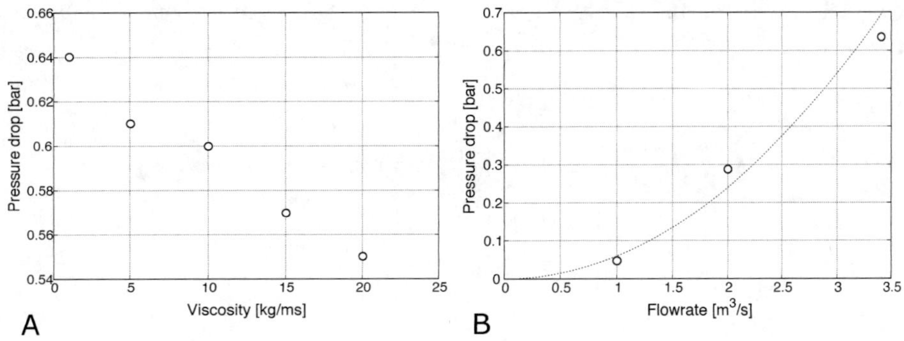

Figure 11. A) Numerical results for the pressure drop over the cyclone as function of the viscosity of the fluid. B) Numerical results for the pressure drop over the cyclone as a function of the liquid (water) flowrate.

Figure 12 shows the static pressure distribution for a CFD simulation with a flowrate of 3.4 m^3/s and a fluid of high viscosity (15 cP). The low-pressure vortex core is much less evident than for the same flow with water, an indication of the less strong vortex motion when the cyclone acts on fluids with high viscosity.

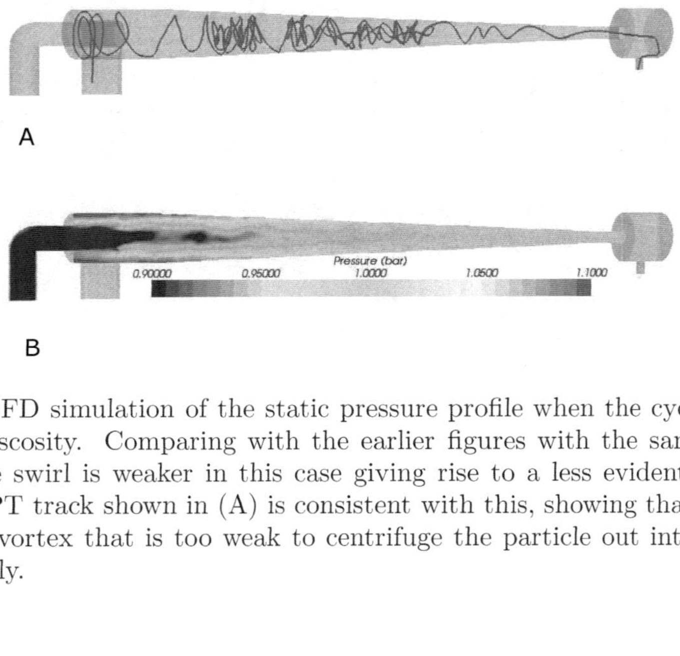

Figure 12. (B) CFD simulation of the static pressure profile when the cyclone is working on a liquid of high viscosity. Comparing with the earlier figures with the same fluid flowrate it is evident that the swirl is weaker in this case giving rise to a less evident core of low static pressure. The PEPT track shown in (A) is consistent with this, showing that the particle track is irregular in the vortex that is too weak to centrifuge the particle out into the outer part of the vortex efficiently.

4.5. Effect of a modified inlet

It is often thought that a dual inlet will lead to more axisymmetric flow in the cyclone and therefore stabilise the vortex motion in the cyclone. To test this, such a configuration was made for the CFD simulations, the height of the inlets were reduced so that the total inlet area was the same as for the single-inlet cyclone, and the results are therefore directly comparable with the simulations for the single-inlet cyclone at the same flowrate. The geometry and the simulation results are shown in Figure 13. The flow pattern can be seen to be very similar to that in the single-inlet cyclone, and the particle path confirms the presence of the 'end of the vortex' in the separation space, causing a disturbance in the particle transport to the particle outlet.

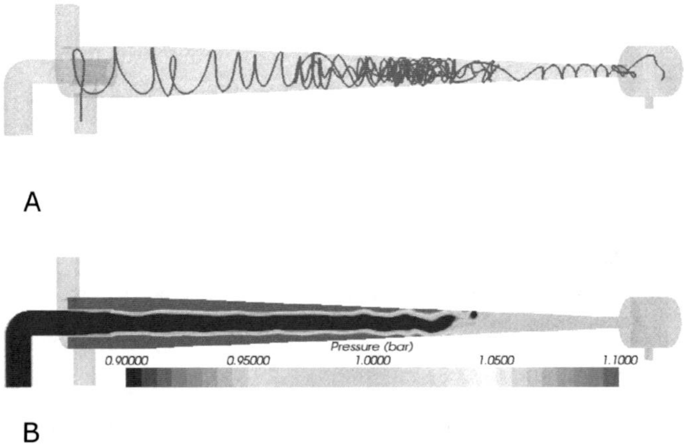

Figure 13. CFD simulations of hydrocyclone with two inlets. A) Pathlines of a particle. B) Distribution of static pressure in a plane containing the cyclone axis.

Lately increasing interest has been paid to the option of angling the cyclone inlet downwards with the aim of enhancing the downward flow in the outer part of the vortex and stabilise the vortex flow in the cyclone separation space and also to reduce the lip-leakage flow short-circuiting the cyclone [26, 52]. In order to add an unbiased assessment of the merits of such a design modification to the published literature, some simulations were performed with this configuration. Also in this case the inlets were configured so that the total inlet areas was equal to the inlet area in the single-inlet cyclone. The geometry and the results of the simulations are shown in Figure 14.

Also here the flow pattern low in the cyclone is seen to be more or less unaffected by the configuration of the inlet. A detailed study of the flows showed that this inlet configuration does lead to a reduction of the lip-leakage flow and therefore to less particles short-circuiting the cyclone in this flow and a slightly increased separation efficiency.

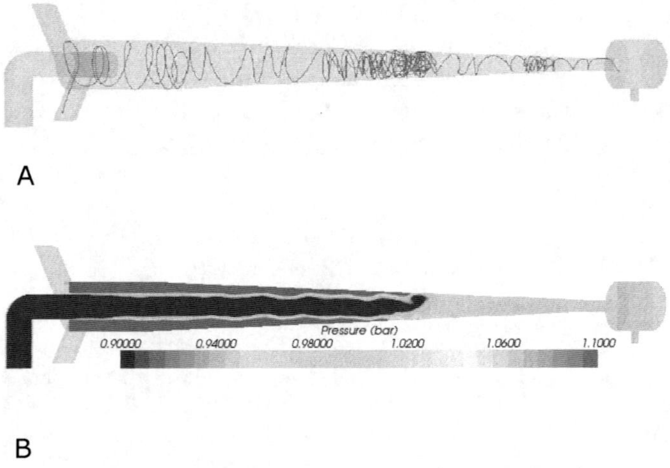

Figure 14. CFD simulations of hydrocyclone with two inlets angled downwards. A) pathlines of a particle. B) Distribution of static pressure in a plane containing the cyclone axis.

5. Conclusions

- The locus of zero axial velocity does extend from the vortex finder wall as assumed in the classical and most-used cyclone models and forms a cylindrical surface in the upper part of the cyclone as assumed in many models. Lower down it forms a conical mantle as some others hold.
- A much weaker secondary vortex below the end of the main vortex in the cyclone, as discussed in ref [48], has been demonstrated by the simulations.
- Higher viscosity fluids give rise to a weaker vortex but also to a lower pressure drop and a longer vortex.
- The classical cyclone models appear to work best at higher Reynolds numbers. At low Reynolds number the high friction factor seems to present a problem.
- A double inlet, angled downward or not, does not affect the flow low in the cyclone significantly, but the angled inlet does reduce the flow from the outer to the inner part of the vortex in the upper part of the cyclone separation space.
- LES was successful in simulating details of the flow in the cyclone.
- The position of the 'end of the vortex' is a crucial feature for the operation and design of cyclones. It is an instability that depends on the wall friction and the details of the flow pattern in the cyclone. Only recently it has been possible to model this correctly numerically using the most advanced computational techniques. The fact that this phenomenon and even the axial position at which it appears are reflected correctly in almost all of the simulations presented in this article is an important quantitative testament to the validity of the simulations.
- The only way of revealing important flow features in cyclone separators affecting the separation is Lagrangian particle tracking, showing features that point measurements, e.g. with laser Doppler anemometry (LDA) or particle image velocimetry (PIV), cannot show. The best experimental way of doing that is PEPT using an accurate cross-triangulation algorithm combined with high radioactivity on the tracer particle.

Acknowledgments
The authors gratefully acknowledge funding from the Norwegian Research Council through the PETROMAKS program. The project was established and the experimental rig built in collaboration with Aker Solutions. The authors thank the staff at Haukeland University Hospital and Siemens Healthineers for their expertise and assistance.

References

[1] Barth W 1956 *Brennstoff-Wärme-Kraft* **8, Heft 1** 1–9 (in German)
[2] Muschelknautz E 1972 *Chemie-Ing.-Techn.* **44** 63–71 (in German)
[3] Mothes H and Löffler F 1985 *Ger. Chem. Eng.* **8** 223–233
[4] Lorenz T 1996 *VDI Forschrittsberichte* **3** 1–113 (In German)
[5] Muschelknautz U 2018 *Gefahrstoffe Reinhaltung der Luft* **78** 453–465 ISSN 0949-8036
[6] Svarovsky L 2000 *Solid-Liquid Separation* 4th ed (Oxford: Butterworth-Heinemann)
[7] Wasilewski M, Brar L S and Ligus G 2020 *Separation and Purification Technology* **239**
[8] Dong S, Jiang Y, Jin R, Dong K and Wang B 2020 *Applied Mathematical Modelling* **80** 683–701 ISSN 0307–904X
[9] Zhao B, Wang D, Su Y and Wang H L 2020 *Separation and Purification Reviews* **49** 112–142 ISSN 1542-2119
[10] Nakhaei M, Lu B, Tian Y, Wang W, Dam-Johansen K and Wu H 2020 *PROCESSES* **8**
[11] Wasilewski M and Brar L S 2019 *Separation and Purification Technology* **213** 19–33 ISSN 1383-5866
[12] Wang J, Guo Y and Pan J 2019 *Powder Technology* **345** 457–467
[13] Bartz-Beielstein T, Zaefferer M and P Q C 2018 *Structural and Multidisciplinary Optimization* **58** 919–933 ISSN 1615-147X
[14] El Ashry Y, Abdelrazek A M and Elshorbagy K A 2018 *Separation Science and Technology* **53** 2500–2516 ISSN 0149-6395
[15] Wasilewski M 2017 *Separation and Purification Technology* **179** 236–247 ISSN 1383-5866
[16] Bogodage S G and Leung A Y T 2016 *Journal of Hazardous Materials* **311** 100–114 ISSN 0304-3894
[17] Singh P, Couckuyt I, Elsayed K, Deschrijver D and Dhaene T 2016 *Applied Mathematical Modelling* **40** 4248–4259 ISSN 0307-904X
[18] Gronald G and Derksen J J 2011 *Powder Technology* **205** 160–171
[19] Hoffmann A C, Vansanten A, Allen R W K and Clift R 1992 *Powder Technology* **70** 83–91
[20] Peng W M, Hoffmann A C, Dries H, Regelink M and Foo K K 2007 *AIChE Journal* **53** 589–597
[21] Chang Y F, Ilea C G, Aasen O L and Hoffmann A C 2011 *Chemical Engineering Science* **66** 4203–4211
[22] Li A, Zhu L, Wang K, Wang G and Wang Z 2020 *Powder Technology* **364** 205–217 ISSN 0032–5910
[23] Gopalakrishnan B and Prakash K A 2019 *International Journal of Advances in Engineering Sciences and Applied Mathematics* **11** 280–287
[24] Tofighian H, Amani E and Saffar-Avval M 2020 *Powder Technology* **360** 1237–1252 ISSN 0032–5910
[25] Svensen E A and Hoffmann A C 2020 *Heliyon* **6** ISSN 2405–8440
[26] Wasilewski M and Brar L S 2019 *Separation and Purification Technology* **213** 19–33 ISSN 1383–5866
[27] Parvaz F, Hosseini S H, Elsayed K and Ahmadi G 2018 *Separation and Purification Technology* **201** 223–237 ISSN 1383–5866
[28] Li G and Lu Y 2018 *Particuology* **39** 55–67 ISSN 1674–2001
[29] Huang A N, Ito K, Fukasawa T, Yoshida H, Kuo H P and Fukui K 2018 *Separation and Purification Technology* **190** 25–32 ISSN 1383–5866
[30] Mousavian S M and Najafi A F 2009 *Archive of Applied Mechanics* **79** 1033–1050
[31] Skorpen Å 2012 *Investigation of Particle Trajectories and Flow Patterns in a Hydrocyclone by Positron Emission Particle Tracking (PEPT)* (Master's thesis, The University of Bergen)
[32] Fangary Y S, Barigou M, Seville J P K and Parker D J 2000 *Chemical Engineering Science* **55** 5969–5979 ISSN 0009–2509
[33] Snieders F F, Hoffmann A C, Cheesman D, Yates J G, Stein M and Seville J P K 1999 *Powder Technology* **101** 229–239
[34] Hoomans B P B, Kuipers J A M, Salleh M A M, Stein M and Seville J P K 2001 *Powder Technology* **116** 166–177 ISSN 0032–5910
[35] Parker D J, Dijkstra A E, Martin T W and Seville J P K 1997 *Chemical Engineering Science* **52** 2011–2022 ISSN 0009–2509
[36] Bailey D L, Maisey M N, Townsend D W and Valk P E 2003 *Positron Emission Tomography: Basic Sciences and Clinical Practice* (London: Springer-Verlag)
[37] Chang Y F, Hoffmann A C and Adamsen T C H 2012 *Proceedings of the 2012 IEEE International Instrumentation and Measurement Technology Conference, I2MCT* (Graz, Austria)

[38] Sovechles J M, Boucher D, Pax R, Leadbeater T, Sasmito A P and Waters K E 2017 *Measurement Science and Technology* **28** 095402
[39] Wiggins C, Santos R and Ruggles A 2017 *Nuclear Instruments and Methods in Physics Research Section A: Accelerators, Spectrometers, Detectors and Associated Equipment* **843** 22–28
[40] Blakemore D M, Govender I, McBride A T and Mainza A N 2019 *Chemical Engineering Science* **207** 780–789
[41] Nicusan A and Windows-Yule C 2020 *Review of Scientific Instruments* **91** 013329
[42] Cole K, Buffler A, van der Meulen N, Cilliers J, Franzidis J P, Govender I, Liu C and van Heerden M 2012 *Chemical Engineering Science* **75** 235–242 ISSN 0009-2509
[43] Parker D J 2017 *Review of Scientific Instruments* **88** ISSN 0034-6748
[44] van Heerden M R, Cole K, van der Meulen N P, Franzidis J and Buffler A 2020 *Applied Radiation and Isotopes* **158** ISSN 0969-8043
[45] Boucher D, Jordens A, Sovechles J, Langlois R, Leadbeater T W, Rowson N A, Cilliers J J and Waters K E 2017 *Minerals Engineering* **100** 155–165 ISSN 0892-6875
[46] Valdesueiro D, Garcia-Trinanes P, Meesters G M H, Kreutzer M T, Gargiuli J, Leadbeater T W, Parker D J, Seville J P K and van Ommen J R 2016 *Nuclear Instruments & Methods in Physics Research Section A-Accelerators Spectrometers Detectors And Associated Equipment* **807** 108–113 ISSN 0168-9002
[47] Cole K, Buffler A, Cilliers J J, Govender I, Heng J Y Y, Liu C, Parker D J, Shah U V, van Heerden M and Fan X 2014 *Powder Technology* **263** 26–30 ISSN 0032-5910
[48] Pisarev G I, Hoffmann A C, Peng W and Dijkstra H A 2011 *Applied Mathematics and Computation* **217** 5016–5022
[49] Hoffmann A C, Skorpen Å and Chang Y F 2019 *Chemical Engineering Science* **200** 310–319
[50] Hoffmann A C and Stein L E 2007 *Gas Cyclones and Swirl Tubes, Principles, Design and Operation* 2nd ed (Berlin, Heidelberg, New York: Springer)
[51] Marthinussen S A, Chang Y F, Balakin B and Hoffmann A C 2014 *Chemical Engineering Science* **108** 169–175
[52] Borhani Jebeli M, Moridi P, Beheshti M H and Yarahmadi R 2020 *International Journal of Environmental Science and Technology* **17** 1075–1086

Mathematical modeling of non-stationary deformation of cylindrical structures

O Korotunova[1], Yu Mastynovsky[1], G Shyshkanova[1, 2*]

[1] Department of Applied Mathematics, National University 'Zaporizhzhia Polytechnic', Zhukovskogo str., 64, Zaporizhzhia, 69063, Ukraine
[2] Department of Mathematics, Humboldt-Universität zu Berlin, Unter den Linden, 6, Berlin, 10099, Germany

[*] Corresponding Author E-mail: mathzntu@meta.ua

Abstract. The paper investigates the dynamic behavior of a hollow elastic cylinder of finite length, placed in a rigid cylindrical shell, under a sharp change in internal pressure. A numerical solution of a two-dimensional dynamic problem is obtained using the method of spatial characteristics. The stress-strain state of an elastic cylinder is described by a system of hyperbolic equations with two circular conical surfaces as characteristic surfaces. The outer cones correspond to longitudinal waves; the inner ones correspond to transverse waves. The calculations were carried out under various conditions at the ends and the outer (contact) surface of the cylinder and shell. An analysis of the results for a cylinder with load-free ends shows that the absence of gluing the outer surface of the cylinder with the shell leads to a significant increase in the velocities of the points of the cylinder. Under the action of internal pressure, the ends move apart, resulting in a significant increase in the radial velocities of the internal channel. Qualitative patterns of behavior of the structures under consideration, which are widely used in mechanical engineering, on impact-type impacts, can be used to improve and optimize them at the design stage.

1. Introduction

Cylindrical structures, consisting of elements with different physical properties, are widely used in various fields of technology, including rocket and aircraft building, surface and submarine ships, the oil and gas industry. During operation, such structures can be affected by high-intensity short-term loads created, for example, by explosions. The study of the stress-strain state that occurs in such structures under non-stationary loading presents significant difficulties. However, this is an important aspect of ensuring the safety and reliability of such structures. Designers and engineers in these areas of mechanical engineering use sophisticated calculation methods [1, 2] and experimental tests [3] to analyze the behavior and optimize these structures under various operating conditions.

If the structure is affected by short-term impulses (for example, from an explosive, a solid body impact, etc.), the pressure distribution can be localized in the form of a spot of limited size [4, 5]. Experimental studies do not allow us to trace the evolution and interaction of stress waves in the material of structural elements, and are mainly limited to recording only the final surface parameters of the process [6].

The study of the deformation of various structures under the action of a shock type load began more than a century ago. In recent years, impulse load actions have received increased attention due to their relevance to modern engineering applications such as the aerospace and automotive industries.

Many theoretical studies of the behavior of the structures under consideration were carried out, as a rule, using the equations of the theory of plates and shells, which do not take into account the wave nature of the load propagation [7]. The propagation of elastic stress waves is of decisive importance in hollow pressure cylinders under the action of explosive or impulse loads [8].

It is impossible not to mention the finite difference method (FDM) for calculating a hollow infinite cylinder with a step-function distribution of material along the radial direction [9] and a functional gradient method for a thin-walled pipe subjected to a combination of longitudinal and torsional stepped impact loading [10]. Here, too, the generalized theory of characteristics was used to analyze the propagation of combined stress waves.

An improved boundary element method (BEM) was proposed in [11] to study the propagation of propagation waves of one-dimensional or multidimensional exponentially distributed properties of materials. In the study of wave processes in three-dimensional viscoelastic three-layer cylindrical bodies under the action of non-stationary loads [6], the methods of separation of variables and orthogonal sweep, truncation in time by the Lagrange interpolation polynomial and BEM were used.

In [12], a method using fractional derivatives and the extended polynomial Legendre technique with the advantage of reducing the solution of complex wave equations in partial derivatives with variable coefficients applied to an eigenvalue problem eliminates the shortcomings of previous iterative approaches.

Most authors used various versions of the finite element method (FEM) to study the propagation of stress waves in cylinders [13, 14]. Gradient FEM is a powerful method for solving multidimensional stress wave propagation equations for a hollow, thick end cylinder made of functionally graded materials [15–17].

The system of equations describing the behavior of an infinitely small volume of an isotropic linearly elastic medium and the system of equations describing acoustic spatially dynamic wave processes are solved jointly using the grid-characteristic method [18]. This made it possible to perform full-wave modeling taking into account contact and boundary conditions, including contact conditions at the boundaries between linear elastic and acoustic media.

In conclusion, we note that the impulse (impact) load is a complex phenomenon that can have a significant impact on the performance of structural elements, and hence the structure itself as a whole. Researchers have made significant advances in understanding the physics of impulse loading, and recent advances in materials science have further expanded the field of research. Future research is likely to focus on developing more sophisticated methods for modeling and predicting the behavior of materials under impulsive loading.

In the present work, we analyze the dynamic motions n of a hollow linearly elastic cylindrical body placed in an absolutely rigid cylindrical shell at a sharply changing radial pressure on the inner surface of the cylinder. The relevance of the study is due to their widespread use in modern technology. Such a problem arises, for example, when detonation of a charge located along its axis deforms a hollow cylinder.

There are more and more publications about research in this area. At the same time, a review of the literature showed that the issues of studying the dynamic behavior of composite structures containing zones of rapid structural change are still insufficiently studied.

Therefore, the development of a mathematical model and a numerical algorithm for studying wave processes in multilayer cylindrical bodies under the action of short-term non-stationary loads, taking into account the possibility of the delamination effect, is an urgent task. The determining factor in assessing the strength of structures is the dynamic stress concentration due to the wave nature of the load propagation. Wave phenomena that occur at or near the contact surface of media with different physical and mechanical properties lead to stress concentration in certain places, which can cause partial or complete destruction of the structure. Thus, the purpose of the study is to analyze the stress-strain state of the cylinder, to identify delamination zones between the outer surface of the cylinder and the shell under various geometric, physical parameters and loading conditions.

2. Mathematical Model and Numerical Method of the Research

To describe the movement of the cylinder, linear equations of the theory of elasticity are used. The two-dimensional dynamic problem is solved numerically using the method of spatial characteristics [3, 19]. For a numerical solution, calculation formulas are obtained for the internal points of the grid area and on its boundaries [19]. The effectiveness of the spatial characteristics method in solving such problems is shown in [19–21]. Numerical-characteristic methods, in comparison with other methods, have a number of advantages, the most important is of them are the following: they bring the area of dependence of differential and difference equations as close as possible; changing the initial and boundary conditions does not require changing the formulas for calculating the internal points of the area. This allows in the process of calculations to check the fulfillment of certain conditions (for example, the conditions for the strength of the connection of structural elements) and to change the boundary conditions, i.e. solve complex interaction problems. Therefore, the method of characteristics can handle different boundary and initial conditions, and nonlinearities. It also helps to understand the physical nature of the wave propagation by tracking the wave fronts and their velocity.

The scheme of the cylinder loading is shown in figure 1, where R_1, R_2 are inner and outer radiuses, l is half length of the cylinder, x, r are longitudinal and radial coordinates, $P(t)$ is a function of loading.

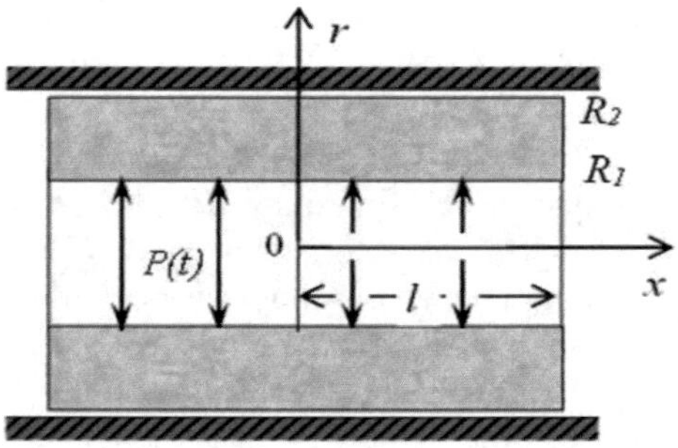

Figure 1. An element of a cylindrical structure under the action of an impulse load.

The domain, where $x \geq 0$ and $r \geq 0$, is only in consideration because of the conditions of radial symmetry. The stress-strain state of an elastic cylinder is described by a dimensionless equation system of the form

$$\frac{\partial w}{\partial t} = \frac{\partial}{\partial r}(p+q) + \frac{\partial \tau}{\partial x} + \frac{1}{r}(p+q-\sigma),$$

$$\frac{\partial u}{\partial t} = \frac{\partial}{\partial x}(p-q) + \frac{\partial \tau}{\partial r} + \frac{\tau}{r},$$

$$\frac{\gamma^2}{\gamma^2-1} \cdot \frac{\partial p}{\partial t} = \frac{\partial w}{\partial r} + \frac{\partial u}{\partial x} + \frac{\gamma^2-2}{\gamma^2-1} \cdot \frac{w}{r}, \qquad \gamma^2 \frac{\partial q}{\partial t} = \frac{\partial w}{\partial r} - \frac{\partial u}{\partial x}, \tag{1}$$

$$\frac{\gamma^2}{\gamma^2-2} \cdot \frac{\partial \sigma}{\partial t} = \frac{\partial w}{\partial r} + \frac{\partial u}{\partial x} + \frac{\gamma^2}{\gamma^2-2} \cdot \frac{w}{r}, \qquad \gamma^2 \frac{\partial \tau}{\partial t} = \frac{\partial u}{\partial r} + \frac{\partial w}{\partial x},$$

$$u = \frac{\partial u_x}{\partial t}; \quad w = \frac{\partial u_r}{\partial t},$$

where dimensionless and dimensional coordinates and functions are related by:

$$\tilde{r}=\frac{r}{R_2};\quad \tilde{x}=\frac{x}{R_2};\quad \tilde{t}=\frac{at}{R_2};\quad \tilde{l}=\frac{\hat{l}}{R_2};\quad p=\frac{\sigma_{rr}+\sigma_{xx}}{2\rho a^2};\quad q=\frac{\sigma_{rr}-\sigma_{xx}}{2\rho a^2};$$

$$\tilde{\tau}=\frac{\sigma_{rx}}{\rho a^2};\quad \tilde{\sigma}=\frac{\sigma_{\theta\theta}}{\rho a^2};\quad \tilde{u}=\frac{1}{a}\frac{\partial u_x}{\partial t};\quad \tilde{w}=\frac{1}{a}\frac{\partial u_r}{\partial t};\tag{2}$$

$$a=\left(\frac{E(1-\nu)}{\rho(1+\nu)(1-2\nu)}\right)^{\frac{1}{2}};\quad b=\left(\frac{E}{2\rho(1+\nu)}\right)^{\frac{1}{2}};\quad \gamma=\frac{a}{b}>1,$$

where $u_r(r,x,t)$ and $u_x(r,x,t)$ are radial and axial components of the displacement vector; $\frac{\partial u_x}{\partial t}$; $\frac{\partial u_r}{\partial t}$ are axial and radial components of the velocity vector, respectively; σ_{rr}, $\sigma_{\theta\theta}$, σ_{xx}, σ_{rx} are components of the stress state tensor; E is elasticity modulus, ν is Poisson's ratio, ρ is density. For ease of notation, the upper sign $\tilde{}$ is omitted from dimensionless quantities in (1) and below. The system (1) depends on one material parameter γ.

Boundary conditions can be as follows

1) on the inner surface of the cylinder (when $|r|=R_1\cdot R_2^{-1}$; $|x|\le l$), $p+q=-P(x,t)$, where $P(x,t)$ is a given function of loading,

$$\tau=0;\tag{3}$$

2) on the outer surface of the cylinder (when $|r|=1$; $|x|\le l$) two cases were considered:

a) a rigid fastening with the absolutely rigid shell

$$u=0,\quad w=0;\tag{4}$$

b) a sliding contact, when the cylinder is freely inserted into a shell with a smooth surface

$$p+q\le 0;\quad \tau=0;\quad w\le 0;\tag{5}$$

3) at the ends of the cylinder, when $R_1\cdot R_2^{-1}\le|r|\le 1$; $x=\pm l$

a) a rigid fastening of the end with an absolutely rigid plate

$$u=0,\quad w=0;\tag{6}$$

b) free ends

$$p+q=0;\quad \tau=0.\tag{7}$$

3. Results and discussions

The calculations have been carried out for different boundary conditions at the ends and at the outer surface of the cylinder for the following initial data values $h=\Delta x=\Delta r=0{,}02$; $k=\Delta t=0{,}01$; $R_1\cdot R_2^{-1}=0{,}7$; $\gamma=1{,}87$; $l=1$ and the loading function $P(x,t)=Ate^{-t}$.

Figure 2-5 shows the time-varying components of the velocity vector and radial stress at x=0.25; x=0.5; x=0.75 points along the interface between the cylinder and an absolutely rigid shell. The graphs shown in the following figures are made considering the symmetry of the structure (figure 1).

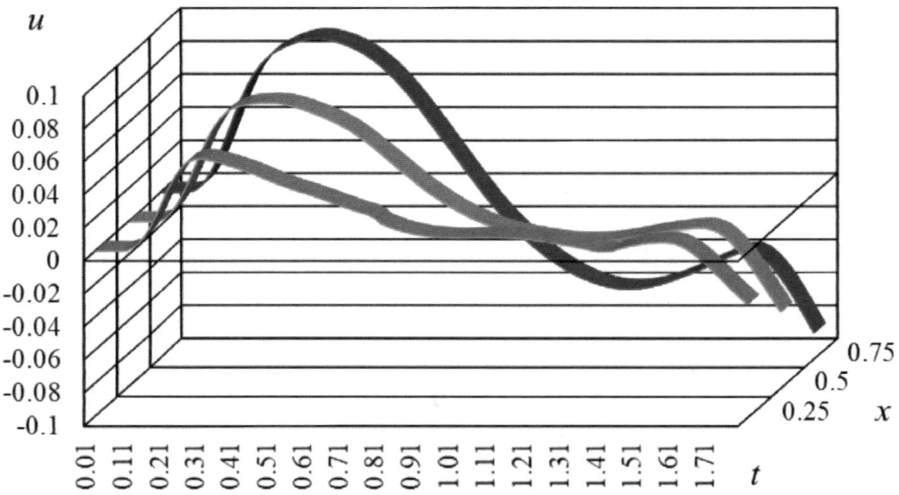

Figure 2. Changing of the longitudinal component of the velocity vector in time.

Damage is observed mainly in the areas adjacent to the cross sections $x = \pm l/R_2$ when the pressure inside the channel changes.

It was shown that the presence of negative values of the axial displacements of the cylinder, caused by its "expansion" (figure 2), leads to an increase in shear stresses (figure 3). Thus, the applied load tends to "stretch" the cylinder in the axial direction.

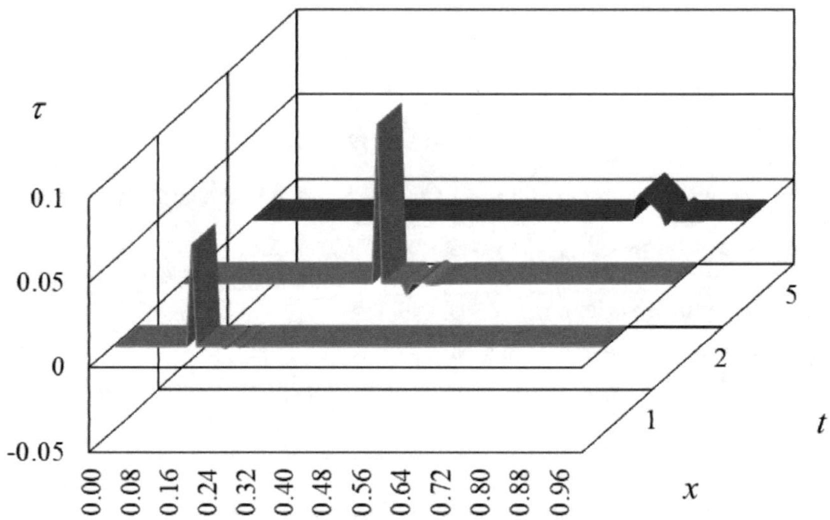

Figure 3. Changing of shear stresses at point x=0.5 in time.

An analysis of the change in time of the radial component of the velocity vector (figure 4) shows that with an increase in the duration, the delamination process occurs with less intensity and at a certain point in time the process becomes steady

The change in time of the radial stress (figure 5) is not as pronounced as it occurs, for example, during an end impact [11, 19].

The cross-sectional deformation contours of a hollow cylinder for different conditions at the ends are shown in Figures 6-7. The solid line corresponds to the cylinder under the joint deformation conditions, and the dashed line shows the sliding contact.

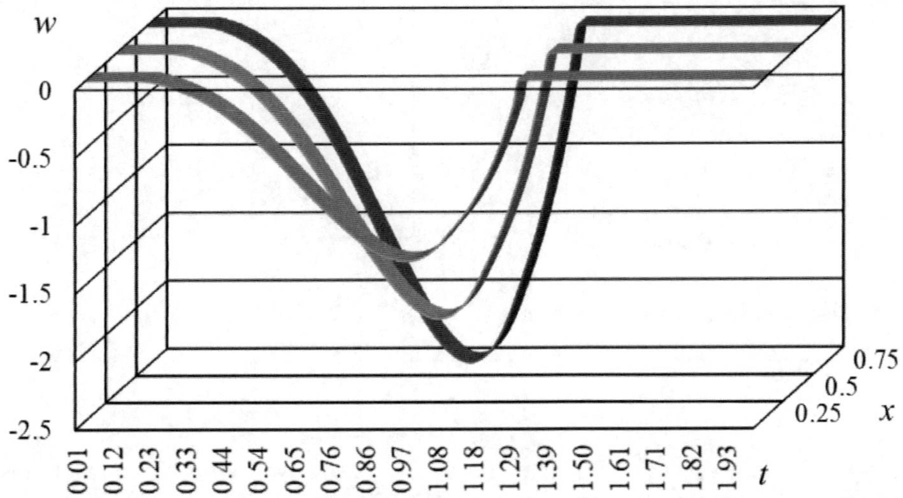

Figure 4. Changing of the radial component of the velocity vector in time.

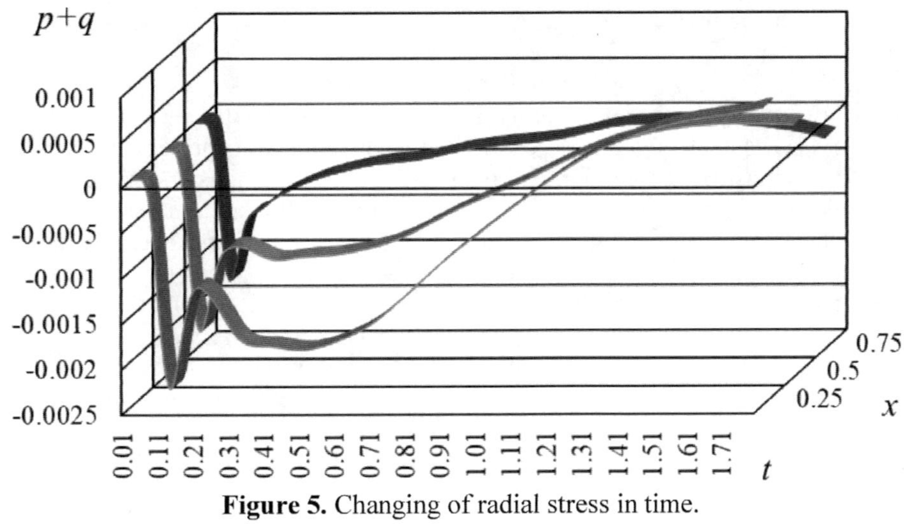

Figure 5. Changing of radial stress in time.

Analysis of the results shows that the lack of adhesion between the outer surface of the cylinder and the shell leads to a significant increase in the velocities of the points of the cylinder. Under the effect of internal pressure, the ends move apart, which leads to a significant increase in the radial velocities of the points on the surface of the internal channel. The free ends contribute to the formation of a rarefaction zone at the end and the generation of longitudinal waves is not observed.

The stress-strain state of a cylinder freely inserted into a rigid shell differs little from that of a cylinder rigidly connected to the shell (figure 7) in the case with fixed ends. However, the chosen dependence of the pressure on the x coordinate caused the appearance of zones of "delaminating" of the cylinder surface from the cavity.

This occurs because at the end, fastened to a rigid plate, there are significant compressive stresses that propagate into the structure. Consequently, the most significant influence on the dynamic behavior of the structure under study is exerted by the conditions that restrict the axial movement of the end face.

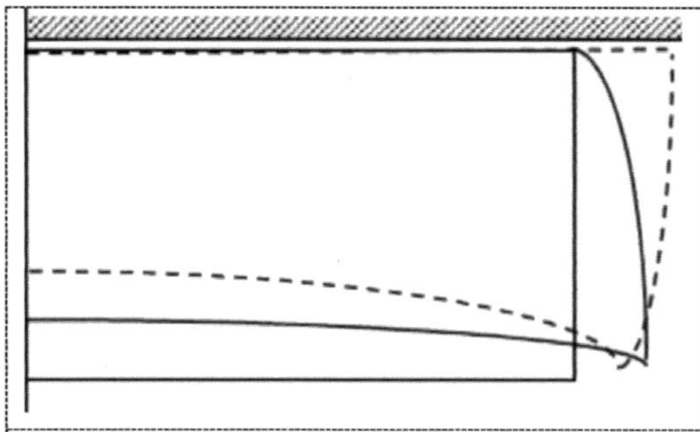

Figure 6. The case of free ends: deformed contours of the cylinder cross-section,

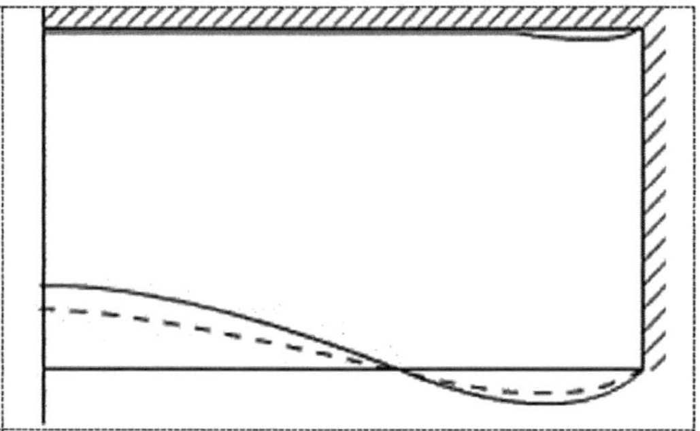

Figure 7. The case of rigidly fixed ends: deformed contours of the cylinder cross-section.

4. Conclusions

In this work, we studied the dynamic response of a hollow elastic cylinder of finite length inserted into a rigid cylindrical shell under a sharp change in internal pressure. The numerical solution of the two-dimensional dynamic problem was obtained by the method of spatial characteristics. An analysis of the results revealed the influence of various boundary conditions at the ends and the outer surface of the cylinder on the stress-strain state and the “separation” zones between the cylinder and the shell. It should be noted that the normal components of the wave field parameters on the contact line undergo the most significant changes.

The obtained results indicate that the introduction of restrictions on the axial movement of the end face plays a decisive role in determining the dynamic behavior of the structure. The lack of adhesion between the outer surface of the cylinder and the shell leads to a significant increase in the velocities of movement of the points of the cylinder, and the presence of restrictions on axial movement causes the formation of delamination zones between the surface of the cylinder and the shell. It is shown that the violation of the integrity of the structure occurs mainly towards the end, and the size of the delamination area increases with time. The presence of delamination zones significantly changes not only the quantitative characteristics of the transient process, but also its qualitative form. Varying the loading conditions changes the nature of the transient process in structures of the type under consideration. It has been established that the boundary conditions at the end have the greatest influence on the movement of the structure. Zones of stable delamination of a cylindrical structure under internal impulse loading are located near the fixed ends. The dimensions of these zones increase with the growth of load amplitude. Over time, the process of delamination becomes less dynamic. Further studies of the influence of other boundary conditions, geometric and mechanical characteristics on the movement of the structure under consideration are planned. In addition, experimental verification of the numerical results presented in this paper will provide valuable information on the reliability and applicability of the method of spatial characteristics for this class of problems.

The proposed calculation method allows to carry out numerical experiments at the design stage, which makes it possible to select design parameters that are resistant to various types of loads encountered in its operation.

Acknowledgments

The authors thank Deutsche Forschungsgemeinschaft (DFG) for their support within "MATH+" (fund: 0502381701 / 19023300)

References

[1] Ghadirian A and Bredmose H 2019 Pressure impulse theory for a slamming wave on a vertical circular cylinder. *J. of Fluid Mech.* **867** R1 https://doi.org/10.1017/jfm.2019.151

[2] Carrer J A M, Pereira W L A and Mansur W J 2012 Two-dimensional elastodynamics by the time-domain boundary element method: Lagrange interpolation strategy in time integration, *Eng. Analysis with Boundary Elements* **36** **(7)** 1164-1172 https://doi.org/10.1016/j.enganabound.2012.01.004

[3] Clifton R J 2000 Response of materials under dynamic loading, *Int. J. of Solids and Struct.* **37** **(1–2)** 105-113 https://doi.org/10.1016/S0020-7683(99)00082-7

[4] Safarov I I, Kulmuratov N R, Teshaev M K and Kuldaschov N U 2019 Interaction of Nonstationary Waves on Cylindrical Body. *Appl. Math.* **10** 435-447 https://doi.org/10.4236/am.2019.106031

[5] Chen Y C and Hwu Ch 2014 Boundary element method for vibration analysis of two-dimensional anisotropic elastic solids containing holes, cracks or inter-faces. *Eng. Analysis with Boundary Elements* **40** 22-35 https://doi.org/10.1016/j.enganabound.2013.11.013

[6] Safarov I I, Teshaev M X and Nuiddinov B Z 2021 Propagation of non-stationary waves in three-layer cylindrical bodies from local loads *AIP Conf. Proc.* **2428** **(1)** 050001 https://doi.org/10.1063/5.0072048

[7] Yang J and Fu Y 2007 Analysis of dynamic stability for composite laminated cylindrical shells with delaminations. *Compos. Struct.* **78** **(3)** 309-315 https://doi.org/10.1016/j.compstruct.2005.10.004

[8] Wu B, Su Y, Liu D, Chen W and Zhang C 2018 On propagation of axisymmetric waves in pressurized functionally graded elastomeric hollow cylinders. *J. of Sound and Vibration* **421** 17–47 https://doi.org/10.1016/j.jsv.2018.01.055

[9] Dorduncu M, Apalak, M K and Cherukuri H P 2015 Elastic wave propagation in functionally graded circular cylinders. *Compos. Part B: Eng.* **73** 35–48 https://doi.org/10.1016/j.compositesb.2014.12.021

[10] Cui S and Ni X 2021 Propagation of combined longitudinal and torsional stress waves in a functionally graded thin-walled tube. *Appl. Math. and Mech.* **42(12)** 1717–32 https://doi.org/10.1007/s10483-021-2805-6

[11] Yang Y and Liu Y 2020 A new boundary element method for modeling wave propagation in functionally graded materials. *European J. of Mech. - A/Solids* **80** 103897 https://doi.org/10.1016/j.euromechsol.2019.103897

[12] Zhang X, Li Z, Wang X, and Yu J 2021 The fractional Kelvin-Voigt model for circumferential guided waves in a viscoelastic FGM hollow cylinder. *Appl. Math. Modelling* **89** 299–313 https://doi.org/10.1016/j.apm.2020.06.077

[13] Xu C and Yu Z 2017 Numerical simulation of elastic wave propagation in functionally graded cylinders using time-domain spectral finite element method. *Advances in Mech. Eng.* **9** **(11)** 168781401773445. https://doi.org/10.1177/1687814017734457

[14] Zhang W and Xu J 2021 Tunable traveling wave properties in one-dimensional chains composed from hollow cylinders: From compression to rarefaction waves, *Int. J. of Mech. Sci.*, **191**, 106073, https://doi.org/10.1016/j.ijmecsci.2020.106073

[15] Santare M H, Thamburaj P and Gazonas G A 2003. The use of graded finite elements in the study of elastic wave propagation in continuously nonhomogeneous materials. *Int. J. of Solids and Struct.* **40(21)** 5621–34 https://doi.org/10.1016/S0020-7683(03)00315-9

[16] Najibi A and Jing G 2021 Two dimensional stress wave propagation analysis of infinite 2D-FGM hollow cylinder. *Waves in Random and Complex Media* 1–18 https://doi.org/10.1080/17455030.2021.1987584

[17] Najibi A and Shojaeefard M H 2023 Stress wave propagation analysis of 2D-FGM axisymmetric finite hollow thick cylinder, *Mech. Based Design of Struct. and Machines* 1-23 https://doi.org/10.1080/15397734.2023.2165099

[18] Favorskaya A V and Petrov I B 2018 Numerical Modeling of Wave Processes in Rocks by the Grid-Characteristic Method. *Math. Models Comput. Simul.* **10**, 639–647. https://doi.org/10.1134/S207004821805006X

[19] Korotunova O, Mastynovsky Yu, Shyshkanova G, Mizerna O, Nechyporenko N, Zaytseva T, Frydman O and Shcherbyna O. 2022 Multilayer Structures under Non-Stationary Loading: Prediction of Damages and Further Operation Possibility. *J. Phys.: Conf. Ser.*, **2346(1)**, 012006. https://dx.doi.org/10.1088/1742-6596/2346/1/012006

[20] Sabodash P F and Naval I K 1975 Numerical solution of the problem of the propagation of stress waves in a cylindrical shell of finite thickness. *Soviet Appl. Mech.* **11** 470–475. https://doi.org/10.1007/BF00886353

[21] Sabodash P F and Cherednichenko R A 1971 Application of the method of spatial characteristics to the solution of axially symmetric problems relating to the propagation of elastic waves. *J. Appl. Mech. Tech. Phys.* **12** 571–577 https://doi.org/10.1007/BF00851864

Moving iso-contour method for solving partial differential equations

Aarne Pohjonen

University of Oulu, Materials and Mechanical Engineering Department, Pentti Kaiteran Katu 1, 90570 Oulu, Finland

E-mail: Aarne.Pohjonen@Oulu.fi

Abstract. The numerical solution of partial differential equations is often performed on a numerical grid, where the grid points are used for estimating the partial derivatives. The grid can be fully static as in Eulerian type of solution method, or the grid points can move during the solution, which is the case in Lagrangian type of method. In the current article, a numerical solution method is presented, where the grid points are located on iso-contours of the two-dimensional field. The method calculates the local movement of the iso-contours according to an evolution equation described by the PDE, and the solution proceeds by moving the grid points towards the calculated direction. Additional stability is obtained by setting the grid points to move along the iso-contour line. To exemplify the application of the method, numerical examples are calculated for the two-dimensional diffusion equation.

1. Introduction

In the previous study [1], the movement of a phase interface was simulated with level-set type method using a physical science based model which takes into account the transformation strains and carbon partitioning and diffusion. In that study the connection to the Allen-Cahn equation [1] was made, which connected it's solution to the level-set approach. As a extension of this idea, the connection between a general partial differential equation (PDE) with first order time derivative and the level set formulation is investigated in the current study.

The current approach is closely connected with the level-set method [2], which is often applied for simulations involving phase boundary movement. Level set methods usually perform the solution of a partial differential equation (PDE) in a grid that is not based on the points contained on the isocontours. The level-set method can be used in any types of grids, even in completely Eulerian framework [3, 4], where the computational grid does not need to move. Sometimes it is important to be able to adapt the mesh to allow for more precision at certain locations that require fine grids for accurate solution, and to allow to the grid become sparse at regions that do not require dense grid for the solution accuracy so that the computational time does not increase too much. For generating adaptive meshes, a class of methods, called deformation methods have been created based on the level-set concept. In the deformation methods, the grid points are moved according to a specific monitor function, which allows to refine the grid locally [5].

In the current study, the mesh points are contained in the iso-contour lines and the focus is on the movement of the grid points, where the function retains it's value. Thus the method calculates the movement of the iso-contour lines, and the points contained in these iso-contours are used as the computational mesh gridpoints. This has certain advantages: first, the the iso-contours and their movement can have actual physical or application specific meaning, and it is often the desired result of the simulation. The current study provides the equations relating the PDE, gradient of the field and the first order time deriative and it provideds the means to calculate the iso-contour movement directly. Secondly, as the iso-contour points are used as the grid-points, this method provides a good basis for generating adaptive meshes.

2. Theory

The method described in this article provides a numerical solution procedure to PDE described by Eq. (1)

$$\frac{\partial}{\partial t}u = L(u) \tag{1}$$

where t is time and $L(u)$ is general PDE operator. For example, in diffusion equation with constant diffusion coefficient D the operator is $L(u) = D(\partial_{xx}u + \partial_{yy}u)$.

The method is based on the fact that advection equation can be used to calculate the movement of isocontours and it contains first order time derivative, which can be equated with the left hand side of Eq. (1). The advection equation is $\partial_t u = -\vec{v} \cdot \nabla u$, which describes the movement of a field u in the direction $\vec{v}$. If the propagation velocity $\vec{v}$ changes as function of position, the shape of the field changes with time. Let us now choose the vector $\vec{v}$ so that it is directed in the negative direction of the gradient, $\vec{v} = -s\hat{n}$, where $\hat{n} = \nabla u/|\nabla u|$ is a unit vector that has the same direction as ∇u ($|\nabla u|$ is the length of the gradient vector). In this case $=$ $\partial_t u = -\vec{v} \cdot \nabla u = s\hat{n} \cdot \nabla u = s\frac{\nabla u \cdot \nabla u}{|\nabla u|} = s|\nabla u|$, so that Eq. (2) is true.

$$\frac{\partial}{\partial t}u = s|\nabla u| \tag{2}$$

Eq. (2) is applied in the Level-Set (LS) method to describe shape evolution of diffuse interface, when the local propagation speed s is first determined based on physical principles. In the level set method, signed distance from the region interface is usually used as the propagating field, and re-calculation of this field can be required.

In the current article, we apply the Eq. (2) to describe the evolution of the iso-contours of a general field whose evolution equations can be calculated with an equation of the form given by Eq. (1). Iso-contour normal vector is parallel to the gradient. Since the local velocity $\vec{v} = s\hat{n}$ of the field u was chosen in the negative direction of the gradient of u, the Eq. (2) describes the local movement of the field iso-contour with speed s.

Now, comparing Eq. (1) and (2), it can be seen that since both equations contain the first order time derivative, they can be equated, which yields the propagation speed of the iso-contour to the negative direction of the gradient, Eq. (3), which is the direction of the iso-contour normal.

$$s = \frac{L(u)}{|\nabla u|} \quad \text{when } |\nabla u| \neq 0 \tag{3}$$

Once the local speed of an iso-contour point $\vec{p}(t) = (p_x, p_y)$ is calculated using Eq. (3), the time derivative of the isocontour point position is obtained using Eq. (4).

$$\frac{\partial \vec{p}}{\partial t} = \vec{v} = -s\hat{n} \tag{4}$$

Further, by noting that $\hat{n} = \nabla u/|\nabla u|$, Eq. (4) can be also written out as Eq. (5)

$$\frac{\partial \vec{p}}{\partial t} = -\frac{L(u)}{|\nabla u|}\hat{n} = -L(u)\frac{\nabla u}{|\nabla u|^2} \tag{5}$$

Now, the movement of the point $\vec{p}$ contained at the iso-contour can be calculated, using the time derivative $\partial_t\vec{p} = (\partial_t p_x, \partial_t p_y)$ which is obtained either from Eqs. (3) and (4), or directly from Eq. (5).

However, as described in the section 3, it appeared that the numerical solution was rather unstable if simple Euler forward time step was used. This is most likely because of the same issues as associated with the solution of the advection equation, which is notorious for its instability in the basic Euler forward time-stepping procedure. To overcome this issue, the time-stepping described by Eqs. (6) and (7) was applied, which is similar to the one used for obtaining stable solution for advection equation in [6], but here the average time derivative $\left\langle \frac{\partial\vec{p}}{\partial t} \right\rangle$ was used instead of $\left.\frac{\partial\vec{p}}{\partial t}\right|_{t=t_i}$. In general, the average time derivative could provide better solution also in the case if variable timesteps were used.

$$\vec{p}(t_i + \Delta t) = \vec{p}(t_i - \Delta t) + \left\langle \frac{\partial\vec{p}}{\partial t} \right\rangle 2\Delta t \tag{6}$$

where

$$\left\langle \frac{\partial\vec{p}}{\partial t} \right\rangle = \frac{1}{2}\left(\left.\frac{\partial\vec{p}}{\partial t}\right|_{t=t_i-\Delta t} + \left.\frac{\partial\vec{p}}{\partial t}\right|_{t=t_i} \right) \tag{7}$$

After moving the isocontour gridpoint in the direction of the isocontour normal (i.e. the direction of the gradient) using either Eqs. (3, 4, 6) or Eq. (5, 6), the gridpoint can be additionally moven along the isocontour line. This is because the value of the function u does not change when the gridpoint is moved along the isocontour - when the gridpoint is moved to a new position along the isocontour, the new position will also belong to the isocontour. This fact was used to maintain the grid shape more regular, as shown in the section 3.

For maintaining better grid shape, the gridpoints were moved along the isocontours after the time evolution step, so that the grid would remain suitably regular. This was achieved by calculating the difference of the current x-coordinate of the gridpoint from it's initial x-coordinate $p_{x(i,j)}(t) - p_{x(i,j)}(0)$. If this value was negative, the gridpoint was moved towards the right hand side gridpoint $(i+1, j)$ which is on the same isocontour. On the other hand, if $p_{x(i,j)}(t) - p_{x(i,j)}(0) > 0$, the gridpoint was moved towards the left gridpoint $(i-1, j)$, located on the same isocontour. More precisely, the vector $\vec{b}$ descibing the movement of the gridpoint along the isocontour is stated Eq. (8).

$$\begin{aligned}
&\vec{b} = \gamma\hat{h} \\
&\text{where} \\
&\hat{h} = \frac{\vec{p}_{(i+1,j)}(t) - \vec{p}_{(i,j)}(0)}{|\vec{p}_{(i+1,j)}(t) - \vec{p}_{(i,j)}(0)|} \text{ if } p_{x(i,j)}(t) - p_{x(i,j)}(0) < 0 \\
&\text{and} \\
&\hat{h} = \frac{\vec{p}_{(i-1,j)}(t) - \vec{p}_{(i,j)}(0)}{|\vec{p}_{(i-1,j)}(t) - \vec{p}_{(i,j)}(0)|} \text{ if } p_{x(i,j)}(t) - p_{x(i,j)}(0) > 0
\end{aligned} \tag{8}$$

where the adjustment parameter γ controls how much the gridpoint moves along the isocontour line during a timestep. The value $\gamma = 0.05$ was used in the current example.

3. Numerical experiments

To test the application of the method in practice, numerical experiments were conducted. To make sure that the method produces correct result, it is useful to compare the numerical solution obtained with the method to an analytical solution. For this purpose, the solution for the one dimensional diffusion equation was calculated for a case that has a simple analytical solution. By using the separation of variables, the usual Fourier solution for the following case can be found: The analytical solution for a diffusion equation $\partial_t u = \partial_{xx} u$ with Dirichlet boundary conditions $u(0) = 1$ and $u(2\pi) = 0$ and initial condition $u(x, 0) = 1 - x/(2\pi) + 0.15\sin(x)$ has the analytical solution $u(x, t) = 1 - x/(2\pi) + 0.15\exp(-t)\sin(x)$.

To apply the method described in the theory section for this one dimensional case, the range of values of the initial function were divided to 100 equally spaced iso-contours (0, 0.01, 0.02, 0,03, ...). The second order spatial derivative $\partial_{xx} u$ was calculated at each point by applying the region weighted average method [7]. In one dimensional case the regions are the lengths of the segments separating the isocontour points. The speed of the movement of the isocontour is calculated applying Eq. (3) (isocontour is a straigh line represented by a point in the 1 dimensional case). The direction of the movement $\hat{n} = \nabla u/|\nabla u|$ is the sign of the gradient, and the velocity of the isocontour is obtained from Eq. (5). For each timestep Δt, the new position of the isocontour is calculated as

$$\vec{p}(t_i + \Delta t) = \vec{p}(t_i) + \frac{\partial \vec{p}}{\partial t}\Delta t \tag{9}$$

The calculated solution compared to the analytical solution for the one dimensional diffusion equation is shown in Fig. 1, where the solid lines represent the numerical solution at different times, and the dashed black line shows the corresponding analytical solution. It can be seen that the method reproduces the analytical solution.

The isocontour method is particularly useful in a context, where the function is defined on one boundary region, and the points contained in this region move. Such condition was defined and the time evolution of the system was calculated. For the test case, the two dimensional diffusion equation with constant diffusivity $D = 1$ was picked, i.e. the PDE is described by Eq. (10). The method [7], which is capable of handling irregular grids, was used to calculate the first and second order partial derivatives for the two dimensional case.

$$\frac{\partial}{\partial t}u = \frac{\partial^2 u}{\partial x^2} + \frac{\partial^2 u}{\partial y^2} \tag{10}$$

The calculation domain was defined as a rectangular grid with grid point spacing in the x and y directions $\Delta x = 0.02 = \Delta y$ and the grid $x_{i,j}$ points were numbered with indices i and j in the horizontal x and vertical y direction. The initial condition of the system was defined as $u(x, y) = 1 - y$, as shown in Fig. 2 a). This is the steady state condition for Eq. (10) for the boundary values described by the function at their initial position. The values of the function on the upper, left and right boundaries were not allowed to change during the simulation, and the positions of the points at these boundaries were held fixed. On the bottom boundary, where initially $y = \Delta y$, the function values were defined as $u = \Delta y$ for all t, but the positions of the gridpoints of this iso-contour were changed during the calculation according to Eq. (11).

$$\begin{aligned} y(x) &= a(t)\exp\left(\frac{-(x - x_0)^2}{b}\right) + \Delta y \\ a(t) &= \frac{t}{2 \cdot 10^{-2}} 0.2 \text{ when } 0 \leq t < 2 \cdot 10^{-2} \\ a(t) &= 0.2 \text{ when } t \geq 2 \cdot 10^{-2} \end{aligned} \tag{11}$$

where $x_0 = 0.48$ and $b = 0.1$. This means that the points contained on the isocontour $u = 0$ were moving in the y-direction, so that the y-coordinates as function of time were described by

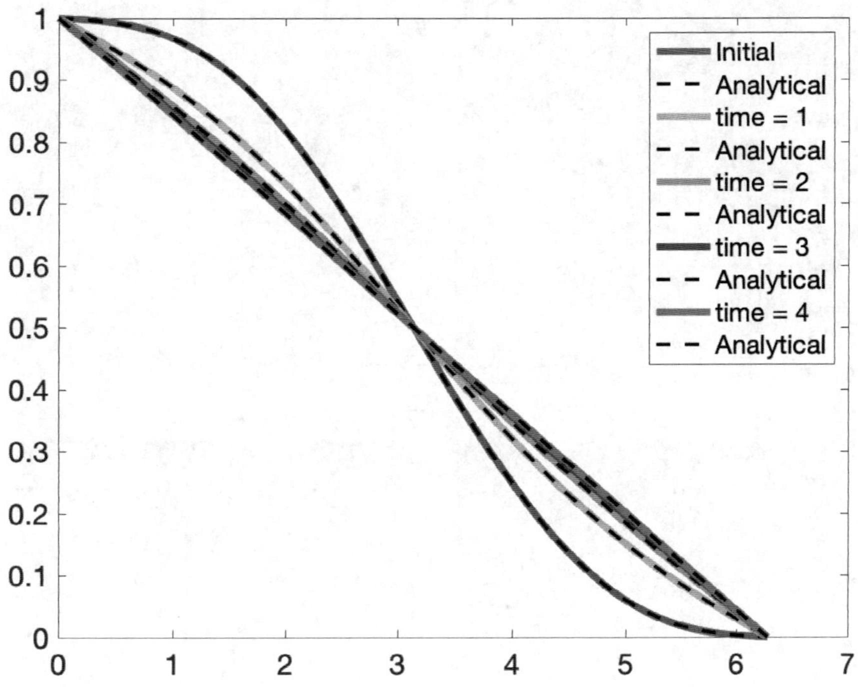

Figure 1. Comparison of the numerical solution to the analytical solution for $\partial_t u = \partial_{xx} u$ with Dirichlet boundary conditions $u(0) = 1$, $u(2\pi) = 0$ and initial condition $u(x, 0) = 1 - x/(2\pi) + 0.15\sin(x)$. The coloured solid lines represent the numerical solution at different times, and the dashed black line shows the corresponding analytical solution, which fully overlaps with the numerical solution for each time instant.

the gaussian function, Eq. (11). The x-coordinates of the boundary gridpoints retained their original value for all t.

As a first instance, the movement of the the internal gridpoints were calculated using Eqs. (3, 4, 6) and they were not moved along the isocontours. It was found that when the boundary changed significantly, the grid became deformed, as shown in Fig. 2.

The approach of moving the gridpoints along the isocontour lines, descried in section 2 helped in maintaining a suitable grid shape, as shown in Fig. 3.

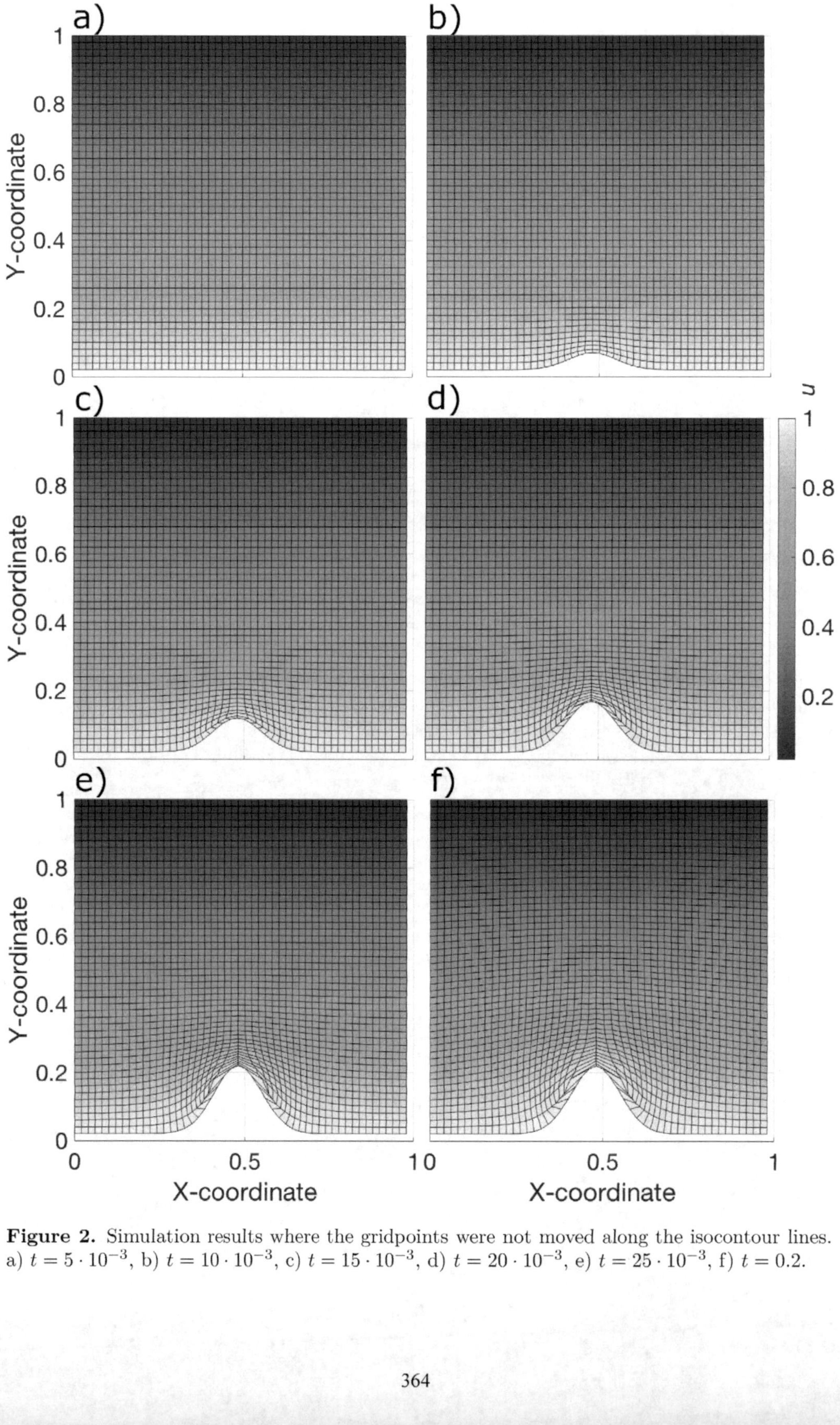

Figure 2. Simulation results where the gridpoints were not moved along the isocontour lines. a) $t = 5 \cdot 10^{-3}$, b) $t = 10 \cdot 10^{-3}$, c) $t = 15 \cdot 10^{-3}$, d) $t = 20 \cdot 10^{-3}$, e) $t = 25 \cdot 10^{-3}$, f) $t = 0.2$.

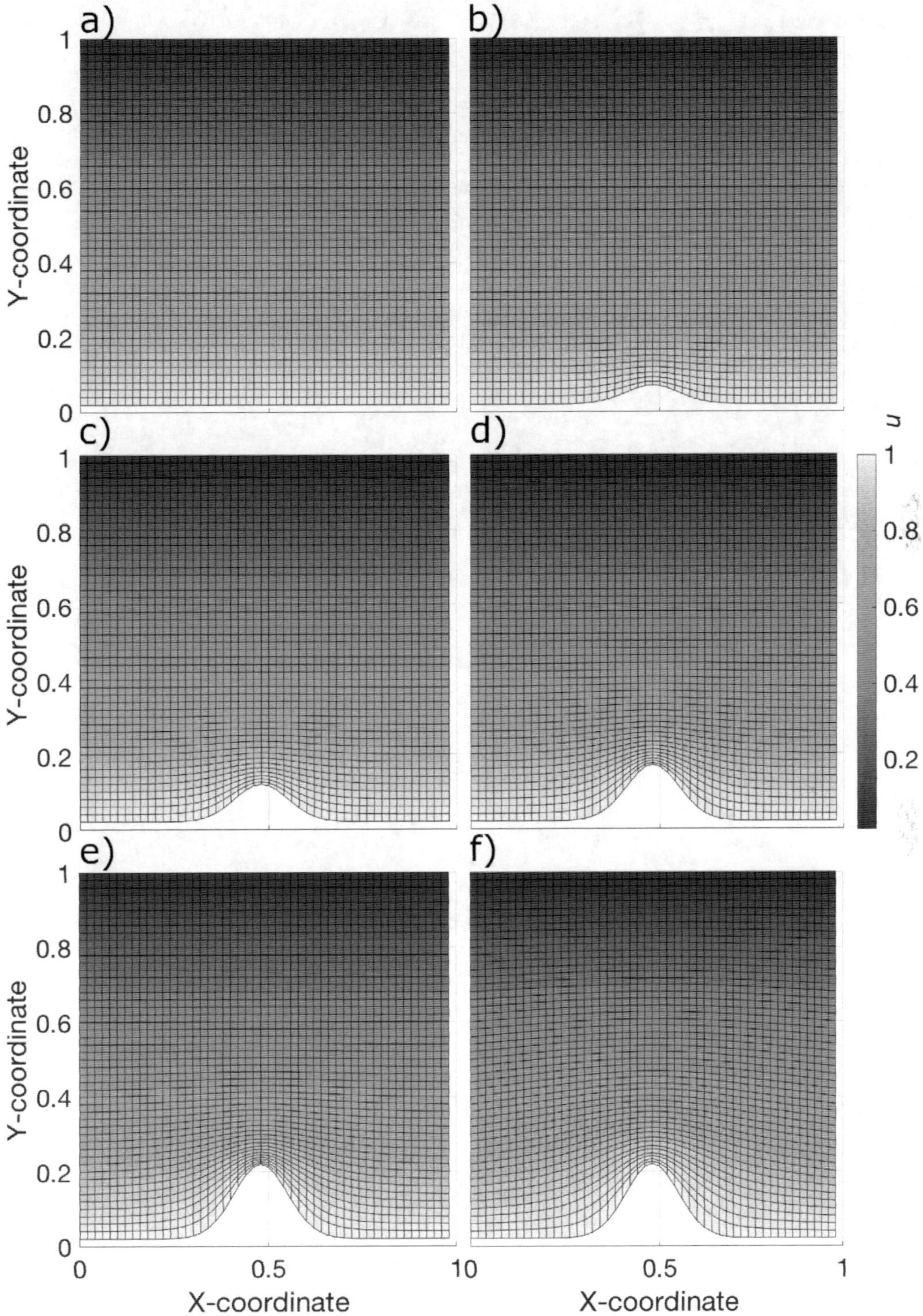

Figure 3. Simulation results where the gridpoints were moved along the isocontour lines to maintain a more regular grid shape. a) $t = 5 \cdot 10^{-3}$, b) $t = 10 \cdot 10^{-3}$, c) $t = 15 \cdot 10^{-3}$, d) $t = 20 \cdot 10^{-3}$, e) $t = 25 \cdot 10^{-3}$, f) $t = 0.2$.

When t increased, $\frac{\partial^2 u}{\partial x^2} + \frac{\partial^2 u}{\partial y^2}$ approached zero, as the system approached steady state, since the boundary movement was stopped at time $t = 2 \cdot 10^{-2}$ as defined in Eq. (11).

4. Conclusions and outlook

A moving isocontour method was described and numerical experiments were performed. The method was compared to an analytic solution for one dimensional diffusion equation and it's applicability was demonstrated also for two dimensional diffusion equation. The method provides a way for constructing a moving mesh, based on the movement on the field iso-contours to calculate the time evolution of a PDE with first order time derivative. The method can also be used for the PDEs containing higher order time derivatives, if the first order time derivative can be calculated numerically. The method is useful for calculating time evolution of the iso-contours when the domain boundary is moving. In future the method could be applied in calculating the solution to a first order time PDEs for a sharp interface problems with moving boundary, such as diffusional growth of precipitates or phases in steels [8, 9].

References

[1] Pohjonen A 2023 *Adv. Theory Simul.* 6(3):2200771
[2] Osher S and Sethian J A 1988 *J. Comput. Phys.* 79(1):12–49.
[3] Sussman M *et al* 1995 Technical report, Lawrence Livermore National Lab (LLNL), Livermore, CA (United States)
[4] Sussman M *et al* 1994 *J. Comput. Phys.* 114(1):146–159.
[5] Liao G *et al.* 2000 *J. Comput. Phys.* 159(1):103–122
[6] Gershenfeld N 2011 *The Nature of Mathematical Modeling* (Cambridge University Press)
[7] Pohjonen A 2021 *J. Phys. Conf. Ser*, 2090(1):012069
[8] Pohjonen A *et al* 2022 *Comput. Mater. Sci.* 209:111413.
[9] Seppälä O *et al* 2023 *Materials & Design*, 230:111956

A structure-activity relationship modelling of opioid compounds by using machine learning

Fatima Sapundzhi[1], Meglena Lazarova[2], Tatyana Dzimbova[1,3], Slavi Georgiev[4,5], Antonina Ivanova[6]

[1] Department of Communication and Computer Engineering, Faculty of Engineering, South-West University "Neofit Rilski", 66 Ivan Myhailov Str., 2700 Blagoevgrad, Bulgaria
[2] Faculty of Applied Mathematics and Informatics, Technical University of Sofia, 8 "St. Kliment Ohridski", Blvd., 1000 Sofia, Bulgaria
[3] Institute of Molecular Biology, Bulgarian Academy of Sciences, Acad. G.Bonchev Str, bl. 21, 1113 Sofia, Bulgaria,
[4] Department of Information Modeling, Institute of Mathematics and Informatics, Bulgarian Academy of Sciences, Acad. Georgi Bonchev Str., bl. 8, 1113 Sofia, Bulgaria
[5] Department of Applied Mathematics and Statistics, Faculty of Natural Sciences and Education, University of Ruse, 8 Studentska Str., 7004 Ruse, Bulgaria
[6] Department of Computer Science, Faculty of International Economics and Administration, Varna Free University "Chernorizets Hrabar", 84 Yanko Slavchev Str., Chaika Resort, 9007 Varna, Bulgaria

sapundzhi@swu.bg

Abstract. Opiates are among the oldest drugs that are used to treat many medical problems. They are analgesic and sedative drugs that contain opium. The morphine is its most active ingredient and it is a widely used pain reliever despite its side effects. The main objective of this study is to construct a model which gives the structure-activity relationship among a series of mu-opioid ligands and molecular docking results. For this purpose, a model of mu-opioid receptors using machine learning is introduced. By obtaining a relationship between the docking results and the in vivo test, we could predict the biological effect of the newly synthesized ligands.

1. Introduction

Opioids are pain relievers that are extracted from the papaver somniferum plant [1]. They are widely used as analgesic drugs in the treatment of pain in humans. Opioids bind to three brain receptors: the mu-opioid receptor (MOR), the kappa-opioid receptor (KOR), and the delta-opioid receptor (DOR) in the central nervous system (CNS) and the peripheral organs [2,3]. In contrast to mu-opioid and kappa-opioid agonists, delta-opioid agonists have limited antinociceptive properties which are measured by morphine-sensitive antinociceptive assays as a result of DOR activation. The DOR system may play an essential role in regulating mood and emotional states [4]. Although there are powerful clinically available analgesics (morphine, oxycodone, and fentanyl), they are highly addictive. It is important for the treatment of pain to develop new opioid drugs that produce analgesia without causing dependence [5,6,7].

With the help of computer-aided drug design or CADD, the drug developers could create more drugs faster by using the state-of-the-art technology. Molecular docking is one of the fundamentals of CADD that analyzes the binding interaction between the target and the small molecules called ligands. These ligands are potential drug candidates for the development of phenotypic and therapeutic models targeted by the CADD proteins [8, 9,10].

In our previous work [11] we have investigated the relationship between the values of the biological activity of delta-opioid analogues that were previously synthesized and biologically tested and have made some investigations over the results of the *in silico* docking. Moreover, the calculation of the minimal energy conformation for each obtained ligand-receptor complex after the docking procedure was investigated.

The main purpose of this study is to create a model for the structure-activity relationship of a series of delta-opioid ligands and molecular docking results by using machine learning where the Delta-opioid receptor (DOR) has a crystal structure (PDBid: 4ej4). This is a continuation of the recent study [12] where the relationship had been sought between the scores and it is obtained by another optimization algorithm where the biological activity of the studied compounds is used. The current paper briefly presents the data while the proposed fitting algorithm as well as the machine learning methods are explained in detail. The paper concludes with a result analysis and with an outline for future research.

2. Experimental remarks

2.1. *Docking procedure*

The docking procedure was performed by using the software for molecular docking GOLD and the ChemScore algorithms, see [11, 13, 14]. The scoring results give information on how good the pose is. Precisely, the goodness of the pose is sketched by the scale of the score.

2.2. *Receptor*

We use the delta-opioid receptor's model with a crystal structure which is published in the RCSB Protein DataBase (PDBid: 4ej4), (http://www.rcsb.org).

2.3. *Ligands*

The Delta-opioid ligands that were previously synthesized by our colleagues were also used in this study, see [11, 15, 16], Table 1. Ligand preparation was performed by the Molegro Molecular Viewer 2.5 (MMV) program, (www.clcbio.com). The prepared structures were utilized for molecular docking by GOLD software. The total energies of the formed ligand–receptor complex after docking were computed by MMV 2.5 where the MolDock scoring algorithm is used, see [17].

3. Results and discussions

In order to establish a relationship between the biological activity of the studied delta-opioid compounds and the results from molecular docking (the values of the scoring functions) the Surface Curve Fitting Toolbox in MATLAB was applied [11].

In our previous research, see [11, 18-20] the experimental data fitting for DOR was carried out by using a polynomial function $z = f(x,y)$ where the values $(z_1, z_2, ..., z_n)$ of the dependent variable z represent the values of the biological activity of the mu-opioid ligands. The values $(x_1, x_2, ..., x_n)$ of the independent variable x represent the results from the docking – these are the values of ChemScore function (calculated by GOLD). The values $(y_1, y_2, ..., y_n)$ of the independent variable y represent the total energies for the ligand-receptor complex which are formed after the docking – the values of MolDock function (calculated by MMV).

Table 1. Data for the biological activity of ligands and docking studies, see [11].

Ligand	*ChemScore*	*Total energy*	*Ligand efficacy*
[Cys(Bzl)^2-Leu^5]-enk	38.91	-170.657	9.3
[Cys(Bzl)^2-Met^5]-enk	35.19	-125.108	3.5
[Cys(O_2NH_2)^2-Leu^5]-enk	28.48	-118.805	29.2
[Cys(O_2NH_2)^2-Met^5]-enk	25.82	-87.343	7.3
[DCys(O_2NH_2)^2-Leu^5]-enk	31.84	-136.187	7.4
[DCys(O_2NH_2)^2-Met^5]-enk	31.55	-139.449	7.1
[HCys(O_2NH_2)^2-Leu^5]-enk	32.75	-100.702	30.2
[HCys(O_2NH_2)^2-Met^5]-enk	26.55	-112.164	3.4
[D-Pen^{2,5}]- enkephalin (DPDPE)	29.23	896.877	4.5
[Leu^5]-enkephalin	31.62	-119.009	5.8
[Met^5]-enkephalin	32.22	-106.792	3.6

The obtained polynomial model of 3rd degree [11] could be interpreted as a surface-fitting function and it analyses the experimental data using a least squares method.

The Surface Fitting Toolbox of MATLAB (http://www.mathworks.com/products/matlab) was applied for analysing the behaviour of one variable that depends on multiple independent variables.

To assess the goodness of fit the following statistical measures were employed: SSE (Sum of squares due to error), $R - Square$ (R^2), $Adjusted\ R^2$, $RMSE$ (*Root Mean Squared Error*).

The results are as follows:

Table 2. Goodness of fit for the polynomial models obtained by the least squares method.

Models Poly (x,y)	Degree of x	Degree of y	SSE	R^2	Adj R^2	RMSE
Poly11	1	1	6988	0.7369	0.6784	27.86
Poly22	2	2	5456	0.7946	0.6234	30.15
Poly33	3	3	0.6290	1.000	0.9999	0.5608

The obtained model shows good fitting properties and significant predictive ability. and was suitable for determining the structure-biological activity relationship: $R^2 = 1.0\ SSE = 0.6290$, $adj\ R^2 = 0.9999$, $RMSE = 0.5608$. Similar studies have been conducted with other compounds, see [11, 18-21].

In this research our aim is to identify a relationship of the from $z = f(x, y)$ where z represents the values of the biological activity of the delta-opioid ligands. The function z is a function of two variables where the independent variables x and y are the values of the GoldScore function and the MolDock function, respectively. Our objective is to discover a nonlinear correlation that utilizes machine learning methodologies, see [22].

The regression models that are employed encompass the k-Nearest Neighbors, Gradient Boosting, Random Forest, and Extra Trees. The latter trio is classified as an ensemble method and is expected to outperform the former, the traditional method which is incorporated primarily for comparison purposes, see [23, 24].

The **k-Nearest Neighbors** (k-NN) algorithm is a multifunctional and easily understood approach that is used across various machine learning applications like regression or classification tasks. In particular, k-NN regression, a derivative of the method, serves as a beneficial instrument for predicting continuous data points.

The core concept of the k-NN regression is founded on the presumption that the data points in the dataset are close in the feature space and they are likely to show analogous outcome values. This algorithm functions by pinpointing ‘k’ data points in the training dataset that are closest to a new, unobserved data point employing a certain distance metric, generally the Euclidean distance in the multi-dimensional feature space.

Once the 'k' nearest data points are ascertained, the algorithm creates a forecast for the new data point. For the regression tasks, this prediction is typically the mean of the dependent variable for the 'k' closest data points. This basically suggests that the outcome for an unobserved data point is the average of the outcomes of its neighboring points. A chief benefit of the k-NN regression is its straightforwardness and logical nature. It does not make any explicit suppositions about the underlying functional form of the data, categorizing it as a non-parametric approach. This adaptability allows the k-NN adequately to fit a sophisticated and non-linear data. Nevertheless, the k-NN regression also presents its own set of difficulties. Selecting an ideal 'k' can be a complicated task – a smaller 'k' may result in a model that is excessively sensitive to noise, whereas a larger 'k' may overgeneralize the model, disregarding crucial trends in the data. In addition, the k-NN regression can find its challenging to deal with high-dimensional datasets which is a problem that is frequently referred to as the "curse of dimensionality".

Gradient Boosting is an influential machine learning method utilized in an array of regression and classification applications. Basically the gradient boosting regression is an ensemble algorithm that constructs a predictive model by sequentially adjusting a collection of weak learners to the data, each striving to correct the errors committed by the one before it. The weak learners are generally - decision trees, although different base models may be employed. The fundamental concept is to amalgamate the outputs of numerous simple models to generate a single very precise forecast. This idea of "boosting" arises from the proposition that a combination of weak learners when they are aptly merged, could evolve into a strong learner.

In gradient boosting, the sequential inclusion of weak learners effectively operates as a method of steepest descent, therefore the term "gradient". The method estimates the gradient of the loss function (the measure that gauges the accuracy of the model's forecasts, compared to the actual values) relative to the model parameters, and incorporates new models that orient in the direction that reduces the loss.

For regression purposes, the objective is to predict a continuous outcome variable. In the context of Gradient Boosting Regression, the ensemble of trees is trained to forecast the residuals or errors of the preceding trees. Thus, each subsequent tree is effectively drawing nearer to the true, unidentified function that we are aiming to approximate.

In general, Gradient Boosting has its advantages and disadvantages. It is an exceptionally potent method, capable of fitting complex, non-linear data, and it often performs well even on datasets composed of a mix of categorical and numerical features. Nevertheless, gradient-boosting models can be susceptible to overtraining, particularly if the data is riddled with noise. They could also be computationally challenging and necessitate meticulous tuning of various hyperparameters. For example the number of estimators, the depth of the trees, and the learning rate.

Random Forest is a renowned machine learning technique, regularly employed in an array of regression and classification tasks. As indicated by its name, the Random Forest model consists of an assembly of individual decision trees. These trees are structured in a way that ensures variety and thus resilience in the collective predictive strength.

In the terms of regression, the Random Forest Regressor operates by producing numerous individual decision trees during training, each tree trained on a random data subset and making its distinct prediction. When a new prediction is necessitated, each tree in the forest generates its prediction and the final output is the mean of these individual forecasts. This procedure enables the Random Forest Regressor effectively to capture the intricate, non-linear associations in data.

The core principle of the Random Forest is the notion that an assembly of "weak learners" could amalgamate to compose a "strong learner". Each decision tree in the forest is a weak learner, trained on a random data subset using a random feature subset at each split. This technique is known as bootstrap aggregating or "bagging" coupled with feature randomness. The strength of the Random Forest lies in its straightforwardness and adaptability. It is relatively unsusceptible to overtraining, due to the randomness incorporated in its construction. It effectively handles both numerical and categorical data and manages missing data efficiently. Moreover, it offers an inherent method for feature importance estimation which can be advantageous for comprehending the model. However, in regard to all models, the Random Forest has its limitations. It can be computationally demanding and slower in both training and forecasting, especially when the tree count gets large. Additionally, it might not perform as well with very high-dimensional sparse data, such as text data and could be less interpretable compared to single decision trees.

Extra Trees Regressor, abbreviated from "Extremely Randomized Trees" is another machine learning method used for regression tasks that belongs to the ensemble learning category. Like the Random Forest the Extra Trees constructs multiple decision trees during the training stage. It introduces an extra degree of randomness, rendering it more resilient and less susceptible to overfitting. In a regular decision tree the optimal split amongst a random subset of features in the node is chosen during the tree growth process. However, in Extra Trees, for each feature being considered, a random value is chosen as the split point, as opposed to the optimal split. Therefore, the Extra Trees Regressor brings in randomness not just at the sample level but also at the level of each decision tree's individual split.

The forecasting process for regression tasks is identical to that of the Random Forest Regressor. Every tree in the ensemble generates a forecast and the final result is the average of these individual forecasts. The randomness in the Extra Trees Regressor aids in capturing complex, non-linear relationships making it an extremely effective tool for regression.

Nevertheless, it is crucial to highlight the distinctions between the Extra Trees and the Random Forest techniques. While they both are using bagging and random feature subsets, the main difference resides in how they split the nodes. The Extra Trees technique is faster due to its randomness at each split, which leads to a quicker training process. However, this could sometimes result in a minor performance decline, thus presenting a trade-off to consider.

The Extra Trees is less likely to overtrain. It handles numerical and categorical data well and could effectively manage missing data. Like Random Forest, it offers an inherent method for feature importance assessment like the Random Forest. The drawback is similar to the other tree-based methods. It might not perform optimally with extremely high-dimensional sparse data such as text data. Its ensemble nature makes it less interpretable compared to the single decision trees.

We have applied the four machine learning methods to build a multivariate regression. The metrics for each model were computed using 10-fold cross-validation and a hyperparameter tuning process was employed to find the optimal parameters. The best hyperparameters can be found in Table 3. Fig. 1 provides the flowchart of the algorithm.

Cross-validation is a tool used to measure how a machine learning model would perform on a novel data set, i. e. data has not been seen during the training. It helps to mitigate the overtraining - a scenario where the model overly memorizes the training data, which usually leads to a poor performance on the unseen data. The most frequently used form of the cross-validation is the k-fold cross-validation where the initial data set is randomly divided into k equally sized subsets. One of these k subsets is set aside for the model validation and the remaining k-1 subsets are utilized for training. This cross-validation procedure is iterated k times ensuring that each subset serves as validation data once. The results from the k folds can then be combined (typically by averaging) to provide a unified measure of the model performance.

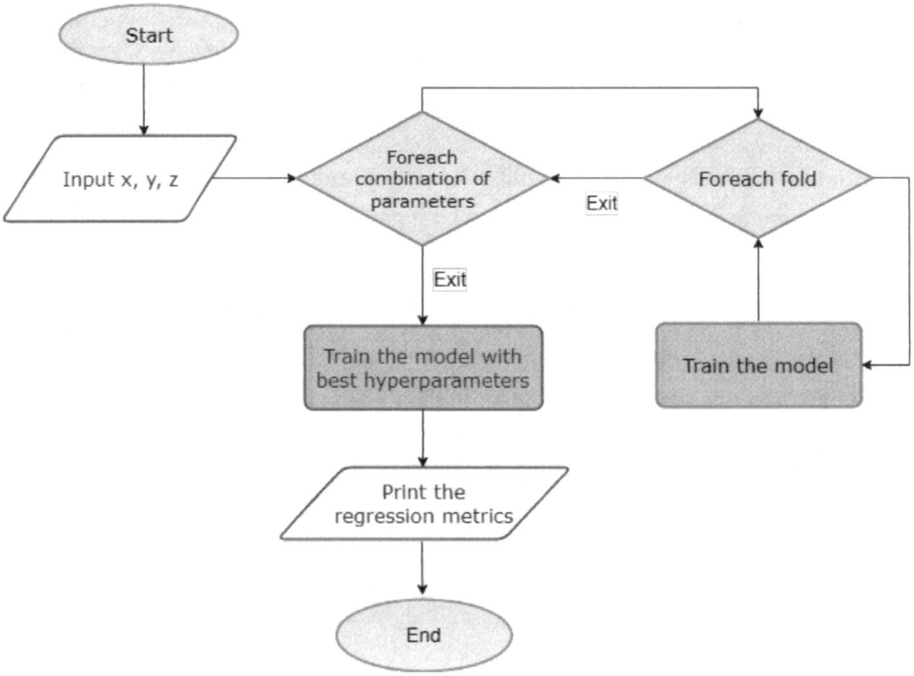

Fig. 1. Flowchart of the regression algorithm (https://app.diagrams.net/).

However, hyperparameter optimization refers to the process of finding the best set of hyperparameters for a machine learning model. Hyperparameters, unlike model parameters, are preset before training and do not change during the learning process. For example, in a Random Forest model, hyperparameters might include the number of trees in the forest or the number of features that each tree evaluates when splitting a node. Hyperparameters significantly affect model performance, making their appropriate selection vital. For the hyperparameter optimization, we apply a 'grid search' strategy, where we define a range of potential values for each hyperparameter and then assess the performance of every possible combination.

Table 3: Modeling by using machine learning regressors.

Regressor	**Best hyperparameters**	**SSE**	**R-square**	**RMSE**
k-Nearest Neighbors	{'n_neighbors': 5, 'weights': 'uniform'}	1223.4328	-0.2559	10.5461
Gradient Boosting	{'max_depth': 2, 'n_estimators': 50}	6.6080	0.9932	0.7751
Random Forest	{'max_depth': 4, 'n_estimators': 200}	228.5116	0.7654	4.5578
Extra Trees	{'max_depth': 2, 'n_estimators': 200}	763.3891	0.2163	8.3306

The optimal hyperparameter combination is presented in the second column of Table 3. The other columns contain the measures, see [14] which establish the comparison between the both approaches. Apparently from the results, the **Gradient Boosting** method indicates an optimal performance. The coefficient of determination reaches one, while the error metrics are almost negligible. In comparison to Table 2, the Gradient Boosting rates as the third-order polynomial approximation Poly33. These findings will agree with the results in [12].

The method can be employed to predict the biological activity of some new ligands via the optimization function ChemScore. The forecast surface is given in Fig. 2, where the white dots are the given data in Table 1. The result could be compared with Fig. 3 in [11].

4. Conclusions

We conducted some docking studies in order to understand the interactions between delta-opioid ligands and the delta-opioid receptor (DOR). These studies allowed us to anticipate the biological activity of newly engineered analogs, which manifested a significantly higher activity compared to the other compounds in the series tested, by creating a correlation between the docking outcomes and in vitro tests.

From this evaluation, we have improved our comprehension of how the biological impacts of compounds correspond to *in silico* experiments. Further, this research opens the avenue to determine if the biological macromolecule models (DOR) are in harmony with the actual three-dimensional structures of the molecules.

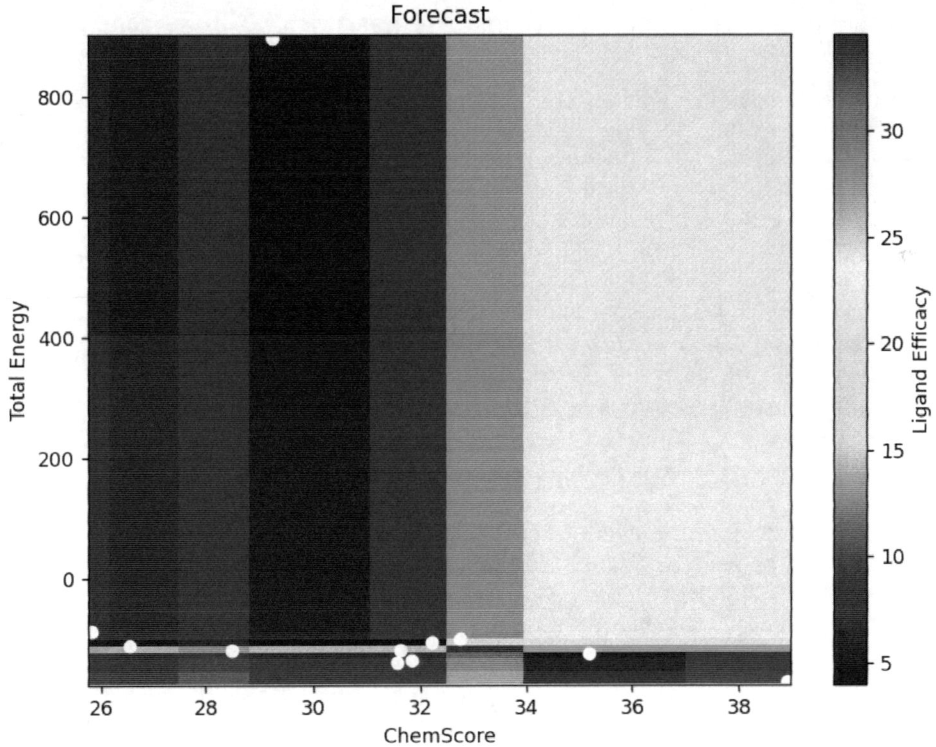

Fig. 2. Forecast surface.

The insights derived from these studies could be instrumental in predicting the potential effectiveness of compounds with known docking scores and total energy calculations. We aim to apply these findings to future research and compound analysis, thereby advancing our understanding of molecular docking and its implications for drug design and development. In essence, our work provides a powerful analytical tool that bridges the gap between in silico predictions and biological effects, accelerating the process of identifying promising compounds for further investigation.

Acknowledgments

This research is supported by the Bulgarian National Science Fund under Project KP-06-M62/1 "Numerical deterministic, stochastic, machine and deep learning methods with applications in computational, quantitative, algorithmic finance, biomathematics, ecology and algebra" from 2022.

References

[1] Ehrlich A., Kieffer B, Darcq E. 2019 *Expert opinion on therapeutic targets* **23(4)** 315–326 https://doi.org/10.1080/14728222.2019.1586882
[2] Valentino R, Volkow N 2018 *Neuropsychopharmacology* **43(13)** 2514-2520 https://doi.org/10.1038/s41386-018-0225-3
[3] Darcq E, Kieffer B. 2018 *Nat Rev Neurosci* **19(8)** 499–514 https://doi.org/10.1038/s41583-018-0028-x
[4] Jutkiewicz E. 2006 *Molecular interventions* **6(3)** 162–169. https://doi.org/10.1124/mi.6.3.7
[5] Morgan M, Christie M 2011 *British journal of pharmacology* **164(4)** 1322–1334 https://doi.org/10.1111/j.1476-5381.2011.01335.x
[6] Reinscheid R, Nothacker H, Bourson A, 1995 *Science* **270** 792-794 https://doi.org/10.1126/science.270.5237.79
[7] Le Merrer J, Becker J, Befort K, Kieffer L 2009 *Physiol Rev.* **89(4)** 1379–1412 https://doi.org/10.1152/physrev.00005.2009
[8] Agarwal R., Singh A., Sen S. 2016 *Applied Case Studies and Solutions in Molecular Docking-Based Drug Design,* 1-28 IGI Global. https://doi.org/10.4018/978-1-5225-0362-0.ch001
[9] Sapundzhi F., Dzimbova T. 2022 *Bulgarian Chemical Communications* **54 (B1)**, 97-102
[10] Sapundzhi F., Popstoilov M., Lazarova M. 2023 *Numerical Methods and Applications. NMA 2022. Lecture Notes in Computer Science (LNCS)* 13858. Springer, Cham.
[11] Sapundzhi F., Dzimbova T., Pencheva N., Milanov P. 2017 *Bulgarian Chemical Communications* **49 (4)** 768 – 774.
[12] Sapundzi F, Lazarova M, Dzimbova T and Georgiev S 2023 Application of machine learning for modelling of biological data *Submitted to J. Phys. Conf. Proc.*
[13] Jones G, Willett P, Glen R, Leach A, Taylor R 1997 *J. Mol. Biol.* **267**, 727-748.
[14] Sapundzhi F, Prodanova K, Lazarova M 2019 *AIP Conference Proceedings*, 2172, 100008 1-6
[15] Pencheva N., Bocheva A, Dimitrov E., Ivancheva C 1996 *Eur. J. Pharmacol.* **304** 99-108.
[16] Pencheva N., Milanov P., Vezenkov L., Pajpanova T., Naydenova E. 2004 *Eur. J. Pharmacol.* **498** 249-256.
[17] R. Thomsen, M. Christensen, J. Med. Chem., 49, 3315(2006).
[18] Sapundzhi F, Dzimbova T. 2019 *International Journal of Online and Biomedical Engineering* **15(15)** 39–56 https://doi.org/10.3991/ijoe.v15i15.11566
[19] Sapundzhi F., Dzimbova T., Pencheva N., Milanov P. 2016 *Bulgarian Chemical Communications* **49 (E)**, 23-30.
[20] Sapundzhi F 2019 *Bulgarian Chemical Communications* **51 (4),** 569-579.
[21] Traykov M, Trenchev I, 2016 *Genetika* **52(9)**, 1089–1096, https://doi.org/10.7868/s0016675816080130
[22] Russel SJ, Norvig P 2020 *Artificial Intelligence: A Modern Approach* 4th Ed Pearson USA
[23] Kelleher JD, Namee BM, D'Arcy A 2020 *Fundamentals of Machine Learning for Predictive Data Analytics: Algorithms, Worked Examples, and Case Studies* 2nd Ed The MIT Press USA
[24] Nedyalkov I, Stefanov A, Georgiev G, 2018 *IX National Conference with International Participation (ELECTRONICA)*, Sofia, Bulgaria, 1-4,

Journal of Physics: Conference Series **2675** (2023) 012033 doi:10.1088/1742-6596/2675/1/012033

Simple layer potential expansion for optimization of contact interaction taking into account friction and adhesion

G Shyshkanova[1, 2*] **and A Walther**[1]

[1] Department of Mathematics, Humboldt-Universität zu Berlin, Unter den Linden, 6, Berlin, 10099, Germany

[2] Department of Applied Mathematics, National University 'Zaporizhzhia Polytechnic', Zhukovskogo str., 64, Zaporizhzhia, 69063, Ukraine

[*] Corresponding Author E-mail: ganna.shyshkanova@hu-berlin.de

Abstract. Friction plays a crucial role in the formation of contact problems, particularly through adhesion. This paper focuses on a quasi-static three-dimensional problem of a punch movement along the boundary of an elastic half-space. The investigation considers friction and adhesion forces, employing a two-term friction law. The objective is to optimize the pressure distribution beneath the punch. The shape of the punch serves as the design variable, while the deviation of the pressure distribution, originating from a given one, is minimized. The optimization problem can be divided into two sequentially solvable sub-problems. The first task involves finding a pressure distribution that minimizes the performance functional, which has a known solution. The second problem entails searching for the optimal shape of the punch to achieve the previously determined pressure distribution. A numeric-analytical solution is developed based on the expansion of the simple layer potential. The coefficients characterizing friction and adhesion act as small parameters. The proposed method gives the ability to obtain closed-form formulas in each approximation, enabling convenient qualitative analysis and practical engineering applications. The calculations and analytical dependence reveal an asymmetric distribution of pressure on the contact area, during the movement of an axisymmetric punch.

1. Introduction

The optimization of contact problems is a challenge in a wide range of engineering applications. Efficient contact interfaces are desirable in structural designs in terms of strength, durability, reliability of connection, load and motion transmission. It is well known that an uneven distribution of contact stress has a profound effect on the wear process and fatigue life, potentially leading to failure [1, 2]. Recent developments in mathematical modeling, optimization techniques, and numerical simulations have enabled the design and optimization of novel devices with improved performance and reduced wear. For example in [3], the profile of the hub edge was optimized to weaken the contact stress concentration. An efficient numerical technique based on direct minimization of the potential energy function using a constrained quadratic programming algorithm was developed in [4] and it was also shown that a Lagrange multiplier method can be used to transform the inequality equilibrium relations found for contact problems to equality relations.

The existence of the optimal shapes in contact problems was proved in [5]. The questions of approximation and numerical realization were investigated in [6]. A nonlinear programming approach to the contact optimization problems for elastic plates with pointed supports were described in [7].

An effective technique to minimize the contact stress concentration and to control the contact stress distribution considering the uncertainties of the loads and wear was proposed in [8], and was also demonstrated on several practical examples. A combined interface shape and material stiffness optimization method based on evolutionary optimization techniques was developed in [1]. The shape and topology optimization of the unilateral contact problems was conducted to minimize the normal contact stress [9, 10]. A topology optimization method was developed in [11] to control the contact stress distribution directly, by simultaneously minimizing the average and variation of the contact stress of an apriori unknown contact area. Although the shape and topology design are effective in homogenizing the contact stress field, there are some limitations [1]. Some of them require complicated operations and may have difficulties in complex geometric situations. For example, in aerospace structures, it is difficult to perform topology design while considering other properties and taking into account friction and adhesion forces [1].

Further research is needed to develop more efficient and accurate models and algorithms for more complex contact problems in various fields. Every day, in the development of advanced technologies and various systems in mechanical engineering, the issue of friction is becoming more and more relevant. This is because contact interaction is accompanied or is only possible with the existence of friction, which depends on the roughness and adhesion of the contacting surfaces.

The optimal limiting contact geometry in fretting wear in the stick-slip mode was researched amply in [12-14]. The role of adhesion in contact mechanics was studied in [15]. A comprehensive overview of recent developments in the area of contact mechanics and friction was done in [16]. Goryacheva and Makhovskaya in [17] present new solutions to contact problems taking into account various effects such, as surface microgeometry and adhesion. The adhesion of indenters was considered in [18-20].

The quasi-static problem taking into account friction and adhesion was considered in [21]. There, it was shown that the application of the proposed approach gives us the possibility to decompose the formulated optimization problem into two problems which are solved successively. The analytic solution of the problem is obtained for a punch of rectangular shape in plan.

In the present work, a similar problem for the punch with a parabolic rotation shape is considered. The purpose of this investigation is to develop an effective approach for a class of quasi-static problems to find the contact pressure distribution, indentations, rotation angle and the unknown area of contact using the small parameter method. Appling the technique proposed in [21], we are going to obtain the optimal pressure distribution, realized in the minimal value of the considered discrepancy functional under constraints on the equilibrium equations and positive pressure condition. Then, to find the optimal punch shape, we develop a simple layer potential expansion and compare the maximum pressure of the initial and optimized cases.

2. Statement of the problem

Let us consider a three-dimensional contact problem for the motion of a rigid punch (indenter) along the boundary of an elastic homogeneous half-space, taking into account friction forces when a two-term friction law applies over the entire contact area [17, 22]. We assume that the speed of movement of the punch is such that it allows us to neglect dynamic phenomena and consider the problem as a quasi-static one. The Cartesian coordinate system is attached to the punch and the movement occurs in the direction of the Ox axis (Fig. 1). The punch is pressed by the vertical force Q which direction coincides with Oz axis. In addition to Q, there is also a moving force T. This is directed parallel to Ox axis and applied at distance d from the punch basis. The tangential forces acting on the punch should be balanced by force T. We suppose that there is no loading out of the contact domain on the elastic half space. We assume that the friction forces on the contact area are directed collinear to the movement along the Ox axis.

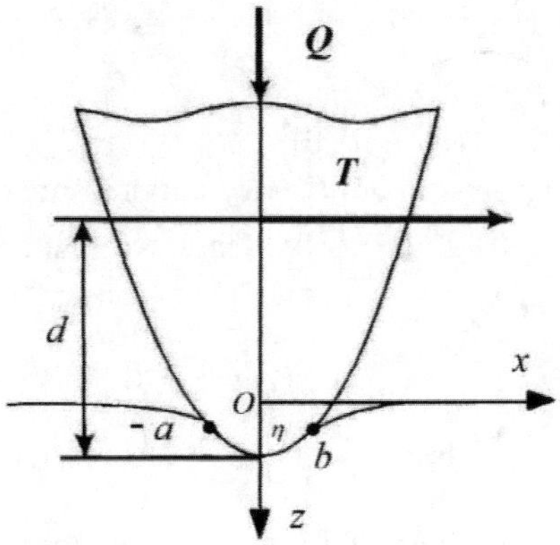

Figure 1. Punch loading scheme.

These forces provide the punch with a state of ultimate equilibrium or uniform motion. Here, the ultimate equilibrium is the state of the system in which a slight increase in external load leads to a violation of the established equilibrium and causes loss of stability.

In the general case, the specified distribution of forces is carried out approximately. It is assumed that the punch under the action of forces applied to it can rotate to a certain angle, the value of which affects the displacement of the contact domain relative to the center of symmetry of the punch.

The domain Ω is a plane projection on $z=0$ of the contact domain. z axis is directed into the elastic half-space. The vertical displacements w of the points of the domain Ω depend on the surface shape of the punch, the penetration value and the sine of the inclination angle γ about Oy axis. Because angle γ is small, in what follows we will assume that $\sin\gamma \approx \gamma$.

The corresponding boundary conditions are written in the form [22]:

$$\begin{aligned} & w=-f(x,y)+\eta+\gamma x, && (x,y)\in\Omega, \\ & \tau_{xz}=\tau_0+\mu\sigma_z,\ \ \sigma_z=-p(x,y),\ \ \tau_{xy}=0, && (x,y)\in\Omega, \\ & \sigma_z=\tau_{xz}=\tau_{yz}=0, && (x,y)\in\overline{\Omega}, \end{aligned} \tag{1}$$

where $f(x,y)$ is a function describing the shape of the punch surface, η is an indentation value of the punch; γ is the punch inclinational angle about Oy axis in Oxz plane; μ is the friction coefficient, τ_0 is the adhesion parameter, σ_z are normal stresses and τ_{xz}, τ_{yz} are shear stresses at the boundary of the elastic half-space ($z=0$). Without adhesion, $\tau_0=0$, whereas in the case of small adhesion, $\tau_0 << 1$.

The main integral equation of the problem for the definition of the pressure $p(x,y)$ under the punch contains integrals with weak singularity of the type of simple layer potentials. In common case, the potentials are taken over an unknown contact domain Ω, depending on friction and adhesion coefficients [22]

$$\iint_\Omega \frac{1+\mu\lambda\cos\phi}{r} p(\rho,\theta)\,\mathrm{d}\Omega + \lambda\tau_0 \iint_\Omega \frac{\cos\phi}{r}\,\mathrm{d}\Omega$$

$$=\frac{\pi E}{1-\nu^2}[-f(\rho_0,\theta_0)+\eta+\gamma\rho_0\cos\theta_0], \tag{2}$$

where (ρ,θ) are polar coordinates, $x=\rho\cos\theta$, $y=\rho\sin\theta$, $r^2=\rho^2+\rho_0^2-2\rho\rho_0\cos(\theta-\theta_0)$, $\mathrm{d}\Omega=\rho\,\mathrm{d}\rho\,\mathrm{d}\theta$, $(\rho_0,\theta_0)\in\Omega$, $\cos\phi=\dfrac{\rho_0\cos\theta_0-\cos\theta}{[\rho^2+\rho_0^2-2\rho\rho_0\cos(\theta-\theta_0)]^{1/2}}$, $\lambda=\dfrac{1-2\nu}{2(1-\nu)}$, ν is the Poisson coefficient, $p(\rho,\theta)$ is an unknown function of the normal pressure distribution.

The contour Γ of the domain Ω of the non-flat punch without angular lines on its surface ($f_x{}', f_y{}'$ are continuous functions) is defined by

$$p(\rho,\theta)\big|_{\Gamma}=0. \tag{3}$$

This condition follows from the continuity of the function that characterizes the normal pressure distribution on the elastic half-space.

In addition, the equilibrium conditions of forces and moments within the punch should be also hold:

$$\iint_{\Omega} p(\rho,\theta)\mathrm{d}\Omega=Q, \quad \iint_{\Omega} p(\rho,\theta)\cdot\rho\cos\theta\,\mathrm{d}\Omega+T(-d+\eta)=0, \tag{4}$$

$$T=\iint_{\Omega} t(x,y)\mathrm{d}\Omega=\mu Q+\tau_0 S, \tag{5}$$

where $S=\iint_{\Omega}\mathrm{d}\Omega$ is the area of the contact domain Ω, $t(x,y)=-\tau_{xz}$; $\tau_0=-\tau_{oxz}$. The equilibrium conditions give the possibility to define unknown punch penetration η and rotational angle γ.

3. Solution method for the general problem

We denote

$$\varepsilon_1=\mu\lambda, \quad \varepsilon_2=\frac{\lambda\tau_0 s_0}{Q},$$

where s_0 is an area of contact domain without friction, $s_0=\pi\cdot b^2$, and ε_1, ε_2 are small parameters in the integral equation (2). As product of the friction coefficient μ by λ and $\lambda\tau_0 s_0/Q$ they are small values ($\lambda=0.286$, when $0<\nu\le 0.5$ [22]). The unknowns of the problem are the normal pressure distribution under the punch, the equation of the boundary of the contact domain, the punch indentation and the rotation value. All are unknown functions of the parameters ε_1 and ε_2. It is assumed that they can be written in the form of series in terms of powers of ε_1 and ε_2:

$$p(\rho,\theta)=\sum_{i,k=0}^{\infty}\varepsilon_1^i\varepsilon_2^k p_{ik}(\rho,\theta),$$

$$\gamma=\sum_{i,k=0}^{\infty}\varepsilon_1^i\varepsilon_2^k\gamma_{ik}, \tag{6}$$

$$\eta=\sum_{i,k=0}^{\infty}\varepsilon_1^i\varepsilon_2^k\eta_{ik}.$$

We propose to reduce the problem for the domain Ω to sequences of similar problems for a circular domain D analogically to the reduction of a problem for a non-circular ring to the circular ring developed in [23, 24]. Because the domain Ω is circular without friction, the equation of the boundary Γ of the contact domain Ω could be assumed as

$$\rho_\Gamma = b\left[1 + g(\varepsilon_1, \varepsilon_2, \theta)\right], \tag{7}$$

$$g(\varepsilon_1, \varepsilon_2, \theta) = \sum_{i,k=0}^{\infty} \varepsilon_1^i \varepsilon_2^k f_{ik}(\theta), \qquad f_{00} = 0.$$

The functions $f_{ik}(\theta)$ are continuous and have continuous derivatives of different order in Ω such that a ray outgoing from the origin crosses the boundary Γ only at one point.

Then, the integrals from (2) are expanded in series in terms of powers of ε_1 and ε_2. Taking into account that the equation of the boundary of the contact domain Ω depends on ε_1 and ε_2, the next formula is obtained

$$\iint_{\Omega} \frac{P(\rho,\theta)}{r} \mathrm{d}\Omega = \sum_{i,k=0}^{\infty} \varepsilon_1^i \varepsilon_2^k \left[\iint_{D} \frac{P_{ik}(\rho,\theta)}{r} \mathrm{d}s + \Phi_{ik}(P) \right], \tag{8}$$

where the operators Φ_{ik} for i,k=0,1 are given by

$$\Phi_{00}(P) = 0;\ \Phi_{10}(P) = L_{10}(P_{00});\ \Phi_{01}(P) = L_{01}(P_{00}) ,$$

and

$$L_{10}(P_{00}) = (1 - \rho_0 \frac{\partial}{\partial \rho_0}) \iint_{D} \frac{P_{00}(\rho,\theta)}{r} f_{10}(\theta)\,\mathrm{d}s,$$

$$L_{01}(P_{00}) = (1 - \rho_0 \frac{\partial}{\partial \rho_0}) \iint_{D} \frac{P_{00}(\rho,\theta)}{r} f_{01}(\theta)\,\mathrm{d}s.$$

Therefore, equation (2) now reads

$$\sum_{i,k=0}^{\infty} \varepsilon_1^i \varepsilon_2^k \left[\iint_{D} \frac{P_{ik}(\rho,\theta)}{r} \mathrm{d}s + \Phi_{ik}(P) + \varepsilon_1 \Phi_{ik}(P\cos\phi) + \right.$$
$$+ \varepsilon_1 \iint_{D} \frac{P_{ik}(\rho,\theta)}{r^2} (\rho_0 \cos\theta_0 - \rho\cos\theta)\mathrm{d}s + \varepsilon_2 \Phi_{ik}\left(\frac{Q}{s_0} \cos\phi \right) +$$
$$\left. + \varepsilon_2 \iint_{D} \frac{Q_{ik}}{s_0 r^2} (\rho_0 \cos\theta_0 - \rho\cos\theta)\mathrm{d}s \right] = \tag{9}$$
$$= \frac{\pi E}{1-\nu^2} \left[-f(\rho_0,\theta_0) + \sum_{i,k=0}^{\infty} (\eta_{ik} + \gamma_{ik}\, \rho_0 \cos\theta_0) \right],$$

where $Q_{00} = Q$, Q_{ik} =0, except when $i = 0$ and k=0 simultaneously and D is given by $\rho \le b$.

The integrals from the equilibrium conditions (4) can also be written in the form of series expansions in terms of powers of ε_1 and ε_2. After writing the pressure distribution under the punch in the form of an expansion in terms of powers of ε_1 and ε_2, the boundary condition (3) is given by

$$P_{ik}(\rho,\theta)\big|_{\rho=b} = 0, \qquad (i,k = 0,1,\ldots). \tag{10}$$

Equating terms with equal powers of parameters ε_1 and ε_2 in (7) and exploiting the equilibrium of forces and moments leads to recurrent systems for the unknown coefficients which appear in (6).

The unknown functions $f_{ik}(\theta)$ describing the contact domain (7) are written as Fourier series from received recurrent systems and conditions (10). Then, functions $P_{ik}(\rho,\theta)$ characterizing the pressure distribution on the contact domain, values η_{ik}, γ_{ik} characterizing the punch penetration and its inclination, and also $f_{ik}(\theta)$ characterizing the shape and sizes of contact domain are defined.

Therefore, the problem for the domain Ω is reduced to sequences of similar problems for circular domain D.

4. Solution of the problem for punch with parabolic shape

Let us study the subcase where

$$f(\rho)=\frac{\rho^2}{2b_0}.$$

Here, only terms that are first order in ε_1 and ε_2 in the desired expansion (6), (9) are taken into account. Thus, the function, characterizing the normal pressure distribution in points $M(\rho,\theta_*)$ of the ray $\theta=\theta_*$ passing point $M_\Gamma(\rho_\Gamma,\theta_*)$ on the boundary of domain Ω, is obtained by

$$p(\rho,\theta_*)=\frac{2bE}{\pi b_0(1-\nu^2)}\left\{\sqrt{1-\frac{\rho^2}{\rho_\Gamma^{\,2}}}+\varepsilon_1\cos\theta_*\left[a_1\frac{\rho}{\rho_\Gamma}\sqrt{\frac{1-\rho/\rho_\Gamma}{1+\rho/\rho_\Gamma}}+\right.\right. \quad (11)$$

$$\left.\left.+\frac{2}{\pi}\sum_{k,n=0}^{\infty}\frac{2k(2n-1)!!}{(2n-2k+1)(2n)!!}\frac{1}{(4k^2-1)}\left(\frac{\rho}{\rho_\Gamma}\right)^{2n+1}\right]+\varepsilon_2\cos\theta_*\left[a_{11}\frac{\rho}{\rho_\Gamma}\sqrt{\frac{1-\rho/\rho_\Gamma}{1+\rho/\rho_\Gamma}}\right]\right\}.$$

The value of the inclination angle γ of the punch, its indentation η and the equation of the boundary $\rho(\varepsilon_1,\varepsilon_2,\theta)$ of the contact domain Ω can be written in the next form

$$\gamma=\frac{\varepsilon_1 b}{b_0}\cdot\left(a_1+\frac{1}{\pi}\right)+\frac{\varepsilon_2 b}{b_0}\cdot\left(-\frac{4}{3\pi}+a_{11}\right),\qquad \eta=\frac{b^2}{b_0},\qquad b^3=\frac{3Qb_0(1-\nu^2)}{4E},$$
$$\rho=b\cdot[1+\varepsilon_1 a_1\cos\theta+\varepsilon_2 a_{11}\cos\theta],\ \ a_1=\frac{1}{\lambda b}\left(d-\frac{b^2}{b_0}\right)+\frac{1}{2\pi},\ \ a_{11}=\frac{-1}{\pi b}\left(d-\frac{b^2}{b_0}\right). \quad (12)$$

5. The optimization problem and the process of finding the solution

The problem of optimizing pressure under a rigid punch is studied using the calculus of variations [8]. The equation describing the punch shape is considered as an unknown function. It is assumed that the desired punch shape $f=\varphi(\rho,\theta)$ is a continuous and smooth function of the coordinates.

The optimization problem [21] is to find the function $\varphi_*(\rho,\theta)$ minimizing the functional which characterizes the mismatch between the pressure distribution $p(\rho,\theta)$ corresponding to a certain shape of the punch and a given target pressure distribution $p_\varphi(\rho,\theta)$

$$J(p(\varphi_*))=\min_{\varphi} J(p(\varphi)), \quad (13)$$

when, in our case, the equilibrium conditions (4) and (5) are satisfied and the normal pressures in the contact area are positive (1)

$$\varphi\in\Lambda_\varphi=\left\{\varphi: p(\varphi)\ge 0,\ \iint_\Omega p(\varphi)\rho d\rho d\theta=Q,\ \iint_\Omega p(\varphi)\cdot\rho^2\cos\theta\, d\rho d\theta=(\mu Q+\tau_0 S)(d-\eta)\right\}. \quad (14)$$

The algorithm for solving such optimization problems is described in detail in [21]. It was shown in [8] that the solution of the formulated optimization problem (13), (14) can be decomposed into two problems that can be solved sequentially. The first problem is to minimize the functional

$$J(p) = \iint_{\Omega} (p_{\varphi} - p)^2 d\Omega \to \min_{p}, \quad (15)$$

where the feasible set is

$$p \in \Lambda_p = \left\{ \begin{array}{l} p : p(\rho,\theta) \geq 0, (\rho,\theta) \in \Omega; \iint_{\Omega} p(\rho,\theta) d\Omega = Q, \\ \iint_{\Omega} p(\rho,\theta) \cdot \rho \cos\theta d\Omega = T(d - \eta). \end{array} \right\}. \quad (16)$$

When searching for the optimal pressure distribution $p_*(\rho,\theta)$ based on the solution of problem (15), (16), an auxiliary variable $\psi(x)$ is introduced as in [21], satisfying the relation

$$p(\rho,\theta) - \psi^2(\rho,\theta) = 0,$$

The solution method that we use here for contact pressure optimization is developed in [8, 21]. An extended Lagrange functional is given by

$$J^L = \iint_{\Omega} \left((p - p_{\varphi})^2 - \lambda_1 p - \lambda_2 p \cdot \rho \cdot \cos\theta - \chi(p - \psi)^2 \right) d\Omega,$$

where the constants λ_i ($i = 1,2$) are the Lagrange multipliers, and $\chi(\rho,\theta)$ is the Lagrange function. The extremum conditions for it with respect to the variables p and ψ are written as follows [21]

$$p = p_{\varphi} + 0.5 \cdot (\lambda_1 + \lambda_2 \rho \cdot \cos\theta + \chi), \quad \psi\chi = 0, \ (\rho,\theta) \in \Omega. \quad (17)$$

The expression for p gives the desired pressure distribution. To determine λ_i, the integral equalities included in (14) are used, and the function χ is found using the second equality in (17).

The second problem is to find normal displacements corresponding to the optimal pressure distribution $p_*(\rho,\theta)$ and determining the optimal shape of the punch $\varphi_*(\rho,\theta)$. It is possible to determine the normal displacements from (2), solving the spatial boundary value problem of the elasticity theory, the solution method for which is described above in the previous paragraph.

6. Numerical results and discussions

The normal pressure distribution surface over the contact domain is shown in Fig. 2, where $p_* = 3Q/2\pi b^2$. Calculations have been made according to formula (11) for $\varepsilon_1 = 0.2$, $\tau_0 = 0$, $d/b = 0.4$; $b/b_0 = 0.1$, assuming that the punch is pressed with force Q such that $b = 1$. The maximal value of the contact pressure is $p_{\max} / p_*(0.2195;\ 0) = 1.0156$.

Figure 3 represents the graphs of normal pressure along the Ox axis at $\theta = \pi$ and $\theta = 0$, for different values of the friction coefficient. Line 1 corresponds to ε_1=0.4; ε_2=0; line 2 – ε_1=0.2; ε_2=0; line 3 – ε_1=0.06; ε_2=0.1; line 4 – ε_1=0.06; ε_2=0.03. It gives a view on the difference in pressure value of head-on and rear punch zones under the motion. Contact domains for the same parameters are shown in Fig. 4.

In the considered model [17, 22], the adhesive component of the frictional force affects the displacement of the contact domain of the punch with elastic half-space, and also significantly reduces the punch rotation angle.

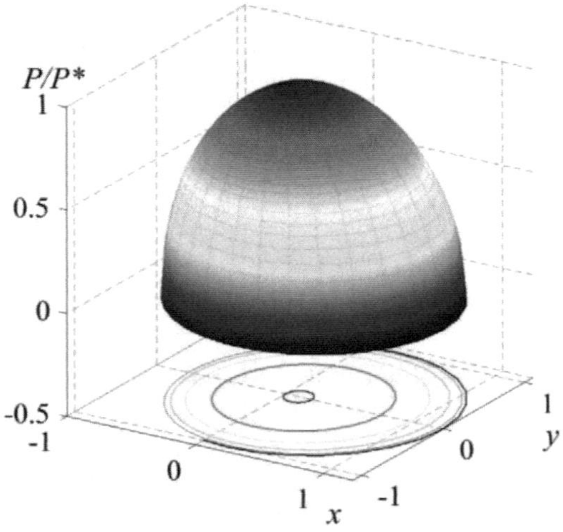

Figure 2. Contact pressure distribution.

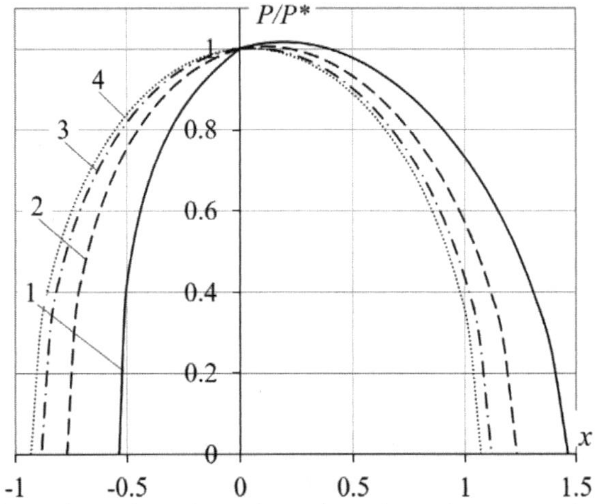

Figure 3. Contact pressure cross section along Ox.

Figure 5 demonstrates the graphs of dependence on the punch rotation angle γ / p_* from distance d/b (the position of the force T application point relative to the punch base), according to (12), line 1 corresponds to the following parameter values: ε_1=0.06; ε_2=0.03; line 2 – ε_1=0.06; ε_2=0; line 3 – ε_1=0.2; ε_2=0.1; line 4 – ε_1=0.2; ε_2=0; line 5 – ε_1=0.4; ε_2=0.2; line 6 – ε_1=0.4; ε_2=0.

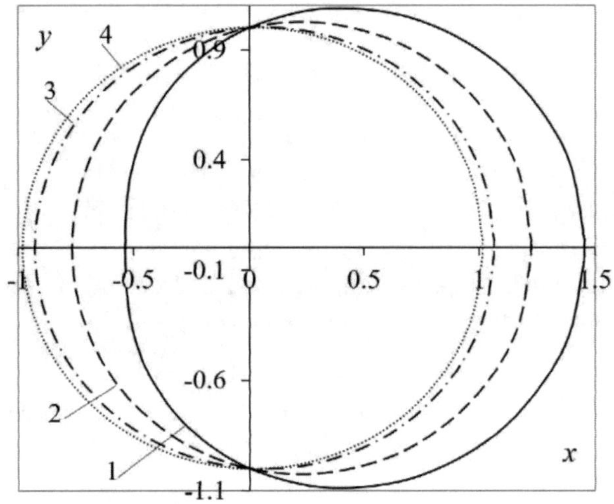

Figure 4. Contact Domains.

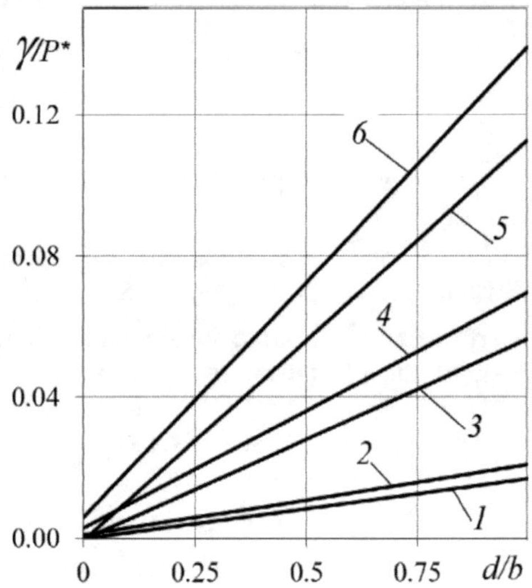

Figure 5. Punch inclination dependence on the height of the application point of the moving force.

The optimal punch shape is shown in Fig. 6, for the case where for a more symmetric distribution the desired pressure is taken as in the case of static loading without moments. The other constants are equal to $\varepsilon_1 = 0.4$, $\tau_0 = 0$, $d/b = 0.4$ and $b = 1$. For such a punch shape the maximal value of contact pressure is $p_{\max} / p_*(0.0873;\ 0) = 0.9328$. It is therefore reduced by 8.15%.

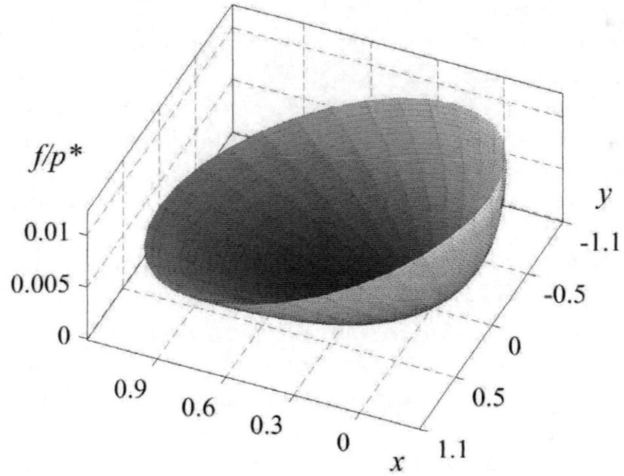

Figure 6. Optimal punch surface

7. Conclusions

An expansion of a simple layer potential is derived. A method is proposed for the analytic solution of a quasi-static three-dimensional contact problem for the motion of a rigid punch along the boundary of an elastic homogeneous half-space, taking into account friction forces when a two-term friction law applies over the entire contact area. The solution is written as a small parameter expansion, and the resulting problem is similar to the investigated problem for circular contact domain in each approximation. The contact domain has a circular area without consideration of friction forces. An example of punch limited by the rotation surface was considered.

The advantages of the proposed method are the possibility to obtain simple formulae in closed form for each approximation, this being convenient for qualitative analysis and feasibility of application in mechanical engineering.

Friction causes asymmetry of pressure distribution that may result in rotation of the punch. In this model, the adhesive component affects the value of the displacement of the contact area and reduces the angle of rotation of the punch. Optimization of the pressure distribution over a contact domain was considered accounting for friction and adhesion. The function to be minimized is chosen as the root mean square deviation of the pressure distribution under the punch from a given optimal distribution. The punch shape optimized to obtain the optimal pressure distribution. The formulated optimization problem can be decomposed into two sequentially solved problems, as it was done in [8] and a similar solution algorithm was used in this work.

The optimal shape of the punch was found using the expansion of the simple layer potential developed in the present work. A numerical example showed the reduction of the maximal value of contact pressure, when the optimal shape punch was applied.

Acknowledgments

The authors thank the Deutsche Forschungsgemeinschaft (DFG) for their support within "MATH+" (fund: 0502381701 / 19023300)

References

[1] Zhou Yi, Lin Q, Hong J and Yang N 2023 Combined interface shape and material stiffness optimization for uniform distribution of contact stress, *Mech. Based Design of Struct. and Machines*, **51:2**, 1001-15, https://doi.org/10.1080/15397734.2020.1860086

[2] Collins J A, Busby H R and Staab G H 2010 *Mechanical design of machine elements and machines: A failure prevention perspective* 2nd ed. Hoboken (New York: John Wiley & Sons)

[3] Hénaff G, Lee D H, Choi H Y, Kwon S J and Seo J W 2018 Optimization of contact edge profile for minimizing contact pressure in a press-fitted shaft *MATEC Web of Conf.* **165** 22029 https://doi.org/10.1051/matecconf/201816522029

[4] Benedict R L and Taylor J E 1981 Optimal design for elastic bodies in contact *Optimization of Distributed Parameter Structures* Haug E J and Cea J eds. – V. II . (Alphen aan den Rijn, Netherlands : Sijthoff and Nordhoff) pp 1553 – 1599

[5] Haslinger J and Neittaanmaki P 1984 On the existence of optimal shapes in contact problems *Numer. Funct. Anal. and Optim.* **7** 107–124

[6] Haslinger J and Neittaanmaki P 1988 *Finite Element Approximation for Optimal Shape Design* (London : John Wiley & Sons)

[7] Banichuk N V 1983 *Problems and Methods of Optimal Structural Design* (New York: Plenum Press)

[8] Banichuk N V and Ivanova S Yu 2017 *Optimal Structural Design: Contact Problems and High-Speed Penetration* (De Gruyter: Berlin, Boston)

[9] Myśliński A and Wróblewski M 2017 Structural optimization of contact problems using Cahn–Hilliard model *Comput. & Struct.* **180** 52–9 https://doi.org/10.1016/j.compstruc.2016.03.013

[10] Niu C, Zhang W H and Gao T 2019 Topology optimization of continuum structures for the uniformity of contact pressures *Struct. and Multidiscip. Optim.* **60 (1)** 185–210 https://doi.org/10.1007/s00158-019-02208-8

[11] Kristiansen H, Poulios K and Aage N 2020 Topology optimization for compliance and contact pressure distribution in structural problems with friction *Comput. Methods in Appl. Mech. and Eng.* **364** 112915 https://doi.org/10.1016/j.cma.2020.112915

[12] Popov V 2014 Analytic solution for the limiting shape of profiles due to fretting wear *Sci. Rep.* **4** 3749 https://doi.org/10.1038/srep03749

[13] Argatov I I and Chai Y S 2018 Limiting shape of profiles in fretting wear *Tribol. Int.* **125** 95-99 https://doi.org/10.1016/j.triboint.2018.04.026

[14] Argatov I I and Chai Y S 2020 A note on optimal design of contact geometry in fretting wear. *Int. J. Mech. Mater. Des.* **16** 415–422 https://doi.org/10.1007/s10999-019-09467-9

[15] Ciavarella M, Joe J, Papangelo A and Barber J R 2019 The role of adhesion in contact mechanics *J. R. Soc. Interface* 162018073820180738 https://doi.org/10.1098/rsif.2018.0738

[16] Goryacheva I G, Paggi M and Popov V L 2021 Contact mechanics perspective of tribology *Front. in Mech. Eng.* **7** 649792 https://doi.org/10.3389/fmech.2021.649792

[17] Goryacheva I and Makhovskaya Y 2022 Discrete Contact Mechanics with Applications in Tribology (Amsterdam: Elsevier) https://doi.org/10.1016/C2019-0-03650-3

[18] Pande S S and Turner K T 2023 Geometric optimization of pillars for enhanced adhesion *Extreme Mech. Letters* **59** 101969 https://doi.org/10.1016/j.eml.2023.101969

[19] Kossa A, Hensel R and McMeeking R M 2023 Adhesion of a cylindrical punch with elastic properties that vary radially *Mech. Res. Commun.* **130** 104123 https://doi.org/10.1016/j.mechrescom.2023.104123

[20] Li P and Liu T J 2022 Axisymmetric adhesive contact of multi-layer couple-stress elastic structures involving graded nanostructured materials. *Appl. Math. Modelling* **111** 501-520 https://doi.org/10.1016/j.apm.2022.06.044

[21] Banichuk N V and Ivanova S Y 2012 Determination of optimal shape of a moving punch with friction taken into account. *Mech. Solids* **47** 591–597 https://doi.org/10.3103/S0025654412050135

[22] Goryacheva I G 2013 *Contact Mechanics in Tribology*. (Dordrecht: Springer)

[23] Shyshkanova G 2018 About stability of first kind equation solving. *IEEE First Int. Conf. on System Analysis & Intelligent Computing (SAIC)* 8516805 https://doi.org/10.1109/SAIC.2018.8516805

[24] Shyshkanova G and Walther A 2023 Optimization of a punch shape with a doubly connected contact domain *Preprint* docId/523 https://opus4.kobv.de/opus4-trr154/frontdoor/index/index/docId/523

Parallel scalability of sparse solvers for sparsified fractional diffusion problems discretized with FEM

D Slavchev[1] and S Margenov[2]

Institute of Information and Communication Technologies at Bulgarian Academy of Sciences, Acad. G. Bonchev St., Block 2 1113 - Sofia, Bulgaria

E-mail: [1]dimitargslavchev@parallel.bas.bg, [2]margenov@parallel.bas.bg

Abstract. Anomalous diffusion is a non-local process that describes a wide spectrum of natural processes and phenomena with many applications in science and technology. It is described mathematically by the fractional Laplace operator. In this work we examine the integral definition of the fractional Laplacian modeled with the Riesz potential and discretized with the finite element method. The thus obtained system of linear algebraic equations is dense due to the non-local nature of the fractional Laplacian and is computationally complex to solve. With LU factorization, for example, solving the problem has an $O(N^3)$ computational complexity, where N is the number of unknowns. However, it can be observed that a large amount of the off-diagonal coefficients have very small absolute values compared to the diagonal coefficients. Those small off-diagonal coefficients can be lumped (set to zero and added to the diagonal coefficient) without significant loss of accuracy. In this work we employ a direct sparse parallel solver to the resultant sparse matrix. We analyze the parallel performance and speed-up, as well as the accuracy, varying the fractional power and the lumping threshold.

1. Introduction

A significant amount of natural processes and phenomena are described by anomalous diffusion. Unlike the normal variety, anomalous diffusion is a non-local process that includes supperdiffusion and subdiffusion. It can be observed, for example, in diffusion through porous media, machine learning, electromagnetostatics, protein diffusion within cells, superconductivity, crime diffusion, just to name a few. See [1, 2] for an overview of applications of anomalous diffusion.

The fractional Laplace operator is used to model anomalous diffusion (sometimes called fractional for this reason). It is a non-local operator and the systems of linear algebraic equations arising from its discretization with a suitable method (like finite elements, finite volumes, etc.) are dense. As a result solving the systems of linear algebraic equations stemming from real life applications is very computationally expensive. With LU decomposition the computational complexity of solving an elliptic problem is $O(N^3)$, where N is the number of unknowns. One way to lower the computational complexity is to use the spectral definition with rational approximations (see for example the BURA method in [3]) or use hierarchical compression (see [4]).

In this work we eliminate the smaller off-diagonal coefficients of the dense fractional stiffness matrix K thus obtaining a sparse approximation K_{sparse}. A lumping scheme with a relative threshold is used as described in [5]. Systems of linear algebraic equations with sparse matrices can be solved with specialized sparse solvers like nested dissection, the frontal method and others. In this work we use the PARDISO sparse solver for an elliptic fractional diffusion problem and study the parallel performance and speed-up.

The dense stiffness matrix is sparsified by lumping the smallest off-diagonal coefficient with their corresponding diagonal coefficient. Here by *lumping* we mean adding the eliminated coefficient to the diagonal coefficient. A relative threshold is used to determine which off-diagonal coefficient can be eliminated, i.e. an off-diagonal coefficient will be eliminated if the ratio of its value and the corresponding diagonals' values is smaller than a suitable threshold. The size of the threshold and sparsification methods are discussed in more length in [5].

2. Fractional Diffusion problem

The Fractional operator of Laplace (also known as Laplacian) is used to mathematically describe anomalous diffusion processes. It can be defined in several ways (see [6, 7] for comparison between some of the definitions) and in this work we will be examining the integral definition through the Riesz potential as described by Acosta et al. in [8]. A modified version of the code from [9] by the same authors is used for the computation of the system of liner algebraic equations arising from the finite element discretization. The fractional Laplacian can be written as

$$(-\Delta)^{\alpha} u(x) = C(d,\alpha)\,\text{P.V.} \int_{\mathbb{R}^d} \frac{u(x)-u(y)}{|x-y|^{d+2\alpha}} dy, \quad \alpha \in (0,1), \tag{1}$$

where α is the fractional power of the fractional Laplacian Δ, $u(x)$ is the unknown, d is the dimension (in the numerical experiments we use $d = 2$), P.V. stands for principal value, Γ is the gamma function and the constant $C(d,\alpha)$ has the following form

$$C(d,\alpha) = \frac{2^{2\alpha}\alpha\Gamma\left(\alpha + \frac{d}{2}\right)}{\pi^{d/2}\Gamma(1-\alpha)}.$$

The finite element method is used to discretize the fractional Laplace operator. An admissible quasi uniform triangulation $\mathcal{T}_h$ is introduced over the problem domain Ω. Continuous piecewise linear elements form the discrete space $\mathbb{B}$. The nodal basis $\varphi_i(x_j) = \delta^i_j$, $\{\varphi_1, \dots, \varphi_N\} \in \mathbb{V}$ corresponds to the internal nodes $\{x_1, \dots, x_N\}$. The fractional stiffness matrix can then be formed as

$$K_{ij} = \frac{C(d,\alpha)}{2}\langle\varphi_i,\varphi_j\rangle_{H^{\alpha}(\mathbb{R})} = \frac{C(d,\alpha)}{2}\sum_{l=1}^{N_{\tilde{\mathcal{T}}}}\left(\sum_{m=1}^{N_{\tilde{\mathcal{T}}}} I^{i,j}_{l,m} + 2J^{i,j}_{l}\right), \; l,m \in [1, N_{\tilde{\mathcal{T}}}].$$

Here H is a Hilbert space, $\tilde{\mathcal{T}}_h$ is an admissible quasi uniform triangulation of an auxiliary ball domain B that includes and extends Ω. The number of nodes in B is denoted with $N_{\tilde{\mathcal{T}}}$ The triangulations for a very small problem are visualized on Fig. 1. The integrals $I^{i,j}_{l,m}$ and $J^{i,j}_{l}$ can be written as

$$I^{i,j}_{l,m} = \int_{T_l}\int_{T_m} \frac{\big(\varphi_i(x)-\varphi_i(y)\big)\big(\varphi_j(x)-\varphi_j(y)\big)}{|x-y|^{d+2\alpha}} dxdy$$

$$J^{i,j}_{l} = \int_{T_l}\int_{B^c} \frac{\varphi_i(x)\varphi_j(x)}{|x-y|^{d+2\alpha}} dydx.$$

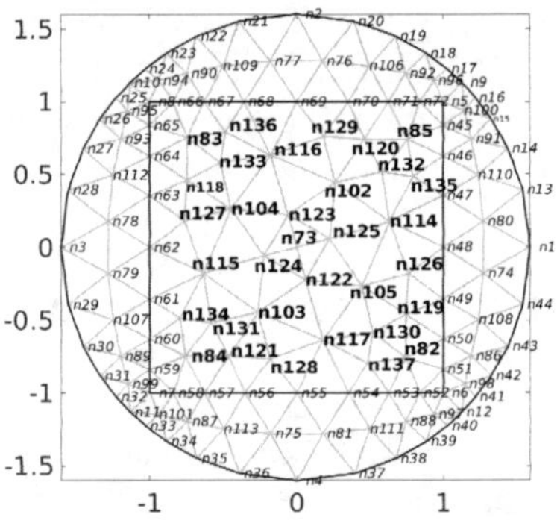

Figure 1: An auxiliary ball B enveloping the square problem domain Ω.

The thus formed fractional stiffness matrix K is dense symmetric positive definite. We should note that a large amount of the off-diagonal coefficients are very small relative to the diagonal coefficients. On Fig. 2 we present the order of magnitude of the matrix coefficients with the diagonal values (top) and without them (bottom). It could be observed that for higher α the difference is more pronounced with larger amount of very small off-diagonal coefficients. Elimination of such coefficients will produce a sparse approximation of K that would allow the application of more efficient sparse solvers when solving systems of linear algebraic equations involving K.

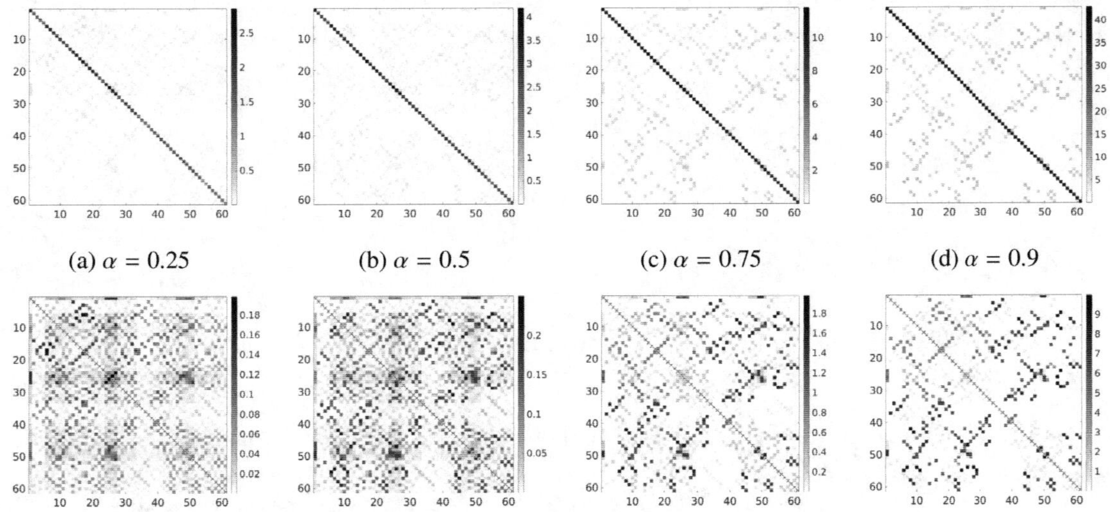

(a) $\alpha = 0.25$ (b) $\alpha = 0.5$ (c) $\alpha = 0.75$ (d) $\alpha = 0.9$

(e) $\alpha = 0.25$ (f) $\alpha = 0.5$ (g) $\alpha = 0.75$ (h) $\alpha = 0.9$

Figure 2: Order of magnitude of the coefficients of the stiffness matrix for an $N = 61$ problem. The diagonals are omitted in the second row (marked off with a red 'x').

3. Lumping Sparsification Scheme

As stated above, a significant amount of off-diagonal coefficients of K have relatively small values. On Fig. 3 we present the means of the diagonal and off-diagonal coefficients (MV^{K_D} and $MV^{K\setminus K_D}$ respectively). They are calculated in the following way

$$MV^{K_D} = \frac{1}{N}\sum_{i=1}^{N} K_{ii} \quad \text{and} \quad MV^{K\setminus K_D} = \frac{1}{N}\sum_{i\neq j}^{N} |K_{ij}|,$$

where K_D is the set of the diagonal coefficients and $K \setminus K_D$ is the set of off-diagonal coefficients of the fractional stiffness matrix. We should note that both depend on the values of α and N. The mean of the off-diagonal coefficients is significantly smaller than that of the diagonals.

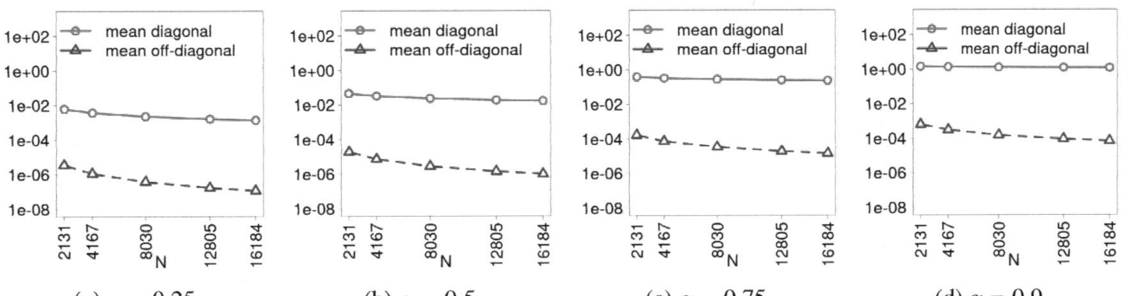

(a) $\alpha = 0.25$ (b) $\alpha = 0.5$ (c) $\alpha = 0.75$ (d) $\alpha = 0.9$

Figure 3: Means of the diagonal and off-diagonal values in K.

On Fig. 4 we present histogram of the coefficients in K normalized by the corresponding diagonal coefficients (each coefficient is divided by its corresponding diagonal one by column) for $N = 2131$ (the

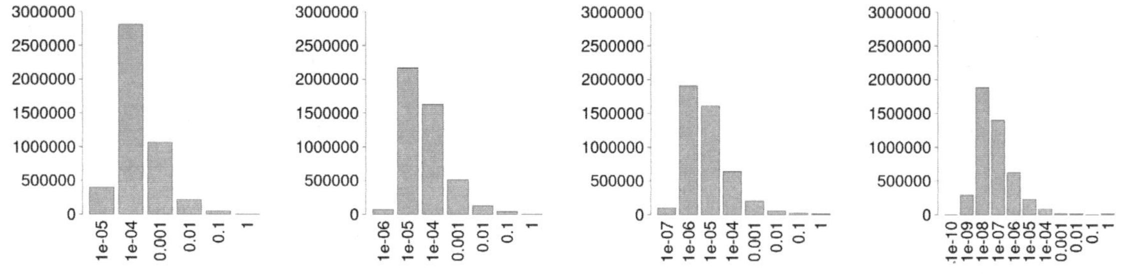

(a) $\alpha = 0.25$ (b) $\alpha = 0.5$ (c) $\alpha = 0.75$ (d) $\alpha = 0.9$

Figure 4: Histograms of the coefficients in K normalized by the diagonal for $N = 2131$.

histograms are similar for other mesh sizes). It should be noted that for the larger values of α more coefficients have smaller values.

In [5] the choice of sparsification threshold $\varepsilon_{\text{sparse}}$ and sparsification method are discussed. In this work we will use the best method from that paper: the lumping scheme, where eliminated off-diagonal coefficients are summed with the corresponding diagonal coefficient. A coefficient $K_{ij,i\neq j}$ is set to zero and added to the diagonal if

$$\frac{|K_{ij}|}{K_{ii}} < \varepsilon_{\text{sparse}},\ j \neq i \quad \text{and} \quad \frac{|K_{ij}|}{K_{jj}} < \varepsilon_{\text{sparse}},\ i \neq j$$

are both satisfied. Here we will present numerical experiments for $\varepsilon_{\text{sparse}} = h^3$ and h^4, where h is a parameter of the triangulation calculated as $h = S_\Omega/N^2$, S_Ω is the surface of Ω.

4. Experimental study of the computational times

4.1. Relative error

For the numerical experiments we consider an elliptic problem $K\mathbf{u} = \mathbf{f}$. The fractional stiffness matrix K is obtained from a finite element discretization in $\Omega = (-1,1) \times (-1,1)$. With $\mathbf{u} \in \mathbb{R}^N$ we denote the nodal unknowns and $\mathbf{f} = (f_j) \in \mathbb{R}^N$, $f_j = \int_\Omega f\varphi_j$ is the right hand side. The presented numerical experiments vary the number of unknowns $N \in (2\,131, 16\,184)$.

On Fig. 5 we present the relative error of the solution with the thus sparsified matrix as well as the number of eliminated coefficients. For thresholds we use $\varepsilon_{\text{sparse}} = h^3$ and h^4. The relative errors R are calculated with the following formula

$$R = \frac{\left\|\mathbf{u}^{\text{dense}} - \mathbf{u}^{\text{sparse}}\right\|_{l_2}}{\left\|\mathbf{u}^{\text{dense}}\right\|_{l_2}} = \frac{\sqrt{\sum_{i=1}^{N}(u_i^{\text{dense}} - u_i^{\text{sparse}})^2}}{\sqrt{\sum_{i=1}^{N}(u_i^{\text{dense}})^2}}.$$

Here $\mathbf{u}^{\text{dense}}$ is the solution obtained with LU factorization. This solution is taken as a reference for calculating the error of the solution with the sparse approximation $\mathbf{u}^{\text{sparse}}$.

We should note that for $\alpha = 0.25$ using the $\varepsilon_{\text{sparse}} = h^4$ threshold eliminates very few of the off-diagonal coefficients (for $n = 8030, 12805$ and 16184 no coefficients were eliminated), therefore the h^4 threshold should be preferred. For the experiments with larger α more off-diagonal coefficients are eliminated by both examined thresholds and the use of h^3 may be preferred. We present numerical experiments with both thresholds.

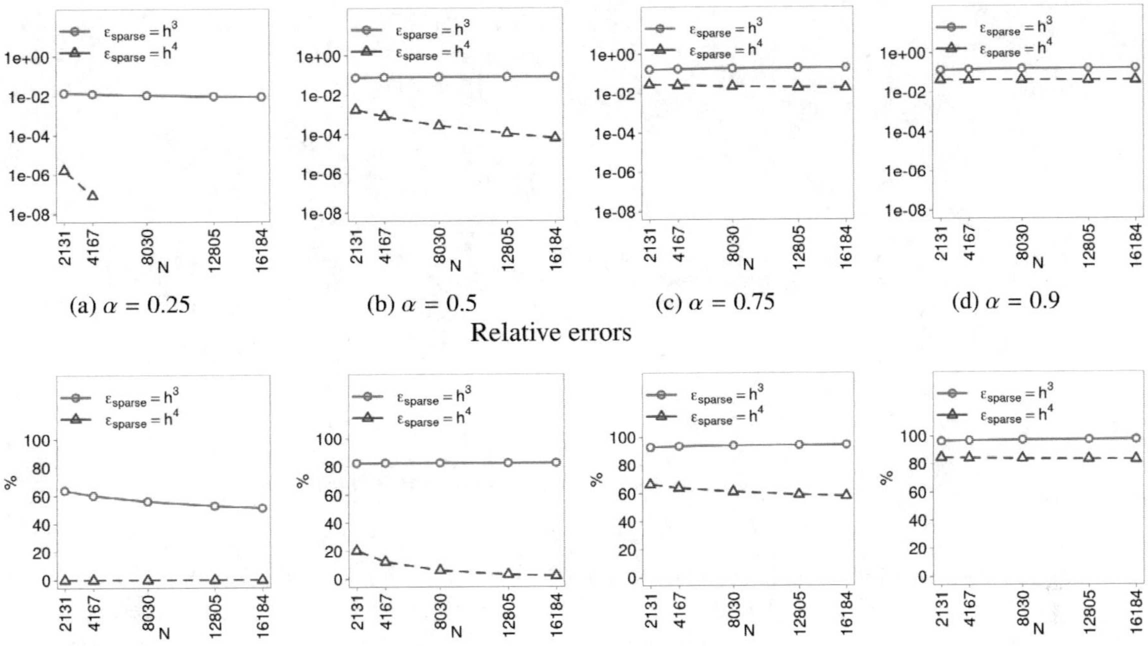

(a) $\alpha = 0.25$ (b) $\alpha = 0.5$ (c) $\alpha = 0.75$ (d) $\alpha = 0.9$

Relative errors

(e) $\alpha = 0.25$ (f) $\alpha = 0.5$ (g) $\alpha = 0.75$ (h) $\alpha = 0.9$

Percent eliminated off-diagonal coefficients

Figure 5: Relative errors of the solution of an elliptic problem after sparsification (top) and percent of eliminated diagonal coefficients (bottom).
NOTE: Missing data on Subfigure a for $\varepsilon_{\text{sparse}} = h^4$ is because no coefficients were eliminated for this threshold.

4.2. Parallel experiments with the PARDISO solver

The performance of the MKL PARDISO Sparse solver for symmetric positive definite matrices is studied for the problem. PARDISO uses nested dissection algorithm from the METIS package, followed by the parallel left-right looking numerical Cholesky factorization $PAP^T = LL^T$. We are using the Intel's MKL PARDISO implementation. The numerical experiments are carried on the AVITOHOL supercomputer[1] at the Institute of Information and Communication Technologies at the Bulgarian Academy of Sciences. We are using one server with 2 Intel Xeon E5-2650 v2 8C 2.6GHz CPUs, each with 8 cores for a total of 16 cores. The presented numerical experiments include five different mesh sizes $N \in (2\,131, 16\,184)$ in a square domain Ω and four values of the fractional power of the Laplace operator $\alpha = 0.25,\ 0.5,\ 0.75$ and 0.9.

The sequential and parallel execution times of the sparse solver PARDISO are presented on Fig. 6, 7, 8 and 9 for $\alpha = 0.25, 0.5, 0.75$ and 0.9. Computational times of dense solvers (using the original K) with the dense LU and Cholesky factorizations from the MKL library are presented for comparison. LU factorization has computational complexity $O(N^3)$ and Cholesky has slightly better computational complexity of $\frac{1}{2}O(N^3)$.

For the smallest value $\alpha = 0.25$ we observe that the PARDISO solver has similar sequential times to the dense solvers, falling between LU and Cholesky for both studied thresholds. This can be explained by the very small amount of eliminated coefficients by the sparsification scheme (percent of eliminated coefficients is presented in Fig. 5e). For the parallel experiments the dense solvers show much better performance, which is explained by the better parallel nature of the algorithms.

[1] http://www.hpc.acad.bg/system-1/

Journal of Physics: Conference Series **2675** (2023) 012034 doi:10.1088/1742-6596/2675/1/012034

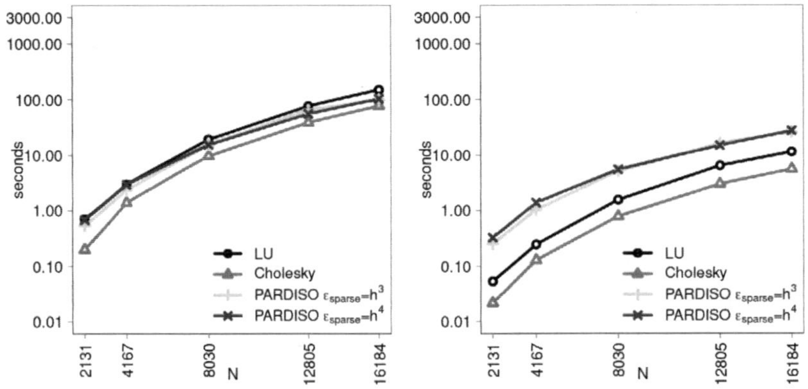

(a) Sequential (b) 16 threads

Figure 6: Execution times for $\alpha = 0.25$

For $\alpha = 0.5$ on Fig. 7 the sequential PARDISO solver with a matrix sparsified with $\varepsilon_{\text{sparse}} = h^3$ has faster computational times than the dense solvers for most sizes of the problem. Experiments with $\varepsilon_{\text{sparse}} = h^4$ have performance between the dense LU and Cholesky factorizations. For the larger values of $\alpha = 0.75$ and 0.9 this trend is more pronounced and for the sequential experiments for $\alpha = 0.90$ the PARDISO solver has better computational times for all experiments. This is because a larger amount of the off diagonal coefficients of K have relatively small values and are eliminated (see Fig. 5gh). This however comes at the cost of larger error of the solution (see Fig. 5cd).

The parallel experiments for $\alpha = 0.5, 0.75$ and 0.9 also show better performance of the sparse experiments with the higher values of α. For $\alpha = 0.5$ and the largest $N = 16184$ parallel experiments the sparse solver with $\varepsilon_{\text{sparse}} = h^3$ has better times than the LU decomposition. For $\alpha = 0.75$ this is more pronounced and for $\alpha = 0.9$ the parallel PARDISO experiments show better times with $\varepsilon_{\text{sparse}} = h^3$ than the Cholesky factorization for $N \geq 8030$, while experiments with $\varepsilon_{\text{sparse}} = h^4$ have similar performance as the LU based solver.

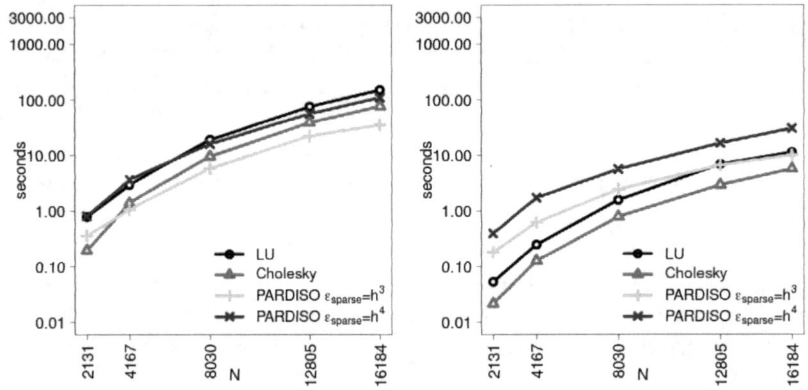

(a) Sequential (b) 16 threads

Figure 7: Execution times for $\alpha = 0.5$

The lower parallel performance of the PARDISO solver can be explained by the more complex and less parallelizable structure of the algorithm. The parallel speed-up of the dense solvers (Cholesky and

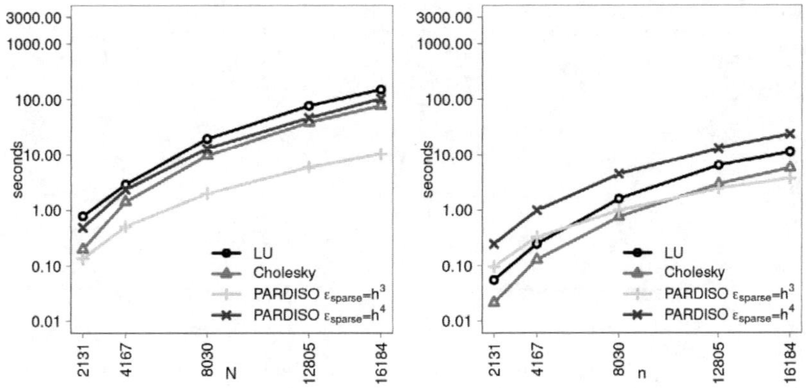

(a) Sequential (b) 16 threads

Figure 8: Execution times for $\alpha = 0.75$

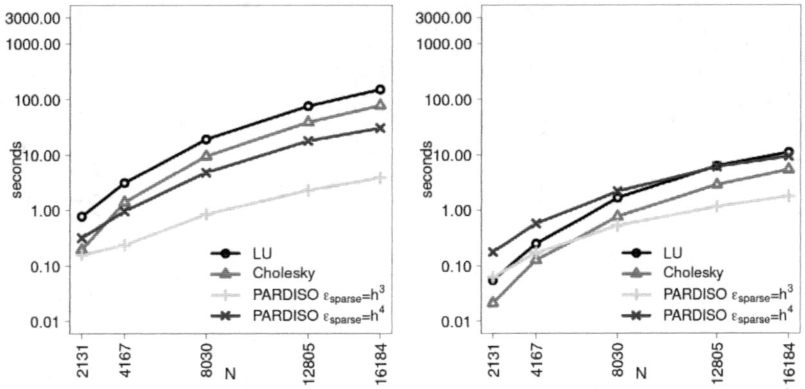

(a) Sequential (b) 16 threads

Figure 9: Execution times for $\alpha = 0.9$

LU) and the sparse PARDISO solver with the tested sparsification thresholds is presented on Fig. 10 for $\alpha = 0.75$. For other values of α the speed-up is similar and not shown for brevity. Both the Cholesky and LU based solvers have very good parallel speed-up with ~14 to ~15 times faster performance with 16 threads. At the same the PARDISO solver achieves a modest ~3 to ~5 speed-up. This is because of the more complex and harder to parallelize structure of the PARDISO solver. This property of the sparse solver explains the better performance of the dense solvers in many of the parallel experiments and especially for the smaller values of α, where less off-diagonal coefficients are eliminated.

It should be noted that for the larger experiments the sparse solver "catches up" and even exceeds the performance of the dense solvers. For larger problems the presented method is likely to show better performance than the LU and Cholesky based solvers.

5. Concluding Remarks

The stiffness matrix K produced by the finite element discretization of the fractional Laplacian has a large amount of relatively small off-diagonal coefficients. They can be eliminated using a suitable chosen sparsification threshold to obtain a sparse approximation of K with some loss of accuracy. In this work we use a relative lumping scheme in order to obtain the sparse approximation of K. Two relative

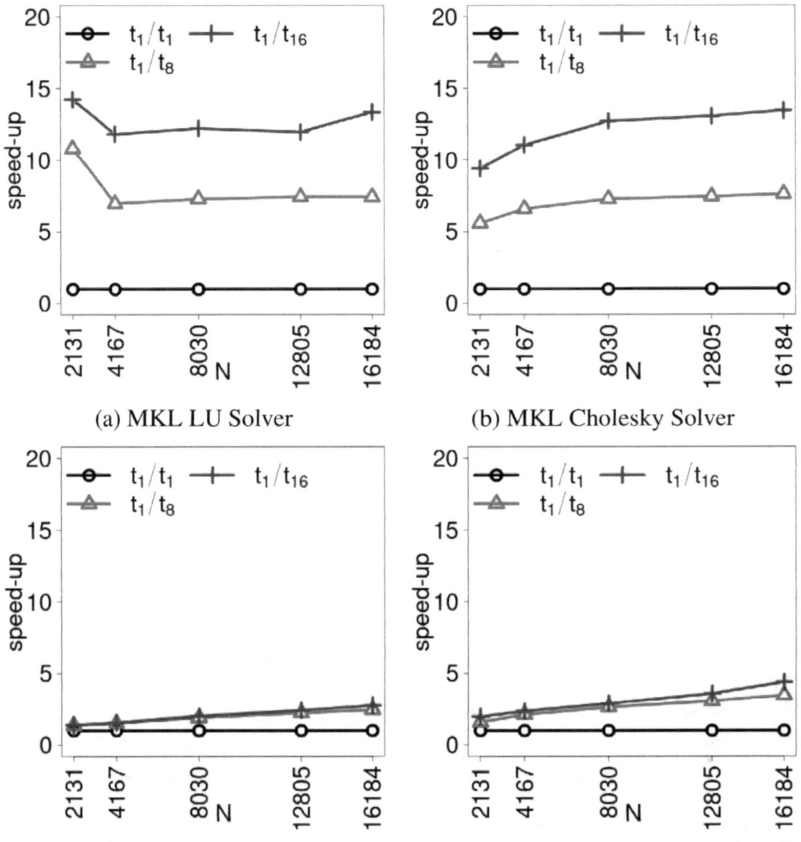

(a) MKL LU Solver (b) MKL Cholesky Solver

(c) MKL PARDISO $\varepsilon_{\text{sparse}} = h^3$ (d) MKL PARDISO $\varepsilon_{\text{sparse}} = h^4$

Figure 10: Parallel speed-up of the MKL PARDISO solver and a comparison with dense solvers based on LU and Cholesky. All figures are for $\alpha = 0.75$.

threshold values based on the properties of the mesh are studied $\varepsilon_{\text{sparse}} = h^3$ and h^4. The accuracy loss induced from the elimination of small coefficients of K is analyzed. Comparative performance analysis is carried out with the MKL PARDISO sparse solver, while the dense solvers based on LU and Cholesky factorizations are used as a reference.

For the larger values of α and for larger sized problems this sparsification method along with the sparse solver PARDISO have better sequential performance than the dense solvers. This is not surprising as at $\alpha = 1$ the stiffness matrix is sparse. As the fractional power approaches 1, the fractional stiffness matrix becomes closer to sparse (in the sense that a large amount of coefficients are very close to zero). This facilitates the sparsification method studied in this work.

The parallel performance of the sparse solver for this problem has lower performance than the dense solver. This is explained with the less parallelism in the algorithm of the sparse solver. As a result only with the larger threshold $\varepsilon_{\text{sparse}} = h^3$ and for the larger α we obtain better parallel performance with the PARDISO solver.

Possible avenue of further research are the use of other sparse solvers, for example iterative solvers based on Krylov subspace methods. A better criteria for the sparsification threshold involving the fractional power α is another possibility.

Acknowledgments
The first author is partially supported by Grant No BG05M2OP001-1.001-0003, financed by the Science and Education for Smart Growth Operational Program (2014–2020) and co-financed by the European Union through the European structural and Investment funds.
This work was partially supported by the Bulgarian Ministry of Education and Science under the National Research Programme "Young scientists and postdoctoral students -2" approved by DCM 206 / 07.04.2022
We acknowledge the provided access to the e-infrastructure of the NCHDC – part of the Bulgarian National Roadmap on RIs, with the financial support by the Grant No D01-387/18.12.2020.

References
[1] Klafter J and Sokolov I 2005 Anomalous diffusion spreads its wings *Physics World* vol 18 (IOP Publishing) pp 29–32 URL `https://doi.org/10.1088/2058-7058/18/8/33`

[2] Metzler R and Klafter J 2004 The restaurant at the end of the random walk: recent developments in the description of anomalous transport by fractional dynamics *Journal of Physics A: Mathematical and General* vol 37 (IOP Publishing) pp R161–R208 URL `https://doi.org/10.1088/0305-4470/37/31/r01`

[3] Harizanov S, Lazarov R, Margenov S, Marinov P and Pasciak J 2020 Analysis of numerical methods for spectral fractional elliptic equations based on the best uniform rational approximation *Journal of Computational Physics* **408** 109285 ISSN 0021-9991 URL `https://www.sciencedirect.com/science/article/pii/S0021999120300590`

[4] Slavchev D, Margenov S and Georgiev I G 2020 On the application of recursive bisection and nested dissection reorderings for solving fractional diffusion problems using hss compression *AIP Conference Proceedings* **2302** 120008 ISSN 0094-243X URL `https://doi.org/10.1063/5.0034506`

[5] Slavchev D and Margenov S 2023 Study of sparsification schemes for the fem stiffness matrix of fractional diffusion problems *Large-Scale Scientific Computing* ed Lirkov I and Margenov S (Cham: Springer International Publishing) in press

[6] Harizanov S, Lazarov R and Margenov S 2020 A survey on numerical methods for spectral space-fractional diffusion problems *Fractional Calculus and Applied Analysis* **23** 1605 – 1646 URL `https://doi.org/10.1515/fca-2020-0080`

[7] Harizanov S and Margenov S 2019 Comparison analysis on two numerical solvers for fractional laplace problems *Studies in Computational Intelligence* **793** 163 – 176 URL `https://doi.org/10.1007/978-3-319-97277-0_13`

[8] Acosta G and Borthagaray J 2017 A Fractional Laplace Equation: Regularity of Solutions and Finite Element Approximations *SIAM Journal on Numerical Analysis* vol 55 pp 472–495 URL `https://doi.org/10.1137/15M1033952`

[9] Acosta G, Bersetche F M and Borthagaray J P 2017 A short FE implementation for a 2d homogeneous Dirichlet problem of a fractional Laplacian *Computers & Mathematics with Applications* **74** 784 – 816 ISSN 0898-1221

Advanced Stochastic Approaches for Fredholm Integral Equations

Venelin Todorov[1,2], Slavi Georgiev[1,3]

[1]Department of Information Modeling, Institute of Mathematics and Informatics, Bulgarian Academy of Sciences, 8 Acad. Georgi Bonchev Str., 1113 Sofia, BG

[2]Department of Parallel Algorithms and Machine Learning with a Laboratory in Neurotechnologies, Institute of Information and Communication Technologies, Bulgarian Academy of Sciences, 25A Acad. Georgi Bonchev Str., 1113 Sofia, BG

[3]Department of Applied Mathematics and Statistics, University of Ruse, 8 Studentska Str., 7004 Ruse, BG

E-mail: vtodorov@math.bas.bg,venelin@parallel.bas.bg, sggeorgiev@math.bas.bg,sggeorgiev@uni-ruse.bg

Abstract. Integral equations are of high applicability in different areas of applied mathematics, physics, engineering, geophysics, electricity and magnetism, kinetic theory of gases, quantum mechanics, mathematical economics, and queuing theory. That is why it is reasonable to develop and study efficient and reliable approaches to solve integral equations. For multidimensional problems the existing biased stochastic algorithms based on evaluation of finite number of integrals will suer more from the effect of high dimensionality, because they are based on quadrature points. So we need advanced unbiased algorithms for solving the multidimensional problem which is developed in this paper. A new unbiased stochastic method for solving multidimensional Fredholm integral equations of second kind is proposed and analysed. We compared the newly proposed unbiased algorithm with the old unbiased stochastic algorithm for the one dimensional problem and multidimensional problem.

1. Introduction

Integral equations stand as fundamental pillars in the realm of mathematical analysis, underpinning a vast range of scientific and engineering pursuits. Particularly, when these equations are expanded into higher dimensions, they become the foundation of several computational paradigms, offering solutions to complex challenges, from the intricacies of quantum physics to the nuances of electromagnetic dynamics. With the ongoing expansion of computational prowess and deepening of theoretical understanding, the importance of deciphering high-dimensional integral equations cannot be overstated, as they provide nuanced perspectives and catalyze breakthroughs across diverse scientific fields.

As expounded in [1, 2, 3, 6], the biased approach is elucidated. Within this context, the objective is to identify a stochastic variable whose expected outcome resonates with the inferred solution of the underlying problem. This inferred solution emerges from an elongated Liouville-Neumann series, encapsulating a considerable number of terms. On the flip side, the unbiased approach is grounded in the construction of a specific stochastic variable, with the overarching principle being that this variable's average converges, over time, to the genuine solution of the problem in question.

Shifting focus to the Fredholm integral equation, particularly of the second kind, it is given by:

$$u(\mathrm{x}) = \int_\Omega k(\mathrm{x}, \mathrm{x}')u(\mathrm{x}')\mathrm{dx}' + f(\mathrm{x}).$$

An alternative representation of this is:

$$u = \mathcal{K}u + f \ \ \text{(where } \mathcal{K} \text{ denotes an integral operator)}.$$

Within these equations, the function $k(\mathrm{x}, \mathrm{x}')$ resides within the space $\mathrm{L}_2(\Omega \times \Omega)$. Concurrently, $f(\mathrm{x})$ is ascribed to $\mathrm{L}_2(\Omega)$. Both functions are predefined. The crux of the problem revolves around determining $u(\mathrm{x})$, which belongs to $\mathrm{L}_2(\Omega)$. Notably, both x and x′ belong to the domain Ω, a subset of $\mathbb{R}^n$. It is imperative to note that in this discourse, Ω is demarcated as a confined region.

One of the primary obstacles in this field is to identify linear functionals related to the solution $u(\mathrm{x})$. Specifically, our focus is on:

$$J(u) = \int \varphi(\mathrm{x})u(\mathrm{x})\mathrm{dx} = (\varphi, u),$$

where $\varphi(\mathrm{x})$, situated in $L_2(\Omega)$, is predetermined.

To embark on our methodology, we formulate a stochastic path, termed a Markov chain, symbolized as T_i. This chain, with a length of i, initiates from the position x_0 and navigates through the confines of Ω:

$$T_i : \mathrm{x}_0 \to \mathrm{x}_1 \to \ldots \to \mathrm{x}_i.$$

The formation of this path heavily relies on the fundamental $\pi(\mathrm{x})$ and the shifting $p(\mathrm{x}, \mathrm{x}')$ probabilities. Both $\pi(\mathrm{x})$ and $p(\mathrm{x}, \mathrm{x}')$ conform to specific benchmarks. They maintain non-negativity and are recognized as suitable, as highlighted in [2], concerning the function $\varphi(\mathrm{x})$ and the kernel $k(\mathrm{x}, \mathrm{x}')$. Furthermore, they adhere to:

$$\int_\Omega \pi(\mathrm{x})\mathrm{dx} = 1$$

and

$$\int_\Omega p(\mathrm{x}, \mathrm{x}')\mathrm{dx}' = 1$$

for any x that belongs to the subset Ω within $\mathbb{R}^n$. Drawing from the revelations in [2], the shifting likelihood of the Markov chain, $p(x, x')$, is chosen to reflect the amplitude of $k(x, x')$.

This suggests that the connection between $p(x, x')$ and $k(x, x')$ is delineated as $p(x, x') = c|k(x, x')|$. The constant c is extracted from the relation:

$$c = \left(\int_\Omega |k(x, x')|\mathrm{d}x'\right)^{-1} = 1 \ \ \text{for each} \ \ x \in \Omega.$$

On the basis of these elementary premises, we deduce:

$$\mathbf{E}\theta_i[\varphi] = (\varphi, u^{(i)}), \quad \text{where} \quad \theta_i[\varphi] = \frac{\varphi(\mathrm{x}_0)}{\pi(\mathrm{x}_0)} \sum_{j=0}^{i} W_j f(\mathrm{x}_j).$$

Additionally:

$$\text{Note that} \quad W_0 = 1, \quad W_j = W_{j-1}\frac{k(\mathrm{x}_{j-1}, \mathrm{x}_j)}{p(\mathrm{x}_{j-1}, \mathrm{x}_j)}, \quad j = 1, \ldots, i.$$

From the above calculations, it is clear that the Monte Carlo technique, when introducing bias, approximates $(\varphi, u^{(i)})$ as:

$$(\varphi, u^{(i)}) \approx \frac{1}{N}\sum_{l=1}^{N} \theta_i[\varphi]_l.$$

Put simply, the Monte Carlo approximation is a potent method to estimate the linkage between φ and $u^{(i)}$, drawing upon the established associations between our functions and variables [7].

Consequently, we can interpret the random variable $\theta_i[\varphi]$ as a skewed approximation of our desired outcome (φ, u) when i reaches a substantial magnitude. The inherent statistical variability of this approximation falls into the category $\mathcal{O}(N^{-1/2})$. Here, N represents the aggregate count of chains under review, and each distinct value $\theta_i[\varphi]_l$ mirrors the assessment of $\theta_i[\varphi]$ over its particular l-th chain.

Shifting to the unbiased viewpoint, the stochastic path, or Markov chain, represented as T_i, which was previously detailed, displays endless traits. The Unbiased Stochastic Approach (USA) emerges as a comprehensive strategy to weave the methodologies tailored for linear algebraic equation systems, as highlighted in [4], into the domain of integral equations. In juxtaposition with the discrete Markov chain presented in [4], it is succeeded by its continuous variant. This seamless nature is profoundly shaped by the transition probabilities anchored in the kernel $k(\mathrm{x}, \mathrm{x}')$. Our exposition will encompass both contexts; the preliminary being the constrained case where $0 \le k(\mathrm{x}, \mathrm{x}') \le 1$ and where the expression $1 - k(\mathrm{x}, \mathrm{x}')$ deciphers the absorption pace, and the subsequent context broaching more sophisticated kernel configurations.

2. Unbiased stochastic approach (USA)

In this section, our focus shifts to the USA as detailed in [3]. The essence of this unprejudiced approach is rooted in expressing the desired functional as an endless series. This method resonates with the approaches adopted in our earlier investigation, as referenced in [4]. Nonetheless, a clear differentiation emerges in the current scenario, where the state space metamorphoses from being discretely categorized to continuously defined. Consequently, the transition density function, alongside the absorption probability, has transitioned from its formerly discrete characterization.

2.1. Traditional unbiased methodology for a basic scenario

Our discussion narrows down to the situation marked by the condition $0 \le k(\mathrm{x}, \mathrm{x}') \le 1$. In this case, the expression $1 - k(\mathrm{x}, \mathrm{x}')$ stands as a depiction for the absorption's rate of occurrence.

Algorithm 1 *Unbiased stochastic approach for* $0 \le k(\mathrm{x}, y) \le 1, \quad \mathrm{x}, \mathrm{y} \in [0, 1]^n$*:*

1. *Setup Phase*
   Input *the kernel* $k(\mathrm{x}, \mathrm{y})$*, the designated function* $f(\mathrm{x})$*, and the count of stochastic trajectories* N*.*
2. *Execution Phase:*

   2.1. ***Assign*** *score=0*
   2.2. ***Assign*** $\mathrm{x}_0 = \mathrm{x}, test = 1$
   For $j = 1, \ldots, N$
   While $(test \ne 0)$
   $score = score + f(\mathrm{x}_0)$
   $U = random[0, 1]^n, \quad V = random[0, 1]$
   If $k(\mathrm{x}_0, U) < V$ ***then***
   $test = 0$
   Else
   $\mathrm{x}_0 = U$

EndIf
EndWhile
EndFor

3. *Derive the solution:*
$u(\mathrm{x}) = \dfrac{score}{N}$

2.2. Traditional unbiased methodology for the universal scenario
A comprehensive exposition of the USA algorithm tailored for broader kernel types is extensively detailed in [3].

Algorithm 2 *Unbiased stochastic technique for the universal scenario:*

1. *Setup Phase*
Input *the kernel* $k(\mathrm{x}, \mathrm{y})$, *the designated function* $f(\mathrm{x})$, *and the count of stochastic trajectories* N.
2. *Execution Phase:*

2.1. ***Assign*** *score=0*
2.2. ***For*** $j = 1, \ldots, N$
2.3. ***Assign*** $\mathrm{x}_0 = \mathrm{x}, test = 1$, $prod = 1$
While $(test \neq 0)$
$score = score + f(\mathrm{x}_0) * prod$
$U = random(0,1)^n$
If $|k(\mathrm{x}_0, U)| > 1$ ***then***
$prod = prod * k(\mathrm{x}_0, U)$
else
$\mathrm{x}_0 = U$
$V = random[0, 1]$
If $|k(\mathrm{x}_0, U)| < V$ ***then***
$test = 0$
Else
$prod = prod * sign\,(k(\mathrm{x}_0, U)), \quad \mathrm{x}_0 = U$
EndIf
EndIf
EndWhile
EndFor

3. *Derive the outcome:*
$u(\mathrm{x}) = \dfrac{score}{N}$

3. Novel unbiased stochastic approach (NUSA)

3.1. New unbiased methodology for a basic scenario
In this section, we introduce a fresh neutral stochastic approach, labeled as NUSA. Starting from a foundational setting, we assume the correlation $0 \leq k(\mathrm{x}, \mathrm{x}') \leq 1$ is consistently true for all combinations of values x, x' contained within the boundary $\Omega \equiv [0,1]^n$, where $\mathrm{x} = (x^1, \ldots, x^n)$ and $\mathrm{x}' = (x^{(1)'}, \ldots, x^{(n)'})$.

Assuming further that $Y \equiv U[0,1]^n$ is a random variable distributed uniformly within the n-dimensional unit cube, we then embrace the stochastic expression:

$$u(\mathrm{x}) = E\{k(\mathrm{x}, Y)u(Y) + (1 - k(\mathrm{x}, Y))\} + f(\mathrm{x}).$$

This equation suggests a blueprint for $u(\mathrm{x})$, reflecting strategies elaborated in [4] designed for linear equation systems. Within this structure, the absorption probability is represented by $1 - k(\mathrm{x}, Y)$. The modifier or 'score' in this scenario is represented by $f(\mathrm{x})$. If the trajectory bypasses absorption, it remains and is pinpointed at location Y. Consider a stochastic journey or a Markov chain, symbolized as T_i, which spans a length of i. This chain originates from a state labeled as $X = \mathrm{x} = \mathrm{x}_0$ within the spatial boundary Ω. Adding to this framework, we introduce a concluding state, represented as ∂, located beyond the boundary Ω and situated within the expansive $\mathbb{R}^n$. This end state symbolically stands for absorption. This forms the trajectory described as:

$$T_i : \mathrm{x}_0 \to \mathrm{x}_1 \to \ldots \to \mathrm{x}_i.$$

The probabilistic dynamics of this procedure can be outlined as:

$$P(\mathrm{x}_{i+1} \in \Omega | \mathrm{x}_i \neq \partial) = k(\mathrm{x}, U_i) P(U_i \in \Omega),$$

$$P(\mathrm{x}_{i+1} = \partial | \mathrm{x}_i \neq \partial) = 1 - k(\mathrm{x}, U_i),$$

and

$$P(\mathrm{x}_{i+1} = \partial | \mathrm{x}_i = \partial) = 1, \quad \text{and} \quad \mathrm{f}(\partial) = 0.$$

From the above, $u(\mathrm{x})$ can be expressed as:

$$u(\mathrm{x}) = E\left(\sum_{i=0}^{\infty} f(\mathrm{x}_i) | \mathrm{x}_0 = \mathrm{x}\right) = E\left(\sum_{i=0}^{\tau} f(\mathrm{x}_i) | \mathrm{x}_0 = \mathrm{x}\right),$$

with

$$\tau = \inf_{i \geq 0}(\mathrm{x}_i) = \partial.$$

Conceptually, this expression can be reformulated and extended as:

$$\begin{aligned} u(\mathrm{x}) = E\left(\sum_{i=0}^{\infty} f(\mathrm{x}_{i+1}) | \mathrm{x}_0 = \mathrm{x}\right) + f(\mathrm{x}) &= E\left(E\left(\sum_{i=0}^{\infty} f(\mathrm{x}_{i+1}) | \mathrm{x}_1\right)\right) + f(\mathrm{x}) \\ &= \int_{\Omega} k(\mathrm{x}, \mathrm{y}) u(\mathrm{y}) d\mathrm{y} + f(\mathrm{x}). \end{aligned} \tag{1}$$

Based on this foundational understanding, we can detail the behavior of the algorithm under these circumstances.

Algorithm 3 *Unbiased method NUSA for the basic scenario:*

1. Setup Phase
  Input *the kernel* $k(\mathrm{x}, \mathrm{y})$, *the function* $f(\mathrm{x})$, *total random paths* N *and evaluation point* x_0.

2. Execution Phase:

3. **Assign** $S = 0$

4. **For** $k = 1$ **to** N **do** *(MC cycle)*
  4.1 weight $= 1$; $x = x_0$; $P_d = 0.5$.
  4.2. test $= 0$.
  4.5. **while** *test* $\neq 1$ **do**:
    4.5.1. U=random(0,1).
    4.5.2. **If** $U < P_d$ **then** *test=1;*
    4.5.3. **Else**
      4.5.3.1. select V *based on the probability distribution function* $|K(x, \cdot)|$.

4.5.3.2. **If** $K(x, V) < 0$ **then** $weight = weight * \frac{(-1)}{(1-P_d)}$; $x = V$;
4.5.3.3. **EndIf**
4.5.4. **EndIf**
4.6. **EndWhile**
4.7. $weight = weight * \frac{f(x)}{P_d}$; $S = S + weight$
5. **EndFor**
6. Derive the outcome $= \dfrac{S}{N}$.

3.2. New unbiased methodology NUSA for the universal scenario
Let us go in-depth into describing the strategy when the kernel $k(\mathrm{x}, \mathrm{y})$ might have negative magnitudes. We also consider scenarios where the absolute strength of $k(\mathrm{x}, \mathrm{y})$ is at least 1. Nevertheless, we maintain the consistent requirement that $k(\mathrm{x}, \mathrm{y})$ belongs to $\mathrm{L}_2(\Omega \times \Omega)$ and its associated Neumann series converges.

A crucial understanding from this context is the intrinsic link between the absorption likelihood at each phase, represented by p_{abs}, and the spans of the trajectories. In essence, as p_{abs} increases, trajectory spans proportionately decrease.

Algorithm 4 *Unbiased method NUSA for the universal scenario:*

1. Setup Phase
Input *the kernel* $k(\mathrm{x}, \mathrm{y})$, *the function* $f(\mathrm{x})$, *total random paths* N, *and assessment point* x_0.
2. Execution Phase:
3. **Assign** $S = 0$
4. **For** $k = 1$ **to** N **do** *(MC cycle)*
4.1 $weight = 1$; $x = x_0$; $P_d = 0.5$.
4.2. $test = 0$.
4.5. **While** $test \neq 1$ **do**:
4.5.1. *U=generate(0,1).*
4.5.2. **If** $U < P_d$ **then** *test=1;*
4.5.3. **Else**
4.5.3.1. pick V *using the probability distribution function* $M(x, V)$;
4.5.3.2. $weight = weight * \frac{K(x,V)}{M(x,V)(1-P_d)}$;
4.5.3.3. $x = V$;
4.5.4. **EndIf**
4.6. **EndWhile**
4.7. $weight = weight * \frac{f(x)}{P_d}$; $S = S + weight$
5. **EndFor**
6. Derive the outcome $= \dfrac{S}{N}$.

4. Numerical simulations
In this section, we discuss the computational tests, divided into two separate subsections. The first one highlights a comparison between the conventional methodology and the newly presented unbiased stochastic algorithm. The following subsection delves deeper into the intricacies of the NUSA algorithm.

4.1. Probabilistic techniques for single-dimensional problems

To determine the effectiveness of our computational techniques, we apply them to a specific case in a one-dimensional framework, as mentioned in [3, 5]:

$$u(x) = \int_{\Omega} k(x, x')u(x')\mathrm{d}x' + f(x), \qquad \Omega = [0, 1]. \tag{2}$$

The kernel along with related functions are described as:

$$k(x, x') = x^2 \mathrm{e}^{x'(x-1)}, \tag{3}$$

$$f(x) = x + (1 - x)\mathrm{e}^{x}, \tag{4}$$

$$\varphi(x) = \delta(x - x_0). \tag{5}$$

This sample problem provides a solution represented by $u(x) = e^x$. Our main goal is to estimate the value of (φ, u), where the function $\varphi(x)$ is denoted by the Dirac delta function centered at $x_0 = 0.5$.

Table 1. Relative error comparison for 1000 trajectories for computing various $u(\mathrm{x}_0)$ values for equations (2)-(5).

x_0	0.1	0.2	0.3	0.4	0.5	0.6	0.7	0.8	0.9
USA	0.4092	0.3426	0.2651	0.1783	0.0839	0.0167	0.1226	0.2326	0.3459
NUSA	8.8545e-5	2.9906e-4	5.6785e-4	8.5177e-4	0.0011	0.0014	0.0015	0.0016	0.0016

Analyzing the compiled results, it is evident that the innovative NUSA algorithm for Fredholm integral equations outperforms its predecessor, the USA method. This advantage becomes increasingly pronounced at data points that are considerably close to the left boundary of the tests 0.1. In this context, the NUSA approach outstrips the USA method by at least two magnitudes.

4.2. Probabilistic techniques for multi-dimensional problems

Tackling the multi-dimensional Fredholm integral equation of the second kind has always posed a formidable challenge in computational strategies. We shift our attention to a particular multi-dimensional issue with dimensions $n > 1$, as highlighted in [3]:

$$u_n(\mathrm{x}) = \int_0^1 \dots \int_0^1 \prod_{j=1}^{n} \{k(x^{(j)}, x^{(j)\prime})\} u(\mathrm{x}') \prod_{j=1}^{n} \mathrm{d}x^{(j)\prime} + f_n(\mathrm{x}), \tag{6}$$

where

$$\mathrm{x} \equiv (x^{(1)}, \dots, x^{(n)}),\ \mathrm{x}' \equiv (x^{(1)\prime}, \dots, x^{(n)\prime}),\ \prod_{j=1}^{n} \mathrm{d}x^{(j)\prime} = \mathrm{dx}',$$

Within this framework, the constituents are expressed as:

$$k(x^{(j)}, x^{(j)\prime}) = \beta_j \left(2x^{(j)\prime} - 1\right) \{x^{(j)}\}^2 \exp\{x^{(j)\prime}(x^{(j)} - 1)\},\ \ \beta = \prod_{j=1}^{n} \beta_j, \tag{7}$$

$$f_n(x) = \mathrm{e}^{\sum_{j=1}^{n} x^{(j)}} + (-1)^{(j-1)}\beta_n(2\mathrm{e}^{x^{(j)}} - 2 - x^{(j)}\mathrm{e}^{x^{(j)}} - x^{(j)}), \tag{8}$$

$$\varphi(\mathrm{x}) = \delta(\mathrm{x} - \mathrm{x}_0), \tag{9}$$

Table 2. Values, relative errors, and processing tikes for deriving $u(\mathrm{x}_0)$ at $(0.75, 0.75, \ldots, 0.75)$ with varying kernel norms for 1000 trajectories

β	Procedure	Metrics	Problem size		
			$n = 3$	$n = 5$	$n = 10$
$\frac{0.1}{\|K\|_{L_2}}$		Accurate value	9.4877	42.5211	1808.0424
	USA method	Percentage error	**0.0020**	**0.0012**	0.0660
		Processing span	0.010465	0.010016	0.011934
	NUSA method	Percentage error	0.0038	0.0088	**0.0133**
		Processing span	0.032336	0.040444	0.072550
$\frac{0.5}{\|K\|_{L_2}}$		Accurate value	9.4877	42.5211	1808.0424
	USA method	Percentage error	0.0396	0.0831	0.2756
		Processing span	0.012185	0.012483	0.011559
	NUSA method	Percentage error	**4.1e-4**	**0.0086**	**0.0133**
		Processing span	0.032047	0.041941	0.067919
$\frac{0.95}{\|K\|_{L_2}}$		Accurate value	9.4877	42.5211	1808.0424
	USA method	Percentage error	0.3005	0.1529	0.1834
		Processing span	0.013337	0.012670	0.012186
	NUSA method	Percentage error	**0.0034**	**0.0085**	**0.0133**
		Processing span	0.032936	0.041976	0.069516

with x_0 chosen as $(1.0, 1.0, \ldots 1.0)$ for our numerical examinations.

The equation's solution is represented by the combination of the continuous function $u(\mathrm{x})$ and the Dirac delta function $\delta(\mathrm{x} - \mathrm{x}_0)$, located at the point x_0. In particular, the values of $u(\mathrm{x})$ at x_0 can be obtained by the convolution of the inhomogeneous term $f(\mathrm{x})$ and the Green function. Furthermore, the calculation procedure becomes notably more complex in higher dimensions due to the problem's inherent complexity.

With the kernel chosen as mentioned in Eq. (7), three separate cases, $n = 3$, $n = 5$, and $n = 10$, are evaluated. Table 2 demonstrates the variance in value acquisition for $u(\mathrm{x}_0)$ for these instances with varying kernel norms and two computational approaches: the USA and the NUSA method. For each n, the kernel norm is varied at three distinct levels: $\frac{0.1}{\|K\|_{L_2}}$, $\frac{0.5}{\|K\|_{L_2}}$, and $\frac{0.95}{\|K\|_{L_2}}$.

From the results summarized in Table 2, the NUSA method consistently outperforms the USA technique in terms of accuracy across the different scenarios. While the NUSA method requires a slightly longer processing time, this marginal delay is more than compensated for by its superior accuracy in estimating $u(\mathrm{x}_0)$. This makes the NUSA method preferable for more precise evaluations, especially in higher dimensions.

5. Conclusions

We have presented and analyzed an advanced stochastic method specifically designed for addressing multi-dimensional Fredholm integral equations of the second kind. Compared to its predecessor, this new approach consistently delivers enhanced results for both one-dimensional and multiple-dimensional contexts. Impressively, as the number of dimensions grows, our refined method increasingly distinguishes itself from the earlier model in terms of proficiency.

A central insight from our research indicates that the rise in computational intricacy aligns linearly with the growth in dimensionality. As we progress, our goal is to probe the resilience of this avant-garde technique in situations marked by extremely high-dimensional computational demands.

Further exploration is possibly focusing on the variability that these unbiased methods might manifest. We are particularly keen on discerning the implications of any asymmetries in the kernel on the accuracy of the results. Moreover, a deep dive into the computational intricacy of the USA, particularly concerning the count of floating-point operations and how they correlate with specific problem variables, is planned. In our future investigative pursuits, we are driven to test the concurrent performance of these algorithms on various technological infrastructures, including multi-CPU configurations and GPUs.

Acknowledgments

Slavi Georgiev is supported by the Bulgarian National Science Fund (BNSF) under Project KP-06-M62/1 "Numerical deterministic, stochastic, machine and deep learning methods with applications in computational, quantitative, algorithmic finance, biomathematics, ecology and algebra" from 2022. Venelin Todorov is supported by BNSF under Project KP-06-N52/5 "Efficient methods for modeling, optimization and decision making" and BNSF under Project KP-06-N62/6 "Machine learning through physics-informed neural networks".

References

[1] Curtiss, J. H.: Monte Carlo methods for the iteration of linear operators, J. Math Phys., vol. 32, 209-232, 1954.
[2] Dimov, I.T.: Monte Carlo methods for applied scientists., *World Scientific*, New Jersey, London, Singapore, World Scientific, 2008, 291p., ISBN-10 981-02-2329-3.
[3] Dimov, I.T., Maire, S.: A new unbiased stochastic algorithm for solving linear Fredholm equations of the second kind. Adv Comput Math 45, 1499-1519 (2019).
[4] Dimov, I.T. , Maire, S., Sellier, J.M.: A New Walk on Equations Monte Carlo Method for Solving Systems of Linear Algebraic Equations, *Appl. Math. Modelling*, 2014, http://dx.doi.org/10.1016/j.apm.2014.12.018
[5] Farnoosh R., Ebrahimi., M.: Monte Carlo Method for Solving Fredholm Integral Equations of the Second Kind, *Appl. Math. and Comput.*, 195 309–315, 2008.
[6] Kalos, Malvin H.; Whitlock, Paula A. (2008). Monte Carlo Methods. Wiley-VCH. ISBN 978-3-527-40760-6.
[7] Sobol., I. M.: *Monte Carlo Numerical Methods*, Nauka, Moscow, 1973, (in Russian).

Air Pollution Modeling of UNI-DEM Model by using Innovative Stochastic Approaches

Venelin Todorov[1,2], Slavi Georgiev[1,3], Ivan Dimov[2]
[1]Department of Information Modeling, Institute of Mathematics and Informatics, Bulgarian Academy of Sciences, 8 Acad. Georgi Bonchev Str., 1113 Sofia, BG
[2]Department of Parallel Algorithms and Machine Learning with a Laboratory in Neurotechnologies, Institute of Information and Communication Technologies, Bulgarian Academy of Sciences, 25A Acad. Georgi Bonchev Str., 1113 Sofia, BG
[3]Department of Applied Mathematics and Statistics, University of Ruse, 8 Studentska Str., 7004 Ruse, BG

E-mail: vtodorov@math.bas.bg, venelin@parallel.bas.bg, sggeorgiev@math.bas.bg, sggeorgiev@uni-ruse.bg, ivdimov@bas.bg

Abstract. In this paper we develop advanced multidimensional sensitivity analysis, based on some innovative stochastic approaches for performing air pollution modelling on a large-scale model of long-range transport of air pollutants. The Unified Danish Eulerian Model (UNI-DEM) is a very important mathematical model, and can be applied in many different studies related to damaging effects caused by high air pollution levels. We shall use it in this paper to get a reliable answer to a some very important questions regrading environmental protection. We develop some advanced Monte Carlo and quasi-Monte Carlo methods, based on special lattice and digital sequences. In this paper we will improve the digital ecosystem modeling by improving the existence stochastic approaches. The computational efficiency (in terms of relative error and computational time) of the advanced Monte Carlo algorithms for multidimensional numerical integration has been studied to analyze the sensitivity of UNI-DEM model output to variation of input emissions of the anthropogenic pollutants and of rates of several chemical reactions. The algorithms will be applied to compute global Sobol sensitivity measures corresponding to the influence of several input parameters on the concentrations of important air pollutants. The study will be done for the areas of several European cities with different geographical locations.

1. Introduction

Sensitivity analysis, as highlighted by various studies [1, 2, 3, 4, 5], plays a pivotal role in comprehending and evaluating the behavior and dependability of mathematical models. These models are utilized for the simulation and representation of intricate systems, encompassing physical processes, biological entities, economic occurrences, and environmental dynamics. Given that these models often encompass numerous input parameters and variables, there's potential for introducing uncertainties and variations in their outputs.

The core objective of sensitivity analysis is to quantitatively assess the impact of alterations in input parameters on model outputs or desired outcomes. Through the methodical manipulation of input parameters and observation of ensuing changes in model outputs, sensitivity analysis yields insights into the relative significance and influence of distinct inputs on the model's behavior. It assists in the identification of key drivers or influential factors that exert substantial effects on the model's predictions or findings.

Certainly, sensitivity analysis is a crucial tool in various fields such as engineering, science, economics, and environmental studies. It plays a vital role in enhancing our understanding of complex models and their behavior under different conditions. Here, let's delve deeper into the concepts and methods associated with sensitivity analysis.

In many real-world situations, mathematical models are used to make predictions, inform decisions, or gain insights into complex systems. However, these models often involve a multitude of input parameters, each contributing to the model's output in a different way. Sensitivity analysis helps address the following questions [6, 7]:

- Identifying Critical Factors: Which input parameters have the most significant influence on the model's outputs? Sensitivity analysis helps identify these critical factors, allowing decision-makers to focus their efforts on controlling or optimizing these factors to achieve desired outcomes.
- Managing Uncertainties: Real-world data is rarely perfect, and there are uncertainties associated with the input parameters. Sensitivity analysis helps quantify the effect of these uncertainties on the model's outputs, providing a more comprehensive understanding of the potential range of outcomes.
- Model Validation: Sensitivity analysis can aid in validating a model's accuracy. By comparing the model's predictions with observed data, analysts can adjust the model's input parameters to ensure it aligns closely with real-world observations.

2. Problem settings

Sobol' indices [8], named after the Russian mathematician Ilya M. Sobol', are quantitative measures used in sensitivity analysis to assess the contribution of individual input parameters and their interactions to the overall variance of a model's output. These indices offer valuable insights into the relative importance of different parameters in influencing the model's behavior, making them a widely employed tool in various fields, including engineering, environmental science, economics, and more.

Sobol' indices [8, 5] provide a comprehensive understanding of the sensitivity of a model to its input parameters by partitioning the total output variance into various components. These components correspond to the variance attributed to individual parameters, interactions between parameters, and higher-order interactions. The indices quantify the proportion of the total variance that each of these components contributes, offering a clear picture of which parameters play significant roles in driving output variability.

First-order Sobol' indices quantify the portion of output variance attributed solely to a single input parameter, independently of other parameters. They reveal the individual impact of each parameter on the model output.

Total Sobol' indices consider the contribution of an input parameter along with its interactions with other parameters. They provide a more comprehensive perspective on the influence of a parameter by accounting for both its direct and indirect effects.

The calculation of Sobol' indices typically involves generating a series of model evaluations by systematically varying the values of the input parameters. This process can be achieved through methods like Monte Carlo simulations or quasi-random sampling techniques. By observing how changes in parameter values affect the output variance, the Sobol' indices can be estimated.

Sobol' indices have numerous applications in various fields [8]:

- Engineering Design: In engineering, Sobol' indices help engineers understand which parameters have the most significant impact on the performance of a design, enabling them to optimize critical factors and achieve desired outcomes.

- Environmental Modeling: In environmental studies, Sobol' indices help assess the contribution of different factors to the uncertainty of predictions related to climate models, pollutant dispersion, and ecological systems.
- Financial Modeling: In finance, Sobol' indices aid in identifying the key drivers of risk and uncertainty in complex financial models, assisting analysts in making better-informed investment decisions.
- Chemical Process Optimization: In chemical engineering, Sobol' indices assist in optimizing processes by highlighting the parameters that affect product quality and process efficiency the most.

Advantages of using Sobol' indices include their ability to capture interactions between parameters, their quantifiable nature, and their suitability for both global and local sensitivity analysis. However, calculating Sobol' indices can be computationally demanding, especially for models with a large number of input parameters or high-dimensional interactions.

Sobol' indices are a valuable tool in sensitivity analysis, providing a systematic and quantitative approach to understanding the influence of input parameters and their interactions on a model's output variance. By deciphering the contributions of different factors to overall variability, Sobol' indices enhance decision-making, model validation, and optimization in various domains.

The numerical sensitivity measure called **t**otal **s**ensitivity **i**ndex (TSI) of input parameter $x_i, \;\; i \in \{1, \ldots, d\}$ is defined in the following way [8]:

$$S_i^{tot} = S_i + \sum_{l_1 \neq i} S_{il_1} + \sum_{l_1, l_2 \neq i, l_1 < l_2} S_{il_1 l_2} + \ldots + S_{il_1 \ldots l_{d-1}}, \quad (1)$$

where S_i is called *the main effect (first-order sensitivity index)* of x_i and $S_{il_1 \ldots l_{j-1}}$ is the j-th order sensitivity index (respectively *two-way interactions* for $j = 2$, and so on) for parameter x_i $(2 \leq j \leq d)$.

The quantities

$$\mathbf{D} = \int_{U^d} f^2(\mathrm{x})\mathrm{dx} - f_0^2, \;\; \mathbf{D}_{l_1 \, \ldots \, l_\nu} = \int_{U^\nu} f_{l_1 \, \ldots \, l_\nu}^2 \mathrm{d}x_{l_1} \ldots \mathrm{d}x_{l_\nu} \quad (2)$$

are called variances (total and partial variances, respectively). The total variance of the model output is partitioned into partial variances [8] in the analogous way as the model function, that is the ANOVA-decomposition:

$$\mathbf{D} = \sum_{\nu=1}^{d} \sum_{l_1 < \ldots < l_\nu} \mathbf{D}_{l_1 \ldots l_\nu}.$$

Based on the above assumptions about the model function and the output variance, the following quantities

$$S_{l_1 \, \ldots \, l_\nu} = \frac{\mathbf{D}_{l_1 \, \ldots \, l_\nu}}{\mathbf{D}}, \quad \nu \in \{1, \ldots, d\} \quad (3)$$

are called Sobol global sensitivity indices [8, 5].

3. The methods

Let I be the desired value of the integral. Let for a given r.v. θ it can be proven that the mathematical expectation satisfies $\mathbf{E}\theta = I$. Suppose that the mean value of n values of θ: $\theta^{(i)}, \; i = 1, \ldots, n$ is considered as a Monte Carlo plain (crude) approximation [?] to the solution: $\bar{\theta}_n = 1/n \sum_{i=1}^{n} \theta^{(i)} \approx I$, where $\theta^{(i)}(i = 1, 2, \ldots, n)$ correspond to values (realizations) of a r.v. θ.

Lattice low discrepancy sequences [10, 11, 12] are a class of deterministic sequences used in numerical simulations, particularly in the field of Monte Carlo methods and quasi-Monte Carlo methods. These sequences play a crucial role in enhancing the efficiency and accuracy of various computational tasks, such as integration, optimization, and solving partial differential equations. They are designed to provide a more uniform and even distribution of points in high-dimensional spaces compared to random sequences, leading to faster convergence and improved precision in numerical computations.

A lattice low discrepancy sequence is generated by transforming a regular grid structure using a lattice, which is essentially a set of linearly independent vectors forming a basis for the grid. The idea is to distribute the points in such a way that they fill the space evenly and minimize the gaps or clusters that can arise with purely random sequences.

The construction of lattice low discrepancy sequences involves the following steps [10, 11, 12]:

- Select a suitable lattice basis. Transform the regular grid points using the lattice basis to obtain the sequence points. Apply a permutation or randomization step to ensure randomness-like properties. Advantages and Applications: Lattice low discrepancy sequences offer several advantages over traditional random sequences:
- Quicker Convergence: These sequences are known to converge faster than random sequences when used in numerical integration or optimization problems. This means that accurate results can be obtained using fewer samples, reducing computational time.
- Lower Variance: The even distribution of points across the space leads to reduced variance in the computed results, resulting in more stable and reliable estimations.
- Deterministic Nature: Lattice sequences are deterministic, meaning that the same sequence will be generated each time the method is applied. This reproducibility is advantageous for validation and debugging.
- Higher Dimensions: Lattice low discrepancy sequences maintain their efficiency even in high-dimensional spaces, where random sequences tend to exhibit the "curse of dimensionality."

While lattice low discrepancy sequences offer significant benefits, they are not universally superior and have some limitations:

- Construction Complexity: Designing an effective lattice and ensuring its quality can be challenging. Selecting an inappropriate lattice can lead to poor results.
- Limited Randomness: Although sequences undergo randomization steps, they may not possess the same level of unpredictability as truly random sequences.

Lattice low discrepancy sequences find application in various fields: In the numerical integration they are used to accurately estimate integrals, especially in high-dimensional spaces, with faster convergence compared to random sequences. In financial mathematics, option pricing and risk assessment, lattice sequences enhance the efficiency of Monte Carlo simulations. In simulation and modeling these sequences aid in solving complex physical, engineering, and scientific problems that involve multiple dimensions and require accurate sampling. In cryptography lattice-based cryptography uses mathematical structures similar to those in lattice low discrepancy sequences to create cryptographic schemes with strong security properties.

In summary, lattice low discrepancy sequences are powerful tools in numerical simulations, offering improved accuracy and efficiency over random sequences, particularly in high-dimensional spaces. Their application extends to a wide range of fields where precision and reliability in numerical computations are essential.

Particularly we will implement lattice low discrepancy sequences based on component by component construction (CBC) method [13] of rank-1 lattices with corresponding generators with prime power of points with product weights and with order dependent weights. The worst case error [13] in the first case (FC) is

$$e^2(Q,K) = -1 + \frac{1}{n}\sum_{k=0}^{n-1}\prod_{j=1}^{d}(1+\gamma_j w(x_j^{(k)})), \gamma_u = \prod_{j\in U}\gamma_j. \tag{4}$$

And the worst case error in the second case (SC) is

$$e^2(Q,K) = \frac{1}{n}\sum_{k=0}^{n-1}\sum_{l=1}^{d}\Gamma_l \sum_{u\subseteq D_d,|u|=l}\prod_{j\in U} w(x_j^{(k)})), \gamma_u = \Gamma_{|u|}. \tag{5}$$

4. Numerical results

The Unified Danish Eulerian Model (UNI-DEM) [14, 15, 16, 17] stands as an advanced and resilient extensive-scale air pollution model utilized for approximating the quantities of multiple contaminants and chemical components present in the atmosphere. It possesses the capability to replicate fluctuations in these quantities across designated time frames, such as months, years, or decades. Encompassing an expansive geographical expanse measuring 4800 × 4800 km, the model covers Europe, the Mediterranean, and portions of Asia and Africa. UNI-DEM takes into account pivotal physical, chemical, and photochemical mechanisms taking place within the atmosphere. Given the dynamic nature of emissions and meteorological conditions, substantial datasets of input information are required.

The model's outcomes furnish valuable insights for a variety of applications spanning environmental conservation, agriculture, and healthcare. The specific focus of our research, which involves UNI-DEM, centers on investigating the model's output sensitivity. We intend to examine how variations in the input emissions of human-made pollutants affect the mean monthly levels of various noteworthy pollutants.

The first goal of our particular research of UNI-DEM, is to study the sensitivity of the output (in terms of mean monthly concentrations of several important pollutants) with respect to variation of input emissions of the anthropogenic pollutants. The anthropogenic emissions are determined as vectors with 4 different components $E = (E^N, E^C, E^S, E^A)$, given separately in the input data flow. These components correspond to 4 different groups of pollutants as follows:

- E^N – nitrogen oxides ($NO + NO_2$);
- E^C – anthropogenic hydrocarbons;
- E^S – sulphur dioxide (SO_2);
- E^A – ammonia (NH_3).

The initial stage of computations includes producing the necessary input information tailored to the particular sensitivity analysis investigation. In our situation, this encompasses executing a sequence of trials with UNI-DEM by introducing particular adjustments to the emissions data.

The commonly employed outcomes from UNI-DEM are the average monthly levels of a specified set of perilous chemical components at different grid locations within the computational region. These levels rely on the specific chemical mechanism applied. In the context of this article, our primary attention is directed toward the subsequent chemical species:

- ammonia (NH_3);
- ozone (O_3);
- ammonium sulphate and ammonium nitrate ($NH_4SO_4 + NH_4NO_3$).

Actually, UNI-DEM produces summarized concentrations of ammonium sulphate and ammonium nitrate. That's why the latter sum of pollutants is considered as one united pollutant in our further study.

With respect to the grid points of the computational domain the three European cities with different climate and level of pollution are selected in [18, 19]: (i) Milan, (ii) Manchester and (iii) Edinburgh.

The quantity f_0 is represented by a 4-dimensional integral and the rest of quantities under consideration are represented by double integrals.

Table 1. Relative error for the evaluation of $f_0 \approx 0.048$.

	PLAIN	**FC**	**SC**
# of samples n	Relative error	Relative error	Relative error
2^{10}	1.02e-02	4.28e-04	6.43e-04
2^{12}	3.42e-03	7.91e-05	1.29e-04
2^{14}	2.51e-03	**1.64e-05**	4.06e-05
2^{16}	1.73e-03	5.21e-06	7.78e-06
2^{18}	4.32e-04	2.57e-06	**1.61e-06**
2^{20}	6.72e-05	5.31e-07	**3.21e-07**
2^{22}	6.46e-05	1.44e-07	4.84e-07
2^{24}	1.63e-05	2.85e-08	4.71e-08

Table 2. Relative error for the evaluation of the total variance $\mathbf{D} \approx 0.0002$.

	PLAIN	**FC**	**SC**
# of samples n	Relative error	Relative error	Relative error
2^{10}	1.15e-01	**1.57e-02**	3.25e-02
2^{12}	2.87e-02	**1.24e-04**	6.05e-03
2^{14}	4.3e-02	7.7e-04	**1.8e-04**
2^{16}	1.76e-02	**2.09e-04**	2.26e-04
2^{18}	1.16e-02	1.19e-04	**9.19e-05**
2^{20}	5.8e-03	3.38e-05	**8.12e-06**
2^{22}	7.36e-04	**7.81e-06**	3.01e-05
2^{24}	2e-03	**9.92e-08**	2.15e-06

Tables 1, 2, 3, 4, and 5 exhibit the relative discrepancies associated with the approximate computation of the values f_0, $\mathbf{D}$, as well as the first and total sensitivity indices. The calculation of f_0 involves a four-dimensional integral, while the other quantities involve eight-dimensional integrals. The most accurate relative error is highlighted in bold font. In general, it can be concluded that our devised lattice low discrepancy sequences FC and SC display significantly superior performance compared to the PLAIN Monte Carlo algorithm.

The second objective of this study is to investigate the sensitivity of ozone concentration concerning variations in the rates of specific chemical reactions: $\#\#\ 1, 3, 7, 22$ (time-dependent)

Table 3. Relative error for estimation of sensitivity indices of input parameters using various Monte Carlo and quasi-Monte Carlo approaches ($n = 2^{16}$).

EQ	RV	**PLAIN**	**FC**	**SC**
S_1	9e-01	8.23e-03	**1.3e-04**	4.52e-04
S_2	2e-04	2.29e+00	**2.28e-02**	1.23e-01
S_3	1e-01	5.32e-02	**6.4e-04**	1.32e-03
S_4	4e-05	3.02e+00	**4.07e-01**	6.38e-01
S_1^{tot}	9e-01	6.98e-03	**6.98e-05**	2.26e-04
S_2^{tot}	2e-04	1.67e+00	**2.96e-02**	1.31e-01
S_3^{tot}	1e-01	6.25e-02	**1.09e-03**	3.19e-03
S_4^{tot}	5e-05	1.14e+00	**3.73e-02**	4.87e-01

Table 4. Relative error for estimation of sensitivity indices of input parameters using various Monte Carlo and quasi-Monte Carlo approaches ($n = 2^{20}$).

EQ	RV	**PLAIN**	**FC**	**SC**
S_1	9e-01	7.87e-04	**6.29e-06**	6.13e-05
S_2	2e-04	5.55e-01	**6.82e-04**	5e-03
S_3	1e-01	3.42e-03	**5.6e-05**	2.38e-04
S_4	4e-05	2.16e-01	**1.59e-02**	3.48e-02
S_1^{tot}	9e-01	5.27e-04	**6.7e-06**	3.29e-05
S_2^{tot}	2e-04	4.43e-01	**4.29e-03**	1.46e-02
S_3^{tot}	1e-01	5.33e-03	**4.53e-05**	4.71e-04
S_4^{tot}	5e-05	1.47e-01	3.9e-03	**1.9e-03**

Table 5. Relative error for estimation of sensitivity indices of input parameters using various Monte Carlo and quasi-Monte Carlo approaches ($n = 2^{24}$).

EQ	RV	**PLAIN**	**FC**	**SC**
S_1	9e-01	2.07e-04	6.69e-07	**4.63e-07**
S_2	2e-04	3.34e-03	1.37e-04	**8.6e-05**
S_3	1e-01	1.36e-03	5.35e-06	**3.2e-07**
S_4	4e-05	1.68e-01	**4.73e-04**	1.27e-03
S_1^{tot}	9e-01	1.67e-04	5.75e-07	**7.62e-08**
S_2^{tot}	2e-04	6.33e-02	1.61e-04	**9.65e-05**
S_3^{tot}	1e-01	1.76e-03	5.92e-06	**2.75e-06**
S_4^{tot}	5e-05	1.53e-02	**9.28e-04**	9.84e-04

and $27, 28$ (time independent) reactions of the condensed CBM-IV scheme ([14]). The simplified chemical equations of these reactions are as follows:

#1 $NO_2 + h\nu \Longrightarrow NO + O$

#3 $O_3 + NO \Longrightarrow NO_2$

#7 $NO_2 + O_3 \Longrightarrow NO_3$

#22 $HO_2 + NO \Longrightarrow OH + NO_2$

#27 $HO_2 + HO_2 \Longrightarrow H_2O_2$

#28 $OH + CO \Longrightarrow HO_2$

It should be noted that ozone might not be directly engaged in all the indicated reactions. Instead, significant compounds that contribute to the creation of ozone are active participants in these reactions.

The central outcome provided by the UNI-DEM model involves computing average monthly levels of different hazardous chemical species or clusters of these species. These levels are calculated according to the particular chemical mechanism and are evaluated at diverse grid locations within the computational area. In this research, the reaction rates of chemicals are regarded as inputs, while the pollutant concentrations are treated as outputs.

Initially, it is important to note that the quantity denoted as f_0 corresponds to a six-dimensional integral, while the remaining sensitivity measures are represented by double integrals.

Table 6. Relative error for the evaluation of $f_0 \approx 0.27$.

# of samples n	**PLAIN** Relative error	**FC** Relative error	**SC** Relative error
2^{10}	6.97e-04	**2.45e-04**	6.74e-04
2^{12}	8.06e-04	1.8300e-04	**8.4e-05**
2^{14}	2.80e-03	**8.47e-06**	3.21e-05
2^{16}	5.95e-04	**4.54e-06**	1.43e-05
2^{18}	7.66e-04	9.69e-07	**9.46e-07**
2^{20}	3.45e-04	**2.14e-07**	2.5e-07
2^{22}	4.7e-05	1.55e-07	**1.1e-07**
2^{24}	8.62e-06	**3.92e-08**	4.56e-08

Table 7. Relative error for the evaluation of the total variance $\mathbf{D} \approx 0.0025$.

# of samples n	**PLAIN** Relative error	**FC** Relative error	**SC** Relative error
2^{10}	7.22e-02	**4.31e-02**	5.27e-02
2^{12}	9.54e-02	**1e-02**	1.22e-02
2^{14}	6.4e-02	1.94e-03	**1.25e-03**
2^{16}	2.97e-02	3.21e-04	**8.28e-05**
2^{18}	7.42e-03	**2.22e-04**	8.04e-04
2^{20}	8.92e-03	**7.24e-05**	1.09e-04
2^{22}	2.21e-03	2.65e-05	**8.92e-06**
2^{24}	1.29e-03	**2.65e-06**	3.80e-06

Tables 6, 7, 8, 9, and 10 display the relative errors associated with the approximate computation of the values f_0, $\mathbf{D}$, as well as the first and total sensitivity indices. The computation of f_0 involves a four-dimensional integral, while the remaining quantities are derived from eight-dimensional integrals. The most accurate relative error is emphasized in bold font. In a broader context, it can be deduced that our developed lattice low discrepancy sequences FC and SC significantly outshine the conventional PLAIN Monte Carlo method.

Table 8. Relative error for estimation of sensitivity indices of input parameters using various MC and QMC approaches ($n \approx 2^{16}$).

EQ	RV	**PLAIN**	**FC**	**SC**
S_1	4e-01	5.61e-03	2.51e-03	**2.19e-03**
S_2	3e-01	4.73e-02	**5.39e-04**	4.13e-03
S_3	5e-02	1.38e-01	**7.61e-03**	9.16e-03
S_4	3e-01	1.05e-02	1.78e-03	**1.13e-04**
S_5	4e-07	8.73e+02	**1.09e+02**	1.52e+02
S_6	2e-02	2.69e-01	2.27e-02	**1.55e-03**
S_1^{tot}	4e-01	2.83e-02	**1.93e-03**	3.66e-03
S_2^{tot}	3e-01	3.63e-02	**1.43e-04**	4.99e-03
S_3^{tot}	5e-02	1.62e-01	5.9908e-03	**5.49e-03**
S_4^{tot}	3e-01	2.47e-02	**1.08e-03**	1.84e-03
S_5^{tot}	2e-04	2.44e+00	5.23e-01	**4.77e-01**
S_6^{tot}	2e-02	2.72e-01	6.19e-02	**2.05e-02**
S_{12}	6e-03	6.31e-01	**3.37e-02**	7.57e-02
S_{14}	5e-03	8.28e-01	4.43e-02	**3.05e-02**
S_{24}	3e-03	2.23e-01	**4.17e-02**	2.31e-01
S_{45}	1e-05	1.89e+01	**2.15e-01**	3.49e-01

Table 9. Relative error for estimation of sensitivity indices of input parameters using various MC and QMC approaches ($n \approx 2^{20}$).

EQ	RV	**PLAIN**	**FC**	**SC**
S_1	4e-01	2.55e-03	1.75e-04	**1.23e-04**
S_2	3e-01	1.91e-02	**1.08e-04**	3.22e-04
S_3	5e-02	3.5e-03	**3.36e-04**	4.56e-04
S_4	3e-01	1.79e-02	3.24e-04	**1.41e-04**
S_5	4e-07	1.87e+02	**6.19e+00**	4.55e+01
S_6	2e-02	5.38e-02	**2.29e-04**	8.7e-04
S_1^{tot}	4e-01	7.95e-03	**1.27e-04**	4.24e-04
S_2^{tot}	3e-01	1.97e-02	2.22e-04	**2.51e-05**
S_3^{tot}	5e-02	3.1e-03	2.75e-04	**8.18e-05**
S_4^{tot}	3e-01	2.01e-02	**8.39e-05**	1.76e-04
S_5^{tot}	2e-04	5.8e-01	2.88e-01	**2.79e-02**
S_6^{tot}	2e-02	7.27e-02	**1.06e-03**	1.46e-03
S_{12}	6e-03	1.33e-01	**6.65e-03**	1.41e-02
S_{14}	5e-03	2.71e-01	1.7e-02	**9.3e-04**
S_{24}	3e-03	3.07e-01	8.44e-03	**4.76e-03**
S_{45}	1e-05	3.15e+00	**9.87e-02**	2.62e-01

5. Conclusions

Meteorological large-scale models, such as UNI-DEM, inherently harbor inherent uncertainties within their input data and parameters. Sensitivity analysis emerges as a potent instrument for scrutinizing and enhancing the dependability of these models. This paper presents specific outcomes from a global sensitivity study carried out on UNI-DEM.

This study involves a series of numerical experiments conducted with a comprehensive air

Table 10. Relative error for estimation of sensitivity indices of input parameters using various QMC approaches ($n \approx 2^{24}$).

EQ	RV	**PLAIN**	**FC**	**SC**
S_1	4e-01	2.43e-03	**3.11e-06**	4.95e-06
S_2	3e-01	1.86e-03	**9.76e-06**	1.44e-05
S_3	5e-02	3.93e-03	**8.87e-06**	3.13e-05
S_4	3e-01	1.73e-03	2.19e-06	**1.21e-06**
S_5	4e-07	6.3e+01	2.11e+00	**2.09e-01**
S_6	2e-02	1.4e-02	3.84e-05	**1.99e-05**
S_1^{tot}	4e-01	1.97e-03	**1.26e-06**	1.58e-05
S_2^{tot}	3e-01	2.6e-03	6.92e-06	**8.77e-07**
S_3^{tot}	5e-02	2.07e-03	4.94e-05	**2.51e-05**
S_4^{tot}	3e-01	1.75e-03	1.09e-05	**2.44e-06**
S_5^{tot}	2e-04	4.3e-02	**2.45e-03**	5.25e-03
S_6^{tot}	2e-02	5.36e-03	**7.3e-06**	8.68e-05
S_{12}	6e-03	2.49e-02	4.39e-04	**2.83e-04**
S_{14}	5e-03	1.21e-02	**2.55e-05**	4.81e-04
S_{24}	3e-03	2.66e-02	4.04e-04	**1.48e-04**
S_{45}	1e-05	1.52e-01	1.44e-03	**4.94e-04**

pollution model known as the Unified Danish Eulerian Model (UNI-DEM) to calculate Sobol sensitivity metrics. We introduce advanced lattice low discrepancy sequences to estimate the Sobol' sensitivity indices of UNI-DEM, which exhibit significant superiority over the conventional Monte Carlo method. The primary aim was to assess the sensitivity of ozone concentrations in relation to changes in chemical reaction rates. The research specifically targeted four European cities: Genova, Milan, Manchester, and Edinburgh, each characterized by distinct geographical attributes.

Furthermore, the investigation encompassed a sensitivity analysis of the output generated by UNI-DEM concerning emission levels. The influence of emission levels on the concentrations of four pivotal air pollutants (ammonia, ozone, ammonium sulphate, and ammonium nitrate) in three European cities: Milan, Manchester, and Edinburgh was taken into account. The predominant strategy for sensitivity analysis is variance-based methods, and we utilized the Sobol variance-based technique for conducting global sensitivity analysis to compute the relevant sensitivity metrics.

Acknowledgments

Slavi Georgiev is supported by the Bulgarian National Science Fund (BNSF) under Project KP-06-M62/1 "Numerical deterministic, stochastic, machine and deep learning methods with applications in computational, quantitative, algorithmic finance, biomathematics, ecology and algebra" from 2022. Venelin Todorov is supported by BNSF under Project KP-06-N52/5 "Efficient methods for modeling, optimization and decision making" and BNSF under Project KP-06-N62/6 "Machine learning through physics-informed neural networks".

References

[1] Homma T, Saltelli A 1996 *Reliability Engineering and System Safety* **52** 1-17
[2] Ishigami T, Homma T 1990 *Proc. First Int. Symp. on Uncertainty Modeling and Analysis* (University of Maryland, USA) 398-403
[3] Kucherenko S, Feil B, Shah N, Mauntz W 2011 *Reliability Engineering and System Safety* **96** 440-449

[4] Saltelli A, Tarantola S, Campolongo F, Ratto M 2004 *Sensitivity analysis in practice: A guide to assessing scientific models* (New York: Halsted Press)
[5] Sobol I M 2001 *Mathematics and Computers in Simulation* **55(1-3)** 271-280
[6] Saltelli A 2002 *Computer Physics Communication* **145** 280-297
[7] Saltelli A, Ratto M, Andres T, Campolongo F, Cariboni J, Gatelli D, Saisana M, Tarantola S 2008 *Global sensitivity analysis, The Primer* (New Jersey: Wiley)
[8] Sobol I M 1993 *Mathematical Modeling and Computational Experiment* **1(4)** 407-414
[9] Dimov I T 2008 *Monte Carlo methods for applied scientists* (London, Singapore: World Scientific)
[10] L'Ecuyer P, Lemieux C 2002 *Modeling Uncertainty: An Examination of Stochastic Theory, Methods, and Applications* (Boston: Kluwer Academic Publishers) 419-474
[11] Levitan Y, Markovich N, Rozin S, Sobol I 1988 *USSR Comput. Math. and Math. Phys.* **28(5)** 755-759
[12] Niederreiter H 1988 *Journal of Number Theory* **30** 51-70
[13] Baldeaux J, Dick J, Leobacher G, Nuyens D, Pillichshammer F 2012 *Numer. Algorithms* **59** 403-431
[14] Zlatev Z 1995 *Computer treatment of large air pollution models* (Dorsrecht-Boston-London: KLUWER Academic Publishers)
[15] Zlatev Z, Dimov I T 2006 *Computational and numerical challenges in environmental modelling* (Amsterdam: Elsevier)
[16] Zlatev Z, Dimov I T, Georgiev K 1994 *IEEE Computational Science and Engineering* **1(3)** 45-52
[17] Zlatev Z, Dimov I T, Georgiev K 1996 *Zeitschrift für Angewandte Mathematik und Mechanik* **76(S4)** 473-476
[18] Dimov I T, Georgieva R, Ivanovska S, Ostromsky Tz, Zlatev Z 2010 *Journal of Computational and Applied Mathematics* **235** 391-402
[19] Dimov I T, Georgieva R, Ostromsky Tz, Zlatev Z 2013 *Central European Journal of Mathematics, "Numerical Methods for Large Scale Scientific Computing"* **11(8)** 1531-1545

New "Walk on Equations" Monte Carlo Algorithm for Linear Systems

Venelin Todorov[1,2], Slavi Georgiev[1,3], Stoyan Apostolov[4]

[1]Department of Information Modeling, Institute of Mathematics and Informatics, Bulgarian Academy of Sciences, 8 Acad. Georgi Bonchev Str., 1113 Sofia, BG
[2]Department of Parallel Algorithms and Machine Learning with a Laboratory in Neurotechnologies, Institute of Information and Communication Technologies, Bulgarian Academy of Sciences, 25A Acad. Georgi Bonchev Str., 1113 Sofia, BG
[3]Department of Applied Mathematics and Statistics, University of Ruse, 8 Studentska Str., 7004 Ruse, BG
[4]Department of Mathematical Analysis, University of Sofia, 8 James Bourchier Blvd., 1164 Sofia, BG

E-mail: vtodorov@math.bas.bg,venelin@parallel.bas.bg, sggeorgiev@math.bas.bg,sggeorgiev@uni-ruse.bg,sapostolov@fmi.uni-sofia.bg

Abstract. A novel version of Monte Carlo algorithm for solving systems of linear algebraic equations is presented and studied. The algorithm is similar to the "Walk on Equations" Monte Carlo method recently developed by Ivan Dimov, Sylvain Maire and Jean Michel Sellier. It is done a comparison with the Gauss-Seidel method for matrices up to size of 2^{12}. The algorithm could be drastically improved by choosing appropriate values for the relaxation parameters, which in turn leads to dramatic reduction in time and lower relative errors for a given number of iterations. What is more, a sequential Monte Carlo method of John Halton based on an iterative use of the control variate method has been applied. Some of the most important numerical applications are the large system, coming from a finite element approximation of problems, describing a beam structure in constructive mechanics, and the block-diagonal matrices, which come from discretization of models in a regime-switching economy.

1. Introduction

Linear Algebraic Equations (LAEs) serve as the foundational framework for a plethora of scientific and engineering endeavors, translating theoretical ideas into tangible real-world applications. Ranging from sophisticated simulations in the realm of computational fluid dynamics to complicated optimizations within the field of structural engineering, LAEs provide a mathematical prism through which we can comprehend and tackle a broad spectrum of intricate issues. With the advent of ever-growing computational power and increasing complexities in problems to be solved, the spotlight has now shifted to large-scale LAEs. These significantly expand the frontiers of both analytical techniques and the computational infrastructure needed to solve them. Such expansive systems encapsulate intricate interrelations and voluminous datasets, rendering their resolution both crucial and demanding. In most cases, addressing these extensive systems effectively is feasible primarily through the deployment of sophisticated stochastic techniques.

Iterative Monte Carlo methods designed to solve Linear Algebraic Equations (LAEs) can be conceptualized as truncated Markov chains [3]:

$$T = \{\alpha_{t_0} \to \alpha_{t_1} \to \alpha_{t_2} \to \cdots \to \alpha_{t_k}\}, \tag{1}$$

where each state α_{t_q} for $q = 1, \dots, i$ represents an absorbing state in the chain.

Matrices A and B are defined as square matrices of dimensions $n \times n$, formally $A, B \in \mathbb{R}^{n\times n}$. We represent these matrices as follows:

$$A = \{a_{ij}\}_{i,j=1}^{n} = (a_1, \dots, a_i, \dots, a_n)^\top ,$$

where each vector a_i is comprised of the elements $(a_{i1}, \dots, a_{in})$ for $i = 1, \dots, n$. Here, the symbol $\top$ signifies the mathematical operation known as transposition.

For vector norms, we employ the following representations:

$$\| b \| = \| b \|_1 = \sum_{i=1}^{n} |b_i|, \quad \| a_i \| = \| a_i \|_1 = \sum_{j=1}^{n} |a_{ij}|$$

and for matrix norms, the following formulations are used:

$$\| A \|_1 = \max_j \sum_{i=1}^{n} |a_{ij}|, \quad \| A \|_\infty = \max_i \sum_{j=1}^{n} |a_{ij}|,$$

where the vector b is an element of $\mathbb{R}^n$.

In this framework, the truncated Markov chains serve as a computational pathway for the iterative Monte Carlo methods, and the norms offer a mathematical means to gauge the magnitudes of vectors and matrices involved. This mathematical framework provides a rigorous foundation for deploying advanced stochastic techniques to tackle LAEs.

2. Novel "Walk on Equations" Monte Carlo approach explained

We explore a specific system of LAEs, which is mathematically defined as follows:

$$Bx = f, \tag{2}$$

Here, the matrix $B = \{b_{ij}\}_{i,j=1}^{n}$ is part of the input, belonging to $\mathbb{R}^{n\times n}$; while $f = (f_1, \dots, f_n)^\top$ and $v = (v_1, \dots, v_n)$ are designated vectors in $\mathbb{R}^{n\times 1}$ and $\mathbb{R}^{1\times n}$, respectively.

We then focus our attention on an associated matrix $A = \{a_{ij}\}_{ij=1}^{n}$, formulated as:

$$A = I - DB.$$

Here, D denotes a diagonal matrix, represented as $D = \mathrm{diag}(d_1, \dots, d_n)$, and each diagonal entry d_i is calculated as:

$$d_i = \frac{\gamma}{b_{ii}}, \; i = 1, \dots, n.$$

In this setup, γ is a relaxation parameter within the range $(0, 1]$ that can be leveraged to expedite the convergence rate.

Subsequently, the system in equation (2) can be recast into the alternative format:

$$x = Ax + b, \tag{3}$$

where the vector b is given by:

$$b = Df.$$

We operate under the assumption that B is a matrix characterized by diagonal dominance. It is important to note, however, that this requirement is not strictly necessary, as the algorithms we present are applicable to a broader range of matrices as well. In the specific case that B is diagonally dominant, the constituents of the A matrix must adhere to the following constraint:

$$\sum_{j=1}^{n} |a_{ij}| \leq 1, \qquad i = 1, \ldots, n.$$

By introducing these mathematical constructs and conditions, we aim to create a robust framework capable of accommodating various forms of matrices and enabling effective solution strategies for linear algebraic equations.

Consider an initial density vector $p = \{p_i\}_{i=1}^n$ in $\mathbb{R}^n$, such that each element p_i satisfies $p_i \geq 0$ for $i = 1, \ldots, n$, and the sum of all elements equals one, $\sum_{i=1}^n p_i = 1$. Also take into account a transition density matrix $P = \{p_{ij}\}_{i,j=1}^n$ in $\mathbb{R}^{n\times n}$, with each element p_{ij} being non-negative ($p_{ij} \geq 0$) for $i, j = 1, \ldots, n$, and the condition $\sum_{j=1}^n p_{ij} = 1$ holds for any $i = 1, \ldots, n$.

Next, let us introduce the concept of *permissible* density vectors and matrices, denoted as $\mathcal{P}_b$ and $\mathcal{P}_A$, respectively.

Definition 1 *The starting density vector $p = \{p_i\}_{i=1}^n$ is termed permissible in relation to the vector $v = \{v_i\}_{i=1}^n$ in $\mathbb{R}^n$, denoted as $p \in \mathcal{P}_b$, if the following conditions are met:*

$$\begin{cases} p_{\alpha_s} > 0 \text{ if } v_{\alpha_s} \neq 0, \\ p_{\alpha_s} = 0 \text{ if } v_{\alpha_s} = 0. \end{cases} \tag{4}$$

Likewise, the transition density matrix $P = \{p_{ij}\}_{i,j=1}^n$ is classified as permissible with respect to the matrix $A = \{a_{ij}\}_{i,j=1}^n$, denoted as $P \in \mathcal{P}_A$, under the conditions:

$$\begin{cases} p_{\alpha_{s-1},\alpha_s} > 0 \text{ if } a_{\alpha_{s-1},\alpha_s} \neq 0, \\ p_{\alpha_{s-1},\alpha_s} = 0 \text{ if } a_{\alpha_{s-1},\alpha_s} = 0. \end{cases} \tag{5}$$

In our computational procedures, we employ these permissible densities $\mathcal{P}_b$ and $\mathcal{P}_A$ (see [3]), thereby ensuring that randomly generated paths utilized to solve the LAEs avoid visiting zero elements within the matrices. This strategy plays a vital role in reducing computational overhead.

Further, we select these allowable densities as follows:

$$p = \{p_\alpha\}_{\alpha=1}^n \in \mathcal{P}_b, \quad p_\alpha = \frac{|v_\alpha|}{\| v \|};$$

$$P = \{p_{\alpha\beta}\}_{\alpha,\beta=1}^n \in \mathcal{P}_A, \quad p_{\alpha\beta} = \frac{|a_{\alpha\beta}|}{\| a_\alpha \|}, \; \alpha = 1, \ldots, n. \tag{6}$$

Opting for such specific forms of $\mathcal{P}_b$ and $\mathcal{P}_A$ paves the way for what we term as an "Almost Optimal Monte Carlo" (MAO) algorithm, as elaborated in [3].

Consider a linear equation system expressed as $x = Ax + b$, where the matrix A with dimensions $n \times n$ has a convergence radius $\rho(A) < 1$. Each element $a_{i,j}$ in A is a real number, and it satisfies the condition

$$\sum_{j=1}^{n} |a_{i,j}| \leq 1, \quad \text{for all } 1 \leq i \leq n.$$

Next, we introduce a Markov process, denoted by T_k, comprising $n + 1$ states $\alpha_1, \ldots, \alpha_n, n + 1$. The conditional probabilities are defined as follows:

$$P(\alpha_{k+1} = j | \alpha_k = i) = |a_{i,j}|$$

when $i \neq n+1$, and

$$P(\alpha_{k+1} = n+1 | \alpha_k = n+1) = 1.$$

In addition, we define a vector c such that $c(i) = b(i)$ for $1 \leq i \leq n$ and $c(n+1) = 0$. Let $\tau = (\alpha_0, \alpha_1, \ldots, \alpha_k, n+1)$ represent a stochastic path that initiates at the starting state $\alpha_0 < n+1$ and traverses through $(\alpha_1, \ldots, \alpha_k)$ until it reaches the terminal or absorbing state $\alpha_{k+1} = n+1$. The likelihood of traversing along this trajectory τ is expressed as $P(\tau) = p_{\alpha_0} p_{\alpha_0 \alpha_1} \cdots p_{\alpha_{k-1},k \alpha_k} p_{\alpha_k}$.

Utilizing the MAO algorithm, see [1, 2, 10], we choose an initial density vector $p = \{p_\alpha\}_{\alpha=1}^n$ as well as a transition density matrix $P = \{p_{\alpha\beta}\}_{\alpha,\beta=1}^n$. We also define weights Q_α as:

$$Q_m = Q_{m-1} \frac{a_{\alpha_{m-1},\alpha_m}}{p_{\alpha_{m-1},\alpha_m}}, \quad m = 1, \ldots, k, \quad Q_0 = \frac{c_{\alpha_0}}{p_{\alpha_0}}.$$

The estimator $X_\alpha(\tau)$ can be formulated as:

$$X_\alpha(\tau) = c_\alpha + Q_k \frac{a_{\alpha_k \alpha}}{p_{\alpha_k}}, \quad \alpha = 1, \ldots$$

and the probability associated with this estimator is $P(\tau) = p_{\alpha_0} p_{\alpha_0 \alpha_1} \cdots p_{\alpha_{k-1},k \alpha_k} p_{\alpha_k}$.

For confirming the convergence properties of this Monte Carlo technique, we invoke a theorem, proved in [4]. The theorem states that the random variable $X_\alpha(\tau)$ serves as an unbiased estimator for x_α, that is,

$$E\{X_\alpha(\tau)\} = x_\alpha.$$

2.1. Novel WoE algorithm for one solution component

In this subsection, we unveil the novel Monte Carlo technique aimed at calculating a specific element x_{i_0} in the solution vector of a linear equation system with real-valued coefficients. The computational journey commences at the i_0-th row of the given system, with b_{i_0} being augmented to the computational score. The algorithmic walk can either terminate or traverse to a different equation, guided by certain probabilities $p_{i_0 j}$ outlined in Algorithm 1. Additionally, the algorithm adjusts the sign of the contributions to the score to mirror the signs of the system's coefficients. The algorithm is elaborated below:

Algorithm 1 Objective: Calculate a specific component x_{i_0} within the solution vector $x_i, i = 1, \ldots, n$

1. **Initialization Stage**
Provide *initial data: The matrix A, the vector* **b**, *and the number of stochastic paths N.*

2. **Precomputation Step**

2.1. **Calculate** *the vector named asum:*

$$asum(i) = \sum_{j=1}^{n} a_{ij}, \ for\ i = 1, 2, \ldots, n.$$

2.2. **Determine** *the cumulative sum of the elements in vector b:*

$$sb = \sum_{i=1}^{n} b_i, \ for\ i = 1, 2, \ldots, n.$$

3. **Initialize** $S := 0$.

4. **For** $k = 1$ **to** N **do**

4.1. Set test $= 0$.
4.2. Select an index j *based on the probability distribution vector* $(b_1, b_2, \ldots, b_n)$*; set* $pj = j$.
4.3 **While** $test \neq 1$ **do**
4.3.1. Select a fresh value for the index j*, aligned with the probability density vector* $\left(a_{pj,1}, a_{pj,2}, \ldots, a_{pj,n}, 1 - asum(pj)\right)$.
4.3.2. **If** $j \neq n+1$ *then set* $pj = j$.
4.3.3. **Else**, *set* $test = 1$*; update* $S = S + b(i_0) + sb \cdot a_{pj,i_0} / \left(1 - asum(pj)\right)$.
4.3.4. **Endif**
4.4 **Endwhile**
5. **Endfor**
6. *The final outcome is* $x_{i_0} = \frac{S}{N}$.

2.2. Novel WoE algorithm for all solution components

In this subsection, we elucidate a refined algorithm designed for the purpose of evaluating every element in the solution set of a linear system of equations. The central theme of this novel method is to calculate scores for each state encountered along a specific computational trajectory, where these states can be regarded as alternative starting points. Initially, one equation is chosen randomly from the n available equations. For every state i, both the cumulative score $S(i)$ and the total visit count $V(i)$ are defined and updated each time state i is encountered in a computational journey. A critical parameter cc is assigned to represent the number of trajectories specific to component x_i of the solution vector. The hallmark feature of this innovative algorithm is the integration of the sequential Monte Carlo (Sequential MC) technique for linear systems, which was originally formulated by John Halton [5, 6, 7, 8], and is grounded on the control variate methodology [9].

Algorithm 2 Objective: Calculation of all individual elements x_i, $i = 1, \ldots, n$ **within the solution vector.**

1. **Initialization Phase**
Provide *beginning data sets: the matrix* A*, the vector* $\mathbf{b}$*, and the total number of stochastic paths* N.
2. **Preparatory Computations**
2.1. **Evaluate** *the asum vector:*

$$asum(i) = \sum_{j=1}^{n} a_{ij}, \ \textit{for } i = 1, 2, \ldots, n.$$

2.2. **Compute** *the overall sum sb of the b vector:*

$$sb = \sum_{i=1}^{n} b_i, \ \textit{for } i = 1, 2, \ldots, n.$$

3. **Set** $S(i) = 0$ *and* $cc(i) = 0$ *for all* i *ranging from* 1 *to* n.
4. **For** $k = 1$ **to** N **do**
4.1 Randomly select component i_0 *to be estimated from the set* $1, 2, \ldots, n$.
4.2 Increment the value of $cc(i_0)$ *by 1.*
4.3 Set test $= 0$.
4.4 Select index j *based on the probability density vector* $(b_1, b_2, \ldots, b_n)$*; let* $pj = j$.
4.5 **While** $test \neq 1$ **do**

4.5.1 Choose a new index j following the probability distribution $(a_{pj,1}, a_{pj,2}, \ldots, a_{pj,n}, 1-asum(pj))$.

4.5.2 **If** $j \neq n+1$ **then update** $pj = j$.

4.5.3 **Else** *set* $test = 1$; *revise* $S(i_0) = S(i_0) + b(i_0) + sb \cdot a_{pj,i_0}/(1 - asum(pj))$.

4.5.4 **Endif**

4.6 **Wndwhile**

5 **Endfor**

6 **Foreach** i *from* 1 *to* n **do** $x_i = \frac{S(i)}{cc(i)}$.

3. Computational simulations

In this section, we delve into a variety of computational tests involving dense matrices with dimensions set at $n = \{2^6, 2^8, 2^{10}, 2^{12}\}$. These are juxtaposed against established deterministic algorithms, specifically Gauss-Seidel and Jacobi methods, for comparative analysis. Several key insights emerged from this study. Primarily, in each of the computational tests conducted, the Gauss-Seidel algorithm exhibited superior accuracy over the Jacobi method by a minimum of three orders of magnitude. This comparison to Gauss-Seidel is a novel addition to this area of study, as earlier comparisons in [4] were exclusively done with the Jacobi method. In the specific scenario involving a 64×64 matrix (depicted in Fig. 1) over a span of 15 iterations, our Monte Carlo strategy displayed a relative error that was approximately five orders of magnitude lower than Gauss-Seidel's and eight orders of magnitude lower than Jacobi's.

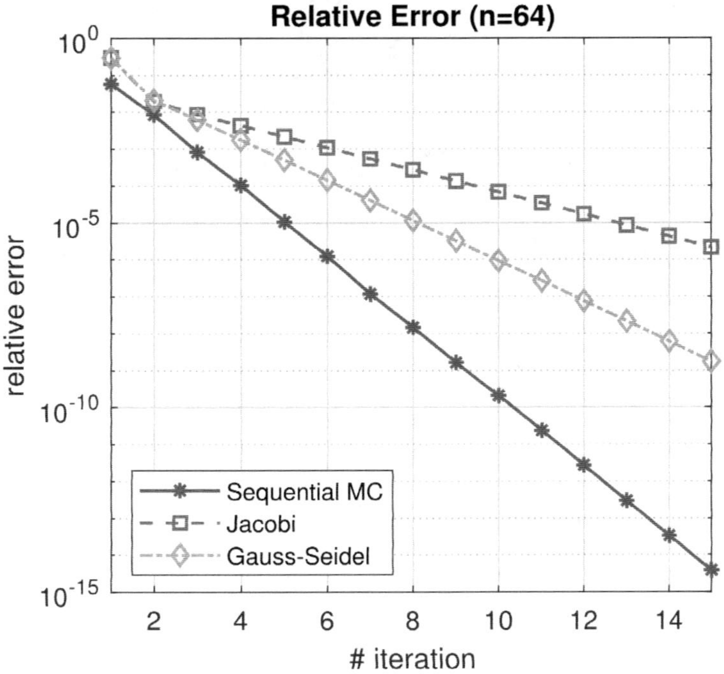

Figure 1. Relative error of a random dense matrix $n \times n$ for $n = 64$ (0.524029 seconds)

Turning our attention to a larger matrix with dimensions 256×256 (as presented in Fig. 2), and limiting the iterations to just 15, we find that the Monte Carlo technique surpasses Gauss-Seidel by approximately six orders of magnitude and outdoes the Jacobi method by about ten orders of magnitude in terms of relative error.

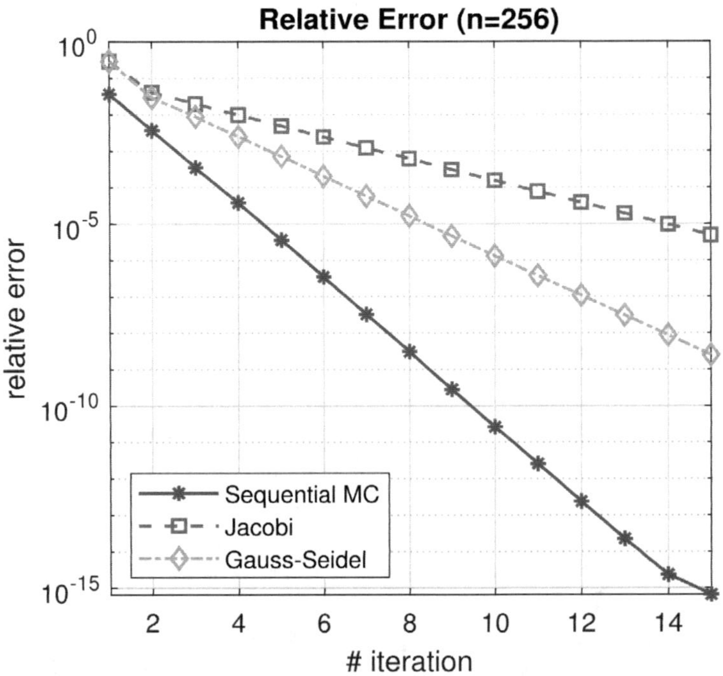

Figure 2. Relative error of a random dense matrix $n \times n$ for $n = 256$ (2.853844 seconds)

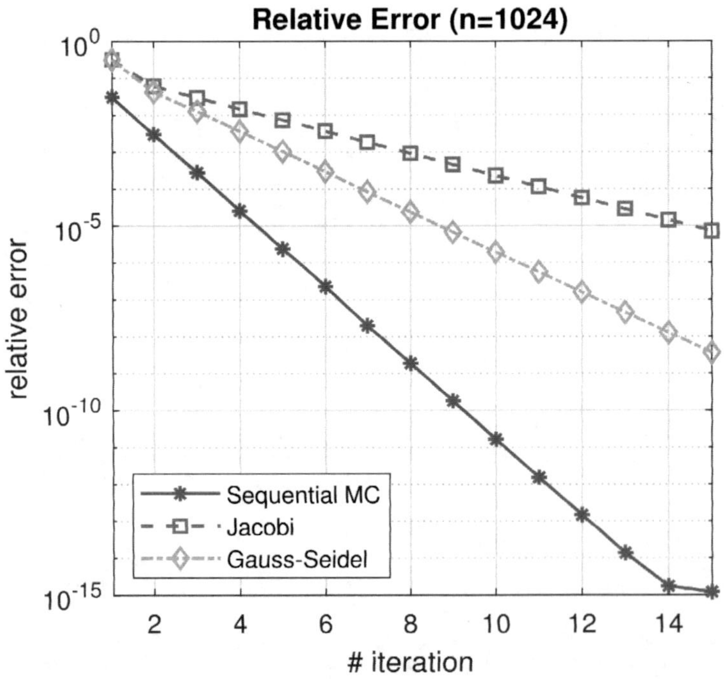

Figure 3. Relative error of a random dense matrix $n \times n$ for $n = 1024$ (40.639203 seconds)

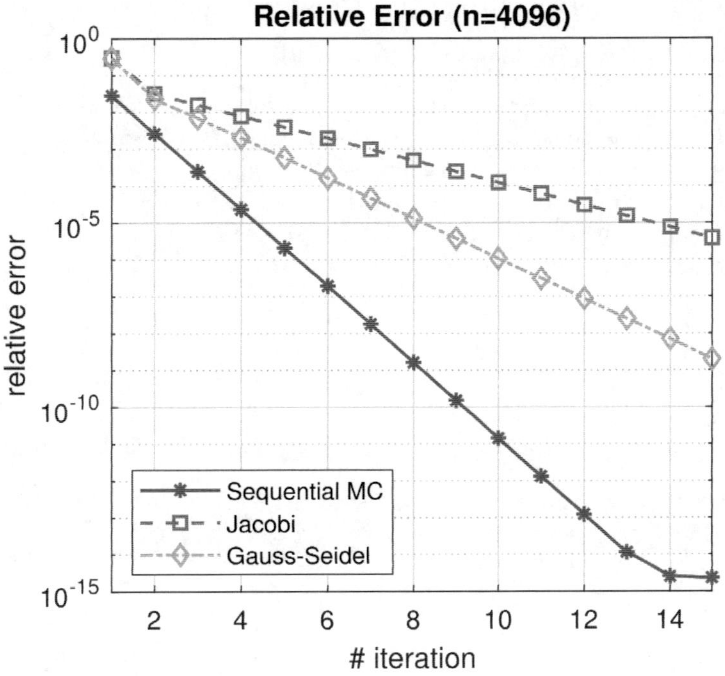

Figure 4. Relative error of a random dense matrix $n \times n$ for $n = 4096$ (2990.56617 seconds)

As we further elevate the matrix size to 1024×1024, the performance differential becomes even more striking. With just 15 iterations (see Fig. 3), the Monte Carlo method exhibits a relative error that is seven orders of magnitude better than Gauss-Seidel and ten orders of magnitude superior to Jacobi. Therefore, we can assert that the performance gap between our Monte Carlo approach and the Gauss-Seidel algorithm widens as we scale up the matrix dimensions.

The apex of our analysis is focused on the colossal 4096×4096 matrix, the outcomes for which are illustrated in Fig. 4. In a mere 14 iterations, the Monte Carlo methodology achieved an impressively low relative error of 10^{-15}. Comparatively, Gauss-Seidel registered a relative error of 10^{-8} and Jacobi languished further behind with 10^{-5}. It is thus evident that for extremely large-scale problems, the Monte Carlo approach we have developed reigns supreme, outclassing the other algorithms by a considerable span of orders of magnitude in terms of accuracy.

4. Conclusions

In the current research, we have suggested and analyzed an innovative Monte Carlo technique specifically designed to tackle systems of linear equations. This innovative approach is derived from the "Walk on Equations" Monte Carlo model, a relatively recent development credited to Ivan Dimov, Sylvain Maire, and Jean Michel Sellier. Notably, our investigation breaks new ground by offering a side-by-side evaluation with the Jacobi and Gauss-Seidel's algorithms, focusing on matrices with dimensions scaling up to 2^{12}. It also shows a significant improvements, compared to the classical WoE method.

Looking toward the future landscape of scholarly pursuits in this domain, we foresee a comprehensive and rigorous examination that pits traditional computational methods against our avant-garde Monte Carlo technique. Moreover, we plan to expand the applicability of our novel algorithm by applying it to a variety of matrices culled from the well-known Harwell-Boeing collection. This initiative aims to highlight the approach's adaptability and efficacy in solving linear algebraic equations that are integral to an array of practical, real-world scenarios.

Acknowledgments
Slavi Georgiev is supported by the Bulgarian National Science Fund (BNSF) under Project KP-06-M62/1 "Numerical deterministic, stochastic, machine and deep learning methods with applications in computational, quantitative, algorithmic finance, biomathematics, ecology and algebra" from 2022. Venelin Todorov is supported by BNSF under Project KP-06-N52/5 "Efficient methods for modeling, optimization and decision making" and Bilateral Project KP-06-Russia/17 "New highly efficient stochastic simulation methods and applications".

References
[1] Curtiss JH 1954 *J Math Phys* **32**(4) 209-32
[2] Curtiss JH 1954
Proc Symp MC Meth 191-233
[3] Dimov IT 2008 *Monte Carlo Methods for Applied Scientists* (New Jersey, London, Singapore: World Scientific)
[4] Dimov IT, Maire S and Sellier JM 2015 *Applied Mathematical Modelling* **39**(15) 4494-510
[5] Halton J 1962 *Proceedings of the Cambridge Philosophical Society* **58** 57-78
[6] Halton, J.: *Sequential Monte Carlo. University of Wisconsin*, Madison, Mathematics Research Center Technical Summary Report No. 816 (1967) 38 pp.
[7] Halton, J. and Zeidman, E. A.: *Monte Carlo integration with sequential stratification.* University of Wisconsin, Madison, Comp. Scien. Departm. Tech. Report No. 61 (1969) 31 pp.
[8] Halton, J.: *Sequential Monte Carlo for linear systems - a practical summary*, Monte Carlo Methods & Applications, 14 (2008) pp. 1–27.
[9] Maire, S.: *Reducing variance using iterated control variates*, Journal of Statistical Computation and Simulation, Vol. 73(1), pp. 1-29, 2003.
[10] I.M. Sobol, Monte Carlo Numerical Methods, Nauka, Moscow (1973).

AUTHOR INDEX